Also of interest

Scientific Leadership.
Niemantsverdriet, Felderhof, 2017
ISBN 978-3-11-046888-5, e-ISBN 978-3-11-046889-2

Teaching Chemistry.
A Course Book
Apotheker, 2019
ISBN 978-3-11-056961-2, e-ISBN 978-3-11-056962-9

Product and Process Design.
Driving Innovation
Harmsen, de Haan, Swinkels, 2018
ISBN 978-3-11-046772-7, e-ISBN 978-3-11-046774-1

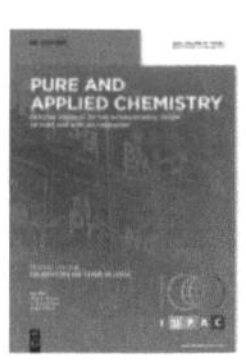

Pure and Applied Chemistry.
The Scientific Journal of IUPAC
Burrows, Stohner (Eds.)
ISSN 0033-4545, e-ISSN 1365-3075

Alexander Senning
The Etymology of Chemical Names

Alexander Senning

The Etymology of Chemical Names

Tradition and Convenience vs. Rationality in Chemical Nomenclature

DE GRUYTER

Author
Alexander Senning
Duevej 24, 3. th
2000 Frederiksberg
Denmark
alexander.senning@gmail.com

ISBN 978-3-11-061106-9
e-ISBN (PDF) 978-3-11-061271-4
e-ISBN (EPUB) 978-3-11-061124-3

Library of Congress Control Number: 2019941581

Bibliographic information published by the Deutsche Nationalbibliothek
The Deutsche Nationalbibliothek lists this publication in the Deutsche Nationalbibliografie; detailed bibliographic data are available on the Internet at http://dnb.dnb.de.

Typesetting: Integra Software Services Pvt. Ltd.
Printing and binding: CPI books GmbH, Leck
Cover images: MicrovOne / iStock / Getty Images Plus (foreground), Gwengoat / iStock / Getty Images Plus (background)

www.degruyter.com

Preface

The thinking and knowledge ensconced in this book are the fruit of more than half a century's university teaching and research [1] as well as extensive nomenclature work [2–5]. Unsurprisingly, an exhaustive discussion of the naming of the 118 known elements, as well as of a few unknown elements, as available in Chapter 6, is a manageable task. On the other hand, other areas of chemical nomenclature can and should only be presented and analyzed with selected name examples as can be seen below. However, an understanding of the way the language of chemistry developed to its present stage should be well served by the examples and comments given.

https://doi.org/10.1515/9783110612714-201

Acknowledgments

A travel grant from *Kemisk Forenings Rejsefond* (The Danish Chemical Society's Travel Fund) is gratefully acknowledged.

Dr. Peter J. T. Morris, Science Museum, London, UK, with the kind assistance of David Allen, Robert Baptista, William H. Brock, Gérard Emptoz, Yasu Furukawa, Urs Leo Gantenbein, Gregory Girolami, the Glasgow University Archives (Laura Burchell), Ernst Homburg, Fyaz Ismail, Alan Rocke, Henry Rzepa, Reiner Salzer, Jeffrey Seeman, Anthony S. Travis, and Mark Whiting, supported the author's work in a most generous manner with the location and evaluation of hard-to-find and sometimes contradictory bibliographic and biographic data as well as error-spotting.

Likewise, Dr. Ture Damhus, Copenhagen, Denmark, with his first-hand knowledge of IUPAC's activities in the area of chemical nomenclature, graciously contributed with a critical check of selected chapters. In this way text passages were identified where errors needed to be corrected or which suffered from a lack of clarity or rigor.

Dr. Mogens Torben Jensen of the University of Southern Denmark, Odense, Denmark, and Dr. Søren Møller of the University of Roskilde, Roskilde, Denmark, gave valuable assistance in the course of some complicated literature searches.

Mrs. Lotte Burkhardt, an independent collaborator of the Botanic Garden and the Botanical Museum of the Freie Universität Berlin, Berlin, Germany, and the author of an encyclopedia of eponymic systematic plant names [6], contributed valuable information concerning taxonomy.

Professor emeritus Sven Harnung, University of Copenhagen, Copenhagen, Denmark generously provided much of the artwork.

References

[1] A. Senning, Sulfur – an Important Element in My Life, *Sulfur Rep.* **24**, 101–253 (2003); *Chem. Abstr.* **140**, 93297 (2004).

[2] A. Senning, Racemic Carbohydrates – Fact or Fiction?, *ARKIVOC* **2007** (ii), 266–271; *Chem. Abstr.* **148**, 33937 (2008).

[3] A. Senning and T. Damhus (Eds.), *Kemisk Ordbog*, Teknisk Forlag, Copenhagen, Denmark, 1996, ISBN 87-571-1976-7; 2nd Ed. 2005; 3rd Ed. 2008.

[4] M. T. Jensen, P. E. Andersen, T. Damhus, A. Senning, and S. Møller, *Dansk Kemisk Nomenklatur*, www.kemisknomenklatur.dk (retrieved August 2, 2018).

[5] A. Senning, *Elsevier's Dictionary of Chemoetymology*, Elsevier, Amsterdam etc., 2007, ISBN 978-0-444-52239-9.

[6] L. Burkhardt, Verzeichnis eponymischer Pflanzennamen Erweiterte Edition, Freie Universität Berlin, Berlin, Germany, 2018, ISBN 978-3-946292-26-5; DOI: https://doi.org/10.3372/epolist2018.

https://doi.org/10.1515/9783110612714-202

Contents

1 Introduction

Both the occasional and the heavy user of chemical nomenclature will at times be baffled to find ambiguous names for a given compound, such as a trivial and a systematic name, or several systematic names, coined over time after changing rules. While an encyclopedia-style compilation of chemical etymological lore is available elsewhere [1] the present book is devoted to providing a coherent picture of how the trivial and systematic names in current use developed over time and especially how a series of conflicting demands had to be reconciled during the creation of the current IUPAC rules in order to make them practically useful.

Chapter 2 discusses the etymology of trivial names, both obsolete ones and those in current use, organized by the method used to coin the names.

Chapters 3 to 5 deal with the development of systematic nomenclature, again with strong emphasis on the etymology of the parent names and affixes used as building blocks. The simple question of whether two different names refer to the same substance or not can be vexing at times and this book should help to demystify the plethora of names connected with the multimillions of chemical substances on record. Chemical names can be more or less specific with regard to connectivity and geometry and both kinds play their own legitimate role, depending on the context in which the substance is to be discussed. Problems and solutions are presented here concerning the optimization of specificity. It is, however, not the intention to provide a crash course in the construction of IUPAC or CAS names. The instructions needed to generate correct systematic names are to be found in the current corresponding official rule books [2–4].

Chapters 6 through 11 are concerned with the special types of nomenclature which have been designed to create names for the elements and several classes of natural products. Also here etymological analysis helps to grasp the principles which have been applied. The development of the Periodic Table is considered in Chapter 6.

Chapter 12 deals with chemical terminology applied to concepts such as named rules and reactions.

In Chapter 13 the naming of minerals is considered together with the annotation of their composition by empirical formulas.

The INN concept for the naming of drugs is dealt with in Chapter 14 while ISO Common Names for Pesticides and Agrochemicals are the subject of Chapter 15.

Finally, common initialisms and acronyms, for chemical substances more often than not based on obsolete systematic names, are discussed in Chapter 16.

Chapter 17 provides an etymological analysis of chemical terms other than element or compound names while Chapter 18 lists miscellaneous prefixes and suffixes used in scientific terms.

Chapters 19 and 20 show examples of names with no or with disputed etymology.

Source languages appearing in this text are listed in the Source Language Index.

https://doi.org/10.1515/9783110612714-001

This book mentions many historical facts of relevance to chemical nomenclature. However, it is not a textbook of the history of chemistry. Much more on the history of alchemy and chemistry can be found in online sources [5,6] and in reference books [7,8].

References

[1] A. Senning, *Elsevier's Dictionary of Chemoetymology*, Elsevier, Amsterdam etc., 2007, ISBN 978-0-444-52239-9 and literature cited there.

[2] 'The Blue Book', H. A. Favre and W. H. Powell, *Nomenclature of Organic Chemistry – IUPAC Recommendations and Preferred Names 2013*, Royal Society of Chemistry, Cambridge, UK, 2013, ISBN 978-0-85404-182-4. Extensive lists with corrections of misprints and errors are available as www.qmul.ac.uk/sbcs/iupac/bibliog/BBerrors.html (retrieved March 14, 2019) and www.qmul.ac.uk/sbcs/iupac/bibliog/BBminor.html (retrieved March 14, 2019).

[3] 'The Red Book', N. G. Connelly, T. Damhus, R. M. Hartshorn, and A. T. Hutton, *Nomenclature of Inorganic Chemistry – IUPAC Recommendations 2005*, Royal Society of Chemistry, Cambridge, UK, 2005, ISBN 0-85404-438-8; free online version, https://www.iupac.org/cms/wp-content/uploads/2016/07/Red_Book_2005.pdf (retrieved August 2, 2018). Lists with corrections of misprints and errors are available as http://www.qmul.ac.uk/sbcs/iupac/bibliog/RBcorrect.html (retrieved March 14, 2019) and http://www.qmul.ac.uk/sbcs/iupac/bibliog/RBcorrect2.html (retrieved March 14, 2019). G.P. Moss, Brief Guide to the Nomenclature of Inorganic Chemistry, IUPAC Technical Report, https://www.qmul.ac.uk/sbcs/iupac/BriefGuide/inorganic.html (retrieved March 22, 2019).

[4] *Naming and Indexing of Chemical Substances for Chemical Abstracts, 2007 Edition*, American Chemical Society, Washington, DC, USA, 2008.

[5] https://en.wikipedia.org/wiki/History_of_chemistry (retrieved August 2, 2018).

[6] http://www.rudolf-werner-soukup.at/Publikationen/Dolumente/Lexikon_der_Natutstoffchemie3_neu.pdf

[7] W. E. Flood, *The Origins of Chemical Names*, Oldbourne, London, UK, 1963.

[8] J. S. Fruton, *A Bio-bibliography for the History of the Biochemical Sciences since 1800*, American Philosophical Society, Philadelphia, PA, USA, 1982, ISBN 0-87169-983-4.

2 Samples of trivial and semitrivial names

A trivial name is a name which does not or only to a limited degree convey information about the structure of the substance in question, with IUPAC's words 'a name having no part used in a systematic sense'. Obviously, only such names could be coined until the development of chemical science provided more and more insight into the exact structures of an ever-increasing multitude of chemical compounds.

A semitrivial name does provide some, but not exhaustive structural information.

Without access to structural information something else had to be relied on to coin these trivial names such as where the compound was first observed or a characteristic, readily observable property.

An extreme case of confusing trivial names was created when two structurally unrelated natural products from the same source were called scopolin and scopoline, respectively. Cf. Section 2.1.

Chapters 7 through 11 deal with important classes of natural products and their naming.

While many trivial and semitrivial names have fallen into disuse over time the remaining trivial names are used ubiquitously and fall into three categories:

1) Trivial names retained in the current IUPAC nomenclature.
2) Trivial names retained in earlier, but not the current IUPAC nomenclature.
3) Trivial names whose use is discouraged by IUPAC.

In the following self-explanatory names are listed without comment.

The common use of systematic names, which has led to the disappearance of many alchemistic trivial names, such as wood spirit (methanol) or muriatic acid (hydrochloric acid), does not prevent new trivial names to be coined, and adopted by the chemical community, on a daily basis. Newly discovered natural products are still assigned trivial names, typically derived from biological species or genus names, but with prefixes and suffixes which do not contradict contemporary practice in systematic names. Molecules with out-of-the-ordinary geometry, such as churchane {homopentaprismane, hexacyclo[$5.4.0.0^{2,6}.0^{3,10}.0^{5,9}.0^{8,11}$]undecane}, are routinely named after their shape, often based on an English source rather than the customary Greek or Latin sources. Also new eponyms, such as buckminsterfullerene {(C_{60}-I_h)[5,6]fullerene}, extend the range of modern trivial names.

2.1 Names based on source

Aaptamine {8,9-dimethoxy-1*H*-benzo[*de*][1,6]naphthyridine}, after the species *Aaptos aaptos* (marine sponge). The genus name and the specific epithet are from Greek *aaptos*, invincible, strong.

https://doi.org/10.1515/9783110612714-002

Abietic acid (abieta-7,13-dien-18-oic acid), from Latin *abies, abiet-*, fir, after the genus *Pinus* (pines). A.k.a. **sylvic acid**, from French *sylvique*, of a forest, from Latin *silva*, forest.

Absinthin {(1*R*,2*R*,5*S*,8*S*,9*S*,12*S*,13*R*,14*S*,15*S*,16*R*,17*S*,20*S*,21*S*,24*S*)-12,17-dihydroxy-3,8,12,17,21,25-hexamethyl-6,23-dioxaheptacyclo[13.9.2.$0^{1,16}$.$0^{2,14}$.$0^{4,13}$.$0^{5,9}$.$0^{20,24}$]hexacosa-3,25-diene-7,22-dione}, after the species *Artemisia absinthum* (common wormwood). The specific epithet is from Greek *apsinthion*, wormwod, possibly from Persian *aspand*, wormwood. Cf. anabsinthin.

Acacic acid (3β,16α,21β-trihydroxyolean-12-en-28-oic acid) after the species *Robinia pseudoacacia* (black locust tree). The specific epithet is ultimately from Greek *akakia*, shittah tree, akin to *ake*, point, thorn.

Aceric acid (3-*C*-carboxy-5-deoxy-L-xylofuranose), after the genus *Acer* (maples). The genus name is from Latin *acer*, maple.

Acetic acid (ethanoic acid), from Latin *acetum*, vinegar, from *acer*, sharp.

Aconitic acid (prop-1-ene-1,2,3-tricarboxylic acid), after the genus *Aconitum* (monkshoods). Suggestions to derive the genus name from Greek *akonitos*, without dust, one referring to these plants' habitat on rocky ground, and another, in terms of able to prevail in a fight without effort, i.e. invincible, bear the hallmark of folk etymology. Another possibility is a derivation from Greek *akon*, dart, javelin, referring to the practice of poisoning the tips of weapons with the poisonous juice of such a plant. Most likely the Greek name of the plant is from a Semitic source.

Actaplanin (a group of antibiotics), with contraction, after the bacterial genus *Actinoplanes*. The genus name is from Greek *aktis, aktin-*, ray, and *planes*, wanderer.

Actinidine {(7*S*)-4,7-dimethyl-6,7-dihydro-5*H*-cyclopenta[*c*]pyridine}, after the species *Actinidia polygama* (silver vine). The genus name is from Greek *aktis*, ray, referring to the rayed styles of the flowers.

Adenine (9*H*-purin-6-amine), from Greek *aden*, gland, referring to this compound's first isolation from bovine pancreas.

Adlumidine {(6*S*)-6-[(5*S*)-6-methyl-7,8-dihydro-5*H*-[1,3]dioxolo[4,5-*g*]isoquinolin-5-yl]-6*H*-furo[3,4-*g*][1,3]benzodioxol-8-one}, after the genus *Adlumia* (vines). The genus name is after the American viticulturist John Adlum (1759–1836).

Aflatoxin (a group of natural toxins), with contraction, after the genus *Aspergillus flavus* (a mold). The genus name is from New Latin *aspergillum*, short-handled brush, from Latin *aspergere*, to sprinkle.

Agaric acid (3-*C*-carboxy-2,4-dideoxy-2-hexadecylpentaric acid), after the genus *Agaricus* (mushrooms). The genus name is from Greek *agarikon*, a mushroom. A.k.a. **agaricic acid, agarinic acid**, and **agaricin**.

Agmatine [*N″*-(4-aminobutyl)guanidine], from Greek *agma*, fragment, referring to this compound's formation from arginine in the course of putrefaction. A.k.a. **argmatine**.

Agroclavine (6,8-dimethyl-8,9-didehydroergoline), from Latin *ager, agr-*, field, and after the fungal genus *Claviceps*. The genus name is from Latin *clava*, club, and *-ceps*, headed, from *caput, capit-*, head.

Agrocybin (8-hydroxyocta-2,4,6-triynamide), after the fungal genus *Agrocybe*. The genus name is from Latin *ager, agr-*, field, and Greek *kybe*, head.

Agropine [1′-deoxy-D-mannitol-1′-yl)-L-glutamine 1′,2′-lactone], after the genus *Agrobacterium* and opine, referring to this amino acid's occurrence in crown galls caused by *Agrobacterium tumefaciens*. The genus name is from Latin *ager, agr-*, field, and bacterium.

Ajaconine [21-(2-hydroxyethyl)-4-methyl-7α,20-epoxyatid-16-en-15β-ol], after the species *Delphinium ajacis* (rocket larkspur). The specific epithet is after Ajax, a Greek mythical hero, possibly akin to Greek *aiastes*, mourner.

Ajoene {(1*E*)-1-[prop-2-en-1-yl)disulfanyl]-3-[(prop-2-en-1-yl)sulfinyl]prop-1-ene}, from Spanish *ajo*, garlic (*Allium sativum*).

Ajugarin {[(1*R*,4a*R*,5*S*,6*R*,8*S*,8a*S*)-8-acetoxy-5,6-dimethyl-5-[2-(5-oxo-2,5-dihydrofuran-3-yl)ethyl]octahydro-8a*H*-spiro[naphthalene-1,2′-oxiran]-8a-yl]methyl acetate}, after the species *Ajuga remota* (a bugleweed). The genus name is either from a(n)- and Latin *iugum*, yoke, referring to this plant's undivided calyx, or from Latin *abiga*, a plant inducing abortion, from *abigere*, to drive away.

Alexine [(1*R*,2*R*,3*R*,7*S*,7a*S*)-3-(hydroxymethyl)hexahydro-1*H*-pyrrolizine-1,2,7-triol], after the species *Alexia leiopetala* (the tropical tree haiariballi). The genus name is after Empress Aleksandra Fyodorovna of Russia (1798–1860).

Alantolactone (8β-hydroxy-4a*H*-eudesm-5-en-12-oic acid γ-lactone), after alant (sneezeweed, *Helenium autumnale*), ultimately from Latin *inula*, sneezeweed. A.k.a. **helenine**.

Alkannin {5,8-dihydroxy-2-[(1*S*)-1-hydroxy-4-methylpent-3-en-1-yl]naphthalene-1,4-dione}, after the species *Alkanna tinctoria* (alkanet). The genus name is from Arabic *al-hinna*, henna. The (*R*)-enantiomer is known as **shikonin**, after the shikonin plant (red-root gromwell, *Lithospermum erythrorhizon*) and the racemate as **shikalkin**.

Allantoin [(2,5-dioxoimidazolidin-4-yl)urea], after allantois. Allantois is from French *allantois*, allantois, ultimately from Greek *allantoeides*, sausage-shaped, from *allas*, *allant-*, sausage.

Alliin 3-[(prop-2-en-1-yl)sulfinyl]-L-alanine, after the genus *Allium* (garlic). The genus name is from Latin *allium*, garlic, ultimately probably from Sanskrit *aluka*, edible root of whitespot giant arum (*Amorphophallus campanulatus*).

Aloe emodin [1,8-dihydroxy-3-(hydroxymethyl)anthracene-9,10-dione], from aloe and after the species *Rheum emodi* (Gilgiti rhubarb). The specific epithet is from Greek *Emodos*, Himalaya. Cf. frangula emodin.

Aloin [(10*S*)-10-β-D-glucopyranosyl-1,8-dihydroxy-2-hydroxymethylanthracen-9(10*H*)-one], after the genus *Aloe* (aloe), ultimately from Sanskrit *agaru*, agalloch. A.k.a. **barbaloin**, with contraction, after the species *Aloe barbadensis* (Barbados aloe).

Alstophylline {1-[(4a*R*,6*S*,13*S*,13a*R*)-9-methoxy-7,14-dimethyl-1,4a,5,6,7,12,13,13a-octahydro-6,13-iminopyrano[3′,4′:5,6]cyclooct[1,2-*b*]indol-4-yl]ethan-1-one}, with contraction, after the species *Alstonia macrophylla* (batino, devil tree). The genus name is after the British botanist Charles Alston (1683–1760).

Amabiline {[(7a*S*)-2,3,5,7a-tetrahydro-1*H*-pyrrolizin-7-yl]methyl (2*S*,3*S*)-2,3-dihydroxy-2-(propan-2-yl)butanoate}, after the species *Cynoglossum amabile* (hound's tongue, Chinese forget-me-not). The specific epithet is from Latin *amabilis*, loveable, from *amare*, to love.

Amanitin (a natural toxin), after the genus *Amanita* (agarics), from the Greek plural *amanitai*, a kind of fungus. A.k.a. **amatoxin.**

Ambrein {(1*R*,2*R*,4a*S*,8a*S*)-1-[(3*E*)-6-[(1*S*)-2,2-dimethyl-6-methylidenecyclohexan-1-yl]-4-methylhex-3-en-1-yl]-2,5,5,8a-tetramethyldecahydronaphthalen-2-ol}, from Latin *ambra*, ambergris.

Ambrettolide [(8*Z*)-oxacycloheptadec-8-en-2-one], from ambrette (*Abelmoschus moschatus*) and 'olide'. The genus name *Abelmoschus* is ultimately from Arabic *habb al misk*, seeds of the musk.

Ampelopsin [(2*R*,3*R*)-3,5,7-trihydroxy-2-(3,4,5-trihydroxyphenyl)-2,3-dihydro-4*H*-1-benzopyran-4-one], after the genus *Ampelopsis* (peppervine, porcelain berry). The genus name is from Greek *ampelos*, grapevine, and *opsis*, appearance, resemblance.

Amygdalin {[6-*O*-(β-D-glucopyranosyl-β-D-glucopyranosyl)oxy](phenyl)acetonitrile}, named after its source, almond, ultimately from Greek *amygdalon*, diminutive of *amygdale*, almond.

Amyl alcohol (pentan-1-ol or any of its isomers), from Greek *amylon*, starch, literally not ground; starch was obtained from unground grain. Thus the name amyl alcohol refers to the first observation of this alcohol as a starch fermentation product.

Anabasine [3-(piperidin-2-yl)pyridine], after the plant genus *Anabasis* (anabasis). The genus name is from Greek *anabasis*, the plant horsetail (*Equisetum arvense*), from Greek *anabasis*, act of going up, from ana- and *bainein*, to go.

Anacardic acid {2-hydroxy-6-[(8*Z*,11*Z*)-pentadeca-8,11,14-trien-1-yl]benzoic acid}, after the genus *Anacardium* (cashew). The genus name is from ana- and Greek *kardia*, heart, referring to the heartlike shape of the top of these trees' fruit stem.

Anagyrine (11,12,13,14-tetradehydrospartein-15-one), after the genus *Anagyris* (oro de risco), from Greek *anagyros*, bean trefoil.

Anemonin (1,7-dioxadispiro[4.0.4.2]dodeca-3,9-diene-2,8-dione), after the genus *Anemone* (anemone, buttercup). The genus name is from Greek *anemone*, anemone, probably of Semitic origin, but influenced by Greek *anemos*, wind.

Anethole {1-methoxy-4-[(1*E*)-prop-1-en-1-yl]benzene}, from Latin *anethum*, dill, from Greek *anethon*, dill, probably a loan from Ancient Egyptian. A.k.a. **anise camphor**.

Angelic acid [(2*Z*)-2-methylbut-2-enoic acid], after the genus *Angelica* (angelica), from Latin *angelicus*, angel-like, from *angelus*, angel, referring to these plants' medicinal use.

Anhalonine {4-methoxy-9-methyl-6,7,8,9-tetrahydro[1,3]dioxolo[4,5-*h*]isoquinoline}, after the genus *Anhalonium*, later renamed *Ariocarpus* (cacti). The former genus name is from a(n)- and Greek *halonion*, diminutive of *halon*, threshing floor. Via Latin *areola*, diminutive of *area*, piece of level ground, threshing floor, this name refers to a perceived lack of areoles in these cacti.

Annotinine {(1*R*,2*S*,4*R*,10*S*,11*R*,14*S*,17*S*)-15-methyl-3,12-dioxa-6-azahexacyclo[8.4.3.$1^{11,14}.0^{1,17}.0^{2,4}.0^{6,17}$]octadecan-13-one}, after the species *Lycopodium annotinum* (stiff clubmoss). The specific epithet is from Latin *annotinus*, one year old.

Aphylline (spartein-10-one), with contraction, after the species *Anabasis aphylla* (anabasis). The genus name is from Greek *anabasis*, the plant horsetail (*Equisetum arvense*), from *anabasis*, act of going up, from *anabanein*, to go up. The specific epithet is from Greek *aphyllos*, leafless, from *a(n)-*, not, and *phyllon*, leaf.

Apiole (4,7-dimethoxy-5-(prop-2-en-1-yl)-1,3-benzodioxole), from Latin *apium*, parsley, celery, ultimately from Latin *apis*, bee, and 'ole'. A.k.a. **apiol**.

Apocynin [1-(4-hydroxy-3-methoxyphenyl)ethan-1-one], after the genus *Apocynum* (dogbane). The genus name is ultimately from Greek *apokynon*, dogbane, from *kyon*, *kyno-*, dog. A.k.a. **acetovanillone**.

Arborescin (1,10-epoxy-6β-hydroxy-1β,5β,7α-guiai-3-en-12-oic acid γ-lactone), after the species *Artemisia arborescens* (tree wormwood). The specific epithet is from Latin *arborescens*, becoming like a tree, tree-like, from *arbor*, tree.

Arbutin (4-hydroxyphenyl β-D-glucopyranoside), after the genus *Arbutus* (strawberry trees). The genus name is from Latin *arbutus*, strawberry tree, of obscure etymology.

Aristolochic acid (8-methoxy-6-nitrophenanthro[3,4-*d*][1,3]dioxole-5-carboxylic acid), after the plant family *Aristolochiaceae* (birthworts). The family name is from Greek *aristos*, best, and *locheia*, childbirth.

Armepavine {4-[(6,7-dimethoxy-2-methyl-1,2,3,4-tetrahydroisoquinolin-1-yl)methyl] phenol}, with transposition and contraction, after the species *Papaver armeniacum* (Armenian poppy). The genus is from Latin *papaver*, poppy, and the specific epithet from Latin *armeniacus*, Armenian. A.k.a. **evoeuropine**.

Asaricin (5-methoxy-6-(propan-2-yl)-1,3-benzodioxole), after the species *Asarum heterotropoides* (the Chinese herb xi sin). The genus name is from Greek *asaron*, wild ginger, probably of Semitic origin. A.k.a. **sarisan**.

Ascochitine [3-(butan-2-yl)-6-hydroxy-4-methyl-8-oxo-8*H*-2-benzopyran-7-carboxylic acid], after the fungal genus *Ascochyta*. The genus name is from New Latin *ascus*, from Greek *askos*, wineskin, bladder, and probably from Greek *chytos*, poured, from *chein*, to pour.

Atranorin [3-hydroxy-4-(methoxycarbonyl)-2,5-dimethylphenyl 3-formyl-2,4-dihydroxy-6-methylbenzoate], with contraction, after the lichen species *Lecanora atra*. The genus name is from Greek *lekane*, basin, and *hora*, beauty, grace, referring to the form and color of the apothecium. The specific epithet is from Latin *ater, atr-*, black.

Atromentin [2,5-dihydroxy-3,6-bis(4-hydroxyphenyl)cyclohexa-2,5-diene-1,4-dione], after the fungal species *Tapinella atrotomentosa* (velvet roll-rim, velvet-footed pax). The specific epithet is from Latin *ater, atr-*, black, and *tomentosus*, having a mass of rough hairs.

Aucubin {(1*S*,4a*R*,5*S*,7a*S*)-5-hydroxy-7-(hydroxymethyl)-1,4a,5,7a-tetrahydrocyclopenta [*c*]pyran-1-yl β-D-glucopyranoside}, after the genus *Aucuba* (Japanese laurel). The genus name is from Japanese *aokuba*, aucuba, from *ao*, green, *ki*, *ko*, tree, and *ba*, leaf.

Azadirachtin {dimethyl (2a*R*,3*S*,4*S*,4a*R*,5*S*,7a*S*,8*S*,10*R*,10a*S*,10b*R*)-10-(acetyloxy)-3,5-dihydroxy-4-[(1*S*,2*S*,6*S*,8*S*,9*R*,11*S*)-2-hydroxy-11-methyl-5,7,10-trioxatetracyclo [6.3.1.$0^{2,6}$.$0^{9,11}$]dodec-3-en-9-yl]-4-methyl-8-[[(2*E*)-2-methylbut-2-enoyl]oxy]octahydro-1*H*-furo[3′,4′:4,4a]naphtho[1,8-*bc*]furan-5,10a(8*H*)-dicarboxylate}, after the genus *Azadirachta* (neem tree), from Persian *azad dirakht*, neem tree, literally free or noble tree.

Azafrin {(2*E*,4*E*,6*E*,10*E*,12*E*,14*E*)-15-[(1*R*,2*R*)-1,2-dihydroxy-2,6,6-trimethylcyclohexan-1-yl]-4,9,13-trimethylpentadeca-2,4,6,8,10,12,14-heptaenoic acid}, from Spanish *azafranillo,* safflower (*Carthamus tinctorius*), ultimately from Arabic *zafaran*, saffron.

Bacitracin (a polypeptide antibiotic), with contraction, after the species *Bacillus subtilis* var. Tracy, so named, however misspelled, after Margaret Treacy (1936–1994), a child in whose infected tissue antibiotics were found.

Baicalein (5,6,7-trihydroxy-2-phenyl-4*H*-1-benzopyran-4-one), after the species *Scutellaria baicalensis* (Baikal scullcap). The specific epithet refers to Lake Baikal, Russia. The toponym Lake Baikal is from Mongolian *baigal nuur*, the nature lake.

Barbatic acid {2-hydroxy-4-[(2-hydroxy-4-methoxy-3,6-dimethylbenzoyl)oxy]-3,6-dimethylbenzoic acid}, after the species *Usnea barbata* (beard lichen). The specific epithet is from Latin *barbatus*, bearded, from *barba*, beard. A.k.a. **coccellic acid,** after the species *Cladonia coccifera* (red pixie cup). The specific epithet is from Latin *coccifer*, scarlet-bearing, from *coccum*, scarlet.

Batyl alcohol (1-*O*-octadecylglycerol), from Greek *batis*, a flat fish. A.k.a. **batilol.**

Bergamottin {4-[(2*E*)-(3,7-dimethylocta-2,6-dien-1-yl)oxy]-7*H*-furo[3,2-*g*][1]benzopyran-7-one}, after the species *Citrus bergamia* (bergamot orange). Bergamot is from Italian *bergamotto*, bergamot orange, influenced by the unrelated name of the Italian city of Bergamo, ultimately from Turkish *beg armudu*, prince's pear or prince of pears.

Bergapten {4-methoxy-7*H*-furo[3,2-*g*][1]benzopyran-7-one}, from bergamot orange (*Citrus bergamia*) and stearopten. Cf. bergamottin. A.k.a. **bergamot camphor.**

Berteroin [1-isothiocyanato-5-(methylsulfanyl)pentane], after the species *Oxalis peridicaria* (Molina) Bertero (lobate oxalis), a.k.a. *Oxalis lobata*. The name refers to the Italian physicist, physician, naturalist, botanist, bryologist, and pteridologist Carlo Luigi Giuseppe Bertero (1789–1831).

Betaine [2-(trimethylazaniumyl)acetate], after the genus *Beta* (beets), from Latin *beta*, beet. Also used as synonym for zwitterion.

Betulin [lup-20(29)-ene-3β,28-diol], as betaine.

Bibrotoxin (a polypeptide), after the species *Atractaspis bibroni* (Bibron's mole viper). The specific epithet is after the French zoologist Gabriel Bibron (1806–1848). A.k.a. **BTX.**

Bicuculline {(6*R*)-6-[(5*S*)-6-methyl-5,6,7,8-tetrahydro[1,3]dioxolo[4,5-*g*]isoquinolin-5-yl]furo[3,4-*e*][1,3]benzodioxol-8(6*H*)-one}, derived from bi-, referring to the two rings of this compound, and after the species *Dicentra cucullaria* (Dutchman's breeches). The specific epithet is from Latin *cucullus*, cap, hood.

Bikhaconitine [8-acetoxy-20-ethyl-13-hydroxy-1α,6α,16β-trimethoxy-4-(methoxymethyl)aconitan-14α-yl 3,4-dimethoxybenzoate], with contraction, from bikh (*Aconitum spicatum*) and its sytematic name. Bikh is from Hindi *bikh*, poison. Suggestions to derive the genus name *Aconitum* from Greek *akonitos*, without dust, one referring to these plants' habitat on rocky ground, and another, in terms of able to prevail in a fight without effort, i.e. invincible, bear the hallmark of folk etymology. Another possibility is a derivation from Greek *akon*, dart, javelin, referring to the practice of poisoning the tips of weapons with the poisonous juice of such a plant. Most likely the Greek name of the plant is from a Semitic source.

(*S*)-Boldine (1,10-dimethoxy-6aα-aporphine-2,9-diol), from boldo (*Peumus boldus*), from Mapudungun *folo*, boldo.

Bombesin (a polypeptide), after the genus *Bombina* (fire-bellied toads), from Latin *bombinare, bombilare*, to buzz, from *bombus*, deep hollow sound.

Bombykol [(10*E*,12*Z*)-hexadeca-10,12-dien-1-ol], after the species *Bombyx mori* (silk worm moth). The genus name is from Greek *bombyx, bombyk-*, silk worm moth, inferred from *bombykion*, cocoon of a moth, probably the Syrian silkworm (*Pachypasa otus*).

Bongkrekic acid [(2*E*,4*Z*,6*R*,8*Z*,10*E*,14*E*,17*S*,18*E*,20*Z*)-20-(carboxymethyl)-6-methoxy-2,5,17-trimethyldocosa-2,4,8,10,14,18,20-heptaenedioic acid], from Javanese *tempe bongkrèk*, an Indonesian molded coconut product.

β-Boswellic acid (3α-hydroxyurs-12-en-24-oic acid), after the genus *Boswellia* (frankincense trees). The genus name is after the British botanist John Boswell (1710–1780).

Botulinum toxin (a polypeptide), after the bacterial species *Clostridium botulinum*. The specific epithet is from Latin *botulus*, sausage, a word likely loaned from Osco-Umbrian. A.k.a. **BTX**.

Brefeldin A {(1*R*,2*E*,6*S*,10*E*,11a*S*,13*S*,14a*R*)-1,13-dihydroxy-6-methyl-1,6,7,8,9,11a,12,13,14,14a-decahydro-4*H*-cyclopenta[*f*]oxacyclotridecin-4-one}, after the fungal species *Penicillium brefeldianum*. The specific epithet is after the German botanist and bacteriologist Oskar Brefeld (1839–1925).

Brevetoxin (a family of cyclic polyethers), after the dinoflagellate *Karenia brevis*. The specific epithet is from Latin *brevis*, short.

Brucine (2,3-dimethoxystrychnidin-10-one), after the genus *Brucea* (Macassar kernels, kosam seed), which was erroneously assumed to contain brucine. The genus name is after the British explorer James Bruce (1730–1794). The real source of brucine was the species *Strychnos nux-vomica* (strychnos tree).

Bulbocapnine {(7a*S*)-11-methoxy-7-methyl-6,7,7a,8-tetrahydro-5*H*-[1,3]benzodioxolo[6,5,4-*de*]benzo[g]quinolin-12-ol}, after the genus *Bulbocapnos* (fumitory). The genus name is from Latin *bulbus*, onion, bulb, and Greek *kapnos*, smoke.

Bungarotoxin (a group of polypeptides), after the genus *Bungarus* (krait), from Sanskrit *bhrugara*, krait.

Butein [(2*E*)-1-(2,4-dihydroxyphenyl)-3-(3,4-dihydroxyphenyl)prop-2-en-1-one], after the genus *Butea* (dhak tree). The genus name is after the British Prime Minister and scholar John Stuart, 3rd Earl of Bute (1713–1792).

Butyric acid (butanoic acid), from Greek *butyron*, butter, from *bous*, cattle, and *tyros*, cheese.

Byssochlamic acid {(4*S*,10*R*)-10-ethyl-4-propyl-5,9,10,11-tetrahydro-1*H*-furo[3′,4′:5,6]cyclonona[1,2-c]furan-1,3,6,8(4*H*)-tetrone}, after the fungal genus *Byssochlamis*. The genus name is from Greek *byssos*, flax, linen, and *chlamys*, a short oblong mantle worn by young men in ancient Greece. Actually a misnomer. This compound is an acid anhydride, not an acid.

Cadaverine (pentane-1,5-diamine), from Latin *cadaver*, cadaver, from *cadere*, to fall, referring to its formation in decaying animal matter.

Caldariomycin [(1*S*,3*S*)-2,2-dichlorocyclopentane-1,3-diol], after the fungal species *Caldariomyces fumago*. The genus name is from Latin *caldarius*, suitable for warming, from *calidus*, warm, hot.

Calicheamicin (a class of antibiotics), after the bacterial species *Micromonospora echinospora* spp. *calichensis*. The subspecific epithet refers to this bacterium's habitat, caliche pits, in Kerrville, TX, USA. Caliche, from Spanish, flake of lime, pebble in a brick, is ultimately from Latin *calx, calc-*, lime.

Calvatic acid {4-[(*Z*)-2-cyano-1-oxidodiazen-1-yl]benzoic acid}, after the genus *Calvatia* (puffball mushrooms). The genus name is from Latin *calvaria*, dome of the skull, from *calvus*, bald.

Calystegine (a class of alkaloids), after the species *Calystegia sepium* (hedge bindweed, Rutland beauty, bugle vine, heavenly trumpets, bellbind, granny-pop-out-of-the-bed). The genus name is from Latin *calix, calic-*, chalice, and Greek *stege*, roof, from *stegein*, to cover. Cf. calystigine a.k.a. palmatine.

Camptothecin {(4*S*)-4-ethyl-4-hydroxy-1*H*-pyrano[3′,4′:6,7]indolizino[1,2-*b*]quinoline-3,14(4*H*,12*H*)-dione}, after the species *Camptotheca acuminata* (happy treem cancer tree, tree of life). The genus name is from Greek *kamptos*, flexible, and *theke*, case, cover.

Canadine {9,10-dimethoxy-5,8,13,13a-tetrahydro-6*H*-[1,3]dioxolo[4,5-*g*]isoquinolino[3,2-*a*]isoquinoline}, after the species *Hydrastis canadensis* (goldenseal). The specific epithet is from New Latin *canadensis*, Canadian.

Capreomycin (an antibiotic), after the bacterial species *Streptomyces capreolus*. The specific epithet is from Latin *capreolus*, wild goat, diminutive of *caper, capr-*, goat.

Cardol [(8*Z*,11*Z*)-5-(pentadeca-8,11-dien-1-yl)benzene-1,3-diol], with contraction, after the genus *Anacardium* (cashew). The genus name is from Greek *ana-*, upwards, and *-kardion*, heart, referring to the heartlike shape of the top of these trees' fruit stem.

Carnegine (6,7-dimethoxy-1,2-dimethyl-1,2,3,4-tetrahydroisoquinoline), after the genus *Carnegiea* (cacti). The genus name is after the British-American industrialist and humanitarian Andrew Carnegie (1835–1919).

L-Carnitine [(3*R*)-3-hydroxy-4-(trimethylazaniumyl)butanoate], from Latin *caro, carn-*, flesh, referring to its isolation from meat extract.

Carotol [(3*R*,3a*S*,8a*R*)-6,8a-dimethyl-3-(propan-2-yl)-2,3,4,5,8,8a-hexahydroazulen-3a(1*H*)-ol], after the species *Daucus carota* (Queen Anne's lace). The specific epithet is from Latin *carota*, carrot.

Carpaine {(1*S*,11*R*,13*S*,14*S*,24*R*,26*S*)-13,26-dimethyl-2,15-dioxa-12,25-diazatricyclo[22.2.2.2^{11,14}]triacontane-3,16-dione}, after the species *Carica papaya* (papaya), from Latin *carica*, a kind of fig, from *caricus*, Carian, and Otomaco *papai*, papaya.

Carvone [*p*-mentha-6,8(9)-dien-2-one], after *Carum carvi* (caraway). The specific epithet is from Medieval Latin *carvi*, caraway.

Cassic acid (4,5-dihydroxy-9,10-dioxoanthracene-2-carboxylic acid), after the genus *Cassia* (cassias). The genus name is ultimately from Hebrew *qesiah*, cassia. A.k.a. **rhein**, after the genus *Rheum* (rhubarb). This genus name is from Greek *rha, rheon*, rhubarb. A connection of the Greek word *rha* with the Scythian name of the Volga River, *Ra*, literally wetness, has been dismissed as folk etymology.

Castanospermine [(1*S*,6*S*,7*R*,8*R*,8a*R*)-octahydroindolizine-1,6,7,8-tetrol], after the genus *Castanospermum* (chestnut). The genus name is from Latin *castanea*, chestnut, and Greek *sperma*, seed, ultimately from *speirein*, to grow.

Catalposide {(1a*S*,1b*S*,5a*R*,6*S*,6a*S*)-2-[(β-D-glucopyranosyl)oxy]-1a-(hydroxymethyl)-1a,1b,2,5a,6,6a-hexahydrooxireno[4,5]cyclopenta[1,2-*c*]pyran-6-yl 4-hydroxybenzoate}, after the genus *Catalpa* (catalpa). The genus name is from Creek Indian *kutuhlpa*, head with wings.

Catechin [(2*R*,3*S*)-2-(3,4-dihydroxyphenyl)-3,4-dihydro-2*H*-1-benzopyran-3,5,7-triol], after the species *Acacia catechu* (black catechu, cutch tree). The specific epithet is from Malay *catechu*, catechu, an extract of this tree's heartwood.

Catechol (benzene-1,2-diol), from catechin. A.k.a. **pyrocatechol**.

Catecholamine, a class name for adrenaline, dopamine, and noradrenaline, from catechol.

Catharanthine [methyl 3,4-didehydro-2α,5β,6α-ibogamine-18β-carboxylate)], after the genus *Catharantus* (periwinkle), ultimately from Greek *katharos*, pure, and *anthos*, flower.

Cathinone (2-amino-1-phenylpropan-1-one), after the genus *Catha* (khat). The genus name is from Arabic *qat*, khat.

Caulophylline {11-methyl-7,11-diazatricyclo[7.3.1.0^{2,7}]trideca-2,4-dien-6-one}, after the genus *Caulophyllum* (cohosh). The genus name is from caul(o)- and phyll(o)-. A.k.a. **methylcytisine**.

Ceanothic acid [2α-carboxy-3β-hydroxy-1-norlup-20(29)-en-28-oic acid]. After the genus *Ceanothus* (perennial shrubs). The genus name is from Greek *keanothos*, thistle. A.k.a. **emmolic** acid, after the genus *Emmenosperma* (tropical trees). The latter genus name is from Greek *emmenein*, to linger, and *sperma*, seed, referring to the seeds remaining after the fruit valves have fallen away.

Cecropin (polypeptides), after the species *Hyalophora cecropi*a (cecropia silkmoth). The genus name is from hyal(o)- and -phore. The specific epithet is from Greek *kekropios*, Athenian, after Cecrops, the mythical first king of Athens. The name Cecrops is probably of non-Greek origin.

Celesticetin {2-[[(5*R*)-[(1*R*,2*R*)-2-methoxy-1-[(1-methyl-L-prolyl)amino]propyl]-β-L-arabinopyranosyl]sulfanyl]ethyl 2-hydroxybenzoate}, after the bacterial species *Streptomyces caelestis*. The specific epithet is from Latin *caelestis*, pertaining to the sky, from *caelum*, sky, referring to this microorganism's blue color.

Centaurein {5-hydroxy-2-(3-hydroxy-4-methoxyphenyl)-3,6-dimethoxy-4-oxo-4*H*-1-benzopyran-7-yl β-D-glucopyranoside}, after the genus *Centaurea* (centaury). The genus name is after the centaur Chiron who, according to Greek myths, discovered these plants.

Cephaeline (7′,10,11-trimethoxyemetan-6′-ol), after the genus *Cephaelis* (tropical shrubs and trees), later renamed *Psychotria*. The genus name is from cepha(lo)- and Greek *eilein*, to compress.

Cephalin (phosphatidylethanolamine), a class name, from cephal(o)-, referring to this compound's occurrence in brain tissue.

Cephalonic acid [(6*Z*)-8-hydroxy-5-oxoophiobola-3,6,19-trien-25-oic acid}, with contraction, after the fungal genus *Cephalosporium*. The genus name is from cephal(o)- and New Latin *spora*, spore, from Greek *spora*, seed, sowing, seed-time, ultimately from *speirein*, to sow, to strew.

Cephalosporin C {7-[(5*R*)-5-amino-5-carboxypentanoyl]amino]cephalosporanic acid]}, with contraction, after the fungal genus *Cephalosporium*. The genus name is from cephal(o)- and New Latin *spora*, spore, from Greek *spora*, seed, sowing, seed-time, from *speirein*, to sow, to strew.

Cephalotaxine {(11b*S*,12*S*,14a*R*)-13-methoxy-2,3,5,6,11b,12-hexahydro-1*H*-[1,3]dioxolo[4′,5′:4,5]benzo[1,2-*d*]cyclopenta[*b*]pyrrolo[1,2-*A*]azepin-12-ol}, after the genus *Cephalotaxus* (plum yew). The genus name is from cephal(o)- and tax-.

Cerane (hexacosane), from ^{2}cer(o)-.

Cerin [2α-hydroxyfriedelan-3-one, (2R,4R,4aS,6aS,6bR,8aR,12aR,12bS,14aS,14bS)-2-hydroxyicosahydropicen-3(2H)-one], from Greek *keros*, wax. So named because cerin was erroneously assumed to be a wax.

Cetane (hexadecane), from cetyl alcohol (hexadecan-1-ol), q.v. This otherwise obsolete name has been institutionalized in the cetane number, used to grade the quality of diesel fuel. The test fuel used in this context is either a mixture of cetane and isocetane or a mixture of cetane and 1-methylnaphthalene, the cetane number being the percentage of cetane in the test fuel.

Cetrarin {9-(ethoxymethyl)-4-formyl-3,8-dihydroxy-1,6-dimethyl-11-oxodibenzo[*b*,*e*][1,4]dioxepine-7-carboxylic acid}, after the species *Cetraria islandica* (Iceland moss). The genus name is from Latin *caetra*, targe, referring to the cup-like shape of these lichen's apothecia.

Cetyl alcohol (hexadecan-1-ol), after its source, from Latin *cetus*, sea monster. A.k.a. **palmityl alcohol**, ultimately from palmitin.

Chaulmoogric acid [13-(cyclopent-2-en-1yl)tridecanoic acid], from chaulmoogra (genus *Hydnocarpus*), from Bengali *caulmugra*, from *caul*, rice (*Oryza sativa*), and *mugra*, bowstring hemp (*Sansevieria zeylanica*).

Chavibetol [2-methoxy-5-(prop-2-en-1-yl)phenol], with contraction, after the species *Chavica betle* (betel pepper). The genus name is from Sanskrit *cavika*, a pepper. The specific epithet is from Portuguese *betle*, betel (*Piper betle*), ultimately from Tamil *verrilai*, betel.

Chavicol [4-(prop-2-en-1-yl)phenol], as chavibetol.

Cherimoline {4*H*-pyrano[3,4-*c*]quinolin-4-one}, after the species *Annona cherimola* (chirimoya, custard apple). The common name is from Quechua *chirimuya*, literally cold seeds, referring to the fact that this plant grows at high altitudes.

Chimaphilin (2,7-dimethylnaphthalene-1,4-dione), after the genus *Chimaphila* (prince's pine). The genus name is from chim(o)- and phil(o)-.

Chimyl alcohol [3-(hexadecyloxy)propane-1,2-diol], after the genus *Chimaera* (ghost sharks, rabbit fish, rat fish, spook fish). The genus name is from Greek *chimaira*, she-goat, mythical monster. A.k.a. **testriol.**

Chlorogenin [(25*R*)-5α-spirostan-3β,6α-diol], after the genus *Chlorogalum* (soap plant, soap root, amole). The genus name is from ^{1}chlor(o)- and gal(a)-, referring to these plants' sap.

Cholic acid (3α,7α,12α-trihydroxy-5β-cholan-24-oic acid), from Greek *chole*, bile.

Choline [(2-hydroxyethyl)trimethylammonium, a cation], from Greek *chole*, bile.

Chondrillasterol [(22*E*,24*R*)-5α-stigmasta-7,22-dien-3β-ol], after the genus *Chondrilla* (sea sponges). The genus name is from Greek *chondrilla*, diminutive of *chondrile*, plant which exudes gum.

Chrysanthenone (pin-2-en-7-one), after the genus *Chrysanthemum* (chrysanthemum), and 'one'. The genus name is from chrys(o)- and anth(o)-.

Chrysarobin [1,8-dihydroxy-3-methylanthracen-9(10*H*)-one], from chrys(o)- and after the species *Andira araroba* (araroba, bee plant), later renamed *Vataireopsis araroba*. The specific epithet is from Tupi *arariba*, araroba.

Cibaric acid [(9*Z*,13*Z*,15*E*)-14,18-dihydroxy-12-oxooctadeca-9,13,15-trienoic acid], after the species *Cantharellus cibarius* (golden chanterelle). The specific epithet is from Latin *cibarius*, of the food, edible, from *cibus*, food.

Cichoriin [7-(β-D-glucopyranosyl)-6-hydroxy-2*H*-1-benzopyran-2-one], after the genus *Cichorium* (chickory). The genus name is from Greek *kichorion*, chickory, possibly from Ancient Egyptian *keksher*, chicory.

Cicutoxin [(8*E*,10*E*,12*E*,14*R*)-heptadeca-8,10,12-triene-4,6-diyne-1,14-diol], after the genus *Cicuta* (hemlock). The genus name is from Latin *cicuta*, poison hemlock, probably *Conium maculatum*.

Ciguatoxin (a group of toxic polycyclic polyethers produced by the dinoflagellate *Gambierdiscus toxicus*), from American Spanish *ciguatera*, sea snail poisoning, from Taino *cigua*, sea snail.

Cimigenol {(1*S*,2*R*,3*S*,4*R*,7*R*,9*S*,12*R*,14*S*,17*R*,18*R*,19*R*,21*R*,22*S*)-22-(2-hydroxypropan-2-yl)-3,8,8,17,19-pentamethyl-23,24-dioxaheptacyclo[19.2.1.0^{1,18}.0^{3,17}.0^{4,14}.0^{7,12}.0^{12,14}]tetracosane-2,9-diol}, after the genus *Cimifuga* (bugbane). The genus name is from the genus *Cimex* (bedbugs), from Latin *cimex, cimic-*, bedbug, and *fugare*, to put to flight.

Cinerolone {2-[(2*Z*)-but-2-en-1-yl]-4-hydroxy-3-methylcyclopent-2-en-1-one}, after the species *Chrysanthemum cinerarifolium* (Dalmatian pyrethrum). The specific epithet is from ciner(o)- and foli(o)-.

Citric acid (2-hydroxypropane-1,2,3-tricarboxylic acid), named after its occurrence in citrus fruits.

Clavulanic acid {(2*R*,5*R*)-3-[(*E*)-2-hydroxyethylidene]-7-oxo-4-oxa-1-azabicyclo[3.2.0]heptane-2-carboxylic acid}, after the bacterial species *Streptomyces clavuligerus*. The specific epithet is New Latin for bearing little clubs, from Latin *clavula*, diminutive of *clava*, club.

Clitocine [*N*-(6-amino-5-nitropyrimidin-4-yl)-β-D-ribofuranosylamine], after the genus *Clitocybe* (mushrooms). The genus name is from Greek *klitos*, slope, and Greek *kybe*, head, meaning sloping head.

Cnicine {3a*R*,4*S*,6*E*,10*Z*,11a*R*)-10-(hydroxymethyl)-6-methyl-3-methylidene-2-oxo-2,3,3a,4,5,8,9,11a-octahydrocyclodeca[*b*]furan-4-yl (3*R*)-3,4-dihydroxy-2-methylidenebutanoate}, after the genus *Cnicus* (thistles). The genus name is from Greek *knekos*, safflower, thistle.

Cobrotoxin (a polypeptide), from cobra (*Naja*) and toxin. Cobra is ultimately from Latin *colubra*, female snake.

Cocaine {methyl (1*R*,2*R*,3*S*,5*S*)-3-(benzoyloxy)-8-methyl-8-azabicyclo[3.2.1]octane-2-carboxylate}, after the species *Erythroxylum coca* (coca). The specific epithet is from Quechua *kuka*, coca.

Coclaurine {1-[(4-hydroxyphenyl)methyl]-6-methoxy-1,2,3,4-tetrahydroisoquinolin-7-ol}, with contraction, after the species *Cocculus laurifolius* (laurel leaf cocculus). The genus name is from New Latin *cocculus*, diminutive of *coccus*, berry. The specific epithet is from Latin *laurus*, laurel, and *folium*, leaf.

Codeine (3-methoxy-17-methyl-7,8-didehydro-4,5α-epoxymorphinan-6α-ol), from Greek *kodeia*, poppy head.

Codlemone [(8*E*,10*E*)-dodeca-8,10-dien-1-ol], after the codling moth (*Cydia pomonella*) and from pheromone.

Cognac lactone [(4*S*,5*R*)-4-methyl-5-pentyloxolan-2-one], from cognac and lactone.

Cohumulone [3,4,5-trihydroxy-4,6-bis(3-methylbut-2-en-1-yl)-2-(2-methylpropanoyl)cyclohexa-2,4-dien-1-one], from con- and humulone, referring to the fact that this compound is a minor hop constituent compared to humulone. Cf. humulone.

Colchicine {(1*E*)-*N*-[(7*S*)-1,2,3,10-tetramethoxy-9-oxo-5,6,7,9-tetrahydrobenzo[*a*]heptalen-7-yl]ethanimidic acid}, after the genus *Colchicum* (autumn crocus, meadow saffron, naked lady). The genus name is after Colchis, an ancient country on the eastern shore of the Black Sea.

Collidine [trimethylpyridine; ethyl(methyl)pyridine; propylpyridine], from coll(a)- and pyridine.

Columbin {(1*R*,2*S*,3*S*,5*S*,8R,11*R*,12*R*)-5-(furan-3-yl)-12-hydroxy-3,11-dimethyl-6,14-dioxatetracyclo[10.2.2.0^{2,11}.0^{3,8}]hexadec-15-ene-7,13-dione}, from calumba (*Jatrorrhiza palmata*), from Hausa *kalumba*, a small tree.

Complicatic acid {3a,5-dimethyl-3-methylidene-2-oxodecahydrocyclopenta[4,5]pentaleno[1,6a-*b*]oxirene-5-carboxylic acid}, after the fungal species *Stereum complicatum*. The specific epithet is from Latin *complicatus*, fuzzy, confused.

Conhydrine {(1*R*)-1-[(2*S*)-piperidin-2-yl]propan-1-ol}, from coniine, q.v.

Coniferyl alcohol {4-[(1*E*)-3-hydroxyprop-1-en-1-yl]-2-methoxyphenol}, from conifer. A.k.a. **coniferol.**

Coniine [(2*S*)-2-propylpiperidin], after the genus name *Conium* (hemlock). The genus name is from Greek *koneion*, hemlock.

Conophthorin {(5*S*,7*S*)-7-methyl-1,6-dioxaspiro[4.5]decane}, after the genus *Conophthorus* (bark beetles). The genus name is from Greek *konos*, cone, and *phthora*, destruction, from *phtheirein*, to destruct.

Conquinamine {(3a*R*,8a*S*)-8a-[(2*R*,4*S*,5*R*)-5-ethenyl-1-azabicyclo[2.2.2]octan-2-yl]-2,3,8,8a-tetrahydro-3a*H*-furo[2,3-*b*]indol-3a-ol}, anagrammatically derived from cinch(ona)- and amine.

Contortin {1,1′-[2,2′-dihydroxy-4,4′,6,6′-tetramethoxy-5,5′-dimethyl-[1,1′-biphenyl]-3,3′-diyl]bisethanone}, after the lichenized fungal species *Psoroma contortum*. The specific epithet is from Latin *contortus*, twisted, from con- and *torquere, tort-*, to twist.

Convallarin (a cardiac glycoside), after the species *Convallaria majalis* (lily of the valley). The genus name is from Latin *convallis*, enclosed valley, from con- and *vallis*, valley.

Convicine (6-amino-2,4-dioxo-1,2,3,4-tetrahydropyrimidin-5-yl β-D-glucopyranoside), after the genus *Vicia* (vetch). The genus name is from Latin *vicia*, vetch.

Coprostane {5β-cholestane, (1*S*,2*S*,7*S*,10*R*,11*S*,14*R*,15*R*)-2,15-dimethyl-14-[(2R)-6-methylheptan-2-yl]tetracyclo[8.7.0.0^{2,7}.0^{11,15}]heptadecane}, from copr(o)- and steroid. A.k.a. **pseudocholestane.**

Coptisine {6,7-dihydro[1,3]dioxolo[4,5-*g*][1,3]dioxolo[7,8]isoquinolino[3,2-*a*]isoquinolin-5-ium, a cation}, after the genus *Coptis* (goldthread). The genus name is from Greek *koptein*, to cut off.

Cordycepin (3′-deoxyadenosine), after the fungal genus *Cordyceps*, from Late Greek *kordyle*, club, and Latin *-ceps*, headed, from *caput, capit-*, head.

Coriamyrtin {(1*S*,2*R*,3*S*,5*R*,6*R*,7*R*,9*S*,12*R*)-2-hydroxy-7-methyl-12-(prop-1-en-2-yl)-11*H*-spiro[4,10-dioxatetracyclo[7.2.1.$0^{2,7}$.$0^{3,5}$]dodecane-6,2′-oxiran]-11-one}, with contraction, after the species *Coriaria myrtifolia* (myrtle-leaved coriaria, Currier's sumac). The genus name is from Latin *coriarius*, useful for tanning leather, from *corium*, leather, the specific epithet is from Latin *myrtus*, myrtle, and *folium*, leaf.

Corybulbine {(13*S*,13a*R*)-2,9,10-trimethoxy-13-methyl-5,8,13,13a-tetrahydro-6*H*-isoquinolino[3,2-*a*]isoquinolin-3-ol}, with contraction, after the species *Corydalis bulbosa* (bulbous corydalis). The genus name is from Greek *korydallis*, crested lark, and the specific epithet from Latin *bulbosus*, bulbous.

Corydine (2,10,11-trimethoxy-6aα-aporphin-1-ol), after the genus *Corydalis* (fumewort), from Greek *korydallis*, crested lark (*Galerida cristata*).

Corynanthine (methyl 17α-hydroxyyohimban-16β-carboxylate), after the genus *Corynanthe* (flowering plants), from coryn(o)- and anth(o)-.

Coumarin (2*H*-1-benzopyran-2-one), from French *coumarou*, tonka bean tree (*Dipteryx odorata*). The French name is ultimately from Tupi *kumarú*, tonka bean tree.

Coumingine {(13*E*)-[2-[2-(dimethylamino)ethoxy]-2-oxoethylidene]-14α-methyl-7-oxo-podocarpan-3β-yl] 3-hydroxy-3-methylbutanoate}, after the species *Erythrophleum couminga* (an African tree). The specific epithet is from Malagasy *komanga*, couminga tree.

Creosol (2-methoxy-4-methylphenol), after creosote, from Greek *kreas*, flesh, and *soter*, preserver, from *sozein*, to preserve, referring to creosol's use as a wood preservative.

Cresol (methylphenol), as creosote.

[1]**Crinine** {1,2-didehydrocrinan-3α-ol, (3*R*,4a*R*,11b*S*)-4,4a-dihydro-3*H*,6*H*,5,11b-ethano[1,3]dioxolo[4,5-*j*]phenanthridin-3-ol}, after the plant species *Crinum moorei*, cf. crinan.

[2]**Crinine** [(4*S*)-4-(4-hydroxyphenyl)-6-methoxy-2-methyl-1,2,3,4-tetrahydroisoquinolin-7-ol], after the plant species *Crinum powellii* (swamp lily).

Crocetin [2*E*,4*E*,6*E*,8*E*,10*E*,12*E*,14*E*)-2,6,11,15-tetramethylhexadeca-2,4,6,8,10,12,14-heptaenoic acid], after the genus *Crocus* (crocus). The genus name is from Greek *krokos*, saffron, crocus, of Semitic origin.

Crotamine (a polypeptide), after the genus *Crotalus* (rattlesnakes). The genus name is from Greek *krotalon*, rattle, castanet, from *krotein*, to rattle.

Cubebin {(2*S*,3*R*,4*R*)-3,4-bis[(1,3-benzodioxol-5-yl)methyl]tetrahydrofuran-2-ol}, after the species *Piper cubeba* (tailed pepper). The specific epithet is from Arabic *kubabah*, tailed pepper.

Cucurbitacin [(4*R*,24*S*)-2,25-dihydroxy-9β,10,14-trimethyl-16α,24-epoxy-4,9-cyclo-9,10-secocholesta-2,5-diene-1,11,22-trione], after the genus *Cucurbita* (gourd). The genus name is from Latin *curcubita*, gourd.

Cupreine {(4-[(*R*)-[(2*S*,4*S*,5*R*)-5-ethenyl-1-azabicyclo[2.2.2]octan-2-yl](hydroxy)methyl] quinolin-6-ol}, from cuprea bark, from American Spanish *cuprea* (*Remijia*), ultimately from Latin *cupreus*, of copper.

Cupressene (abieta-6,13-diene), from Latin *cupressus*, cypress.

Curarine (7′,12′-dihydroxy-6,6′-dimethoxy-2,2,2′,2′-tetramethyl-1α-tubocuraran-2,2′-diium, a dication), after the poison curare, from Tupi *kurari*, he to whom it comes, falls.

Curcumin [(1*E*,6*E*)1,7-bis(4-hydroxy-3-methoxyphenyl)hepta-1,6-dien-3,5-dione], after the genus *Curcuma* (turmeric). The genus name is ultimately from Arabic *kurkum*, saffron, crocus. A.k.a. **diferuloylmethane.**

Curvularin {(5*S*)-13,15-dihydroxy-5-methyl-4-oxabicyclo[[10.4.0]hexadeca-1(16),12,14-triene-3,11-dione}, after the fungal genus *Curvularia*. The genus name is from Latin *curvus*, curved, referring to the boomerang shape of the conidium.

Cuscohygrine {1-[(2*R*)-1-methylpyrrolidin-2-yl]-3-[(2*S*)-1-methylpyrrolidin-2-yl]propan-2-one}, after cusco bark (from *Cinchona pubescens*), named after the city of Cuzco, Peru, and hygrine, q.v., referring to its close structural relationship with this alkaloid.

Cycasin {[(*Z*)-2-methyl-2-oxidodiazenyl]methyl β-D-glucopyranoside}, after the genus *Cycas* (cycad). The genus name is ultimately from Greek *koix, koik-*, doom palm.

[1]**Cyclamin** (malvidin 3-*O*-β-D-glucopyranoside chloride), after the genus *Cyclamen* (Eurasian plants). The genus name is from Greek *kyklaminos*, bulbous plant, from *kyklos*, circle.

[2]**Cyclamin** (a saponin), as [1]**cyclamin.**

Cyclobuxine [(20*S*)-14-methyl-3β,20-bis(methylamino)-4-methylidene-9β,19-cyclopregnan-16α-ol], from cyclopregnane and after the genus *Buxus* (boxwood). The genus name is ultimately from Greek *pyxos*, box.

Cynarine {(1*R*,3*R*,4*S*,5*R*)-1,3-bis[[(2*E*)-3-(3,4-dihydroxyphenyl)prop-2-enoyl]oxy]-4,5-dihydroxycyclohexane-1-carboxylic acid}, after the genus *Cynara* (herbs), from Greek *kynara*, a kind of artichoke.

Cytisine {(1*R*,5*S*)-1,2,3,4,5,6-hexahydro-1,5-methano-8*H*-pyrido[1,2-*a*][1,5]diazocin-8-one}, after the genus *Cytisus* (shrubs). The genus name is ultimately from Greek *kytisos*, moon trefoil, shrub medick, alfalfa arborea, tree medick (*Medicago arborea*).

Daidzein [7-hydroxy-3-(4-hydroxyphenyl)-4*H*-1-benzopyran-4-one], from Japanese *daidzu*, soybean (*Glycine max*).

Damascenine [methyl 3-methoxy-2-(methylamino)benzoate], after the species *Nigella damascena* (nigella, devil-in-a-bush, love-in-a-mist). The specific epithet is from Latin *damascenus*, of Damascus, Syria. A.k.a. **nigelline**, after the genus name. The genus name is from Late Latin *nigella*, a black-seeded plant, from Latin *nigellus*, diminutive of *niger*, black.

Damascone {(2*E*)-1-(2,6,6-trimethylcyclohex-1-en-1-yl)but-2-en-1-one], after the species *Rosa damascena* (damask rose). The specific epithet is from Latin *damascenus*, of Damascus, Syria.

Daphnin (8-hydroxy-2-oxo-2*H*-1-benzopyran-7-yl β-D-glucopyranoside), after the genus *Daphne* (European shrubs). The genus name is from Greek *daphne*, laurel.

Datiscetin [3,5,7-trihydroxy-2-(2-hydroxyphenyl)-4*H*-1-benzopyran-4-one], after the genus *Datisca* (herbs, trees). The genus name is from Late Latin *datisca*, an unrelated plant, assumed to belong to the genus *Catananche*.

Daucol {(1*S*,2*R*,5*R*,7*S*,8*S*)-5,8-dimethyl-2-(propan-2-yl)-11-oxatricyclo[6.2.1.0^{1,5}]undecan-7-ol} after the genus *Daucus* (wild carrots). The genus name is ultimately from Greek *daukos*, a kind of parsnip or wild carrot.

Dauricine {4-[[(1*R*)-6,7-dimethoxy-2-methyl-1,2,3,4-tetrahydroisoquinolin-1-yl]methyl]-2-[4-[[(1*R*)-6,7-dimethoxy-2-methyl-1,2,3,4-tetrahydroisoquinolin-1-yl]methyl]phenoxy] phenol}, after the species *Menispermum dauricum* (Asiatic moonseed). The specific epithet is from Daur, a Manchu-Tungus people.

Deguelin {(7a*S*,13a*S*)-9,10-dimethoxy-3,3-dimethyl-13,13a-dihydro-3*H*-[1]benzopyrano[3,4-*b*]pyrano[2,3-*h*][1]benzopyran-7(7a*H*)-one}, after the genus *Deguelia* (derris). The genus name is from Galibi *assa-ha-pagara undeguélé*, derris.

Dendrobane [(2a*S*,4aS,5*R*,8*R*,8a*S*,8b*R*,9*S*)-1,8b-dimethyl-9-(propan-2-yl)decahydro-5,8-methano-1*H*-7-oxa-1-azacyclopent[*cd*]azulene], after the genus *Dendrobium* (orchids). The genus name is from Greek *dendron*, tree, and *bios*, life.

Dendrotoxin (a polypeptide), after the genus *Dendroaspis* (snakes). The genus name is from Greek *dendron*, tree, and *aspis*, asp.

Derritol {2-[4-hydroxy-2-(prop-1-en-2-yl)benzofuran-5-yl]-1-(2-hydroxy-4,5-dimethoxyphenyl)ethan-1-one}, after the genus *Derris* (derris). The genus name is from Greek *derris*, leather covering, skin.

Detoxin (a group of antibiotics), after the bacterial species *Streptomyces caespitosus* var. *detoxicus*. The subspecific epithet is from New Latin *detoxicus*, detoxifying.

Deutzioside {(1a*R*,1b*S*,2*S*,5a*S*,6*S*,8a*S*)-6-hydroxy-5-methyl-1a,1b,2,5a,6,6a-hexahydrooxireno[4,5]cyclopenta[12-*c*]pyran-2-yl β-D-glucopyranoside}, after the genus *Deutzia* (flowering plants). The genus name is after the Dutch councilman and benefactor of botanists Johan van der Deutz (1743–1788).

Dhurrin {(2*S*)-[(β-D-glucopyranosyl)oxy](4-hydroxyphenyl)acetonitrile}, from durra (*Sorghum bicolor*), from Arabic *dhurrah*, durra.

Diaboline {1-[(17*R*)-17-hydroxy-19,20-didehydro-17,18-epoxycuran-1-yl]ethan-1-one}, after the species *Strychnos diaboli* (devil-doer). The specific epithet is from Greek *diabolos*, devil, from di(a)- and *ballein*, to throw.

Diatretyne I [(2*E*)-8-amino-8-oxooct-2-ene-4,6-diynoic acid], after the species *Clitocybe diatreta* (a mushroom). The specific epithet is from Latin *diatretum*, openwork vessel, ultimately from *rete*, net.

Dicentrine (1,2-methylenedioxy-9,10-dimethoxyaporphine), after the genus *Dicentra* (bleeding heart). The genus name is from di- and Greek *kentron*, sharp point.

Dictamnine (4-methoxyfuro[2,3-*b*]quinoline), after the genus *Dictamnus* (burning bush, dittany, gas plant, fraxinella). The genus name is probably from Greek *diktamnon*, from Mount Dikte, Crete, Greece. A.k.a. **dictamine.**

Dictyopterene C′ (6-butylcyclohepta-1,4-diene), after the genus *Dictyopteris* (brown algae). The genus name is from dicty(o)- and Greek *pteris*, fern, from *pteron*, feather, wing.

Didemnin (depsipeptides), after the genus *Didemnum* (tunicates). The genus name is from di- and Greek *demnion*, bed, nest, home, den.

Digenic acid [(3*S*,4*S*)-3-(carboxymethyl)-4-(prop-1-en-2-yl)-L-proline], after the species *Digenea simplex* (red algae). The genus name is from di- and Greek *genos*, race. A.k.a. **kainic acid**, from Japanese *kainin-sou* (*Digenea simplex*).

Dillapiole (4,5-dimethoxy-6-(prop-2-en-1-yl)-1,3-benzodioxole), from dill (*Anethum graveolens*) and Latin *apium*, parsley, celery, ultimately from Latin *apis*, bee, and 'ole'. A.k.a. **dillapiol.**

Dill ether [(3*S*,3a*S*,7a*R*)-3,6-dimethyl-2,3,3a,4,5,7a-hexahydro-1-benzofuran], cf. dillapiole. A.k.a. **anethofuran.**

Dinosterol [(22*E*)-4α,23-dimethyl-5α-ergost-22-en-3β-ol}, from dinoflagellate, referring to its occurrence in the dinoflagellate *Gonyaulax tamarensis*. Dinoflagellate is from Greek *dinos*, rotation, eddy, and New Latin *flagellum*, whip, shoot of a plant, diminutive of *flagrum*, whip. A.k.a. **Black Sea sterol.**

Dioscin {(25*R*)-14β-spirost-5-en-3β-yl 6-deoxy-α-L-mannopyranosyl-(1→2)-[6-deoxy-β-L-mannopyranosyl-(1→4)]-β-D-glucopyranoside}, after the genus *Dioscorea* (yams). The genus name is after the Greek physician Pedanius Dioscorides (40–90).

Diosmetin [5,7-dihydroxy-2-(3-hydroxy-4-methoxyphenyl)-4*H*-1-benzopyran-4-one], after the genus *Diosma* (shrubs). The genus name is from Greek *dios*, heavenly, and -*osme*, one having an odor.

Diploicin {2,4,7,9-tetrachloro-3-hydroxy-8-methoxy-1,6-dimethyl-11*H*-dibenzo[*b,e*][1,4]dioxepin-11-one}, after the genus *Diploicia* (lichens). The genus name is from Greek *diploos*, double, referring to these lichens' two-celled ascospores.

Discodermolide (a polyketide), after the genus *Discodermia* (marine sponges). The genus name is from Greek *diskos*, quoit, and *derma*, skin.

Disparlure [2-decyl-3-(5-methylhexyl)oxirane], after the species *Lymantria dispar* (gypsy moth) and from lure. The specific epithet being from New Latin *dispar*, unlike.

Djenkolic acid [(2*R*,2′*R*)-3,3′-[methylenebis(sulfanyl)]bis(2-aminopropanoic acid)], after Javanese *djenkol*, velvet bean (*Pithecolobium lobatum*).

Dolabellane (a class of macrocyclic diterpenes), after the genus *Dolabella* (sea hares). The genus name was chosen without stated reason, possibly after the cognomen of a branch of the Cornelia gens in ancient Rome.

Domesticine [2-methoxy-9,10-(methylenedioxy)-6aα-aporphin-1-ol], after the species *Nandina domestica* (heavenly bamboo). The specific epithet is from Latin *domesticus*, of the house, from *domus*, house.

Domoic acid [(3*S*,4*S*)-4-[(2*Z*,4*E*,6*R*)-6-carboxyhepta-2,4-dien-2-yl]-3-(carboxymethyl)-L-proline], from Japanese *domoi*, the red alga *Chondria armata*.

Drosophilin A (2,3,5,6-tetrachloro-4-methoxyphenol), after the genus Drosophila (fruit flies). The genus name is from Greek *drosos*, dew, and -phile.

Dryophantin (a gall pigment), after the genus *Dryophanta* (oak galls). The genus name is from Greek *drys*, tree, oak, and -*phantes*, showing, from *phainein*, to reveal, to show, to make known.

Dysidiolide {(5*S*)-5-hydroxy-4-[(1*S*)-1-hydroxy-2-[(1*S*,2*S*,5R,8a*R*)-1,2,5-trimethyl-5-(4-methylpent-4-en-1-yl)-1,2,3,5,6,7,8,8a-octahydronaphthalen-1-yl]furan-2(5*H*)-one}, after the genus *Dysidea* (marine sponges). The genus name is from Greek *dysis*, immersion, diving.

Dysoxysulfone {2,4,5,7,9-pentathiadecane 2,2,9,9-tetraoxide, [[(methylsulfonyl)methyl]disulfanyl][(methylsulfonyl)methyl]sulfanyl]methane}, after the genus *Dysoxylum* (tropical trees). The genus name is from dys- and xyl(o)-, referring to the unpleasant smell of the wood of these trees.

Ecgonine (3β-hydroxy-1α*H*,5α*H*-tropane-2β-carboxylic acid), from Greek *ekgonos*, born of, sprung from, referring to the fact that this compound can be obtained from cocaine.

Echinochrome (6-ethyl-1,4,5,7,8-pentahydroxynaphthalene-2,3-dione), from Greek *echinos*, hedgehog, sea urchin, and -chrome.

Echinuline {(3*S*,6*S*)-3-methyl-6-[[2-(2-methylbut-3-en-2-yl)-5,7-bis(3-methylbut-2-en-1-yl)-1*H*-indol-3-yl]methyl]piperazine-2,5-dione}, after the fungal species *Aspergillus echinulatus*. The specific epithet is from Latin *echinulatus*, set with small prickles, from *echinulus*, small prickle, diminutive of *echinus*, hedgehog, sea urchin.

Echitamine [(16*R*)-3β,17-dihydroxy-16-(methoxycarbonyl)-4-methyl-2,4(1*H*)-cyclo-3,4-secoakuammilanium, a cation], ultimately from Greek *echis*, viper. A.k.a. **echetamine**.

Elatol {(2*R*,3*S*,6*R*)-2-bromo-8-chloro-1,1,9-trimethyl-5-methylidenespiro[5.5]undec-8-en-3-ol}, after the species *Laurencia elata* (red alga). The specific epithet is from Latin *elatus*, high, lofty, proud, from *efferre, elat-*, to make proud.

Eledoisin (an oligopeptide), after the genus *Eledone* (octopi). The genus name is from Greek *eledone*, a kind of octopus.

Elemicin (1,2,3-trimethoxy-5-(prop-2-en-1-yl)benzene), after elemi, an oleoresin from the species *Canarium luzonicum* (elemi), ultimately from Arabic *al lāmi*, elemi.

Eleutherobin (a diterpene glycoside), after the genus *Eleutherobia* (soft corals). The genus name is from Greek *eleutheros*, free, and *bios*, life.

Ellipticine {5,11-dimethyl-6*H*-pyrido[4,3-*b*]carbazole}, after the species *Ochrosia elliptica* (elliptic yellowwood). The specific epithet is from Latin *ellipticus*, elliptic, from Greek *ellipsis*, omission, deficiency.

Elliptone {(6a*S*,12a*S*)-8,9-dimethoxy-12,12a-dihydrofuro[2′,3′:7,8][1]benzopyrano[2,3-*c*][1]benzopyran-6(6a*H*)-one}, after the species *Derris elliptica* (tuba root), and ‘one’. The specific epithet is from Latin *ellipticus*, elliptic, from Greek *ellipsis*, omission, deficiency.

Elymoclavine [6-methyl-8,9-didehydroergolin-8-yl)methanol], after the genus *Elymus* (dunegrass) and agroclavine. The genus name is from Greek *elymos*, millet.

Embelin (2,5-dihydroxy-3-undecylcyclohexa-2,5-diene-1,4-dione), after the genus *Embelia* (Old World tropical woody vines). The genus name is after a Sri Lankan indigenous plant name.

Enanthotoxin [(2*E*,8*E*,10*E*)-heptadeca-2,8,10-triene-4,6-diyne-1,14-diol], after the genus *Oenanthe* (water dropwort). The genus name is from en(o)- and anth(o)-.

Endiandric acid A {(1*S*,1a*S*,2a*S*,5*R*,5a*R*,7a*R*,7b*S*,7c*S*)-[5-phenyl-1a,2,2a,5,5a,7a,7b,7c-octahydro-1*H*-cyclobuta[*bc*]acenaphthylen-1-yl]acetic acid}, after the genus *Endiandra* (trees), from end(o)- and Greek *aner, andro-*, stamen, referring to these plants' fertile stamens.

Enniatin (a group of depsipeptides). After the fungal species *Fusarium orthoceras* var. *enniatum*. The subspecific epithet is New Latin nine-fold, from Greek *ennea*, nine.

Ephedrine [(1*R*,2*S*)-2-(methylamino)-1-phenylpropan-1-ol], after the genus *Ephedra* (joint-pine, jointfir, Mormon-tea, Brigham tea). The genus name is from Greek *ephedra*, the horsetail plant, from *ephedros*, sitting upon, from *hedra*, chair.

Epibatidine {(1*S*,2*S*,4*R*)-2-(6-chloropyridin-3-yl)-7-azabicyclo]2.2.1]heptane}, after the genus *Epipedobates* (frogs). The genus name is from Greek *epipedos*, on the ground, from ep(i)-, Greek *pous*, foot, and *-bates*, walker, from *bainein*, to walk.

Epilachnene [(5*S*,10*Z*)-5-propyl-1-oxa-4-azacyclopentadec-10-en-15-one], after the genus *Epilachna* (ladybirds). The genus name is from ep(i)- and Greek *lachne*, soft woolly hair.

Equilin [3-hydroxyestra-1,3,5(10),7-tetraen-17-one], from Latin *equus*, horse, referring to its source, mare urine.

Erabutoxin (a class of polypeptides), after Japanese *erabu umi hebi*, the black banded sea krait (*Laticauda semifasciata*).

Ergocristine [12′-hydroxy-2′-(1-methyl-5′α-(phenylmethyl)ergotaman-3′,6′-18-trione], from erg(ot)- and crystalline.

Ergothioneine [(2*S*)-3-(2-sulfanylidene-2,3-dihydro-1*H*-imidazol-4-yl) 2-(trimethylazaniumyl)propanoate], from erg(ot)- and 'thione'. A.k.a. **thioneine**, short for ergothioneine.

Eriodictyol [(2*S*)-2-(3,4-dihydroxyphenyl-5,7-dihydroxy-2,3-dihydro-4*H*-1-benzopyran-4-one), after the genus *Eriodictyon* (North American shrubs). The genus name is from Greek *erion*, wool, and *diktyon*, net, referring to the undersurface of these plants' leaves.

Erucin [1-isothiocyanato-4-(methylsulfanyl)butane], after the genus *Eruca* (arugula, rocket). The genus name is from Latin *eruca*, garden rocket (*Eruca sativa*).

Erysimin {3β-[(2,6-dideoxy-β-D-ribo-hexopyranosyl)oxy]-5β,14-dihydroxy-19-oxocard-20(22)-enolide}, after the genus *Erysimum* (wallflower). The genus name is from Greek *erysimon*, a kind of mustard, from *erysthai*, to protect, to defend, to save. A.k.a. **helveticoside**, after the species *Erysimum helveticum* (Swiss wallflower). The specific epithet is from New Latin *helveticus*, Swiss, from *Helvetia*, Switzerland.

Erythrocentaurin (1-oxo-3,4-dihydro-1*H*-2-benzofuran-5-carbaldehyde), after the species *Erythrea centaurium* (European centaury). The genus name is from erythr(o)- and

Greek *kentauros*, centaur, referring to the ancient belief that this herb was discovered by the centaur Chiron.

β-Erythroidine {(4a*S*,6*R*)-6-methoxy-1,4,6,10,12,13-hexahydro-3*H*,5*H*-pyrano[4′,3′:3,4]pyrido[2,1-i]indol-3-one}, after the genus *Erythrina* (coral tree, flame tree). The genus name is from erythr(o)-, referring to the color of the flowers.

Erythropterin [(3*Z*)-3-(2-amino-4,6-dioxo-4,5,6,8-tetrahydropteridin-7(1*H*)-ylidene)-2-oxopropanoic acid], from erythr(o)- and Greek *pteron*, feather, wing, i.e. named as red pigment from butterfly wings.

Escin (a group of saponins), after the genus *Aesculus* (buckeye, horse chestnut). The genus name is from Latin *aesculus*, a variety of oak tree, of obscure etymology.

Estragole [1-methoxy-4-(prop-2-en-1-yl)benzene], after estragon, a.k.a. tarragon (*Artemisia dracunculus*). The common name estragon is from Arabic *tarkhun*, from Greek *drakon, drakont-*, dragon, snake. The specific epithet is from Latin *dracunculus*, diminutive of *draco, dracon-*, dragon, small serpent, and possibly referring to this plant's twisted roots or to the ancient belief that it could heal snake bites. The unrelated name of the Spanish city of Tarragona, on the other hand, is from the Latin toponym *Tarraco*, according to Catalan legend ultimately of Egyptian origin.

Eugenol [2-methoxy-4-(prop-2-en-1-yl)phenol], after the genus *Eugenia* (tropical trees, shrubs). The genus name is after the Austrian Prince Eugen of Savoy (1663–1736).

Euparin {1-[6-hydroxy-2-(prop-1-en-2-yl)-1-benzofuran-5-yl)ethan-1-one}, after the genus *Eupatorium* (bonesets, thoroughworts, snakeroots). The genus name is from Greek *eupatorion*, hemp agrimony (*Eupatorium cannabinum*), named for Mithridates VI Eupator, king of Pontus (132 BC–63 BC), from Greek *eupator*, born of a noble father.

Evernic acid {2-hydroxy-4-[(2-hydroxy-4-methoxy-6-methylbenzoyl)oxy]-6-methylbenzoic acid}, after the lichen genus *Evernia* (oakmoss). The genus name is from Greek *euernes*, well-sprouting, from *ernos*, sprout.

Evodiamine {(13b*S*)-14-methyl-8,13,13b,14-tetrahydroindolo[2′,3′:3,4]pyrido[2,1-*b*]quinazolin-5(7*H*)-one}, after the genus *Euodia* (shrubs, trees), following its common misspelling as *Evodia*. The genus name is from Greek *euodia*, fragrance, from *eu*, well, and *ozein*, to smell.

Fagarine {4,8-dimethoxyfuro[2,3-*b*]quinoline}, after the genus *Fagara* a.k.a. *Zanthoxylum*. The genus name is from Arabic *fagara*, Sichuan pepper (*Zanthoxylum*).

Faranal [(3*S*,4*R*,6*E*,10*Z*)-3,4,7,11-tetramethyltrideca-6-10-dienal], after the pharaoh ant (*Monomorium pharaonis*). This name refers to one of the ten biblical plagues of Egypt.

Febrifugine {3-[3-[(2*S*,3*R*)-3-hydroxypiperidin-2-yl]-2-oxopropyl]quinazolin-4(3*H*)-one}, after the species *Dichroa febrifuga* (antifeverile dichroa). The specific epithet is from Latin *febrifugus*, antifeverile, from *febris*, fever, and *fugare*, to put to flight.

Fecapentaene 12 {(2*S*)-3-[[(1*E*,3*E*,5*E*,7*E*,9*E*)-dodeca-1,3,5,7,9-pentaen-1-yl]oxy]propane-1,2-diol}, from feces, penta-, and 'ene'.

Felinine [*S*-(4-hydroxy-2-methylbutan-2-yl)-L-cysteine], from Latin *felis*, cat, after its occurrence in cat urine.

Ferulic acid [(2*E*)-3-(4-hydroxy-3-methoxyphenyl)prop-2-enoic acid], after the genus *Ferula* (fennel). The genus name is from Latin *ferula*, rod, of obscure etymology, possibly from *festuca*, cf. festucine.

Fervenulin {6,8-dimethylpyrimido[5,4-*e*][1,2,4]triazine-5,7(6*H*,8*H*)-dione}, after the bacterial species *Streptomyces fervens*. The specific epithet is from Latin *fervens*, boiling, glowing, from *fervere*, to boil, to glow.

Festucine {(1*R*,3*S*,7*S*,8*R*)-*N*-methyl-2-oxa-6-azatricyclo[4.2.1.$0^{3,7}$]nonan-8-amine}, after the genus *Festuca* (fescue). The genus name is from Latin *festuca*, stalk, straw, rod. A.k.a. **loline**, after the genus *Lolium* (grasses). The genus name is from Latin *lolium*, darnel.

Filbertone [(2*E*)-5-methylhept-2-en-4-one], named after filbert (nut), the nut of *Corylus avellana* (hazel). The name filbert refers to the fact that these nuts are ripe by August 20, St. Philbert's day. St. Philbert is the French abbot Philibert de Jumièges (617/18–684).

Filicinic acid (2,2-dimethylcyclohexane-1,3,5-trione), after the genus *Filix* (fern). The genus name is from Latin *filix, filic-*, fern.

Filipin (a group of macrolide polyenes), after the bacterial species *Streptomyces filipinensis*. The specific epithet is from New Latin *filipinensis*, Filipino.

Filixic acid {4,4′-[(5-butanoyl-2,4,6-trihydroxy-1,3-phenylene)bis(methylene)]bis(2-butanoyl-3,5-dihydroxy-6,6-dimethylcyclohexa-2,4-dien-1-one}, after the species *Dryopteris filixmas* (male fern). The specific epithet is from Latin *filix mas*, male fern.

Fisetin [3,7-dihydroxy-2-(3,4-dihydroxyphenyl)-4*H*-1-benzopyran-4-one], from German *Fisetholz*, European smoketree (*Cotinus coggygria*).

Flindersine {2,2-dimethyl-2,6-dihydro-5*H*-pyrano[3,2-*c*]quinolin-5-one}, after the genus *Flindersia* (Australian trees). The genus name is after the British mariner Matthew Flinders (1774–1814).

Flustramine A {(3aR,8aR)-1,2,3,3a,8,8a)-1-methyl-3a-(2-methylbut-3-en-2-yl)-8-(3-methylbut-2-en-1-yl)-6-bromohexahydropyrrolo[2,3-*B*]indole}, after the genus *Flustra* (bryozoans). The genus name is from Latin *flustra*, quiet sea.

Fomannosin {1-(4,4-dimethyl-2-oxocyclopentan-1-yl)-5-(hydroxymethyl)-3-oxabicyclo[4.2.0]octa-5,7-dien-4-one}, after the fungal species *Fomes annosus*. The genus name is from Latin *fomes*, touchwood, tinder, from *fovere*, to warm. The specific epithet is from Latin *annosus*, long lived.

Formic acid (methanoic acid), from New Latin *acidum formicicum*, from Latin *formica*, ant, ultimately from Greek *myrmex*, ant.

Formycin {(1*S*)-1-(7-amino-1*H*-pyrazolo[4,3-*d*]pyrimidin-3-yl)-1,4-anhydro-D-ribitol}, after the bacterial species *Nocardia interformis*. The specific epithet is from inter- and Latin *forma*, shape, form.

Fragranol {2-[(1*R*,2*R*)-1-methyl-2-(prop-1-en-2-yl)cyclobutan-1-yl]ethan-1-ol}, after the species *Artemisia fragrans* (wormwood). The specific epithet is from Latin *fragrans, fragrant-*, fragrant.

Frangula emodin (1,3,8-trihydroxy-6-methylanthracene-9,10-dione), after the genus *Frangula* (buckthorn) and from Greek *Emodos*, Himalaya. The genus name is from Latin *frangere, fract-*, to break, referring to the plant's brittle wood. A.k.a. **emodin** and **frangulic acid**. Cf. aloe emodin.

Fraxetin (7,8-dihydroxy-6-methoxy-2*H*-1-benzopyran-2-one), after the genus *Fraxinus* (ash tree). The genus name is from Latin *fraxinus*, ash tree.

Frequentin {(1*R*,3*R*,4*R*)-6-[(1*E*,3*E*)-hepta-1,3-dien-1-yl]-3,4-dihydroxy-2-oxo-cyclohexane-1-carbaldehyde}, after the fungal species *Penicillium frequentans*. The specific epithet is from Latin *frequentans*, crowding, from *frequentare*, to crowd, from *frequens*, crowded, frequent, ultimately from *farcire*, to cram.

Frontalin {(1*S*,5*R*)-1,5-dimethyl-6,8-dioxabicyclo[3.2.1]octane}, after the species *Dendroctonus frontalis* (southern pine beetle). The specific epithet is from Latin *frontalis*, frontal, from *frons, front-*, forehead.

Fuchsin {4-[(4-aminophenyl)(4-iminocyclohexa-2,5-dien-1-ylidene)methyl]-2-methylaniline hydrochloride (1/1)}, after the genus *Fuchsia* (lady's eardrops). The genus name is after the German botanist Leonhard Fuchs (1501–1566). A.k.a. **fuchsine, magenta**, after the battle of Magenta, Italy in 1859, shortly before the discovery of this dye, and **rosaniline**, after its color and its preparation from aniline.

Fucoserratene [(3*Z*,5*E*)-octa-1,3,5-triene)], after the species *Fucus serratus* (brown seaweed, toothed wrack). The genus name is ultimately from Greek *phykos*, seaweed, rouge, of Semitic origin, and Latin *serratus*, jagged, from *serra*, saw.

Fulvoplumierin {methyl (7*E*)-7-[(2*E*)-but-2-en-1-ylidene]-1-oxo-1,7-dihydrocyclopenta[*c*]pyran-4-carboxylate}, from fulv(o)- and after the genus *Plumeria* (shrubs, trees). The genus name is after the French botanist Charles Plumier (1646–1704).

Fumagillin {(2*E*,4*E*,6*E*,8*E*)-10-[[(3*R*,4*S*,5*S*,6*R*)-5-methoxy-4-[(2*R*,3*R*)-2-methyl-3-(3-methylbut-2-en-1-yl)oxiran-2-yl]-1-oxaspiro[2.5]octan-6-yl]oxy]-10-oxodeca-2,4,6,8-tetraenoic acid}, with anagrammatic contraction, after the fungal species *Aspergillus fumigans*. The genus name is from aspergillum, from Latin *aspergillum*, a brush for sprinkling holy water in a liturgical service, from *aspergere*, to dampen, to sprinkle, to humidify. The specific epithet is from Latin *fumigans, fumigant-*, smoking, from *fumigare*, to smoke, from *fumus*, smoke.

Fumigatin (3-hydroxy-2-methoxy-5-methylcyclohexa-2,5-diene-1,4-dione), after the fungal species *Aspergillus fumigatus*. The specific epithet is from Latin *fumigatus*, smoked, from *fumigare*, to smoke, from *fumus*, smoke.

Fumonisin (a group of mycotoxins), with contraction, after the fungal species *Fusarium moniliforme*. The genus name is from Latin *fusus*, spindle. The specific epithet is from Latin *monile*, necklace, and *-formis*, shaped.

Funtumine (3α-amino-5α-pregnan-20-one), after the genus *Funtumia* (tropical African trees). The genus name is from Ewe *funtum, Funtumia*.

Furfural (furan-2-carbaldehyde), earlier called furfurol, from Latin *furfur*, bran, referring to its formation by degradation of bran.

Fusaric acid (5-butylpyridine-2-carboxylic acid), after the fungal genus *Fusarium*. The genus name is from Latin *fusus*, spindle.

Fuscin {5-hydroxy-4,8,8-trimethyl-9,10-dihydro-2*H*,4*H*-benzo[1,2-*b*:4,3-*c′*]dipyran-2,6(8*H*)-dione}, after the fungal species *Oidiodendron fuscum*. The specific epithet is from Latin *fuscus*, dark, tawny, brownish orange.

Fusidic acid [(17*Z*)-16β-acetoxy-3α,11α-dihydroxyfusida-17(20),24-dien-21-oic acid], after the fungal genus *Fusidium*. The genus name is from Latin *fusus*, spindle.

Fustin [(2*R*,3*R*)-2-(3,4-dihydroxyphenyl)-3,7-dihydroxy-2,3-dihydro-4*H*-1-benzopyran-4-one], from fustet (*Rhus cotinus*), ultimately from Arabic *fustaq, fustuq*, fustet.

Galangin (3,5,7-trihydroxy-2-phenyl-4*H*-1-benzopyran-4-one), from galangal, resin of lesser galangal (*Alpinia officinarum*), ultimately from Arabic *khalanjan*, galangal.

Galegine [*N*-(3-methylbut-2-en-1-yl)guanidine], after the genus *Galega* (goat's rue). The genus name is from Medieval Latin *Gallica herba*, Gallic herb.

Gallic acid (3,4,5-trihydroxybenzoic acid), from French *galle*, gall, from Latin *galla*, gall nut.

Gambogic acid (a xanthonoid), from gamboge, a resin from the gamboge tree (*Garcinia hanburyi*), from Cambodia. Gamboge is an older name for Cambodia.

Geiparvarin {7-[[(2*E*)-3-(4,5-dihydro-5,5-dimethyl-4-oxofuran-2-yl)but-2-en-1-yl]oxy]-2*H*-1-benzopyran-2-one}, after the species *Geijera parviflora* (wilga, native willow, sheepbush, dogwood). The genus name is after the German physician and naturalist Johann Daniel Geier (Geyer) (1660–1735). The specific epithet is from Latin *parvus*, small, and *flos, flor-*, flower.

Genistein [5,7-dihydroxy-3-(4-hydroxyphenyl)-4*H*-1-benzopyran-4-one)], after the genus *Genista* (the plant broom). The genus name is from Latin *genista*, the plant broom.

Gentamicin (an antibiotic), from gentian violet, referring to the color of the producing microorganism *Micromonospora pupurea*.

Gentianin (1,7-dihydroxy-3-methoxy-9*H*-xanthen-9-one), after the genus *Gentiana* (gentian), probably named after its discoverer, Gentius, 2nd century BC king of Illyria. A.k.a. **gentiin** and **gentisin**.

Gentisic acid (2,5-dihydroxybenzoic acid), cf. gentianin.

Geronic acid (2,2-dimethyl-6-oxoheptanoic acid), from geran(i)- and 'one'.

Gigantine [(1*S*)-6,7-dimethoxy-1,2-dimethyl-1,2,3,4-tetrahydroisoquinolin-5-ol], after the species *Carnegia gigantea* (saguaro, giant cactus). The specific epithet is from Greek *gigas, gigant-*, giant.

Gingerol [(5*S*)-5-hydroxy-1-(4-hydroxy-3-methoxyphenyl)decan-3-one], from ginger (*Zingiber officinale*).

Ginkgolide (a group of terpenic lactones), from ginkgo (*Ginkgo biloba*), from Japanese *ginkyo*, ginkgo, from *gin*, silver, and *kyo*, apricot.

Ginsenoside (a group of steroid glycosides and triterpene saponins), from ginseng (*Panax shinseng*), from Chinese *jen-shen*, ginseng. Chinese *jen* means man, the meaning of *shen* is obscure.

Gitoxigenin [3β,14,16β-trihydroxy-5β-card-20(22)-enolide], with anagrammatical contraction, after the genus *Digitalis* (foxglove). The genus name is from Latin *digitalis*, of a finger, from *digitus*, finger, toe; and toxin and -gen.

Glaucine {(6a*S*)-1,2,9,10-tetramethoxy-6-methyl-5,6,6a,7-tetrahydro-4*H*-dibenzo[*de*,*g*]quinoline}, after the genus *Glaucium* (horned poppy). The genus name is from Greek *glaukion*, juice of a papaveraceous plant, from *glaukos*, blue-grey.

Glomerine [1,2-dimethylquinazolin-4(1*H*)-one], after the species *Glomeris marginata* (a millipede). The genus name is from Latin *glomus, glomer-*, globule.

Glycosine [2-benzyl-1-methyl-4(1*H*)-quinazoline]. after the genus *Glycosmis* (remote citroid fruit trees), from Greek *glykys*, sweet, and *osme*, smell, referring to these plants' fragrance. A.k.a. **arborine.**

Glycyrrhizic acid [20β-carboxy-11-oxo-30-norolean-12-en-3β-yl 2-*O*-β-D-glucopyranuronosyl-α-D-glucopyranosiduronic acid], after the genus *Glycyrrhiza* (licorice), from Greek *glykys*, sweet, and *rhiza*, root. A.k.a. **glycyrrhizin.**

Gomphidic acid {[(2*Z*)-3-hydroxy-5-oxo-4-(3,4,5-trihydroxyphenyl)furan-2(5*H*)-ylidene](4-hydroxyphenyl)acetic acid}, after the genus *Gomphidius* (spike mushrooms). The genus name is from Greek *gomphos*, tooth, peg, bolt, bond.

Gonyauline [(1*R*,2*R*)-2-(dimethylsulfaniumyl)cyclopropane-1-carboxylate], after the species *Gonyaulax polyedra* (a dinoflagellate), later renamed *Lingulodinium polyedra*. The former genus name is from Greek *gony*, knee, and *aulax, aulak-*, furrow.

Gossyplure (hexadeca-7,11-dien-1-yl acetate), after the species *Pectinophora gossypiella* (pink bollworm) and lure. The specific epithet is from Greek *gossypion*, cotton plant, ultimately from Sanskrit *karpasa*, cotton.

Gougerotin {(2*S*,3*S*,4*S*,5*R*,6*R*)-6-(4-amino-2-oxopyrimidin-1(2*H*)-yl)-4,5-dihydroxy-3-[[(2*R*)-3-hydroxy-2-[[(methylamino)acetyl]amino]propanoyl]amino]tetrahydro-2*H*-pyran-2-carboxamide}, after the bacterial species *Streptomyces gougerotii*. The specific epithet is probably after the French dermatologist Henri Gougerot (1881–1955).

Gramicidin (a cyclic polypeptide), from Gram-positive (bacteria), a property named after the Danish physician and bacteriologist Hans Christian Joachim Gram (1853–1938).

Gramine [1-(1*H*-indol-3-yl)-*N,N*-dimethylmethanamine)], from Latin *gramen, gramin-*, grass.

Grandisol {2-[(1*R*,2*S*)-1-methyl-2-(prop-1-en-2-yl)cyclobutyl]ethan-1-ol}, after the species *Anthonomus grandis* (boll weevil) and 'ol'. The specific epithet is from Latin *grandis*, large, great, grand.

Grapefruit mercaptan {2-[(1*R*)-4-methylcyclohex-3-en-1-yl]propane-2-thiol}, named after its source.

Grayanotoxin [(14*R*)-3β,5,6β,10,16-pentahydroxygrayanotaxan-14-yl acetate], after the species *Leucothoe grayana* (*hana-hirinoki*, a Japanese shrub) and toxin. The specific epithet is after the American botanist Asa Gray (1810–1888).

Grindelic acid (9,13-epoxylabd-7-en-15-oic acid), after the genus *Grindelia* (gumweed). The genus name is after the Latvian chemist and physician David Hieronymus Grindel (1777–1836).

Guaiacol (2-methoxyphenol), after the genus *Guaiacum* (guaiac tree). The genus name is from Taino *guayacan*, guaiac.

Guvacine (1,2,5,6-tetrahydropyridine-3-carboxylic acid), from Old Indo-Aryan *guvaca*, betel palm tree (*Areca cathechu*).

Gymnemic acid (a group of triterpenoid glycosides), after the genus *Gymnema* (tropical herbs). The genus name is from Greek *gymnos*, naked, and *nema*, thread, referring to these plants' hairless staminal filaments.

Hachimycin (a polyene macrolide antibiotic), after the bacterial species *Streptomyces hachijoensis*. The specific epithet is after the island of Hachijo Jima, Japan. A.k.a. **trichomycin**, referring to this antibiotic's action against trichomoniasis.

Halostachine [(1*R*)-2-(methylamino)-1-phenylethan-1-ol], after the genus *Halostachys* (shrubs), later renamed *Allenrolfea*, after the British gardener and botanist Robert Allen Rolfe (1855–1921). The former genus name is from Greek *hals, halo-*, salt, and *stachys*, ear of grain.

Haplophytine {(1*S*,12*R*)-12-(16,17-dimethoxy-1-methyl-21-oxo-3,4-didehydro-19,21-epoxyaspidospermidin-15-yl)-7-hydroxy-15-methyl-5,15-diazatetracyclo[10.3.1.0^{1,5}.0^{6,11}]hexadeca-6,8,10-triene-4,16-dione}, after the genus *Haplophyton* (cockroach plant). The genus name is from Greek *haploos*, single, simple, and *phyton*, plant.

Harmalol (1-methyl-4,9-dihydro-3*H*-β-carbolin-7-ol), after harmal (*Peganum harmala*), from Arabic *harmalah*, harmal.

Hecogenin [(25*R*)-3β-hydroxy-5α-spirostan-12-one], after the genus *Hechtia* (bromeliads). The genus name is after the counselor to the king of Prussia Julius Gottfried Conrad Hecht (1771–1837).

Helenien [*O*,*O*-dipalmitoyllutein, (3*R*,3′*R*,6*R*)-4,5-didehydro-5,6-dihydro-β,β-carotene-3,3′-diyl dihexadecanoate], after the species *Helenium autumnale* (common sneezeweed). The genus name is after the Greek mythical figure Helen of Troy. The ending 'en' (instead of 'ene') is unusual.

Helodermin (a polypeptide), after the genus *Heloderma* (lizards). The genus name is from helo- and derm(o)-.

Hematoxylin {(6aS,11bR)-7,11b-dihydrobenz[*b*]indeno[1,2-*d*[pyran-3,4,6a,9,10(6*H*)-pentol}, after the genus *Haematoxylum* (blackwood, bloodwood tree, bluewood, campeachy tree, Jamaica wood, logwood). The genus name is from hemat(o)- and xyl(o)-.

Hepaxanthin (5,6-epoxy-5,6-dihydroretinol), from Latin *hepar, hepat-*, liver, and xanthin.

Hernandulcin {(6*S*)-6-[(2*S*)-2-hydroxy-6-methylhept-5-en-2-yl]-3-methylcyclohex-2-en-1-one}, a natural sweetener, obtained from *Phyla dulcis* (Aztec sweet herb, bushy lippia, honeyherb, hierba dulce, tzopelic-xihuitl) and named after the Spanish physician and naturalist Francisco Hernandez (1514–1587).

Hesperetin [(2*S*)-5,7-dihydroxy-2-(3-hydroxy-4-methoxyphenyl)-2,3-dihydro-4*H*-1-benzopyran-4-one], after the genus *Hesperidium* (citrus fruit trees). The genus name is from the Greek Hesperides, nymphs who guarded a mythical paradisiacal garden with golden apples. Hesperides is ultimately from Greek *hespera*, evening, i.e. west.

Hierridin B (2,4-dimethoxy-6-pentadecylphenol), after the species *Ramalina hierrensis* (a lichen). The specific epithet is after the island of El Hierro, Canary Islands, Spain.

Hippadine {7*H*-[1,3]dioxolo[4,5-*j*]pyrrolo[3,2,1-*de*]phenanthridin-7-one}, after the genus *Hippeastrum* (amaryllis). The genus name is from Greek *hippeus*, horseman, and *astron*, star. A.k.a. **pratorine**, after the species *Crinum pratense* (amaryllis). The specific epithet is from Latin *pratensis*, of the meadow, from *pratum*, meadow.

Hippocasine {8-methyl-1,2,3,3aα,4,5,6,6aβ,7,9aα-decahydropyrido[2,1,6-*de*]quinolizine}, after the species *Hippodamia caseyi* (Casey's lady beetle). The genus name is after Hippodamia, the mythical queen of Pisa in Ancient Greece. The name means she who masters horses, from Greek *hippos*, horse, and *damazein*, to tame. The specific epithet is after the American entomologist Thomas Lincoln Casey (1857–1925).

Hippuric acid (*N*-benzoylglycine), from Greek *hippos*, horse, and urine, referring to its occurrence in horse urine.

Hirsutic acid {(1a*R*,2*R*,3a*R*,3b*R*,5*S*,6a*R*,7a*S*)-2-hydroxy-3a,5-dimethyl-3-methylidenedecahydrocyclopenta[4,5]pentaleno[1,6a-*b*]oxirene-5-carboxylic acid}, after the species *Stereum hirsutum* (false turkey tail). The specific epithet is from Latin *hirsutus*, hirsute, ultimately from *horrere*, to bristle.

Histamine [2-(1*H*-imidazol-4-yl)ethan-1-amine], from histidine and -amine.

Histrionicotoxin {(2*S*,6*R*,7*S*,8*S*)-7-[(1*Z*)-but-1-en-3-yn-1-yl]-2-[(2*Z*)-pent-2-en-4-yn-1-yl]-1-azaspiro[5.5]undecan-8-ol}, after the species *Dendrobates histrionicus* (poison arrow frog). The specific epithet is from Latin *histrio, histrion-*, actor, of Etruscan origin.

Homarine (1-methylpyridinium-2-carboxylate), after the genus *Homarus* (lobster). The genus name is ultimately from Old Norse *humarr*, lobster.

Hordenine {4-[2-(dimethylamino)ethyl]phenol}, after the genus *Hordeum* (barley). The genus name is from Latin *hordeum*, barley. A.k.a. **anhaline**, after the genus *Anhalonium* (cacti). *Anhalonium* is from a(n)- and Greek *halonion*, diminutive of *halon*, threshing floor. Via Latin *areola*, diminutive of *area*, piece of level ground, threshing floor, this name refers to a perceived lack of areoles in these cacti. A.k.a. **eremursine**, after the genus *Eremurus* (Asian herbs). The genus name is from Greek *eremos*, solitary, and *oura*, tail. A.k.a. **peyocactine**, after peyote (*Lophophora williamsii*). Peyote is from Nahuatl *peyotl*, peyote.

Humulone [(6*R*)-3,5,6-trihydroxy-2-(3-methylbutanoyl)-4,6-bis(3-methylbut-2-en-1-yl)cyclohexa-2,4-dien-1-one], after the genus *Humulus* (hops) and with 'one'. The genus name is ultimately of Germanic origin, cf. Old Norse *humli*, hop. Cf. cohumulone.

Hydnocarpic acid {11-[(1*R*)-cyclopent-2-en-1-yl]undecanoic acid}, after the genus *Hydnocarpus* (trees). The genus name is from Greek *hydnon*, truffle, and *karpos*, fruit.

Hydrastinine {6-methyl-5,6,7,8-tetrahydro[1,3]dioxolo[4,5-*g*]isoquinolin-5-ol}, after the genus *Hydrastis* (goldenseal, turmeric root, yellow root). The genus name is from Greek *hydor*, water and Latin *sistere*, to bring to a standstill, referring to these plants' hemostatic properties.

Hyodeoxycholic acid (3α,6α-dihydroxy-5β-cholan-24-oic acid), from hy(o)-, deoxy-, and cholic acid.

Hypaphorine [(2*S*)-3-(1H-indol-3-yl)-2-(trimethylazaniumyl)propanoate], after the genus *Hypaphorus* (bean plants), a.k.a. *Erythrina*. The former genus name is from hyp(o)- and Greek *aphoros*, sterile, referring to the poor fertility of these plants.

Hypericin {1,3,4,6,8,13-hexahydroxy-10,11-dimethylphenanthro[1,10,9,8-*opqra*]perylene-7,14-dione}, after the genus *Hypericum* (St. John's wort). The genus name is from Greek *hyperikon* (St. John's wort), from hyper- and *ereike*, heath, heather.

Ibotenic acid [amino(3-oxo-2,3-dihydro-1,2-oxazol-5-yl)acetic acid], after the species *Amanita ibutengutake* (a toadstool). The specific epithet is from the Japanese name of this mushroom, *ibutengutake*.

Illudin (a group of natural sesquiterpenes), after the species *Omphalotus illudens* (the mushroom Jack O'Lantern). The specific epithet is from Latin *illudens, illudent-*, mocking, from *illudere*, to mock, from *ludere*, to play.

Imperatorin {9-[(3-methylbut-2-en-1-yl)oxy]-7*H*-furo[3,2-g][1]benzopyran-7-one}, after the genus *Imperatoria* (masterwort). The genus name is from Latin *imperator*, emperor.

Imperialine (3β,20-dihydroxy-5α,17β-cevan-6-one), after the species *Fritillaria imperialis* (crown imperial). The genus name is from Latin *imperialis*, imperial, from *imperator*, emperor, from *imperare*, to rule. A.k.a. **fritillarine**, after the genus *Fritillaria*. The genus name is from Latin *fritillus*, dice cup, ultimately from *fritinnio, fritinnion-*, twittering. A.k.a. **peiminine**, after Chinese pei mu (*Fritillaria roylei*), **si-peimine**, and **zhebeinone**.

Indospicine [6-amidino-L-norleucine, (2*S*)-2,7-diamino-7-iminoheptanoic acid], with contraction, after the species *Indigofera spicata* (creeping indigo, trailing indigo). The genus name is from indigo and Latin *ferre*, to bear, to carry, and the specific epithet from Latin *spicatus*, spiked.

Iodinin (phenazine-1,6-diol 5,10-dioxide), after the bacterial species *Brevibacterium iodinum*. The specific epithet is from New Latin *iodinus*, iodine-colored. A.k.a. **iodinine.**

Ipsdienol [(4*S*)-2-methyl-6-methylideneocta-2,7-dien-4-ol], after the species *Ips confusus* (bark beetle). The genus name is from Greek *ips*, ship worm (*Teredo navalis*), actually a clam with worm-like features.

Iridoid (a group of natural monoterpenes), with contraction, after the genus *Iridomyrmex* (ants). The genus name is from [1]irid(o)- and Greek *myrmex*, ant.

Irone (a group of natural methylionone odorants), after the genus *Iris* (iris)- The genus name is from Greek *iris*, rainbow.

Isatin (1*H*-indole-2,3-dione), after the genus *Isatis* (woad). The genus name is from Greek *isatis* (woad).

Isocomene {(1*R*,3a*S*,5a*S*,8a*R*)-1,3a,4,5a-tetramethyl-1,2,3,3a,5a,6,7,8-octahydrocyclopenta[*c*]pentalene}, after the genus *Isocoma* (jimmyweed, goldenweed). The genus name is from Greek *isokome*, equal hair, referring to these plants' pappus on the seeds. A.k.a. **berkheyaradulene**, after the genus *Berkheya*. The genus names is after the Dutch physician and botanist Johannes Le Franck van Berkhey (1729–1812).

Japonilure {(5*R*)-5-[(1*Z*)-dec-1-en-1-yl]dihydrofuran-2(3*H*)-one}, after the species *Popillia japonica* (Japanese beetle). The specific epithet is from New Latin *japonicus*, Japanese; and lure.

***cis*-Jasmone** {3-methyl-2-[(2*Z*)-pent-2-en-1-yl]cyclopent-2-en-1-one}, after the genus *Jasminum* (jasmine), ultimately from Persian *yasmin, yasman*, jasmine.

Jatrophone {(1*R*,3*R*,6*E*,9*E*)-3,7,11,11,14-pentamethyl-16-oxatricyclo[11.2.1.$0^{1,5}$]hexadeca-4,6,9,13-tetraene-8,15-dione}, after the genus *Jatropha* (physic nut, nettlespurge). The genus name is from iatr(o)- and Greek *trophe*, food, from *trephein*, to nourish, and 'one'.

Jerva acid (4-oxo-4*H*-pyran-2,6-dicarboxylic acid), with contraction, after jervina (white hellebore, *Veratrum album*). Jervina is from Latin *ervium*, vetch, chick-pea. A.k.a. **jervaic acid, jervasic acid**, and **chelidonic acid.**

Juglone (5-hydroxynaphthalene-1,4-dione), after the genus *Juglandacea* (walnut tree). The genus name is from Latin *iuglans, iugland-*, walnut, from *Iupiter*, Jupiter, and *glans, gland-*, acorn.

Justicidin A {9-(1,3-benzodioxol-5-yl)-4,6,7-trimethoxynaphtho[2,3-*c*]furan-1(3*H*)-one}, after the genus *Justicia* (water herbs, shrubs). The genus name is named after the British botanist James Justice (1698–1763).

Kadsuphilactone (a group of natural pentacyclic triterpenoids), with contraction, after the species *Kadsura philippinensis* (kadsura, katsura tree), and with lactone. The genus name is from Japanese *katsura*, kadsura.

Kanamycin {(1*S*,2*R*,3*R*,4*S*,6*R*)-4,6-diamino-3-[(6-amino-6-deoxy-α-D-glucopyranosyl)oxy]-2-hydroxycyclohexan-1-yl 3-amino-3-deoxy-α-D-glucopyranoside}, after the fungal species *Streptomyces kanamyceticus*. The specific epithet is from Japanese *kana-*, golden.

Karanjin {3-methoxy-2-phenyl-4*H*-furo[2,3-*h*][1]benzopyran-4-one}, from the Hindi name *karanj* (Indian beech, *Millettia pinnata*).

Kassinin (a polypeptide), after the genus *Kassina* (running frogs, kassinas). The genus name is most likely after the American ornithologist John Cassin (1813–1869), if not a toponym from Burkina Faso.

Kavatin {(6*R*)-6-[(*E*)-2-(1,3-benzodioxol-5-yl)ethen-1-yl]-4-methoxy-5,6-dihydro-2*H*-pyran-2-one}, from kava (*Piper methysticum*). The specific epithet is from Latinized Greek *methysticus*, intoxicating. Kava is from Tongan & Marquesan *kava*, literally bitter. A.k.a. **kavahin** and **methysticin**.

Kermesic acid (3,5,6,8-tetrahydroxy-1-methyl-9,10-dioxoanthracene-2-carboxylic acid), from kermes (*Kermes*), ultimately from Arabic *qirmiz*, vivid red.

Khellin {4,9-dimethoxy-7-methyl-5*H*-furo[3,2-g][1]benzopyran-5-one}, from khella (toothpick plant, *Ammi visnaga*), from Arabic *akhillah*, khellah.

Kistrin (a polypeptide), after the species *Agkistrodon rhodostoma* (pit viper). The genus name is, irregularly, from Greek *angkistron*, fishhook.

Kojic acid [5-hydroxy-2-(hydroxymethyl)-4*H*-pyran-4-one], from koji, a yeast prepared from rice inoculated with the spores of the mold *Aspergillus oryzae*, from Japanese *koji*, koji.

Kynurenic acid (4-hydroxyquinoline-2-carboxylic acid), from Greek *kynos ouron*, dog's urine, and Latin *ren*, kidney.

Laccaic acid {7-[5-(2-acetamidoethyl)-2-hydroxyphenyl]-3,5,6,8-tetrahydroxy-9,10-dioxoanthracene-1,2-dicarboxylic acid}, after the species *Kerria lacca* (Oriental lac insect). The specific epithet is ultimately from Sanskrit *laksa*, lac.

Lactic acid (2-hydroxypropanoic acid), from Latin *lac, lact-*, milk.

Lactucin {(3a*R*,4*S*,9a*S*,9b*R*)-4-hydroxy-9-(hydroxymethyl)-6-methyl-3-methylidene-3,3a,4,5,9a,9b-hexahydroazuleno[4,5-*b*]furan-2,7-dione}, after the genus *Lactuca* (lettuce). The genus name is from Latin *lactuca*, lettuce, ultimately from *lac, lact-*, milk.

Lapachol [2-hydroxy-3-(3-methylbut-2-en-1-yl)naphthalene-1,4-dione), from lapacho, timber of the genera *Tabebuia* (roble) and *Tecoma* (trumpetbush) from Spanish *lapacho*, a word loaned from a native South American, most likely Argentinian, language.

Lardolure [(1*R*,3*R*,5*R*,7*R*)-3,5,7-trimethylundecan-2-yl formate], after the species *Lardoglyphus konoi* (fish mite) and from lure. The genus name is from Latin *lardum*, bacon, and Greek *glyphe, glypt-*, carved work.

Laserpitine (a sesquiterpene), after the species *Laserpitium latifolium* (broad-leaved sermountain). The genus name is from Latin *laserpicium*, the extinct plant *Ferula historica* (silphium, laserwort).

Lasiocarpine {1,5-dideoxy-2-*C*-methyl-4-*O*-methyl-3-*C*-[[[(1*S*,7a*R*)-1-[[(2*Z*)-2-methylbut-2-enoyl]oxy]-2,3,5,7a-tetrahydro-1*H*-pyrrolizin-7-yl]methoxy]carbonyl]-D-arabinitol}, after the species *Heliotropium lasiocarpum* (an oriental weed). The specific epithet is from Greek *lasios*, shaggy, and *karpos*, fruit.

Latrotoxin (a group of toxic proteins), after the genus *Latrodectus* (widow spiders). The genus name is from Latin *latro, latron-*, mercenary soldier, brigand, and Greek *-dektes*, a biter, from *daknein*, to bite.

Latrunculin (a group of mostly toxic macrolides), after the genus *Latrunculus* (marine snails), later renamed *Babylonia*. The former genus name is from Latin *latrunculus*, diminutive of *latro, latron-*, mercenary soldier, brigand.

Laudanin {5-[(6,7-dimethoxy-2-methyl-1,2,3,4-tetrahydroisoquinolin-1-yl)methyl]-2-methoxyphenol}, after obsolete laudanum, an alcoholic solution of opium, thus referring to its nature as a degradation product of opium. A.k.a. **laudanidine** and **tritopin**.

Laurepukine {7-methyl-6,7,7a,8-tetrahydro-5*H*-[1,3]benzodioxolo[6,5,4-*de*]benzo[*g*]quinolin-12-ol 7-oxide}, with contraction, after the species *Laurelia novae-zeelandiae* (pukatea) and its common name. The genus name is from Latin *laurus*, laurel. The common name is from Maori *pukatea*, the forest tree pukatea.

Laurotetanine {(6a*S*)-1,2,10-trimethoxy-5,6,6a,7-tetrahydro-4*H*-dibenzo[*de,g*]quinolin-9-ol}, with contraction, after the family *Lauraceae* (laurel) and tetanus, from Greek *tetanos*, rigid, stretched, from *teinein*, to stretch. The name of the plant family is from Latin *laurus*, laurel.

Lawsone (2-hydroxynaphthalene-1,4-dione), after the species *Lawsonia inermis* (hina, henna tree, mignonette tree, Egyptian privet). The genus name is after the British physician, mineralogist, and amateur botanist Isaac Lawson (1704–1747).

Leainafulvene (a sesquiterpenoid), after the fungal species *Mycena leaiana* (orange mycena, Lea's mycena) and from fulvene. The specific epithet is after the American amateur botanist Thomas Gibson Lea (1785–1844).

Ledol {(1a*R*,4*R*,4a*S*,7*R*,7a*S*,7b*S*)-11,4,7-tetramethyldecahydro-1*H*-cyclopropa[*e*]azulen-4-ol}, after the obsolete genus *Ledum* (Labrador tea), now regarded as a subsection of the genus *Rhododendron*. The former genus name is from Greek *ledon*, rockrose (*Cistus*). A.k.a. **ledum camphor.**

Lenthionine (1,2,3,5,6-pentathiepane), after the genus *Lentinus* (mushrooms) and with thi(o)-. The genus name is from Latin *lentus*, tough, pliable. Not to be confused with lanthionine.

Leonurine {4-[(aminoiminomethyl)amino]butyl 4-hydroxy-3,5-dimethoxybenzoate}, after the genus *Leonurus* (motherwort). The genus name is from Greek *leon*, lion, and *oura*, tail.

Lepiochlorin [5-(chloromethyl)-5-hydroxy-3-methylfuran-2(5*H*)-one], after the genus *Lepiota* (basidiomycetes) and from chlorine. The genus name is from Greek *lepion*, diminutive of *lepos*, rind, husk, scale, from *lepein*, to peel.

Leucodrin {(4*R*,5*S*,8*R*,9*R*)-8-[(1*S*)-1,2-dihydroxyethyl]-9-hydroxy-4-(4-hydroxyphenyl)-1,7-dioxaspiro[4.4]nonane-2,6-dione}, after the genus *Leucadendron* (conebush, sunshinebush, yellowbush). The genus name is from leuc(o)- and dendr(o)-.

Lichesterol [(22*E*)-ergosta-5,8,22-trien-3β-ol], after lichen and from sterol. Lichen is ultimately from Greek *leichen*, tree moss, lichen, literally what eats around itself, probably from *leichein*, to lick.

Licochalcone {3-[4-hydroxy-2-methoxy-5-[(2*E*)-2-methylbut-3-en-2-yl]phenyl]-1-(4-hydroxyphenyl)prop-2-en-1-one}, with contraction, from licorice (*Glyzyrrhiza*) and chalcone.

Lignin (a complex natural organic polymer), from Latin *lignum*, wood.

Limettin (5,7-dimethoxy-2*H*-1-benzopyran-2-one), after the species *Citrus limetta* (bitter orange). The specific epithet is from French *limette*, lime fruit. A.k.a. **citropten** and **citroptene**, from citrus and stearopten.

Limonin {(4a*S*,6a*R*,8a*R*,8b*R*,9a*S*,12*S*,12a*S*,14a*R*,14b*R*)-12-(furan-3-yl)-6,6,8a,12-tetramethyldecahydro-1*H*,3*H*-oxireno[c]pyrano[4″,3″:2′,3′]furo[3′,4′:5,6]naphtho[1,2-*d*]pyran-3,8,10(6*H*,9a*H*)-trione}, after Latin *limonum*, lemon. A.k.a. **evodin.**

Limulus test (for bacterial endotoxins), after the species *Limulus polyphemus* (horseshoe crab). The genus name is from Latin *limulus*, diminutive of *limus*, slanting.

Lithocholic acid (3α-hydroxy-5β-cholan-24-oic acid), from Greek *lithos*, stone, and cholic acid, referring to this acid's occurrence in gall stones.

Lobeline {2-[(2*R*,6*S*)-6-[(2*S*)-2-hydroxy-2-phenylethyl]-1-methylpiperidin-2-yl]-1-phenylethan-1-one}, after the genus *Lobelia* (lobelia). The genus name is after the Flemish botanist Matthias de Lobel (1538–1616).

Lochnericine (methyl 2,3-didehydro-6α,7α-epoxy-5α,12β,19α-aspidospermidine-3-carboxylate), after the genus *Lochneria* (periwinkle). The genus name is after the German botanist and physician Michael Friedrich Lochner von Hummelstein (1662–1720).

Locustol (5-ethyl-2-methoxyphenol), from locust and 'ol'.

Loganin {methyl (1*S*,4a*S*,6*S*,7*R*,7a*S*)-1-[(β-D-glucopyranosyl)oxy]-6-hydroxy-7-methylhexahydrocyclopenta[*c*]pyran-4-carboxylate}, after the family *Loganiaceae* (herbs, shrubs, trees). The family name is after the Irish-American botanist James Logan (1674–1751).

Looplure [(7*Z*)-dodec-7-en-1-yl acetate], after the cabbage looper moth (*Trichoplusia ni*) and lure. The specific epithet corresponds to the Greek letter ν (lower-case nu), referring to the shape of the markings on the moth's wings.

Lumisterol [(22*E*)-9β,10α-ergosta-5,7,22-trien-3β-ol], from Latin *lumen, lumin-*, light, and sterol, referring to the photochemical synthesis of this compound.

Lunacridine {3-[(2*R*)-2-hydroxy-3-methylbutyl]-4,8-dimethoxy-1-methylquinolin-2(1*H*)-one}, after the genus *Lunasia* (tropical trees and shrubs) and acridine. The genus name is from Tagalog *lunas*, lunasia.

Lunarine {(20a*R*,24a*S*)-tetradecahydro-17,18-etheno-22*H*-benzofura[3a,3-*n*][1,5,10]triazacycloicosine-3,14,22-trione}, after the plant genus *Lunaria* (honesty). The genus name is from Latin *lunarius*, of the moon, from *luna*, moon, and refers to its decorative seedpods.

Lunularic acid {2-hydroxy-6-[2-(4-hydroxyphenyl)ethyl]benzoic acid}, after the genus *Lunularia* (liverwort). The genus name is from Latin *lunula*, a crescent-shaped ornament worn by women, diminutive of *luna*, moon, and refers to its moon-shaped gemmacups.

Lupanine (2-oxosparteine), after the genus *Lupinus* (lupin, lupine). The genus name is from Latin *lupinus*, lupin, literally of a wolf. The motive for the Latin name is obscure.

Lyophyllin (2,*N*,*N*-trimethyldiazenecarboxamide 2-oxide), after the genus *Lyophyllum* (mushrooms), later renamed *Calocybe*. The genus name is from ly(o)- and phyll(o)-.

Lysergic acid (6-methyl-9,10-didehydroergoline-8β-carboxylic acid), from lys(o)- and ergot (*Claviceps*), referring to its formation upon hydrolysis of ergot.

Maclurin [(3,4-dihydroxyphenyl)(2,4,6-trihydroxyphenyl)methanone], after the genus *Maclura* (fustic). The genus name is after the British-American geologist William Maclure (1763–1840).

Magnolol {5,5′-di(prop-2-en-1-yl)-[1,1′-biphenyl]-2,2′-diol}, after the genus *Magnolia* (magnolia). The genus name is after the French botanist Pierre Magnol (1638–1715).

Maitotoxin (a polycyclic toxin), after Tahitian *maito*, striated surgeonfish (*Ctenochaetus striatus*). A.k.a. **MTX**.

Malic acid (2-hydroxybutanedioic acid), from Latin *malum*, apple.

Maltol (3-hydroxy-2-methyl-4*H*-pyran-4-one), from malt, referring to its formation upon roasting of barley.

Malvidin [3,5,7-trihydroxy-2-(4-hydroxy-3,5-dimethoxyphenyl)-1-benzopyrylium, a cation], after the genus *Malvia* (mallow). The genus name is from Latin *malva*, mallow.

Mandelic acid (2-hydroxy-2-phenylethanoic acid), from German *Mandelsäure*, mandelic acid, from *Mandel*, almond, ultimately from Greek *amygdalon*, diminutive of *amygdale*, almond.

Mangostin [1,3,6-trihydroxy-7-methoxy-2,8-bis(3-methylbut-2-en-1-yl)-9*H*-xanthen-9-one], after the species *Garcinia mangostana* (mangosteen tree). The specific epithet is from Malay *manggisutan*, mangosteen.

Manicone [(4*E*,6*S*)-4,6-dimethyloct-4-en-3-one], after the genus *Manica* (ants). The genus name could be from Latin *manica*, long sleeve, possibly referring to an anatomical feature of these animals.

Margatoxin (a polypeptide), after the species *Centruroides margaritatus* (a scorpion). The specific epithet is from Latin *margaritatus*, adorned with pearl, ultimately from Greek *margaron*, pearl.

Marginalin {(3*E*)-5-hydroxy-3-[(4-hydroxyphenyl)methylidene]-1-benzofuran-2(3*H*)-one}, after the species *Dytiscus marginalis* (great diving beetle). The specific epithet is from Latin *marginalis*, marginal.

Marmesin {2,3-dihydro-2-(2-hydroxypropan-2-yl)-7*H*-furo[3,2-*g*][1]benzopyran–7-one}, after the species *Aegle marmelos* (bael, Bengal quince, golden apple, Japanese bitter orange, stone apple, wood apple). The specific epithet is ultimately from Greek *melimelon*, literally honey apple, an apple tree grafted upon a quince.

Marrubiin {(2a*S*,5a*S*,6*R*,7*R*,8a*R*,8b*R*)-6-[2-(3-furyl)ethyl]-6-hydroxy-2a,5a,7-trimethyldecahydro-2*H*-naphtho[1,8-*bc*]furan-2-one}, after the genus *Marrubium* (horehound). The genus name is from Latin *marrubium*, horehound. A.k.a. **marrubin**.

Matatabiether {1,4-dimetthyl-9-methylidene-2-oxabicyclo[3.3.1]nonane}, after the Japanese plant *matatabi*, silver vine, cat powder (*Actinidia polygama*), and ether.

Medicagol {3-hydroxy-6*H*-[1,3]dioxolo[5,6]benzofuro[3,2-*c*][1]benzopyran-6-one}, after the genus *Medicago* (medick). The genus name is from Latin *medica*, medick, from *medicus*, from the ancient country of Media, and *plantago*, plantain.

Megacin (an antibacterial protein), after the bacterial species *Bacillus megaterium*. The specific epithet is from mega- and Greek *teras, terat-*, monster.

Megaphone {(4*R*,6*R*)-6-[(1*R*,2*S*)-1-hydroxy-1-(3,4,5-trimethoxyphenyl)propan-2-yl]-4-methoxy-6-(prop-2-en-1-yl)cyclohex-2-en-1-one}, after the species *Aniba megaphylla* (a tropical tree) and 'one', a name no doubt crafted tongue-in-cheek. The specific epithet is from mega- and Greek *phyllon*, leaf.

Mellitic acid (benzenehexacarboxylic acid), after the mineral mellite, an aluminum salt of mellitic acid.

Melittin (a polypeptidic bee venom), from Greek *meli*, honey.

Menthiafolin {6-[β-D-allopyranosyl)oxy]-5-ethenyl-4a,5-dimethyl-1-oxo-4,4a,5,6-tetrahydro-1*H*,3*H*-pyrano[3,4-*c*]pyran-3-yl (2*E*)-6-hydroxy-2,6-dimethylocta-2,7-dienoate}, after the species *Menyanthes trifoliata* (buckbean). The genus name is from Greek *menyein*, to disclose, and anth(o)-, referring to this plant's successively opening raceme. The specific epithet is from Latin *trifoliatus*, three-leaved.

Mescaline [2-(3,4,5-trimethoxyphenyl)ethan-1-amine], after the intoxicating beverage mescal, obtained from agave (*Agave*), ultimately from Nahuatl *mexcalli*, mescal.

Meteloidine {6,7-dihydroxy-8-methyl-8-azabicyclo[3.2.1]octan-3-yl (2*E*)-2-methylbut-2-enoate}, after the species *Datura meteloides* (angel's trumpet, downy thorn apple, hairy datura, hairy thorn apple, Indian apple, lily-weed, recurved thorn apple, sacred datura, white thorn apple), now renamed *Datura innoxia*. The former specific epithet is New Latin *meteloides*, resembling *Datura metel* (devil's trumpet, downy datura). The latter specific epithet is from metel nut, from Arabic *jawz matil*, metel nut. Arabic *jawz* means nut, *matil* is of obscure meaning.

Mimosine {3-[3-hydroxy-4-oxopyridin-1(4*H*)-yl]-l-alanine}, after the genus *Mimosa* (mimosa). The genus name is from Greek *mimos*, mime, imitator, referring to the notion that these plants imitate animal life by their visible reactions to touch.

Miraculin (a glycoprotein), after the miracle fruit (*Synsepalum dulcificum*).

Miriquidic acid {2-hydroxy-4-[[2-hydroxy-4-methoxy-6-(3-oxopentan-1-yl)benzoyl] oxy]-6-pentylbenzoic acid}, after the fungal genus *Miriquidica*. The genus name is from Proto-Germanic *Miriquidi*, literally dark forest, referring, inter alia, to the German mountain range *Erzgebirge*, literally ore mountains, referring to the preference of this genus for such high-altitude habitats.

Mitragynine [methyl (16*E*)-9,17-dimethoxy-20β-coryn-16-en-16-carboxylate], after the species *Mitragyna speciosa* (kratom). The genus name is from Greek *mitra*, cap, and *gyne*, woman.

Mizoribine [5-hydroxy-1-(β-D-ribofuranosyl)-1*H*-imidazole-4-carboxamide], with contraction, after its discoverer, the Japanese chemist Kimio Mizuno (born 1937), and from ribose. A.k.a. **bredinin**, after the fungal species *Penicillium brefeldianum*. The specific epithet is after the German botanist and bacteriologist Oskar Brefeld (1839–1925).

Mompain (2,5,7,8-tetrahydroxynaphthalene-1,4-dione), after the fungal species *Helicobasidium mompa*. The specific epithet is from Japanese *mompa-byo*, a disease of mulberry trees caused by this fungus, from *mompa*, a kind of cotton cloth, and *byo*, disease.

Monensin (a polyether antibiotic), with drastic contraction, after the bacterial species *Streptomyces cinnamonensis*. The specific epithet is from cinnamon.

Monorden (a macrolide), with contraction, after the fungal species *Monosporium bonorden*. The specific epithet is after the German mycologist Hermann Friedrich Bonorden (1801–1884). A.k.a. **radicicol**, after the fungal species *Stemphylium radicinum*. The specific epithet is from Latin *radicinus*, of consistence like a root.

Morin [2-(2,4-dihydroxyphenyl)-3,5,7-trihydroxy-4*H*-1-benzopyran-4-one], after the genus *Morus* (mulberry tree), formerly including old fustic (*Maclura tinctoria*), from which morin was obtained.

Morindin {1,5-dihydroxy-2-methyl-6-[(6-*O*-β-D-xylopyranosyl-β-D-glucopyranosyl) oxy]anthracene-9,10-dione}, after the genus *Morinda* (Indian mulberry). The genus name is from Latin *morus indica*, Indian mulberry tree.

Multifidine {(1*Z*)-4-[(β-D-glucopyranosyl)oxy]-2-methylbut-2-enenitrile}, after the species *Jatropha multifida* (coral plant, coral tree, physic nut). The specific epithet is from Latin *multus, multi-*, much, and *findere, -fidus*, to cleave.

Muscarine {[(4-hydroxy-5-methyltetrahydrofuran-2-yl)methyl]trimethylammonium, a cation}, after the species *Amanita muscaria* (fly agaric). The specific epithet is from Latin *muscarius*, of a fly, from *musca*, fly.

Muscimol [5-(aminomethyl)-1,2-oxazol-3(2*H*)-one], cf. muscarine. A.k.a. **agarin.**

Muscone (3-methylcyclopentadecan-1-one), from Latin *muscus*, musk, and 'one'.

Mycelianamide {(3*E*,6*S*)-3-[[4-[[(2*E*)-3,7-dimethylocta-2,6-dien-1-yl]oxy]phenyl]methylidene]-1,4-dihydroxy-6-methylpiperazine-2,5-dione}, from mycelium, from myc(o)- and Greek *helos*, nail, wart, callus.

Mycenone {3,6-dichloro-4-[(3*E*)-4-chloro-3-methylbut-3-en-1-yn-1-yl]-5-hydroxycyclohexa-3,5-diene-1,2-dione}, after the fungal genus *Mycena*. The genus name is from Greek *mykes*, mushroom.

Myricyl alcohol (triacontan-1-ol), after the genus *Myrica* (myrtle). The genus name is from Greek *myrike*, tamarisk (*Tamarix*), probably of Semitic origin.

Myxalamide A {(2*E*,4*E*,6*Z*,8*E*,10*E*,12*R*,14*E*,16*S*)-13-hydroxy-*N*-[(2*S*)-1-hydroxypropan-2-yl]-2,10,12,14,16-pentamethyloctadeca-2,4,6,8,10,14-hexaenamide}, after the bacterial species *Myxococcus xanthus*. The genus name is from Greek *myxa*, mucus, and *kokkos*, grain, seed, berry.

Myxothiazol {(2*E*,81,2-d4*R*,5*S*,6*E*)-3,5-dimethoxy-4-methyl-7-[2′-[(3*E*,5*E*)-7-methylocta-3,5-dien-2-yl]-2,4′-bi-1,3-thiazol-4-yl]hepta-2,6-dienamide}, after the bacterial genus *Myxococcus*, from myx(o)-, and thiazole.

Nandinine {10-methoxy-5,8,13,13a-tetrahydro-6*H*-[1,3]dioxolo[4,5-*g*]isoquinolino[3,2-*a*]isoquinolin-9-ol}, after the genus *Nandina* (nandina, heavenly bamboo, sacred bamboo). The genus name is from Japanese *nandin*, nandina.

Napelline {(1*R*,2*R*,4*S*,7*R*,8*R*,9*R*,13*R*,16*S*,17*R*)-11-ethyl-6-methylidene-11-azahexacyclo [7.7.2.1^{5,8}.0^{1,10}.0^{2,8}.0^{13,17}]nonadecane-4,7,16-triol}, after the species *Aconitum napellus* (monkshood). The specific epithet is from Latin *napellus*, diminutive of *napus*, turnip.

Naringenin [(2*S*)-2,3-dihydro-5,7-dihydroxy-2-(4-hydroxyphenyl)-4*H*-1-benzopyran-4-one], after Sanskrit *naranga*, orange tree (*Citrus*).

Navenone A [(3*E*,5*E*,7*E*,9*E*)-10-(pyridin-3-yl)deca-3,5,7,9-tetraen-2-one], after the species *Navanax inermis* (the sea slug California aglaja). The genus name is a variation of the genus name *Navarchus*, from Greek *nauarchos*, fleet commander.

Nepetalactone {(4a*S*,7*S*,7a*R*)-4,7-dimethyl-5,6,7,7a-tetrahydrocyclopenta[*c*]pyran-1(4a*H*)-one}, after the genus *Nepeta* (mints). The genus name is from Latin *nepeta*, catnip.

Nereistoxin (*N*,*N*-dimethyl-1,2-dithiolan-4-amine), after the species *Lumbriconereis heteropoda* (an annelid), later renamed *Kuwaita heteropoda*. Latin *lumbricus*, of obscure etymology, means earthworm. Nereis is the name of Greek mythical sea nymphs.

Neriifolin {3β-[(6-deoxy-3-*O*-methyl-α-L-glucopyranosyl)oxy]-14-hydroxy-5β-card-20(22)-enolide}, after the species *Thevetia neriifolia* (yellow oleander). The specific epithet is from New Latin *neriifolius*, oleander-leaved, from *Nerium* (oleander), from Greek *nerion*, oleander, and Latin *folium*, leaf.

Neuraminic acid (5-amino-3,5-dideoxy-D-*glycero*-D-*galacto*-non-2-ulosonic acid), from Greek *neuron*, nerve, sinew, referring to its formation by degradation of compounds obtained from brain lipids.

Neurine (*N*,*N*,*N*-trimethylethenaminium, a cation), from neur(o)-, referring to its preparation by treatment of cattle brain with barium hydroxide.

Neurosporene (7,8-dihydrolycopin, 7,8-dihydro-ψ,ψ-carotene), after the fungal genus *Neurospora*. The genus name is from Greek *neurospora*, nerve spore.

Nicotine {3-[(2*S*)-1-methylpyrrolidin-2-yl]pyridine}, after the genus *Nicotiana* (tobacco plants). The genus name is after the French diplomat Jean Nicot (1530–1600).

Nimbiol (12-hydroxy-13-methylpodocarpa-8,11,13-trien-7-one), from Sanskrit *nimba*, neem tree (*Azadirachta indica*).

Nitramine {2-azaspiro[5.5]undecan-7-ol}, after the genus *Nitraria* (nitre bush, dillon bush). The genus name is from nitre bush, referring to the fact that these plants were first found near Siberian nitre deposits.

Nivalenol [(12*R*)-3α,4β,7α,15-tetrahydroxy-12,13-epoxytrichothec-9-en-8-one], after the fungal species *Fusarium nivale*. The specific epithet is from Latin *nivalis*, snow white, from *nix, niv-*, snow.

Nocardamine (1,12,23-trihydroxy-1,6,12,17,23,28-hexaazacyclotritriacontane-2,5,13,16,24,27-hexone), after the bacterial genus *Nocardia*, named after the French veterinarian and biologist Edmond Isidore Etienne Nocard (1850–1903).

Nodakenin {(2*R*)-2-[1-[(β-D-glucopyranosyl)oxy]-1-(propan-2-yl)-2,3-dihydro-7*H*-furo[3,2-*g*][1]benzopyran-7-one}, from Japanese *nodake*, peucedanum, hogfennel root (*Peucedanum decursivum*).

Nogalamycin (an anthracycline antibiotic), after the bacterial species *Streptomyces nogalater*. The specific epithet is from New Latin *nogalater*, walnut black, from Spanish *nogal*, walnut tree, and Latin *ater, atr-*, black.

Nostocine A {7-methyl-2,7-dihydro-3*H*-pyrazolo[4,3-*e*][1,2,4]triazin-3-one}, after the bacterial genus *Nostoc*. The genus name is after *Nostoch*, a name given to the conspicuous black and brittle exudates of colonies of the cyanobacterium *Nostoc commune*, often symbiotic with algae and higher plants, a.k.a. troll's butter, mare's eggs, star jelly, and witch's jelly, which can often be seen on soils after thunderstorms. The name *Nostoch* was invented by the Swiss-German alchemist and physician Philippus Aureolus Paracelsus (Theophrastus Bombast von Hohenheim) (1493–1541).

Nudic acid B [(2*E*)-7-cyanohept-2-ene-4,6-diynoic acid], after the fungal species *Tricholoma nudum* (wood blewit), later renamed *Clitocybe nuda*. The specific epithet is from Latin *nudus*, naked. A.k.a. **diatretyne II**, after the fungal species *Clitocybe diatreta*. The specific epithet is from Latin *diatretus*, pierced.

Nupharidine [(1*R*,4*S*,5*R*,7*S*,9a*S*)-4-(3-furyl)octahydro-1,7-dimethyl-2*H*-quinolizine 5-oxide], after the genus *Nuphar* (water lilies). The genus name is from Sanskrit *nilotpala*, nenuphar, from *nila*, dark blue, and *utpala*, nenuphar blossom. English nenuphar is from the same source.

Ochratoxin A {*N*-[[(3*R*)-5-chloro-3,4-dihydro-8-hydroxy-3-methyl-1-oxo-1*H*-2-benzopyran-7-yl]carbonyl]-L-phenylalanine}, after the fungal species *Aspergillus ochraceus* and from toxin The specific epithet is from Latin *ochraceus*, ocher colored, from *ochra*, ocher.

Octopine {N^2-[(1*R*)-1-carboxyethyl]-L-arginine}, after the genus *Octopus* (octopus). The genus name is from Greek *oktopous*, octopod, from *okto*, eight, and *pous, pod-*, foot.

Oenin [5,7-dihydroxy-2-(4-hydroxy-3,5-dimethoxyphenyl)-1-benzopyrylium-3-yl β-D-glucopyranoside], from Greek *oinos*, wine, referring to this compound's role as the coloring matter in blue grapes.

Officinalic acid (4,4,8,17,17,21-hexamethyl-14,23-dioxo-24,25-dioxahexacyclo[$11.11.1.0^{1,10}.0^{3,8}.0^{13,22}.0^{16,21}$]pentacosane-9-carboxylic acid), after the fungal species *Laricifomes officinalis*. The specific epithet, from Latin *officina*, the storeroom of a monastery where medicines are kept, which this species shares with many others, means that this fungus is used in medicine and herbalism.

Olean {1,7-dioxaspiro[5.5]undecane}, after the species *Dacus oleae* (olive fly). The specific epithet is from Latin *olea*, olive, olive tree.

Oleanolic acid (3β-hydroxyolean-12-en-28-oic acid), after the genus *Olea* (olive trees), and 'ol'. The genus name is from Latin *olea*, olive tree.

Olivacine {1,5-dimethyl-6*H*-pyrido[4,3-b]carbazole}, after the species *Aspidosperma olivaceum* (peroba-guatambu, guatambu-vermelho, pequiá-branco). The specific epithet is from Latin *olivaceus*, olive green, from *oliva*, olive, olive tree.

Omphalone [2-(4-methylfuran-2-yl)cyclohexa-2,5-diene-1,4-dione], after the species *Lentinellus omphalodes* (the mushroom stalked lentinellus), later renamed *Lentinellus micheneri*. The former specific epithet is from Greek *omphaloeides*, navel-like, from *omphalos*, navel, referring to the characteristic indentation of this mushroom's cap.

β-Onocerin (8,14-secogammacera-8,13-diene-3β,21α-diol), after the genus *Ononis* (restharrow), and *keros*, wax. The genus name is from Greek *ononis*, restharrow.

Ooporphyrin [3,3′-(7,12-diethenyl-3,8,13,17-tetramethylporphyrin-2,18-diyl)dipropanoic acid], from Greek *oon*, egg, referring to this compound's role as the brown pigment in egg shells, and prophyrin. A.k.a. **protoporphyrin IX**, from prot(o)- and porphyrin.

Orcinol (5-methylbenzene-1,3-diol), from New Latin *orcina* (archil).

β-Orcinol (2,5-dimethylbenzene-1,3-diol), named as homolog of orcinol.

Oripavine (6-methoxy-17-methyl-6,7,8,14-tetradehydro-4,5α-epoxymorphinan-3-ol), with contraction, after the species *Papaver orientale* (oriental poppy). The genus name is from Latin *papaver*, poppy, the specific epithet from Latin *orientalis*, oriental.

Orotic acid (1,2,3,6-tetrahydro-2,6-dioxopyrimidine-4-carboxylic acid), from Greek *orrhos, orrhot-*, whey.

Orotidine (6-carboxyuridine), cf. orotic acid.

Oroxylin A (5,7-dihydroxy-6-methoxy-2-phenyl-4*H*-1-benzopyran-4-one), after the species *Oroxylum indicum* (midnight horror, oroxylum, Indian trumpet flower, broken bones, Indian caper, tree of Damocles). The genus name is from Greek *oros, orot-*, whey, and xyl(o)-.

Orsellinic acid (2,4-dihydroxy-6-methylbenzoic acid), from French *orseille*, archil.

Osajin {5-hydroxy-3-(4-hydroxyphenyl)-8,8-dimethyl-6-(3-methylbut-2-en-1-yl)-4*H*,8*H*-benzo[1,2-*b*:3,4-*b′*]dipyran-4-one}, after Osage orange (*Maclura pomifera*), named after the American Indian people Osage.

Osthole [7-methoxy-8-(3-methylbut-2-en-1-yl)-2*H*-1-benzopyran-2-one], with contraction, after the species *Peucedanum osthrutium* (masterwort, hogfennel). The specific epithet is from New Latin *ostruthius*, born under a lucky star, lucky, useful, from Old Occitan *astruc*, born under a lucky star, lucky, useful.

Otobain {(7*S*,8*R*,9*R*)-9-(1,3-benzodioxol-5-yl)-7,8-dimethyl-6,7,8,9-tetrahydronaphtho[1,2-*d*][[1,3]dioxole}, after the species *Dialyanthera otoba* (cuangare, otoba, saosa) a.k.a. *Otoba novogranatensis*. The former specific epithet and the latter genus name are from Spanish *otoba*. A.k.a. **otobite**.

Ouabain {3β-[(6-deoxy-α-L-mannopyranosyl)oxy]-1β,5β,11α,14,19-pentahydroxycard-20(22)-enolide}, after the species *Acocanthera ouabaio* (ouabaio). The specific epithet is from Somali *waba yo*, ouabaio.

Oudenone {2-[(5*S*)-5-propyldihydrofuran-2(3*H*)-ylidene]cyclopentane-2,5-dione}, after the fungal genus *Oudemansiella*. The genus name is after the Dutch physician and botanist Cornelis Antoon Jan Abraham Oudemans (1825–1906).

Palitantin {(2*R*,3*S*,5*R*,6*R*)-3-[(1*E*,3*E*)-hepta-1,3-dien-1-yl]-5,6-dihydroxy-2-(hydroxymethyl)cyclohexan-1-one}, after the fungal species *Penicillium palitans*. The specific epithet is from Latins *palitans, palitant-*, vagary, from *palitari*, to wander.

Palmatine {2,3,9,10-tetramethoxy-5,6-dihydroisoquinolino[3,2-*a*]isoquinolinium, a cation}, after the species *Jatrorhiza palmata* (columba). The specific epithet is from Latin *palmatus*, hand-shaped. The common name columba refers to St. Columba, the Irish abbot Colm Cille (521–597), who is said to have used this plant as a remedy. A.k.a. **calystigine**, cf. calystegine.

Palustric acid (abieta-8,13-dien-18-oic acid), after the species *Pinus palustris* (longleaf pine). The specific epithet is from Latin *palustris*, marshy, from *palus, palud-*, marsh.

Palytoxin (a polycyclic toxin), after the genus *Palythoa* (corals), and toxin. The genus name has been coined without any stated reason. A.k.a. **PTX.**

Panamine [(18*S*,20*R*)-12,20-cyclo-5α-ormosanine], after the species *Ormosia panamensis* (coronil, sur espino). The specific epithet is from New Latin *panamensis*, Panamanian.

Panaxoside (a group of glycosides), after the genus *Panax* (ginseng). The genus name is from Greek *panax, panak-*, all-healing, panacea. **A.k.a. ginsenoside**, from ginseng.

Papaverine {1-[(3,4-dimethoxyphenyl)methyl]-6,7-dimethoxyisoquinoline)}, after the genus *Papaver* (poppies). The genus name is from Latin *papaver*, poppy.

Parthenolide {(1a*R*,4*E*,7a*S*,10a*S*,10b*S*)-1a,5-dimethyl-8-methylidene-2,3,6,7,7a,8,10a,10b-octahydrooxireno[9,10]cyclodeca[1,2-*b*]furan-9(1a*H*)-one}, after the genus *Parthenium* (feverfew). The genus name is from Greek *parthenion*, feverfew, from *parthenios*, maidenly, from *parthenos*, maiden, virgin.

Paviine [8-[(β-D-glucopyranosyl)oxy]-7-hydroxy-6-methoxy-2*H*-1-benzopyran-2-one], after the genus *Pavia* (buckeyes). The genus name is after the Dutch botanist Peter Paaw (latinized Pavius) (1564–1617). A.k.a. **fraxin**, after the genus *Fraxinus* (ash tree).

Pavine {1-[(3,4-dimethoxyphenyl)methyl]-6,7-dimethoxyisoquinoline}, with contraction, after the genus *Papaver* (poppies), from Latin *papaver*, poppy.

Pavoninine (a group of saponins), after the species *Pardachirus pavoninus* (Pacific sole). The specific epithet is from Latin *pavoninus*, of the peacock, from *pavo, pavon-*, peacock.

Pellitorine [(2*E*,4*E*)-*N*-(2-methylpropyl)deca-2,4-dienamide], from pellitory, ultimately from Greek *pyrethron*, pyrethrum.

Pellotine (6,7-dimethoxy-1,2-dimethyl-1,2,3,4-tetrahydroisoquinolin-8-ol), from Spanish *pellote*, peyote (*Lophophora williamsii*), ultimately from Nahuatl *peyotl*, peyote.

α-Peltatin {(5*R*,5a*R*,8a*R*)-10-hydroxy-5-(4-hydroxy-3,5-dimethoxyphenyl)-5,8,8a,9-tetrahydrofuro[3′,4′:6,7]naphtho[2,3-*d*][1,3]dioxol-6(5a*H*)-one}, after the species *Podophyllum peltatum* (mayapple, American mandrake, wild mandrake, ground lemon). The specific epithet is from New Latin *peltatus*, peltate, ultimately from Greek *pelte*, small shield.

Peonidin [3,5,7-trihydroxy-2-(4-hydroxy-3-methoxyphenyl)-1-benzopyrylium, a cation), after the genus *Paeonia* (peony). The genus name is after its mythical discoverer Paeon, the physician of the Greek gods.

Peramine {2-[3-[2-methyl-1-oxo-1,2-dihydropyrrolo[1,2-*a*]pyrazin-3-yl]propyl]guanidine}, after the species *Lolium perenne* (rye grass). The specific epithet is from Latin *perennis*, perennial, from per- and *annus*, year.

Perezone {3-hydroxy-5-methyl-2-[(2*R*)-6-methylhept-5-en-2-yl]cyclohexa-2,5-diene-1,4-dione}, after the species *Perezia microcephala*, a.k.a. *Acourtia microcephala*. The former genus name is after the 16th century Spanish pharmacist Lázaro Pérez. A.k.a. **pipitzahoic acid**, after Nahuatl *pipitzahuac* (*Trixis pipitzahuac*).

Perillaldehyde (*p*-mentha-1,8-dien-7-al), after the genus *Perilla* (mints). The genus name is perhaps from Latin *perilla*, diminutive of *pera*, leather bag, wallet, of Greek origin.

Periplanone A {(3*S*,6*S*,7*E*,11*Z*)-9-methylidene-6-(propan-2-yl)-1-oxaspiro[2.9]dodeca-7,11-dien-4-one}, after the genus *Periplaneta* (cockroaches). The genus name is from peri- and Greek *planetes*, wanderer.

Periplocin {3β-[(2,6-dideoxy-4-*O*-β-D-glucopyranosyl-3-*O*-methyl-β-D-*ribo*-hexopyranosyl)oxy]-5β,14-dihydroxycard-20(22)-enolide}, after the genus *Periploca* (milkweed). The genus name is from Greek *periploke*, a twining, an interlacing.

Petunidin [2-(3,4-dihydroxy-5-methoxyphenyl)-3,5,7-trihydroxy-1-benzopyrylium, a cation], after the genus *Petunia* (petunia). The genus name is from Medieval French *petun*, tobacco, from Tupi *petyn*, tobacco, referring to the petunia's close relationship to the tobacco plant.

Peucedanin {3-methoxy-2-(propan-2-yl)-7*H*-furo[3,2-*g*][1]benzopyran-7-one}, after the genus *Peucedanum* (sulfurweed). The genus name is from Greek *peukedanon*, sulfurweed, from *peukedanos*, bitter tasting, from *peukos*, sharp.

Phalloidin (a bicyclic heptapeptide), after the species *Amanita phalloides* (death cap). The specific epithet is from Greek *phalloides*, resembling a phallus. It either refers to the shape of these mushrooms or to their resemblance of the mushroom genus *Phallus* (stinkhorn), a rather subtle distinction.

Phaseolin {(6b*R*,12b*R*)-3,3-dimethyl-6b,12b-dihydro-3*H*,7*H*-pyrano[2′,3′:6,7]benzofuro [3,2-*c*][1]benzopyran-10-ol}, after the species *Phaseolus vulgaris* (French bean). The genus name is from Latin *phaseolus*, diminutive of *phaselus*, ultimately from Greek *phaselos*, cowpea (*Vigna unguiculata*). A.k.a. **pterocarpan**.

Phenicin [2,2′-dihydroxy-4,4′-dimethyl-1,1′-bi(cyclohexa-1,4-dien-1-yl)-3,3′,6,6′-tetrone)], after the fungal species *Penicillium phoeniceum*. The specific epithet is from Latin *phoeniceus*, purple-red, literally from Phoenicia.

Phillyrin {4-[(1*R*,3a*R*,4*S*,6a*R*-4-(3,4-dimethoxyphenyl)tetrahydro-1*H*,3*H*-furo[3,4-*c*] furan-1-yl]-2-methoxyphenyl β-D-glucopyranoside)}, after the genus *Phillyrea* (mock privet). The genus name is from Greek *philyrea*, mock privet.

Phloionic acid (9,10-dihydroxyoctadecanedioic acid), from phlo(ro)-, referring to its isolation from cork.

Phloretin [3-(4-hydroxyphenyl)-1-(2,4,6-trihydroxyphenyl)propan-1-one], from phlo (ro)-, referring to its isolation from bark.

Phlorizin {1-[2-[(β-D-glucopyranosyl)oxy]-4,6-dihydroxyphenyl]-3-(4-hydroxyphenyl) propan-1-one}, as phloretin. A.k.a. **phloridzin.**

Phloroglucinol (benzene-1,3,5-triol), from phlorizin and Greek *glykys, glyker*-, sweet.

Phlorotannin (oligomeric phenolic natural products), from Greek *phloios, phlor*-, bark, and tannin.

Phorbol {4a,7b,9,9a-tetrahydroxy-3-(hydroxymethyl)-1,1,6,8-tetramethyl-1,1a,1b,4, 4a,7a,7b,8,9,9a-decahydro-5*H*-cyclopropa[3,4]benz[1,2-*e*]azulen-5-one), after the genus *Euphorbia* (spurge), The genus name is after the 1st century AD physician Euphorbus at the court of Juba, king of Mauretania, who discovered the use of these plants' latex as a remedy.

Phthiocol (2-hydroxy-3-methylnaphthalene-1,4-dione), from Greek *phthisis*, pulmonary tuberculosis, from *phthiein, phthinein*, to decay, to wane, referring to this compound's formation in the microorganism *Mycobacterium tuberculosis*.

Physalaemin (an oligopeptide), after the genus *Physalaemus* (dwarf frogs, foam frogs). The genus name is from phys(o)- and Greek *laima*, throat, gorge.

Physodic acid {3,8-dihydroxy-11-oxo-1-(2-oxoheptyl)-6-pentyl-11*H*-dibenzo[*b,e*][1,4] dioxepine-7-carboxylic acid}, after the lichen species *Parmelia physodes* a.k.a. *Hypogymnia physodes*. The specific epithet is from Greek *physodes*, full of wind. A.k.a. **physodalin**.

Physostigmine {(3a*R*,8a*S*)-1,3a,8-trimethyl-1,2,3,3a,8,8a-hexahydropyrrolo[2,3-*b*]indol-5-yl *N*-methylcarbamate}, after the genus *Physostigma* (Calabar bean). The genus name is from Greek *physa*, bellows, and *stigma*, mark, from *stizein*, to tattoo. A.k.a. **eserine**, after French *éséré*, Calabar bean (*Physostigma venenosum*).

Phytic acid [(1*R*,2*S*,3*r*,4*R*,5*S*,6*s*)-cyclohexane-1,2,3,4,5,6-hexayl hexakis(dihydrogen phosphate)], from phyt(o)-.

Picein {1-[4-[(β-D-glucopyranosyl)oxy]phenyl]ethan-1-one), after the genus *Picea* (spruce). The genus name is from Latin *picea*, pitch pine, from *pix, pic*-, pitch.

Picrocrocin {(4*R*)-4-[(β-D-glucopyranosyl)oxy]-2,6,6-trimethylcyclohex-1-ene-1-carbaldehyde}, from picr(o)- and the genus name *Crocus* (crocus). The genus name is from Greek *krokos*, saffron, crocus, of Semitic origin.

Picoline (methylpyridine), from Latin *pix*, *pic-*, pitch.

Pilocarpine {(3*S*,4*R*)-3-ethyl-4-[(1-methyl-1*H*-imidazol-5-yl)methyl]dihydrofuran-2(3*H*)-one}, after the genus *Pilocarpus* (jaborandi). The genus name is from pil(o)- and carp(o)-.

Pimpinellin {5,6-dimethoxy-2*H*-furo[2,3-*h*][1]benzopyran-2-one}, after the genus *Pimpinella* (herbs). The genus name is from Medieval Latin *pimpinella*, possibly a diminutive of Latin *pepo*, *pepon-*, pumpkin, or Latin *pampinus*, tendril or young shoot of a vine, vine leaf.

Pinonic acid [(3-acetyl-2,2-dimethylcyclobutyl)acetic acid], after the genus *Pinus* (pines), from Latin *pinus*, pine. Cf. nopinic acid.

Pinosylvin {5-[(E)-2-phenylethen-1-yl]benzene-1,3-diol}, after the species *Pinus sylvestris* (Scotch pine). The genus name is from pin(o)-, the specific epithet from Latin *silvestris*, of the forest, from *silva*, forest.

Piperic acid [(2*E*,4*E*)-5-(1,3-benzodioxol-5-yl)penta-2,4-dienoic acid], after the genus *Piper* (peppers), from Latin *piper*, pepper, ultimately from Sanskrit *pippali*, long pepper (*Piper longum*).

Piperine [(2*E*,4*E*)-5-(2*H*-1,3-benzodioxol-5-yl)-1-(piperidin-1-yl)penta-2,4-dien-1-one], as piperic acid.

Piperitone [3-methyl-6-(propan-2-yl)cyclohex-2-en-1-one], after the species *Eucalyptus piperita* (Sydney peppermint) and 'one'. The specific epithet is from Latin *piperitus*, seasoned with pepper.

Piperonal (1,3-benzodioxole-5-carbaldehyde), from piperine and 'al'. A.k.a. **heliotropin**, after the genus *Heliotropium* (heliotrope). The genus name is from heli(o)- and -trope.

Pisatin {3-methoxy-6*H*-[1,3]dioxolo[5,6]benzofuro[3,2-*c*][1]benzopyran-6a(12a*H*)-ol}, after the species *Pisum sativum* (field and garden pea). The genus name is from Latin *pisum*, pea.

Plastoquinone [5-(*all-trans*-nonaprenyl)-2,3-dimethyl-1,4-benzoquinone], named as natural quinone found in chloroplasts. A.k.a. **Kofler's quinone**, after the Austrian pharmcognosist Ludwig Kofler (1891–1951), **plastoquinone A, plastoquinone 9**, and **plastoquinone 45**.

Platyphyllin [(15*Z*)-12-hydroxy-1α,2-dihydrosenecionan-11,16-dione], after the species *Senecio platyphyllus* (a groundsel) a.k.a. *Caucasalia macrophylla*. The former specific epithet is from platy- and phyll(o)-.

Plaunotol {(2*Z*,6*E*)-2-[(3*E*)-4,8-dimethylnona-3,7-dien-1-yl]-6-methylocta-2,6-diene-1,8-diol}, after Thai *plau-noi* (*Croton sublyratus*).

Pleuromutilin [(3a*S*,4*R*,5*S*,6*S*,8*R*,9*R*,9a*R*,10*R*)-6-ethenyl-5-hydroxy-4,6,9,10-tetramethyl-1-oxodecahydro-3a,9-propano-3a*H*-cyclopentacycloocten-8-yl 2-hydroxyacetate}, after the species *Pleurotus mutilus* (a mushroom), later renamed *Omphalina mutila*. The former genus name, meaning side ear, is from pleur(o)- and Greek *ous, ot-* ear, and the specific epithet from Latin *mutilus*, mutilated.

Plicatic acid [(1*S*,2*S*,3*R*)-1-(3,4-dihydroxy-5-methoxyphenyl)-2,3,7-trihydroxy-3-(hydroxymethyl)-6-methoxy-1,2,3,4-tetrahydronaphthalene-2-carboxylic acid], after the species *Thuja plicata* (western red cedar). The specific epithet is from Latin *plicatus*, folded, from *plicare*, to fold.

Plumbagin (5-hydroxy-2-methylnaphthalene-1,4-dione), after the genus *Plumbago* (leadwort). The genus name is from Latin *plumbago, plumbagin-*, galena, referring to the lead-blue color of the flowers of these plants.

Plumericin (an iridoid lactone), after the genus *Plumeria* (frangipani). The genus name is after the French botanist Charles Plumier (1646–1704).

Pododacric acid (12,16,17-trihydroxyabieta-8,11,13-trien-19-oic acid), with contraction, after the species *Podocarpus dacrydioides* (Kahikatea white pine). The genus name is from pod(o)- and carp(o)-, referring to the length of the fleshy stalks, and the specific epithet from Greek *dakrydion* (also *skammonia*), scammony, from *dakry*, a tearing, referring to its purgative action..

Polygodial [(1*R*,4a*S*,8a*S*)-5,5,8a-trimethyl-1,4,4a,5,6,7,8,8a-octahydronaphthalene-1,2-dicarbaldehyde], with contraction, after the genus *Polygonum* (knotweed), and with 'di' and 'al'. The genus name is from Greek *polygonon* (knot grass).

Polyzonimine {6,6-dimethyl-2-azaspiro[4.4]non-1-ene}, after the species *Polyzonium rosalbum* (a millipede) and with 'imine'. The genus name is from poly- and Greek *zona*, belt, referring to the many segments of this arthropod.

Ponasterone A [(22*R*)-2β,3β,14,20,22-pentahydroxy-5β-cholest-7-en-6-one], with contraction, after the species *Podocarpus nakaii* (Nakai podocarp). The specific epithet is after the Japanes plant collector S. Nakai.

Populin [2-(hydroxymethyl)phenyl 6-*O*-benzoyl-β-D-glucopyranoside], after the genus *Populus* (poplars). The genus name is from Latin *populus*, poplar.

Pratensein [5,7-dihydroxy-3-(3-hydroxy-4-methoxyphenyl-4*H*-1-benzopyran-4-one), after the species *Trifolium pratense* (red clover). The specific epithet is from Latin *pratensis*, of the meadow, from *pratum*, meadow.

Pristane (2,6,10,14-tetramethylpentadecane), from Latin *pristis*, shark, sawfish.

Protocatechuic acid (3,4-dihydroxybenzoic acid), from prot(o)- and catechu.

Protokosin {1-[3-[[2,6-dihydroxy-4-methoxy-3-methyl-5-(2-methylpropanoyl)phenyl] methyl]-2,4,6-trihydroxy-5-[[2-hydroxy-4,6-dimethoxy-3-methyl-5-(2-methylpropanoyl) phenyl]methyl]phenyl]-2-methylpropan-1-one}, from prot(o)- and kosin, a crude anthelminthic extract from koso, the Ethiopian tree *Brayera anthelminthica*, from Oromo *kosso*, koso.

Psilocybin {[3-[2-(dimethylamino)ethyl]-1*H*-indol-4-yl] dihydrogen phosphate}, after the mushroom genus *Psilocybe* (psilocybe). The genus name is Greek for bare-headed.

Psoralen {7*H*-furo[3,2-g][1]benzopyran-7-one}, after the genus *Psoralea* (breadroot). The genus name is from Greek *psoraleos*, scabby, itchy, from *psora*, itch, akin to *psen*, to rub.

Psychotrine (7′,10,11-trimethoxy-1′,2′-didehydroemetan-6′-ol), after the genus *Psychotria* (trees, shrubs). The genus name is from Greek *psychotria*, vivification, from *psyche*, soul, life, and *atria*, therapy, medicine.

Ptaquiloside [(2′*R*,3a′*R*,4′*S*,7a′*S*)-4′-hydroxy-2′,4′,6′-trimethyl-3′-oxo-2′,3′,3a′,4′-tetrahydrospiro[cyclopropane-1,5′-inden]-7a′(1′*H*)-yl β-D-glucopyranoside], with contraction, after the species *Pteridium aquilinum* var. *latiusculum* (bracken, common bracken, brake, eagle fern, Eastern brackenfern) and named as glycoside. The genus name is from pter(o)- and the specific epithet from Latin *aquilinus*, eagle-like, from *aquila*, eagle, from *aquilus*, dark-colored, swarthy.

Pterin (2-aminopteridin-4(3*H*)-one, and tautomers), from Greek *pteron*, feather, wing, named as parent of natural compounds found in butterfly wings.

Ptomaine, from ptoma-, a class name for cadaverine and putrescine, referring to their formation in decaying animal and vegetable matter.

Puberulic acid (4,5,6-trihydroxy-3-oxocyclohepta-1,4,6-triene-1-carboxylic acid), after the fungal species *Penicillium puberulum*, later renamed *Penicillium aurantiogriseum*. The former specific epithet is from Latin *puberulus*, diminutive of *puber*, downy.

Pukateine {(7a*R*)-7-methyl-6,7,7a,8-tetrahydro-5*H*-[1,3]benzodioxolo[6,5,4-*de*]benzo [g]quinolin-12-ol}, after the pukatea tree (*Laurelia novo-zealandiae*), from Maori *pukatea*, pukatea tree.

Putrescine (butan-1,4-diamine), from Latin *putrescere*, to rot, from *puter, putr-*, rotten, referring to its formation in decaying animal matter.

Pyocyanine (5-methylphenazin-1-one), after the bacterial species *Bacillus pyocyaneus*. The specific epithet is from py(o)- and cyan(o)-.

Pyrethrin (a class of natural insecticides), from pyrethrum (*Chrysanthemum cinerariaefolium*). Pyrethrum is from Latin *pyrethrum*, pellitory, from Greek *pyrethron*, pellitory, from *pyr*, fire, referring to the spicy taste of the root.

Quassin (2,12-dimethoxypricrasa-2,12-diene-1,11,16-trione), after Surinam quassia (*Quassia amara*). The genus name is after the Surinamese healer, botanist, and former slave Graman Quassi (1692–1787). The name Quassi is from Kwa *kwasi*, Sunday or the first day of the week, and a name often given to children born on a Sunday.

Queen substance [(2*E*)-9-oxodec-2-enoic acid], named after this compound's secretion by queen honey bees (*Apis mellifera*).

Quebrachamine [(7*R*)-7-ethyl-1,4,5,6,7,8,9,10-octahydro-2*H*-3,7-methanoazacycloundecino[5,4-*b*]indole)], from quebracho (*Aspidosperma quebracho blanco*), from American Spanish *quiebrahacha*, from Spanish *quiebrar hacha*, ax-breaker. A.k.a. **kamassin**, from *Gonioma kamassi* (Knysna boxwood, East London boxwood). The specific epithet is from Khoisan *kamassi*, Knysna boxwood. The Khoisan toponym Knysna refers to a town in South Africa and means place of wood or maybe fern leaves.

Quercetagetin [2-(3,4-dihydroxyphenyl)-3,5,6,7-tetrahydroxy-4*H*-1-benzopyran-4-one], from quercetin, referring to its color, and the genus name *Tagetes* (marigold). The genus name is probably after the Etruscan deity Tages, referring to these plants' proliferous growth, both from seeds and from stems.

Quinine [(9*R*)-6-methoxy-8α-cinchonan-9-ol], from Spanish *quinquina*, the bark of the cinchona tree (*Cinchona officinalis*), ultimately from Quechua *quinaquina*, literally bark of bark or holy bark, cinchona bark. The corresponding German names *Chinin, Chinon, Chinolin*, etc. are derived from German *Chinarinde*, cinchona bark, a misnomer wrongly suggesting a relationship with China.

Radicinin {(2*S*,3*S*)-3-hydroxy-2-methyl-7-[(1*E*)-prop-1-en-1-yl]-2,3-dihydro-4*H*,5*H*-pyrano[4,3-*b*]pyran-4,5-dione}, after the fungal species *Stemphylium radicinum*. The specific epithet is from Latin *radicinus*, of consistence like a root.

Ramalic acid [2-hydroxy-4-[(2-hydroxy-4-methoxy-3,6-dimethylbenzoyl)oxy]-6-methylbenzoic acid], after the genus *Ramalia* (lichens). The genus name is from Latin *ramalia*, undergrowth, plural of *ramale*, twig, shoot, referring to their narrowly divided forms.

Rapamycin (an immunosuppressant macrolide), after Rapanui *Rapa Nui* (Easter Island, Chile), referring to the habitat of the bacterium *Streptomyces hydroscopicus*. A.k.a. **sirolimus**.

Raphanin [4-isothiocyanato-1-(methylsulfinyl)but-1-ene], after the species *Raphanus sativus* (radish). The genus name ultimately is from Greek *raphanis*, radish. A.k.a. **sulforaphene** and **sativin**.

Raubasine (methyl 19α-methyl-16,17-didehydro-18-oxayohimban-16-carboxylate), after the genus *Rauwolfia* (devil peppers), and base. The genus name is after the German botanist Leonhard Rauwolf (1535–1586).

Rebaudioside (a steviol glycoside), after the species *Stevia rebaudiana* (candy leaf) and 'oside'. The specific epithet is after the Paraguayan chemist, writer, and occultist Ovidio Rebaudi (1860–1931).

Reserpine {methyl 11,17α-dimethoxy-18β-[(3,4,5-trimethoxybenzoyl)oxy]-3β,20α-yohimban-16β-carboxylate}, anagrammatially derived from the specific epithet of *Rauwolfia serpentina* (Indian snakeroot). The specific epithet is from Latin *serpentinus*, of the snake, from *serpens, serpent-*, snake, from *serpere*, to creep.

Resorcinol (benzene-1,3-diol), a contraction of ammoniated resin gum and orchin, and 'ol'.

Retamine (7α,9α-spartein-5β-ol), after the genus *Retama* (broom bush). The genus name is ultimately from Arabic *retem* or *retam*, broom bush.

Retene [1-methyl-7-(propan-2-yl)phenanthrene], from Greek *rhetine*, resin.

Retinoic acid (15-apo-β-caroten-15-oic acid), from retina. A.k.a. **tretinoin**, by contraction of *all-trans*-retinoic acid.

Reticuline {(1*S*)-1-[(3-hydroxy-4-methoxyphenyl)methyl]-6-methoxy-2-methyl-1,2,3,4-tetrahydroisoquinolin-7-ol}, after the species *Annona reticulata* (custard apple). The specific epithet is from Latin *reticulatus*, consisting of or provided with a small net, from *reticulum*, diminutive of *rete*, net.

Retrorsine (12,18-dihydroxysenecionan-11,16-dione), after the species *Senecio retrorsus* (bushweed). The specific epithet is from Latin *retrorsus*, bent backwards, from *retroversus*, bent backwards.

Rhapontin {3-hydroxy-5-[(*E*)-2-(3-hydroxy-4-methoxyphenyl)ethen-1-yl] β-D-glucopyranoside}, after the species *Rheum rhaponticum* (rhubarb). The specific epithet is from Late Latin *rha ponticum*, Pontic rhubarb, from *rha*, rhubarb, and *Pontus*, the Black Sea. The toponym Black Sea has its roots in an archaic color scheme for the cardinal points of the known world. Black meant north, as in Black Sea, red meant south, as in Red Sea, originally the Indian Ocean, white meant west, and green or light blue meant east.

Rhizonic acid (2-hydroxy-4-methoxy-3,6-dimethylbenzoic acid), after the species *Rizocarpon geographicum* (map lichen). The genus name is from Greek *rhiza*, root, and *karpos*, fruit.

Rhododendrin [(2*R*)-4-(4-hydroxyphenyl)butan-2-yl β-D-glucopyranoside], after the genus *Rhododendron* (shrubs, trees). The genus name is from ^{1}rhod(o)- and dendr(o)-

Rhynchophylline [methyl (16*E*)-2-hydroxy-17-methoxy-1,2-didehydro-7β,20α-corynox-16-en-16-carboxylate, after the species *Uncaria rhynchophylla* (the Chinese medicinal herb gou teng). The specific epithet is from rhynch(o)- and phyll(o)-.

Rhynchophorol [(2*E*,4*S*)-6-methylhept-2-en-4-ol], after the species *Rhynchophorus palmarum* (palm weevil). The genus name is from rhynch(o)- and -phore.

Rimocidin (a macrolide antifungal agent), after the bacterial species *Streptomyces rimosus*. The specific epithet is from Latin *rimosus*, fissured, from *rima*, slit, crack, fissure.

Robinin {3-[[6-*O*-(6-deoxy-α-L-mannopyranosyl)-β-D-galactopyranosyl]oxy]-7-[(6-deoxy-α-L-mannopyranosyl)oxy]-5-hydroxy-2-(4-hydroxyphenyl)-4*H*-1-benzopyran-4-one}, after the genus *Robinia* (locust tree). The genus name is after the French gardener Jean Robin (1550–1629) and his son, the French botanist Vespasien Robin (1579–1662).

Roccellic acid [(2*R*,3*S*)-2-dodecyl-3-methylbutanedioic acid], after the genus *Roccella* (lichens). The genus name is a New Latin variation of Italian *oricello*, archil.

Rohitukin {6-[4-acetoxy-5,9a-dimethyl-2,7-dioxooctahydro-2*H*-pyrano[3,4-b]oxepin-5-yl]-5-(formyloxy)-3-(furan-3-yl)-7a-hydroxy-3a-methyl-7-methylidene-1-oxooctahydro-1*H*-inden-4-yl 3-methylbutanoate}, after the species *Amoora rohituka* (rohituka tree, pitraj tree), later renamed *Aphanamixis polystachia*. The former specific epithet is from Kannada *roheethaka*, rohituka tree, pitraj tree. Not to be confused with rohitukine.

Rohitukine {5,7-dihydroxy-8-[(3*S*,4*R*)-3-hydroxy-1-methylpiperidin-4-yl]-2-methyl-4*H*-1-benzopyran-4-one}, after the species *Amoora rohituka* (rohituka tree, pitraj tree), later renamed *Aphanamixis polystachia*. The former specific epithet is from Kannada *roheethaka*, rohituka tree, pitraj tree. Not to be confused with rohitukin.

Rotenone {(2*R*)-8,9-dimethoxy-2-(prop-1-en-2-yl)-1,2α,12,12*aa*-tetrahydro-[1]benzopyrano[3,4-b]furo[2,3-*h*][1]benzofuran-6(6a*H*)-one}. after Japanese *roten*, derris plant (*Derris elliptica*).

Rottlerin {(2*E*)-1-[6-[(3-acetyl-2,4,6-trihydroxy-5-methylphenyl)methyl]-5,7-dihydroxy-2,2-dimethyl-2*H*-1-benzopyran-8-yl]-3-phenylprop-2-en-1-one}, after the genus *Rottleria* (East Indian trees), later renamed *Mallotus*. The former genus name is after the German-Danish missionary Johann Peter Rottler (1749–1836).

Ruberythric acid (1-hydroxy-9,10-dioxoanthracen-2-yl 6-*O*-β-D-xylopyranosyl-β-D-glucopyranoside), after the genus *Rubia* (Old World herbs). The genus name is from Latin *rubia*, madder, and erythr(o)-. A.k.a. **rubian**.

Rugulovasine A {(4*S*,5*R*)-4-methyl-4-(methylamino)-3,4-dihydro-1*H*,5′*H*-spiro[benzo[*cd*]indole-5,2′-furan]-5′-one}, after the fungal species *Penicillium concavo-rugulosum*, and from cardiovascular activity The latter part of the specific epithet is from Latin *rugulosus*, wrinkled, from *ruga*, wrinkle.

Rutecarpine {8,13-dihydroindolo[2′,3′:3,4]pyrido[2,1-*b*]quinazolin-5(7*H*)-one}, after the species *Evodia rutaecarpa* (the Chinese medicinal herb wu zhu yu). The specific epithet is from Latin *ruta*, rue, and carp(o)-.

Rutin {3-[[6-*O*-(6-deoxy-α-L-mannopyranosyl)-β-D-glucopyranosyl]oxy]-2-(3,4-dihydroxyphenyl)-5,7-dihydroxy-4*H*-1-benzopyran-4-one}, after the genus *Ruta* (rue). The genus name is from Latin *ruta*, rue.

Ryanodine {(1*R*,2*R*,3*S*,6*S*,7*S*,9*S*,10*R*,11*S*,12*R*,13*S*,14*R*)-2,6,9,11,13,14-hexahydroxy-3,7,10-trimethyl-11-(propan-2-yl)-15-oxapentacyclo[7.5.1.$0^{1,6}$.$0^{7,13}$.$0^{10,14}$]pentadec-12-yl 1*H*-pyrrole-2-carboxylate}, after the species *Ryania speciosa* (a South American shrub). The genus name is after the British physician and plant collector John Ryan (deceased 1800/1808).

Sabadine (4α,12,14,16β,17,20-hexahydroxycevan-3β-yl acetate), from sabadilla (*Schoenocaulon officinale*). Sabadilla is from Spanish *sabadilla*, sabadilla, diminutive of *cebada*, barley, ultimately from Latin *cibus*, food, meal.

Saccharic acid (glucaric acid), from Latin *saccharum*, sugar, referring to the fact that this acid can be obtained by oxidation of glucose and other sugars.

Safrole [5-(prop-2-en-1-yl)-1,3-benzodioxol], with contraction, from sassafras (*Sassafras*), maybe, via Spanish, from Latin Latin *saxifraga herba*, rock-breaking herb, more likely an American Indian word.

Safynol [(2*R*,3*E*,11*E*)-trideca-3,11-dien-5,7,9-triyne-1,2-diol], after its source, diseased safflower (*Carthamus tinctorius*) and with ‘yne’ and ‘ol’.

Sakuranetin [5-hydroxy-2-(4-hydroxyphenyl)-7-methoxy-2,3-dihydro-4*H*-1-benzopyran-4-one], from Japanese *sakura*, cherry tree (*Prunus*).

Salbostatin {1,5-anhydro-2-deoxy-2-[[(1*S*,4*S*,5*S*,6*S*)-4,5,6-trihydroxy-3-(hydroxymethyl)cyclohex-2-en-1-yl]amino]-D-allitol}, with contraction, after the bacterial species *Streptomyces albus*, and from statin, referring to this compound’s enzyme inhibition. The genus name is from strepto- and Greek *mykes*, fungus. The specific epithet is from Latin *albus*, white.

Salicylic acid (2-hydroxybenzoic acid), after the genus *Salix* (willows). The genus name is from Latin *salix, salic-*, willow. A.k.a. **spiraeic acid**, after the genus name *Spiraea* (shrubs), from Greek *speira* (coil).

Salsoline (7-methoxy-1-methyl-1,2,3,4-tetrahydroisoquinolin-6-ol), after the genus *Salsola* (saltwort). The genus name is from Italian *salso*, salty, from Latin *salsus*, salty, from *sal*, salt.

Salutaridine (4-hydroxy-3,6-dimethoxy-17-methyl-5,6,8,14-tetradehydromorphinan-7-one), after the species *Croton salutaris* (a tropical herb). The specific epithet is from Latin *salutaris*, wholesome, from *salus, salut-*, health, safety. A.k.a. **floripavine**, after the species *Papaver floribundum* (Armenian poppy). The genus name ist from Latin *papaver*, poppy, and the specific epithet from Latin *floribundus*, prosperous, from *flos, flor-*, flower.

Samaderin {(1*R*,4*S*,5*R*,8*S*,9*R*,10*R*,11*R*)-9-hydroxy-4,11,14-trimethyl-3,7-dioxapentacyclo[8.7.0.0^{1,5}.0^{4,8}.0^{11,15}]heptadeca-13,15-diene-6,12,17-trione}, after the genus *Samadera* (tropical trees). The genus name is from Sinhalese *samadara*, a tree of Ceylon.

Samandarine (1α,4α-epoxy-3-aza-4a-homo-5β-androstan-16β-ol), with contraction, after the genus *Salamander* (lizards). The genus name is ultimately from Greek *salamandra*, lizard, of uncertain origin.

Sanguinarine {13-methyl[1,3]benzodioxolo[5,6-*c*][1,3]dioxolo[4,5-*i*]phenanthridin-13-ium, a cation}, after the genus *Sanguinaria* (bloodroot). The genus name is from Latin *sanguinarius*, blood-thirsty, from *sanguis, sanguin-*, blood, referring to these herbs stanching of blood.

α-Santalol [(2*Z*)-5-[2,3-dimethyltricyclo[2.2.1.0^{2,6}]hept-2-yl]-2-methylpent-2-en-1-ol, after the genus *Santalum* (sandalwood). The genus name is ultimately from Arabic *zandal*, sandalwood.

Santonin {(3*S*,3a*S*,5a*S*,9b*S*)-3,5a,9-trimethyl-3a,5,5a,9b-tetrahydronaphtho[1,2-*b*]furan-2,8(3*H*,4*H*)-dione}, after the species *Artemisia santonica* (wormseed). The specific epithet is from Latin *santonicus*, pertaining to the Santoni, an ancient people of Aquitania.

Sarafotoxin (polypeptides), toxins obtained from the snake *Atractaspis engaddensis*. The name is after Hebrew *saraf*, a biblical poisonous snake. A.k.a. **SRTX**.

Sarsasapogenin [(25*S*)-5β-spirostan-3β-ol], from Spanish *zarzaparilla*, sarsaparilla (*Smilax ornata*), from *zarza*, bush, and *parilla*, diminutive of *parra*, vine, and sapogenin.

Saxitoxin {[(3a*S*,4*R*,10a*S*)-2,6-diamino-10,10-dihydroxy-3a,4,9,10-tetrahydro-1*H*,8*H*-pyrrolo[1,2-*c*]purin-4-yl]methyl carbamate}, after the species *Saxidomus gigantea* (butter clam, Washington clam, smooth Washington clam, money shell). The genus name is from saxi- and Latin *domus*, house.

Scabiolide {(3a*R*,4*R*,5*Z*,9*E*,11a*S*)-6-(acetoxymethyl)-10-methyl-3-methylidene-2-oxo-2,3,3a,4,7,8,11,11a-octahydrocyclodeca[*b*]furan-4-yl 2,3-dihydroxy-2-methylpropanoate}, after the species *Centaurea scabiosa* (greater knapweed), and 'olide'. The specific epithet is from Latin *scabiosus*, mangy, from *scabies*, mange.

Scillaren (a bufadienolide), after the genus *Scilla* (squill). The genus name is from Latin *scilla*, squill.

Sclareolide {(3a*R*,5a*S*,9a*R*,9b*R*)-3a,6,6,9a-tetramethyldecahydronaphtho[2,1-*b*]furan-2(1*H*)-one}, after the species *Salvia sclarea* (clary sage). The specific epithet is from Medieval Latin *sclareia*, clary sage.

Scoparin [8-β-D-glucopyranosyl-5,7-dihydroxy-2-(4-hydroxy-3-methoxyphenyl)-4*H*-1-benzopyran-4-one], after the species *Spartium scoparium* (Scotch broom), later renamed *Cytisus scoparius*. The former genus name is from Latin *spartum*, the herb broom, and the specific epithet from Latin *scopae*, the tool broom.

Scopolin {7-[(β-D-glucopyranosyl)oxy]-6-methoxy-2*H*-1-benzopyran-2-one], after the genus *Scopolia* (herbs), named after the Italian-Austrian naturalist Giovanni Antonio Scopoli (1723–1788). Not to be confused with scopoline.

Scopoline (3α,6α-epoxy-7β-hydroxytropane), cf. scopolin. Not to be confused with scopolin.

Scutellarin {7-[(β-D-glucopyranuronosyl)oxy]-5,6-dihydroxy-2-(4-hydroxyphenyl)-4*H*-1-benzopyran-4-one}, after the genus *Scutellaria* (skullcap). The genus name is from Latin *scutella*, drinking bowl, diminutive of *scutra*, shallow bowl. A.k.a. **breviscapin**, after the species *Erigeron breviscapus* (a Chinese flowering plant). The specific epithet is New Latin for short-stalked, from Latin *brevis*, short, and *scapus*, stalk.

5β-Scymnol [(24*R*)-5β-cholestane-3α,7α,12α,24,26,27-hexol], after the species *Scymnus borealis* (Greenland shark), later renamed *Somniosus microcephalus*. The former genus name is from Latin *scymnus*, young animal, and refers to the slow swimming, low activity level, and non-aggressive nature of these sharks.

Selachyl alcohol {3-[(9*Z*)-[(octadec-9-en-1-yl)oxy]propane-1,2-diol}, after the genus *Selachus* (sharks). The genus name is from Greek *selachos*, cartilaginous fish, akin to *selas*, light, brightness.

Selagine {(1*R*,9*R*,13*E*)-1-amino-13-ethylidene-11-methyl-6-azatricyclo[7.3.1.0^{2,7}]trideca-2(7),3,10-trien-5-one}, after the species *Lycopodium selago* (fir club moss). The specific epithet is from Latin *selago*, a plant resembling the savin. A.k.a. **huperzine A**, after the genus *Huperzia* (firmosses, fir clubmosses), named after the German botanist Johann Peter Huperz (1771–1816).

Sempervirine (3,4,5,6,14,15,20,21-octadehydroyohimban-4-ium-1-ide), after the species *Gelsemium sempervirens* (Carolina jessamine). The specific epithet is from Latin *sempervirens, sempervirent-*, evergreen, from *semper*, always, and *virere*, to green. A.k.a. **sempervirene.**

Senecioic acid (3-methylbut-2-enoic acid), after the genus *Senecio* (ragwort, groundsel). The genus name is from Latin *senecio, senecion-*, old man, groundsel, from *senex, senec-*, old man, referring to the gray hair on the seeds of these plants.

Securinine (securinan-11-one), after the genus *Securinega* (tropical trees). The genus name is from Latin *securis*, hatchet, axe, and *negare*, to refuse, to negate, referring to the hardness of the wood of these trees.

Sesamin {5,5′-(1*S*,3a*R*,4*S*,6a*R*)-tetrahydro-1*H*,3*H*-furo[3,4-*c*]-1,4-diylbis(1,3-benzodioxole}, after sesame (*Fagara*), ultimately from Arabic *simsim*, sesame. A.k.a. **fagarol,** after the genus name. The genus name is from the Arabic name of an unknown plant.

Shellolic acid (10β,14-dihydroxycedr-8-ene-12,15-dioic acid), from shellac.

Shikimic acid [(2*R*,4*S*,5*R*)-3,4,5-trihydroxycyclohex-1-ene-1-carboxylic acid], from Japanese *shikimi*, star anise (*Illicium verum*).

Shionone (D:A-friedo-18,19-secolup-19-en-3-one), after Japanese shi-on, a drug prepared from the roots of the Tatarian aster (*Aster tataricus*).

Showdomycin (3-β-D-ribofuranosyl-1*H*-pyrrole-2,5-dione), after the bacterial species *Streptomyces showdoensis*. The specific epithet is from New Latin *showdoensis*, of the island of Shodo, Kagawa Prefecture, Japan, the source of the soil from which this bacterium was isolated.

Sialic acid [*N*-acetylneuraminic acid, (4*S*,5*R*,6*R*,7*S*,8*R*)-5-acetamido-4,6,7,8,9-pentahydroxy-2-oxononanoic acid], from Greek *sialon*, saliva.

Siccanin {(4a*S*,6a*S*,11b*R*,13b*S*)-4,4,6a,9-tetramethyl-1,2,3,4,4a,5,6,6a,11b,13b-decahydrobenzo[*a*]furo[2,3,4-*mn*]xanthen-11-ol}, after the fungal species *Helminthosporium siccans*. The specific epithet is from Latin *siccans, siccant-*, drying, from *siccare*, to dry, from *siccus*, dry.

Silymarin {(2*R*,3*R*)-2-[(2*S*,3*S*)-3-(4-hydroxy-3-methoxyphenyl)-2-(hydroxymethyl)-2,3-dihydro-1,4-benzodioxin-6-yl]-3,5,7-trihydroxy-2,3-dihydro-4*H*-1-benzopyran-4-one}, with contraction, after the species *Silybum marianum* (milk thistle). The genus name is from Greek *silybos*, thistle, and the specific epithet from Latin *marianus*, of Maria.

Sinalbin {1-*S*-[(1*E*)-2-(4-hydroxyphenyl)-*N*-(sulfooxy)ethanimidoyl]-1-thio-β-D-glucopyranose}, with contraction, after the species *Sinapis alba* (white mustard). The genus name is from Latin *sinapis*, mustard, and the specific epithet from Latin *albus*, white. Cf. sinigrin.

β-Sinensal [(2*E*,6*E*)-2,6-dimethyl-10-methylidenedodeca-2,6,11-trienal], after *Citrus*× *sinensis* (oranges). The specific epithet is from Latin *sinensis*, Chinese.

Sinigrin {potassium 1-*S*-[(1*Z*)-*N*-(sulfonatooxy)but-3-enimidoyl]-1-thio-β-D-glucopyranose}, with contraction after the species *Sinapis nigra* (black mustard). Cf. sinalbin.

Sinomenine (4-hydroxy-3,7-dimethoxy-17-methyl-7,8-didehydro-9α,13α,14α-morphinan-6-one), after the genus *Sinomenium* (orient vine). The genus name is from Greek *sinai*, the Chinese, and *mene*, moon.

Sitophilure [(4*S*,5*R*)-5-hydroxy-4-methylheptan-3-one], after the genus *Sitophilus* (weevils), and lure. The genus name is from Greek *sitos*, grain, and -phile.

Skatole (3-methyl-1*H*-indole), from Greek *skor, skat-*, excrement. A.k.a. **scatole.**

Skimmin {7-[(β-D-glucopyranosyl)oxy]-2*H*-1-benzopyran-2-one}, after the genus *Skimmia* (evergreen shrubs). The genus name is from Japanese *shikimi*, star anise (*Illicium verum*).

Solanine (a glycoalkaloid), named after its source, the genus *Solanum* (nightshade). The genus name is from Latin *solanum*, nightshade, of dubious etymology.

Songorine {(1*R*,2*R*,7*R*,8*R*,9*R*,13*R*,16*S*,17*R*)-11-ethyl-7,16-dihydroxy-13-methyl-6-methylidene-11-azahexacyclo[$7.7.2.1^{5,8}.0^{1,10}.0^{2,8}.0^{13,17}$]nonadecan-4-one}, after the species *Aconitum songoricum* (a monkshood). The specific epithet is from New Latin *songoricus*, of the Songhua River, Manchuria, China. A.k.a. **napellonine**, after the species *Aconitum napellus* (monkshood, aconite, wolfsbane). The specific epithet is from Latin *napellus*, diminutive of *napus*, tuber.

Sorbic acid [(2*E*,4*E*)-hexa-2,4-dienoic acid], after the genus *Sorbus* (whitebeam, rowan, service tree, mountain-ash), from Latin *sorbus* (service tree).

Sparassol (methyl 2-hydroxy-4-methoxy-6-methylbenzoate), after the genus *Sparassis* (cauliflower mushroom). The genus name is from Greek *sparassein*, to tear, to rend.

Spermine [N^1,N^4-bis(3-aminopropyl)butane-1,4-diamine], from sperm. Sperm is from Greek *sperma*, sperm, literally seed, from *speirein*, to sow. A.k.a. **gerontine**, from Greek *geron, geront-*, old man, referring to the fact that this compound has been found in the liver and kidney of aged dogs, and **neuridine**, from neur(o)-.

Sphagnum acid [(2*E*)-3-(4-hydroxyphenyl)pent-2-enedioic acid, after the genus *Sphagnum* (peat mosses). The genus name is ultimately from Greek *sphagnos*, a moss.

Spinasterol [(22*E*)-5α-stigmasta-7,22-dien-3β-ol], after the genus *Spinacia* (spinach) and from sterol. The genus name is ultimately from Persian *isfanakh*, spinach.

Spinosyn (a class of macrocyclic insecticides), after the bacterial species *Saccharopolyspora spinosa*. The specific epithet is from Latin *spinosus*, thorny, from *spina*, thorn, spine. The suffix -syn is probably from 'systemic action'.

Spinulosin (2,5-dihydroxy-3-methoxy-6-methylcyclohexa-2,5-diene-1,4-dione), after the fungal species *Penicillium spinulosum*. The specific epithet is from Latin *spinulosus*, having many little thorns or spines, from *spinula*, diminutive of *spina*, thorn, spine.

Spiraein {2-[4-[(β-D-glucopyranosyl)oxy]-3-hydroxyphenyl]-3,5,7-trihydroxy-4*H*-1-benzopyran-4-one}, after the genus *Spiraea* (shrubs). The genus name is from Greek *speira*, coil.

Splendipherin (an oligopeptide), after the species *Litoria splendida* (magnificent tree frog), and from pheromone. The specific epithet is from Latin *splendidus*, magnificent, from *splendere*, to shine.

Sporidesmin (a class of mycotoxins), after the fungal genus *Sporidesmium*. The genus name is from spor(o)- and Greek *desme*, bundle, from *dein*, to bind.

Stachydrine [(2*S*)-1,1-dimethylpyrrolidinium-2-carboxylate], after the genus *Stachys* (hedgenettle, woundwort, betony, lamb's ears) and from hygric acid (1-methylproline), in turn named after hygrine. The genus name is from Greek *stachys*, ear of grain.

Stegobinone {(2*S*,3*R*)-2,3,5-trimethyl-6-[(2*R*)-3-oxopentan-2-yl]-2,3-dihydro-4*H*-pyran-4-one}, after the genus *Stegobium* (beetles). The genus name is from Greek *stegein*, to cover, and *-bium*, New Latin suffix for living organisms.

Stentorin {2,5-di(propan-2-yl)-1,3,4,6,8,10,11,13-octahydrophenanthro[1,10,9,8-*opqra*]perylene-7,14-dione}, after the species *Stentor coeruleus* (a protist). The genus name refers to its trumpet shape, after the Greek mythical herald Stentor with a loud voice.

Stenusine [1-ethyl-3-(2-methylbutyl)piperidine), after the genus *Stenus* (rove beetles). The genus name is from Greek *stenos*, narrow.

Steviol (13-hydroxy-4α-kaur-16-en-18-oic acid), after the genus *Stevia* (sunflower). The genus name is after the Spanish botanist and physician Petrus Jacobus Stevus (Pedro Jaime Esteve) (1500–1556).

Stictic acid {1,4-dihydroxy-10-methoxy-5,8-dimethyl-3,7-dioxo-1,3-dihydro-7*H*-2,6,12-trioxabenzo[5,6]cyclohepta[1,2-*e*]indene-11-carbaldehyde}, after the genus *Sticta* (lungwort, lung moss), now renamed *Lobaria*. The former genus name is from Greek *stiktos*, spotted.

Stipitatic acid (3,6-dihydroxy-5-oxocyclohepta-1,3,6-triene-1-carboxylic acid), after the fungal species *Penicillium stipitatum*. The specific epithet is from Latin *stipitatus*, having or born on a stipe, from *stipes*, stipe, akin to *stipare*, to press together.

Strigol {(3*E*,3a*R*,5*S*,8b*S*)-5-hydroxy-8,8-dimethyl-3-[[[(2*R*)-4-methyl-5-oxo-2,5-dihydrofuran-2-yl]oxy]methylidene]-3,3a,4,5,6,7,8,8b-octahydro-2*H*-indeno[1,2-*b*]furan-2-one}, after the genus *Striga* (witchweed). The genus name is from Latin *strix, strig-*, a bird-like demon feeding on flesh and blood, referring to the parasitic nature of these plants.

Strobilurin A [methyl (2*E*,3*Z*,5*E*)-2-(methoxymethylidene)-3-methyl-6-phenylhexa-3,5-dienoate], after the genus *Strobilurus* (toadstool). The genus name is from Greek *strobilos*, whirling or twisted object, from *strephein*, to turn, and *oura*, tail.

Strophanthidine [3β,5β,14-trihydroxy-19-oxocard-20(22)-enolide]. after the genus *Strophanthus* (flowering plants). The genus name is from Greek *strophos anthos*, twisted flower.

Strychnine (strychnidin-10-one), after the genus *Strychnos* (trees, vines). The genus name is from Greek *strychnos*, nightshade.

Stylophorine {(5b*R*,6*S*,12b*S*)-13-methyl-5b,6,7,12b,13,14-hexahydro[1,3]benzodioxolo[5,6-*c*]-1,3-dioxolo[4,5-*i*]phenanthridin-6-ol}, after the genus *Stylophorum* (celandine-poppy). The genus name is from Greek *stylos*, stalk, stylus, and phor(o)-, referring to these plants' unusually long style. A.k.a. **chelidonine**.

Styrene (ethenylbenzene), earlier called styrol. After the genus *Styrax* (storax, snowbell). The genus name is from Greek *styrax*, storax, a corruption of Semitic *assthirak* (*Styrax officinalis*). A.k.a. **cinnamol**.

Sulforaphene [(1*E*)-4-isothiocyanato-1-(methanesulfinyl)but-1-ene], from sulfur and after the genus *Raphanus* (radish). The genus name is from Greek *raphanos*, radish, akin to *rhapys, rhaphys*, turnip. A.k.a. **raphanin**.

Sulphurenic acid (3β,15α-dihydroxy-24-methylidenelanost-8-en-21-oic acid), after the fungus *Laetoporus sulphureus* (crab-of-the-woods). The specific epithet is from New Latin *sulphureus*, sulfur-colored, referring to the sulfur color of this fungus.

Surinamine (*N*-methyl-L-tyrosine), after the species *Andira surinamensis* (angelin tree). The specific epithet is from New Latin *surinamensis*, Surinamese. A.k.a. **andirine**, after *Andira*, **angeline**, after angelin tree, **geoffroyine**, after the botanical synonym *Geoffroea surinamensis*, with the genus name after the French botanist, chemist, and pharmacist Claude-Joseph Geoffroy (1685–1752), and **ratanhine**, after rhatany (*Krameria triandra*). The latter plant name is from Quechua *ratánya*, rhatany.

Swainsonine [(1*S*,2*R*,8*R*,8a*R*)-octahydroindolizine-1,2,8-triol], after the genus *Swainsona* (trees, vines, shrubs, and herbs bearing bean pods). The genus name is after the British botanist Isaac Swainson (1746–1812).

Swertiamarin {(4a*R*,5*R*,6*S*)-5-ethenyl-6-[(β-D-glucopyranosyl)oxy]-4a-hydroxy-4,4a,5,6-tetrahydro-1*H*,3*H*-pyrano[3,4-c]pyran-1-one}, after the genus *Swertia* (felworts) and from marine. The genus name is after the Dutch botanist Emanuel Sweerts (1552–1612).

Syringin {4-[(1*E*)-3-hydroxyprop-1-en-1-yl]-2,6-dimethoxyphenyl β-D-glucopyranoside}, named after *Syringa vulgaris* (lilac). The genus name is from Greek *syrinx, syring-*, panpipe.

Tabernanthine (13-methoxyibogamine), after the genus *Tabernanthe* (West African shrubs). The genus name is after the German herbalist Jacobus Theodorus Tabernaemontanus (Jakob Theodor von Bergzabern) (1522–1590) and from anth(o)-.

Tanacetin {(3a*S*,5a*S*,6*R*,9a*R*,9b*S*)-6,9a-dihydroxy-5a-methyl-3,9-dimethylidenedecahydronaphtho[1,2-*b*]furan-2(3*H*)-one}, after the genus *Tanacetum* (tansy). The genus name is ultimately from Greek *athanasia*, immortality, from a(n)- and *thanatos*, death.

Tannin (oligomeric phenolic natural products), from Anglo-French *tanner*, to tan, ultimately from Medieval Latin, *tannare*, to tan, from *tannum*, crushed oak bark. A.k.a. **tannic acid.**

Taraxacin {3,5,8-trimethyl-9,9a-dihydroazuleno[6,5-*b*]furan-2,7-dione}, after the genus *Taraxacum* (dandelion). The genus name, via Latin, is from Arabic *tarakhshaqun*, wild chicory (*Cichorium intybus*), in turn from Persian *tarkhashaqun*, wild endive (*Cichorium pumilum*).

Taurine (2-aminoethane-1-sulfonic acid), from Latin *taurus*, bull, ox, referring to its first isolation from bovine bile.

Tecomanine {(4*R*,7*S*,7a*S*)-2,4,7-trimethyl-1,2,3,4,7,7a-hexahydro-6*H*-cyclopenta[*c*]pyridin-6-one}, after the genus *Tecoma* (ginger thomas, shrubby trumpet flower, yellow bells, yellow cedar, yellow elder). The genus name is from Nahuatl *tecomaxochitl*, tecoma, from *tecomatl*, clay pot, and *xochitl*, flower. A.k.a. **tecomine**.

Tectorigenin [5,7-dihydroxy-3-(4-hydroxyphenyl)-6-methoxy-1-benzopyran-4-one], after the species *Iris tectorum* (wall iris). The specific epithet is from Latin *tector*, wall painter, from *tectum*, roof, house, building, from *tegere, tect-*, to cover. A.k.a. **tectorigenine**.

Tenuazonic acid [3-acetyl-5-(butan-2-yl)-4-hydroxy-1,5-dihydro-2*H*-pyrrol-2-one], after the fungal species *Alternaria tenuis* and with az(o)-. The specific epithet is from Latin *tenuis*, thin, slight, tenuous.

Tephrosin {(7a*R*,13a*R*)-7a-hydroxy-9,10-dimethoxy-3,3-dimethyl-13,13a-dihydro-3*H*-[1]benzopyrano[3,4-*b*]pyrano[2,3-*h*][1]benzopyran-7(7a*H*)-one}, after the genus *Tephrosia* (hoary pea). The genus name is from Greek *tephros*, ash-gray, referring to the color of these plants' leaves.

Terebic acid (2,3-dimethyl-5-oxotetrahydrofuran-3-carboxylic acid), from terebene, a mixture of terpenes obtained from the terebinth tree (*Pistacia terebinthus*). The specific epithet is from Greek *terebinthos*, terebinth tree, probably of Creto-Minoic origin.

Terreic acid {(1*R*,6*S*)-3-hydroxy-4-methyl-7-oxabicyclo[4.1.0]hept-3-ene-2,5-dione}, after the fungal species *Aspergillus terreus*. The specific epithet is from Latin *terreus*, of the earth, from *terra*, earth.

Tetrahymanol (gammaceran-3β-ol), after the ciliate protozoan genus *Tetrahymena* and with 'ane' and 'ol'. The genus name is from tetra- and Greek *hymen*, membrane.

Tetrandrine (6,6′,7,12-tetramethoxy-2,2′-dimethyl-1β-berbaman), after the species *Stephania tetrandra* (the Chinese medicinal herb han fang ji). The specific epithet is from New Latin *tetrandrus*, having four stamens, from tetra- and Greek *aner, andr-*, man, stamen.

Tetrodotoxin {(1*R*,5*R*,6*R*,7*R*,9*S*,11*S*,12*S*,13*S*,14*S*)-3-amino-14-(hydroxymethyl)-8,10-dioxa-2,4-diazatetracyclo[7.3.1.1^{7,11}.0^{1,6}]tetradec-2-ene-5,9,12,13,14-pentol}, after the family *Tetraodontidae* (pufferfish). The family name is from tetra- and Greek *odontos*, toothed, from *odon, odont-*, tooth. A.k.a. **fugu toxin**, from Japanese *fugu*, pufferfish (genus *Takifugu*), and **TTX**.

Thalicarpine {(6a*S*)-9-[2-[[(1*S*)-6,7-dimethoxy-2-methyl-1,2,3,4-tetrahydroisoquinolin-1-yl]methyl]-4,5-dimethoxyphenoxy]-1,2,10-trimethoxy-6-methyl-5,6,6a,7-tetrahydro-4*H*-dibenzo[*de*,*g*]quinoline}, with contraction, after the species *Thalictrum dasycarpum* (purple meadow rue). The genus name is from Greek *thaliktron*, meadow rue (genus *Thalictrum*), and the specific epithet from Greek *dasys*, thick with hair or leaves, and carp(o)-.

Theaflavin {1,8-bis[(2*R*,3*R*)-3,5,7-trihydroxy-3,4-dihydro-2*H*-1-benzopyran-2-yl]-3,4,5-trihydroxy-6*H*-benzocyclohepten-6-one}, after the genus *Thea* (tea plant). The genus name is from New Latin *thea*, tea plant.

Thebaine (3,6-dimethoxy-17-methyl-6,7,8,14-tetradehydro-4,5-epoxy-5α-morphinan), from New Latin *thebaia*, opium produced in the ancient town of Thebes, Egypt.

Theobromine (3,7-dimethyl-1*H*-purine-2,6-dione), after the species *Theobroma cacao* (cocoa tree). The genus name is from Greek *theos*, god, and *broma*, food.

Theophylline (1,3-dimethyl-3,7-dihydro-1H-purine-2,6-dione), from New Latin *thea*, tea plant, and phyll(o)-.

Thevetin (a group of cardiac glycosides), after the genus *Thevetia* (tropical American trees and shrubs). The genus name is after the French traveler and author Frère André Thevet (1516–1590).

Thiolutin {*N*-(4-methyl-5-oxo-4,5-dihydro[1,2]dithiolo[4,3-*b*]pyrrol-6-yl)acetamide}, after the bacterial species *Streptomyces luteosporeus*, and with thi(o)-. The specific epithet is from New Latin *luteosporeus*, with yellow spores.

Thujaplicin [2-hydroxy-3-(prop-1-en-2-yl)cyclohepta-2,4,6-trien-1-one), after the species *Thuja plicata* (red cedar). The genus name is from Greek *thyia*, a kind of cedar, and the specific epithet from Latin *plicatus*, plaited, folded, from *plicare*, to pleat, to fold.

Thujopsene {(1a*S*,4a*S*,8a*S*)-2,4a,8,8-tetramethyl-1,1a,4,4a,5,6,7,8-octahydrocyclopropa [*d*]naphthalene}, after the genus *Thujopsis* (asunaro, hiba, false arborvitae, hiba arborvitae). The genus name is from *Thuja* and -opsis, referring to this genus's similarity to *Thuja*. A.k.a. **sesquichamene**, after the genus *Chamaecyparis* (false cypress), from Greek *chamai*, on the ground, and *kyparissos*, cypress, and **widdrene**, after the genus *Widdringtonia* (cedars, African cypresses), after the British naval officer, author, and plant collector Samuel Edward Widdrington (born Samuel Edward Cook) (1787–1856).

Tiglic acid [(2*E*)-2-methylbut-2-enoic acid], named after the species *Croton tiglium* (purging croton). The origin of the specific epithet is obscure, possibly after pharmacists' name for the seeds of the croton plant. According to one suggestion it might be derived from Greek *tilos*, diarrhea, according to another it might refer to one of the Maluku Islands, Indonesia.

Tigonin {(25*R*)-5α-spirostan-3β-yl β-D-glucopyranosyl-(1→3)-β-D-galactopyranosyl-(1→2)-[β-D-xylopyranosyl-(1→3)]-β-D-glucopyranosyl-(1→4)-β-D-galactopyranoside}, with anagrammatic contraction, after the genus *Digitalis* (foxgloves). The genus name is from Latin *digitalis*, of a finger, from *digitus*, finger, toe. The genus name refers to the ease with which a flower of foxglove can be fitted over a fingertip.

Tomatine {(25*S*)-5α-spirosolan-3β-yl β-D-glucopyranosyl-(1→2)-[β-D-xylopyranosyl-(1→3)]-β-D-glucopyranosyl-(1→4)-β-D-galactopyranoside}, from tomato plant (*Lycopersicon esculentum*), from Nahuatl *tomatl*, tomato plant.

Tonghaosu {2-(hexa-2,4-diyn-1-ylidene)-1,6-dioxaspiro[4.4]non-3-ene}, after the Chinese vegetable plant *tonghao*, corn marigold (*Chrysanthemum segetum*).

Torularhodin [(3′*E*)-3′,4′-didehydro-β,ψ-caroten-16-oic acid], after the fungal genus *Torula*. The genus name is from Latin *torulus*, diminutive of *torus*, protuberance, bulge, cushion, couch, torus molding, and from rhod(o)-.

Trachylobane {(4a*R*,6a*R*,7a*R*,8*S*,8a*R*,9a*R*,9b*R*)-4,4,7a,9b-tetramethyltetradecahydro-6a,8-methano-6a*H*-cyclopropa[*b*]phenanthrene}, after the species *Acalypha trachyloba* (copperleaf, three-seeded mercury). The specific epithet is from trachy- and Greek *lobos*, lobe.

Trehalostatin {*N*-[(3a*R*,4*R*,5*S*,6*S*,6a*S*)-4,5,6-trihydroxy-4-(hydroxymethyl)-4,5,6,6a-tetrahydro-3a*H*-cyclopenta[*d*][1,3]oxazol-2-yl]-α-D-glucopyranosylamine}, after the fungal species *Amycolatopsis trehalostatica*. The specific epithet is from trehalase and statin, referring to this antibiotic's trehalase-inhibiting properties. A.k.a. **trehazolin.**

Trichione (propanedioic acid mono[(3*E*)-4-(3,8-dihydroxy-1,4-dioxonaphthalen-2-yl)but-3-en-1-yl] ester}, after the fungal genus *Trichia* (slime mold). The genus name is from trich(o)-.

Trigonelline (1-methylpyridinium-3-carboxylate), after the genus *Trigonella* (fenugreek). The genus name is from Latin *trigonellum*, diminutive of *trigonum*, triangle, referring to these plants' trifoliolate leaves. A.k.a. **caffearine** and **gynesine.**

Trilobine (6,12′-dimethoxy-2′-methyl-6′,7-epoxy-1α-oxyacanthan), after the species *Cocculus trilobus* (the Chinese medicinal herb mu fang ji). The specific epithet is New Latin for three-lobed.

Truxilline {methyl (1*R*,2*R*,3*S*,5*S*)-8-methyl-3-[[(*E*)-3-phenylprop-2-enoyl]oxy]-8-azabicyclo[3.2.1]octane-2-carboxylate}, from Truxillo coca (*Erythroxylum truxillense*). The specific epithet is after the town of Truxillo, Peru.

Tsuduranine {(6a*R*)-1,2-dimethoxy-5,6,6a,7-tetrahydro-4*H*-dibenzo[*de,g*]quinolin-10-ol}, with contraction, from Japanese *tudurafuzi*, Chinese moonseed (*Simonenium acutum*). A.k.a. **tuduranine.**

Tubercidine {7-(β-D-ribofuranosyl)-7*H*-pyrrolo[2,3-*d*]pyrimidin-4-amine}, after the bacterial species *Streptomyces tubercidicus*. The specific epithet is from New Latin *tubercidicus*, antitubercular.

Tuckolide [(4*S*,5*S*,6*Z*,8*S*,10*R*)-4,5,8-trihydroxy-10-methyl-3,4,5,8,9,10-hexahydro-2*H*-oxecin-2-one], after the stone fungus tuckahoe (*Polyporus tuberaster*) and with 'olide'. Tuckahoe is from Virginia Algonquian *tockawhoughe* and can also mean *Orontium aquaticum* (golden-club, floating arum, never-wets, tawkin), *Peltandra virginica* (green arrow arum), and *Wolfiporia extensa* (hoelen, poria, China root) a.ka. *Poria cocos.*

Turmerone [(6*S*)-2-methyl-6-(4-methylphenyl)hept-2-en-4-one], after turmeric (*Curcuma longa*), possibly from Medieval Latin *terra merita*, worthy earth.

Tutin {(1*S*,2*R*,3*S*,5*R*,6*R*,7*R*,8*S*,9*R*,12*R*)-2,8-dihydroxy-7-methyl-12-(prop-1-en-2-yl)-11*H*-spiro[4,10-dioxatetracyclo[7.2.1.0^{2,7}.0^{3,5}]dodecane-6,2′-oxiran]-11-one}, from tutu (*Coriaria arborea* var. *arborea*), from Maori *tutu*.

Tylocrebrine {2,3,5,6-tetramethoxy-9,11,12,13,13a,14-hexahydrodibenzo[*f*,*h*]pyrrolo[1,2-*b*]isoquinoline}, after the species *Tylophora crebriflora* (coast tylophora), later renamed *Tylophora benthamii*. The genus name is from Greek *tylos*, knob, lump, callus, pad, and -phore. The specific epithet is from Latin *creber*, *crebr-*, crowded, numerous, full.

Uliginosin A {3,5-dihydroxy-4,4-dimethyl-2-(2-methylpropanoyl)-6-[[2,4,5-trihydroxy-3-(3-methylbut-2-en-1-yl)-5-(2-methylpropanoyl)phenyl]methyl]cyclohexa-2,5-dien-1-one}, after the species *Hypericum uliginosum* (St. John's wort). The specific epithet is from Latin *uliginosus*, growing in wet or swampy ground, ultimately from *udus*, *uvidus*, damp, moist.

Umbelliferone (7-hydroxy-2*H*-1-benzopyran-2-one), after the family *Umbelliferae* (celery, carrot, parsley family). The genus name is from Latin *umbella*, parasol, umbrella, diminutive of *umbra*, shade, and *ferre*, to bear, to carry.

Urea (carbonic diamide), from French *urée*, urea, ultimately from Greek *ouron*, urine, referring to its occurrence in urine.

Uric acid [7,9-dihydro-1*H*-purine-2,6,8(3*H*)-trione], from urine.

Urothion {2-amino-7-(1,2-dihydroxyethyl)-6-methylsulfanyl-3*H*-thieno[3,2-*g*]pteridin-4-one}, from urine and 'thio'. A.k.a. **urothione.**

Ursocholic acid (3α,7β,12α-trihydroxy-5β-cholan-24-oic acid), from Latin *ursus*, bear, after its first source, ursine bile.

Urushiol I (3-pentadecylbenzene-1,2-diol), from Japanese *urushi*, lacquer.

Uscharin (a cardenolide glycoside), from *uschari*, a native African word for the arrow poison obtained from rubberbush, apple of Sodom (*Calotropis procera*), ultimately probably from Arabic *usher*, rubberbush, apple of Sodom.

Usnic acid [2,6-diacetyl-7,9-dihydroxy-8,9b-dimethyldibenzo[*b*,*d*]furan-1,3(2*H*,9*H*)-dione], after the genus *Usnea* (mosses), ultimately from Arabic *ushnah*, moss.

Ustilagic acid A {16-[(6-*O*-acetyl-β-D-glucopyranosyl)oxy]-2,15,-dihydroxyhexadecanoic acid}, after the fungal genus *Ustilago* (corn smut) The genus name is from New Latin *ustilago*, smut, from Latin *ustus*, burnt, from *urere*, to burn, referring to the scorched appearance of these fungi.

Uzarine [3β,5α-card-20(22)-enolide], from uzara root (*Xysmalobium undulatum*) a.k.a. fever bush and stomach bush. The etymology of uzara is uncertain.

Valeric acid (pentanoic acid), after the genus *Valeriana* (valerian). The genus name is from Latin *Pannonia Valeria*, a Roman province, now part of Hungary.

Vanillin (4-hydroxy-3-methoxybenzaldehyde), after the genus *Vanilla* (vanilla). The genus name is from Spanish *vanilla*, vanilla, diminutive of *vaina*, sheath, pod, ultimately from Latin *vagina*, sheath.

Varacin [2-(6,7-dimethoxy-1,2,3,4,5-benzopentathiepin-9-yl)ethan-1-amine], after the species *Lissoclinum vareau* (a sea squirt). The specific epithet is from Tahitian *vareau*, violet.

Vasicine.{1,2,3,9-tetrahydropyrrolo[2,1-*b*]quinazolin-3-ol}, after the species *Adhatoda vasica* (Malabar nut, adulsa), later renamed *Justicia adhatoda*. The former specific epithet is from Sanskrit *vasika, vasaka*, Malabar nut, from *vasayati*, it perfumes, it makes fragrant. A.k.a. **peganine**, after *Peganum harmala* (African rue, harmel). The genus name is from Greek *peganon*, African rue. The specific epithet is from Arabic *harmal*, African rue.

Veatchine {(1*S*,2*R*,5*R*,7*S*,8*R*,11*R*,12*R*)-12-methyl-6-methylidene-17-oxa-14-azahexacyclo[10.6.3.$1^{5,8}$.$0^{1,11}$.$0^{2,8}$.$0^{14,18}$]docosan-7-ol}, after the species *Garrya veatchii* (canyon silktassel, Veatch silktassel). The specific epithet is after the American naturalist John Allen Veatch (1808–1870) and his son, the American mining engineer Andrew Allen Veatch (1832–1871).

Vellosimine [(19*E*)-sarpagan-17-al], after the species *Geissospermum vellosii* (pereira tree), later renamed *Geissospermum laeve*. The former specific epithet is after the Brazilian botanist José Mariano da Conceição Velloso (1742–1811).

Venturicidine (a macrolide), after the fungal genus *Venturia*, and from -cide, referring to this antibiotic's activity against apple scab (*Venturia inaequalis*). The genus name is after the Italian botanist Antonio Venturi (1805–1864).

Veralkamine {(20*S*)-17α-methyl-20-[(2*S*,5*S*)-5-methylpiperidin-2-yl]-18-norpregna-5,12-diene-3β,16-diol}, with contraction, after the species *Veratrum album* (white hellebore) and with -amine. Cf. veratraman.

Veratrole (1,2-dimethoxybenzene). Cf. veratraman.

Verbenalin [(1*S*,4a*S*,7*S*,7a*R*)-1-[(β-D-glucopyranosyl)oxy]-7-methyl-5-oxo-1,4a,5,6,7,7a-hexahydrocyclopenta[*c*]pyran-4-carboxylic acid methyl ester], after the genus *Verbena* (verbena). The genus name is from Latin *verbenae*, sacred boughs of laurel, olive, or myrtle, from *verber*, whip.

Vernolepin {(3a*R*,4*S*,5a*R*,9a*R*,9b*R*)-5a-ethenyl-4-hydroxy-3,9-dimethylideneoctahydro-2*H*-furo[2,3-*f*][2]benzopyran-2,8(3*H*)-dione}, after the species *Vernonia hymenolepis* (sweet bitterleaf). The genus name is after the British plant collector William Vernon (1666/67–1711/15) and the specific epithet from Greek *hymen*, membrane, and *lepis, lepid-*, flake, scale.

Verpacrocin [(2*E*,4*E*,6*E*,8*E*,10*E*,12*E*,14*E*)-hexadeca-2,4,6,8,10,12,14-heptaenedial], after the fungal species *Verpa digitaliformis* (foxglove verpa, false morel) and from Latin *croceus*, saffron-colored. The genus name is from Latin *verpa*, penis, little rod.

Verrucarin (a macrolide), after the fungal species *Myrothecium verrucaria*. The specific epithet is from *verrucarius*, of the wart, from Latin *verruca*, wart.

Verticillin (a group of antibiotics), after the fungal genus *Verticillium*. The genus name is from Latin *verticillatus*, arranged in verticils, from *verticillus*, whorl of a spindle, diminutive of *vertex*, *vertic-*, whirl.

Verticine (5α-cevane-3β,6α,20-triol), after the species *Fritillaria verticillata* (fritillary). Cf. verticillin. A.k.a. **peimine**.

α-Vetivone [(4*R*,4aS)-4,4a-dimethyl-6-(propan-2-ylidene)-4,4a,5,6,7,8-hexahydronaphthalen-2(3*H*)-one], after the genus *Vetiveria* (vetiver). The genus name is from Malayalam *vettiver*, vetiver (*Vetiveria zizanioides*), from *veti*, to cut, and *ver*, root, referring to the method of propagation.

Vicine (2,6-diamino-4-oxo-1*H*-pyrimidin-5-yl β-D-glucopyranoside), after the genus *Vicia* (vetch), from Latin *vicia*, vetch, ultimately from *vincire*, to bind, to tie.

Vincaleukoblastine (2β-vincaleukoblastine), after the genus *Vinca* (periwinkle) and from leukoblast, referring to its action on white blood cells. The genus name is from Latin *pervinca*, periwinkle, from per- and *vincire*, to bind, A.k.a. **vinblastine**.

Violacein {(3*E*)-3-[5-(5-hydroxy-1*H*-indol-3-yl)-2-oxo-1,2-dihydro-3*H*-pyrrol-3-ylidene]-1,3-dihydro-2*H*-indol-2-one}, after the bacterial species *Chromobacterium violaceum*. The specific epithet is from Latin *violaceus*, bluish, from *viola*, the flower violet.

Viridicatin [3-hydroxy-4-phenylquinolin-2(1*H*)-one], after the fungal species *Penicillium viridicatum*. The specific epithet is from Latin *viridicatus*, colored green, from *viridis*, grass green.

Viridin {1β-hydroxy-2β-methoxy-18-norandrosta-5,8,11,13-tetraeno[6,5,4-*bc*]furan-3,7,17-trione}, after the fungal species *Gliocladium virens*. The specific epithet is from Latin *virens*, *virent-*, greening, from *virere*, to green, from *viridis*, grass green.

Viscosin {*N*-[(3*R*)-3-hydroxydecanoyl]-L-leucyl-*N*-[(3*S*,6*R*,9*S*,12*R*,15*S*,18*R*,21*R*,22*R*)-3-[(2*S*)-butan-2-yl]-6,12-bis(hydroxymethyl)-9,15-di(2-methylpropyl)-22-methyl-2,5,8,11,14,17,20-heptaoxo-18-(propan-2-yl)-1-oxa-4,7,10,13,16,19-hexaazacyclodocosan-21-yl]-D-α-glutamine}, after the bacterial species *Pseudomonas viscosa*. The specific epithet is from Latin *viscosus*, viscous, from *viscum*, mistletoe, birdlime.

Visnadine {4-acetoxy-3-[(2-methylbutanoyl)oxy]-3,4-dihydroseline}, with contraction, after the species *Ammi visnaga* (bishop's weed, pick-tooth, Spanish carrot, tooth-pick). The genus name is from Greek *ammi*, an umbelliferous plant. The specific epithet is from Spanish *bisnaga*, tooth-pick, from Latin *pastinaca*, parsnip, ultimately from Akkadian *pastum*, adze, ax cleaver.

Voacamine {12-methoxy-13-[(16*S*,19*E*)-17-methoxy-17-oxovobasan-3α-yl]-2α,4α-ibogamine-18β-carboxylic acid methyl ester}, after the genus *Voacanga* (voacanga) and with -amine. The genus name is from Malagasy *voacanga*, voacanga.

Vomicine (4-hydroxy-19-methyl-16,19-secostrychnidine-10,16-dione), after the species *Strychnos nux-vomica* (strychnine tree). The specific epithet is from Latin *nux*, *nuc-*, nut, and *vomicus*, emetic, from *vomere*, to vomit.

Vulgaxanthin I {5-amino-2-[(*E*)-[(2Z)-2-(2,6-dicarboxy-2,3-dihydropyridin-4(1*H*)-ylidene)ethylidene]azaniumyl]-5-oxopentanoate}, after the species *Beta vulgaris* (common beet). The specific epithet is from Latin *vulgaris*, of the mob, from *vulgus*, common people.

Vulpinic acid [methyl (3-hydroxy-5-oxo-4-phenylfuran-2(5*H*)-ylidene)(phenyl)acetate], after the species *Letharia vulpina* (wolf lichen, wolf moss). The specific epithet is from Latin *vulpinus*, of the fox, from *vulpes*, fox, referring to the use of this plant as a poison for wolves and foxes.

Warburganal [(1*S*,4a*S*,8a*S*)-1-hydroxy-5,5,8a-trimethyl-4a,6,7,8-tetrahydro-4*H*-naphthalene-1,2-dicarbaldehyde], after the genus *Warburgia* (East African green heart, fever tree, pepper-bark tree, pepper root) and with 'al'. The genus name is after the German botanist Otto Warburg (1859–1938).

Withaferin (4a,27-dihydroxy-1-oxo-5b,6b-epoxywitha-2,24-dienolide), with contraction, after the species *Withania somnifera* (ashwagandha, Indian ginseng, poison gooseberry, winter cherry). The genus name is after the British naturalist Henry Thomas Maire Witham (1779–1844). The specific epithet is from Latin *somnifer*, sleep-inducing, from *somnus*, sleep, and *ferre*, to bear, to carry.

Xanthatin [(3a*R*,7*S*,8a*S*)-7-methyl-3-methylidene-6-(3-oxobut-1-en-1-yl)-2*H*-cyclohepta[*b*]furan-2-one], after the genus *Xanthium* (cocklebur). The genus name is from Greek *xanthion*, a plant used to color the hair yellow, from *xanthos*, yellow.

Xanthoxyletin {5-methoxy-8,8-dimethyl-2*H*,8*H*-benzo[12-*b*:5,4-*b*′]dipyran-2-one}, after the genus *Zanthoxylum* (prickly ash, yellowwood). The genus name is (irregularly) from xanth(o)- and xyl(o)-. A.k.a. **xanthoxyloin**. from xanth(o)-

Xerocomic acid {[(2*E*)-4-(3,4-dihydroxyphenyl)-3-hydroxy-5-oxofuran-2(5*H*)-ylidene](hydroxyphenyl)acetic acid}, after the genus *Xerocomus* (mushrooms). The genus name is from xer(o)- and Greek *kome*, hair, crest.

Xerulin {(5*Z*)-5-[(2*E*,4*E*,6*E*,12*E*)-tetradeca-2,4,6,12-tetraene-8,10-diyn-1-ylidene]furan-2(5*H*)-one}, after the genus *Xerula* (gilled mushrooms).

Yangonin {4-methoxy-6-[(*E*)-2-(4-methoxyphenyl)ethenyl]-2*H*-pyran-2-one}, from Fijian *yangona*, kava (*Piper methysticum*).

Yingzhaosu A {(3*S*,4*E*)-5-[(1*S*,4*S*,5*S*,8*R*)-4,8-dimethyl-2,3-dioxabicyclo[3.3.1]nonan-4-yl]-2-methylpent-4-ene-2,3-diol}, after the Chinese medicinal plant ying zhao, ying zhua hua (*Artabotrys hexapetalus*).

Yohimbine (methyl 17α-hydroxyyohimban-16α-carboxylate), after the species *Corynanthe yohimbe*, a.ka. *Pausinystalia yohimbe* (yohimbe). The specific epithet is from Bantu *yohimbe*, yohimbe.

Yuehchukene {(6*S*,6a*S*,10a*R*)-6-(1*H*-indol-3-yl)-7,7,9-trimethyl-5,6,6a,7,8,10a-hexahydroindene[2,1-*b*]indole}, from Chinese *chiu li hsiang tsao, jiu li xiang, qian li xiang*, orange jessamine (*Murraya paniculata*).

Zearalenone [(3*S*,11*E*)-14,16-dihydroxy-3-methyl-3,4,5,6,9,10-hexahydro-1*H*-2-benzoxacyclotetradecine-1,7(8*H*)-dione], from ze(a)-, 'ene', and 'one'.

Zeatin {(2*E*)-2-methyl-4-[(7*H*-purin-6-yl)amino]but-2-en-1-ol}, named after the genus *Zea* (maize, corn, Indian corn). The genus name is from Greek *zea*, single-grained wheat.

Zeaxanthin (β,β-carotene-3,3′-diol), from ze(a)- and xanthin.

Zoapatanol {(6*R*)-9-[(2*S*,3*R*,6*E*)3-hydroxy-6-(2-hydroxyethylidene)-2-methyloxepan-2-yl]-2,6-dimethylnon-2-en-5-one}, from zoapatle (*Montanoa tomentosa*). Zoapatle is from Nahuatl *zoapatl*, zoapatle.

Zymosterol (5α-cholesta-8,24-dien-3β-ol), from zym(o)- and sterol, referring to its occurrence in yeast fat.

Zyzzine [4-(1*H*-indol-3-yl)-5-sulfanylidene-1,5-dihydro-2*H*-imidazol-2-one], after the genus *Zyzzya* (sponges). The genus name is widely regarded as a practical joke meant to secure a place at the bottom of any alphabetical order. However, in this race toward the bottom of the alphabet the sponges were overtaken by the weevil genus *Zyzzyva*, and by the hydrozoan genus *Zyzzyzus*.

2.2 Names based on properties

Abscisic acid {(2*Z*,4*E*)-5-[(1*S*)-1-hydroxy-2,6,6-trimethyl-4-oxocyclohex-2-en-1-yl]-3-methylpenta-2,4-dienoic acid}, after Latin *abscissio- abscission-*, abscission, referring to this compound's perceived role in plant abscission. A.k.a. **ABA**.

L-Ascorbic acid {(5*R*)-5-[(1*S*)-1,2-dihydroxyethyl]-3,4-dihydroxyfuran-2(5*H*)-one}. Named as antiscurvy agent. From New Latin *scorbutus*, scurvy, ultimately from

Old Norse *skyr*, curdled milk, and *bjugr*, edema, referring to the ancient belief that scurvy was caused by the sailors' diet of old curdled milk. A.k.a. **vitamin C**.

Durene (1,2,4,5-tetramethylbenzene), from Latin *durus*, hard, referring to this compound's crystallinity at ambient temperature.

Fulminic acid (oxidoazaniumylidynemethane), from Latin *fulmen, fulmin-*, lightning, referring to this compound's explosive properties.

Isofulminic acid [(hydroxynitrilio)methanide], from fulminic acid. A.k.a. **carboxime**, from carbon and oxime.

Laughing gas (dinitrogen monoxide), named after its euphoric properties.

Pinacol (2,3-dimethylbutane-2,3-diol), earlier called pinacone, from Greek *pinakion*, diminutive of *pinax, pinak-*, board, tablet, picture, referring to this compound's characteristic crystal shape.

Prehnitic acid (benzene-1,2,3,4-tetracarboxylic acid), named after the similarity between its crystal shape and that of the mineral prehnite, named after the Dutch colonel and governor of Cape of Good Hope (Cape Colony) Hendrik von Prehn (1733–1785).

Propionic acid (propanoic acid), with contraction, derived from Greek *proteon pion*, the first fat, referring to the fact that propionic acid is to be regarded as the first member of the homologous series of fatty acids judged by the soapy feel of its salts.

Stilbene (1,1′-ethene-1,2-diyldibenzene), from Greek *stilbein*, to glitter, referring to the appearance of this compound's crystals.

Styphnic acid (2,4,6-trinitrobenzene-1,3-diol), named irregularly from Greek *stryphnos*, astringent.

Traumatic acid [(2*E*)-dodec-2-enedioic acid), from Greek *trauma*, wound, named as a plant hormone stimulating the healing of wounds of the plant.

Ubiquinone [6-(*all-trans*-decaprenyl)-2,3-dimethoxy-5-methyl-1,4-benzoquinone], named as natural quinone found 'everywhere', from Latin *ubique*, everywhere. A.k.a. **ubiquinone 10** and **coenzyme Q_{10}**.

2.2.1 Names based on molecular shape

Self-explanatory names are without comment.

Acyclovir {2-amino-1,9-dihydro-9-[(2-hydroxyethoxy)methyl]-3*H*-purin-6-one}, from a(n)-, cycl(o)-, and virus, referring to this compound's less cyclic nature compared to the natural lead compound guanosine.

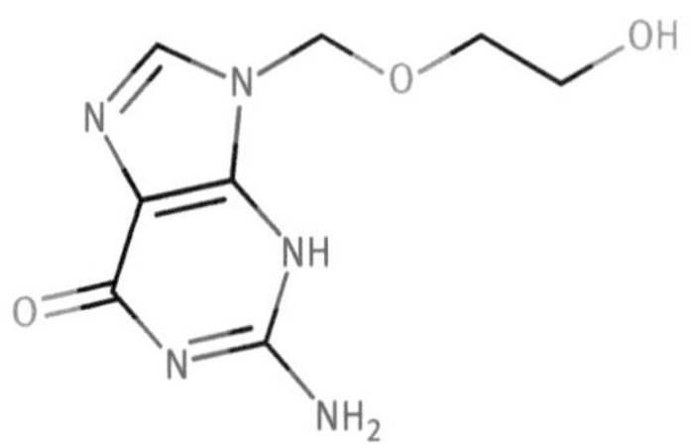

acyclovir

Adamantane {tricyclo[3.3.1.1^{3,7}]decane}, referring to this hydrocarbon's diamond-like structure, from Greek *adamantinos*, of steel, of diamond, from *adamas, adamant-*, steel, diamond.

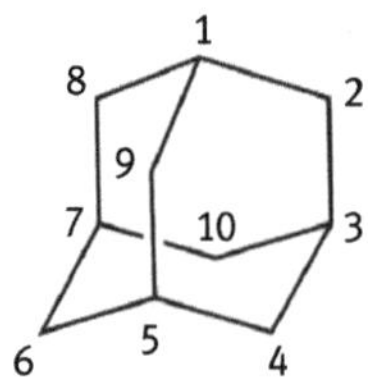

adamantane

Annulene, as in [5]annulene (cyclodeca-1,3,5,7,9-pentaene), a class name for completely conjugated unsaturated monocyclic hydrocarbons, from Latin *anulus*, diminutive of *anus*, ring.

[5]annulene

[3]Asterane {tetracyclo[3.3.1.0^{2,8}.0^{4,6}]nonane}. From Greek *aster, astro-*, star. A.k.a. **triasterane**.

[4]Asterane {pentacyclo[6.4.0.0^{2,7}.0^{4,11}.0^{5,10}]dodecane}, as [3]asterane. A.k.a. **tetraasterane**.

[5]Asterane {hexacyclo[9.2.2.0^{2,7}.0^{4,12}.0^{5,10}.0^{8,14}]pentadecane}, as [3]asterane. A.k.a. **pentaasterane**.

Barrelene {bicyclo[2.2.2]octa-2,5,7-triene}.

barrelene

Barrettane {hexacyclo[4.4.0.0^{2,4}.0^{3,10}.0^{5,8}.0^{7,9}]decane} from Latin *barretum, birretum*, beret, from *birrus*, short hooded cape.

Basketane {pentacyclo[4.4.0.0^{2,5}.0^{3,8}.0^{4,7}]decane}.

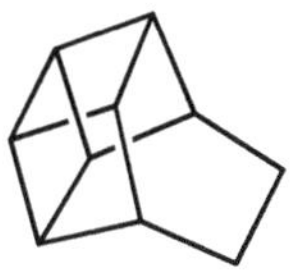

basketane

Betweenanene, as in [10.10]betweenanene {bicyclo[10.10.0]docos-1(12)-ene}, a class name, referring to the bridging and shielding of a central C=C double bond in these compounds by two crossing alkane chains.

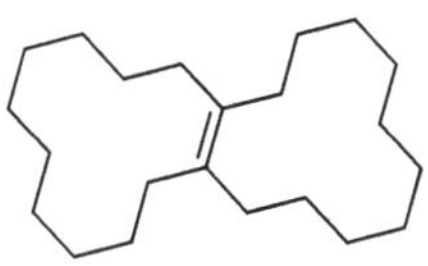

betweenane

Birdcage hydrocarbon {hexacyclo[5.4.1.0^{2,6}.0^{3,10}.0^{4,8}.0^{9,12}]dodecane}.

Broken window {[4.4.4]fenestrane, tricyclo[4.2.0.0^{1,4}]octane}.

Buckminsterfullerene {(C_{60}-I_h)[5,6]fullerene}, referring to the shape of this molecule which corresponds to the geodesic domes created by the American architect Richard Buckminster Fuller (1895–1983). A.k.a. **footballene**, from football (soccer).

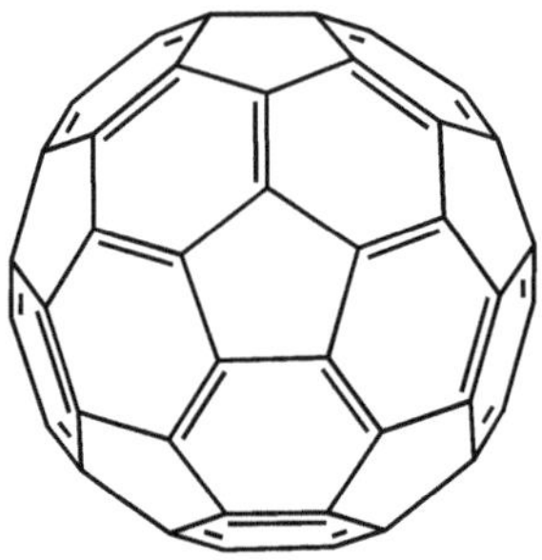

buckminsterfullerene

Calixarene, as in calix[4]arene, from Greek *kalix*, chalice, and arene.

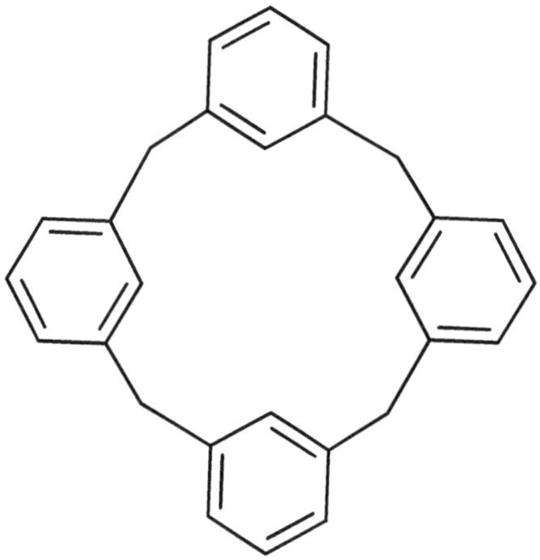

calix[4]arene

Catenane, as in [7]catenane (compounds consisting of seven linked rings), a class name, from Latin *catena*, chain.

Churchane {homopentaprismane, hexacyclo[5.4.0.0^{2,6}.0^{3,10}.0^{5,9}.0^{8,11}]undecane}.

Circulene, as in [6]circulene {coronene, dibenzo[*ghi,pqr*]perylene}, a class name for hydrocarbons consisting of a central polygon completely surrounded by and fused to benzenoids.

Corannulene {[5]circulene, dibenzo[*ghi,mno*]fluoranthene}, from Latin *cor, cord-*, heart and annulene.

Coronene {[6]circulene, dibenzo[*ghi,pqr*]perylene}. From Latin *corona*, crown. This compound also occurs as the mineral carpathite, cf. Section 13.3.1.

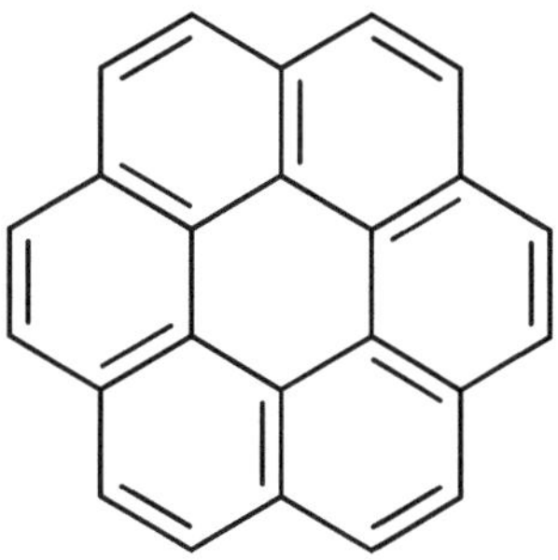

coronene

Cubane {pentacyclo[4.2.0.$0^{2,5}$.$0^{3,8}$.$0^{4,7}$]octane}. A.k.a. **quadriprismane** and **tetraprismane**.

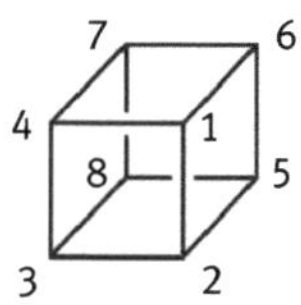

cubane

Deltic acid (2,3-dihydroxycyclopent-2-en-1-one), after the Greek letter Δ (upper-case delta), corresponding to the shape of the molecule.

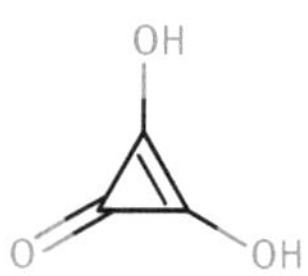

deltic acid

Dendralene, as in [5]dendralene (3,4-dimethylidenehexa-1,5-diene), a class name, by contraction of dendrimer, linear, and alkene.

Dendrimer, a class name for repeatedly branched molecules, from Greek *dendron*, tree, and *meris*, part, share. A.k.a. **arborol**, from Latin arbor, tree.

Diademane {hexacyclo[4.4.0.$0^{2,4}$.$0^{3,9}$.$0^{5,7}$.$0^{8,10}$]decane}. A.k.a. **congressane** and **mitrane**, from mitre, from Greek *mitra*, headband, turban.

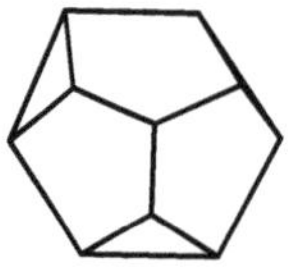

diademane

Diamantane {pentacyclo[7.3.1.1^{4,12}.0^{2,7}.0^{6,11}]tetradecane}, also named **congressane** as the logo of the 1963 IUPAC meeting in London, UK.

Diamondoid, such as in diamantane, a class name for hydrocarbons with their carbon atoms forming a part of the diamond lattice.

Diasterane {tricyclo[3.1.1.1^{2,4}]octane}, from asterane. Not to be confused with the geochemical class name diasterane, cf. Section 4.4. A.k.a. **[2]asterane.**

Dodecahedrane {[5]fullerane-C_{20}-I_h}.

dodecahedrane

Dolichol (a class of long-chain isoprenoid alcohols), from Greek *dolichos*, long.

Fenestrane, as in [4.4.4.4]fenestrane {tetracyclo[3.3.1.0^{3,9}.0^{7,9}]nonane}, a class name, from Latin *fenestra*, window.

Fullerene, a class name for carbon oligomers whose structure is analogous to buckminsterfullerene.

Garudane {heptacyclo[9.3.0.0^{2,5}.0^{3,13}.0^{4,8}.0^{6,10}.0^{9,12}]tetradecane}, named after Garuda, a legendary bird or bird-like creature in Hindu, Buddhist, and Jain mythology.

Gephyrotoxin {2-[(1*R*,3a*R*,5a*R*,6*R*,9a*S*)-6-[(2*Z*)-pent-2-en-4-yn-1-yl]dodecahydropyrrolo[1,2-*a*]quinolin-1-yl]ethan-1-ol}, from Greek *gephyra*, bridge, referring to this compound's bridged structure.

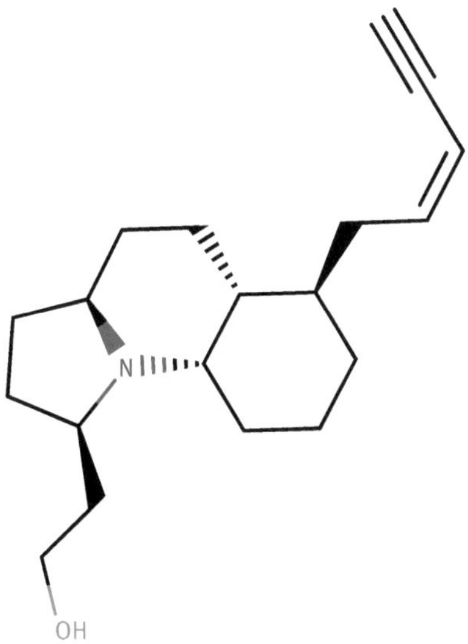

gephyrotoxin

Helicene, as in [6]helicene, a class name for helically shaped molecules consisting of fused benzene rings, from Greek *helix*, spiral.

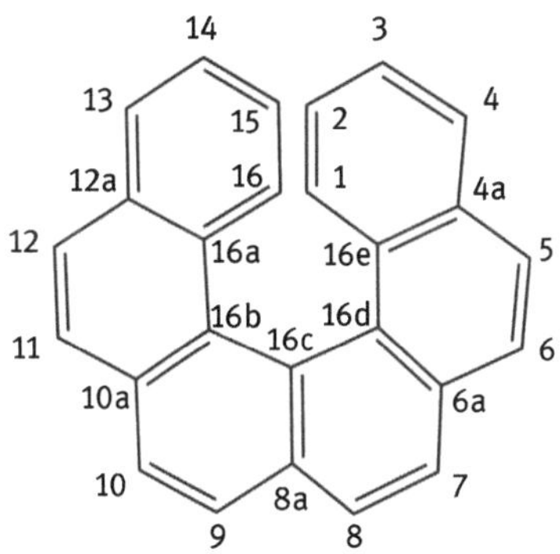

hexahelicene

Heterofullerene, as in diaza[60]fullerene, a class name.

[1]**Housane** {bicyclo[2.1.0]pentane}. The poor man's housane, a compound with a two-dimensional house-shaped carbon skeleton.

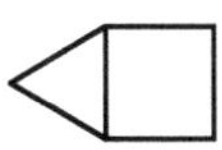

[1]housane

[2]**Housane** {hexacyclo[4.4.0.0^{2,5}.0^{3,9}.0^{4,8}.0^{7,10}]decane}. The rich man's housane, a compound with a three-dimensional house-shaped carbon skeleton. A.k.a. **pentaprismane.**

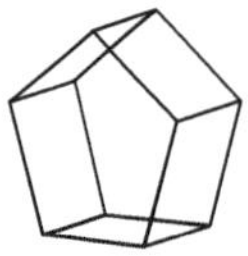

pentaprismane

Iceane (wurtzitane, tetracyclo[5.3.1.1^{2,6}.0^{4,9}]dodecane). Named after its ice-like structure and wurtzitane after its similarity to the crystal structure of the mineral wurtzite. The mineral name is after the French chemist Charles Adolphe Wurtz (1817–1884).

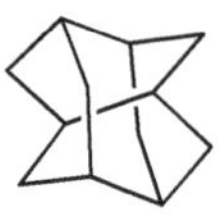

iceane

Ladderane, as in [3]ladderane {tricyclo[4.2.0.0^{2,5}]octane}, a class name for hydrocarbons consisting of fused cyclobutane rings.

(–)-Modhephene {(1*R*,2*R*,5*S*)-2,6,6,8-tetramethyltricyclo[5.3.3.0^{1,5}]undec-7-ene}, from Hebrew *modhephe*, propeller. A.k.a. **(–)-modephene.**

Nortricyclene {tricyclo[2.2.1.0^{2,6}]heptane}.

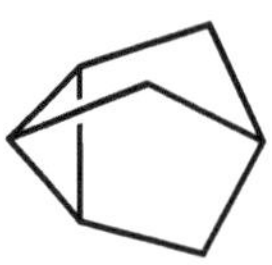

nortricyclene

Olympiadane, a catenane with five interlocking rings.

Olympicene {6*H*-benzo[*c*,*d*]pyrene}, a five-ring hydrocarbon in the shape of the Olympic rings.

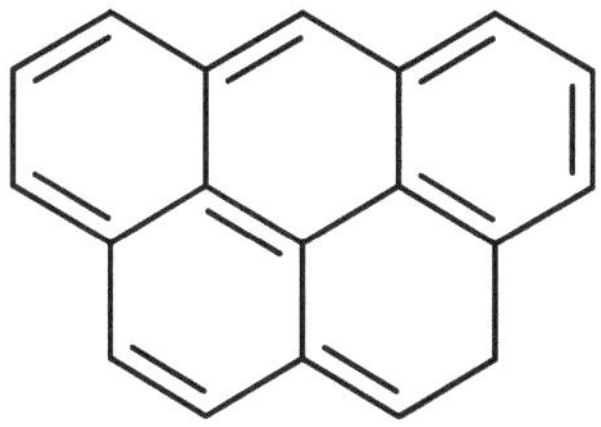

olympicene

Paddlane {tricyclo[*m.n.o.p*]alkane}, a class name.

Penguinone (3,4,4,5-tetramethylcyclohexa-2,5-dien-1-one). The word penguin has a disputed etymology. Around 1570 this name was coined for the (since 1844 extinct) great auk (*Pinguinus impennis*). A suggested Welsh origin, from *pen*, head, and *gwyn*, white, while fitting the appearance of the great auk, and/or its habitat on White Head Island (Welsh *Pen Gwyn*), Newfoundland, Canada, is not universally accepted, due to linguistic objections. From around 1580 the name penguin was used for the Antarctic birds (*Spheniscidae*) which look similar, without being related, to the great auk. Another line of reasoning, equally disputed, relates the name penguin to Latin *pinguis*, fat, in keeping with the Dutch synonyms for penguin, *pinguïn* and *vetgans*, literally fat goose. Among dedicated etymologists this discussion is still ongoing [3].

[3]Peristylane {tetracyclo[3.3.1.0^{2,4}.0^{3,7}]nonane}, from peristyle, from Greek *peristylos*, surrounded by columns. A.k.a **triaxane**.

[4]Peristylane {pentacyclo[5.4.1.0^{3,10}.0^{5,9}.0^{8,11}]dodecane}, as [3]peristylane.

[5]Peristylane {hexacyclo[7.5.1.0^{3,13}.0^{5,12}.0^{7,11}.0^{10,14}]pentadecane}, as [3]peristylane.

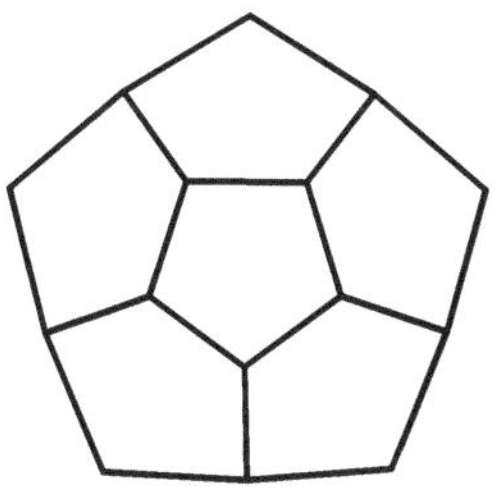

[5]peristylane

Pleiadene {benzo[5,6]cyclohepta[1,2,3-*de*]naphthalene}, named after the Pleiades, in antiquity considered as a constellation of seven stars, a.k.a. Seven Sisters (in modern terms the star cluster Messier 45), referring to the central seven-membered ring of pleiadene. The name Pleiades is most likely derived from Greek *plein*, to sail, referring to the fact that the sight of this constellation was delimiting the sailing season in the Mediterranian Sea. The ancient Greek mythical figures Pleione, an Oceanid nymph, and the Pleiades, seven divine sisters and companions of Artemis, were most likely named after the constellation.

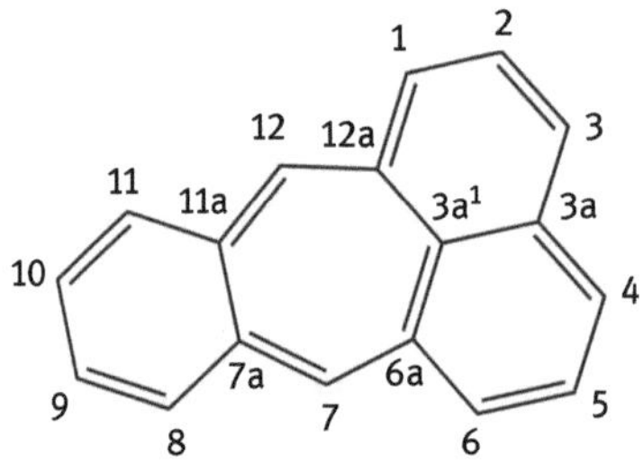

pleiadene

Prismane {tetracyclo[2.2.0.0^{2,6}.0^{3,5}]hexane}. A.k.a. **Ladenburg benzene**, after the German chemist Albert Ladenburg (1842–1911), **triprismane**, and **[3]prismane.**

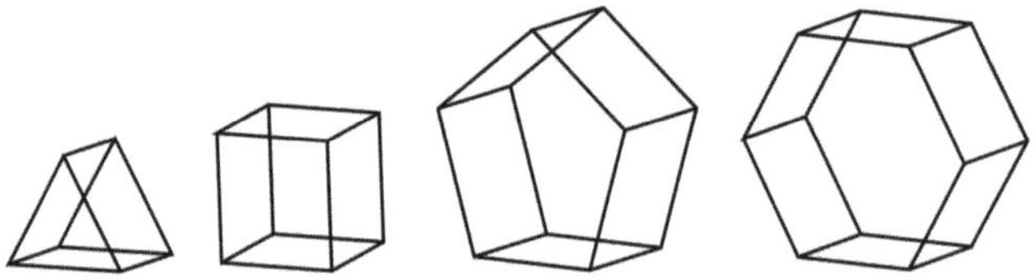

[3]prismane [4]prismane [5]prismane [6]prismane

Propellane, as in [2.2.2]propellane {tricyclo[2.2.2.0^{1,4}]octane}, a class name for tricyclic hydrocarbons with a shared carbon-carbon bond.

Quadricyclane {tetracyclo[3.2.0.0^{2,7}.0^{4,6}]heptane}. From quadr(o)- and cycle.

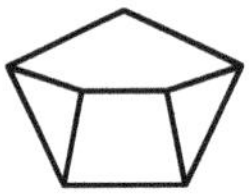

qquadricyclane

Radialene, as in [6]radialene (hexamethylidenecyclohexane), a class name for peralkylidene substituted cycloalkanes, from Latin *radius*, ray, spoke.

[3]Rotane {trispiro[2.0.2.0.2.0]nonane}, from Latin *rota*, wheel.

[4]Rotane {tetraspiro[2.0.2.0.2.0.2.0]dodecane}, as [3]rotane.

Rotaxane, as in [2]rotaxane (compounds consisting of a dumbbell threaded through a macrocycle), a class name, from Latin *rota*, wheel, and *axis*, axle.

Sandwich compound, after sandwich, named after the British statesman John Montagu, Fourth Earl of Sandwich (1718–1792), a devoted card player who did not want his game compromised by a formal meal.

Snoutane {pentacyclo[4.4.0^{2,4}.0^{3,8}.0^{5,7}]decane}.

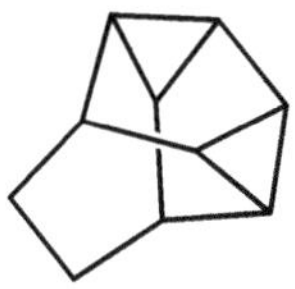

snoutane

Squaric acid (3,4-dihydroxycyclobut-3-ene-1,2-dione).

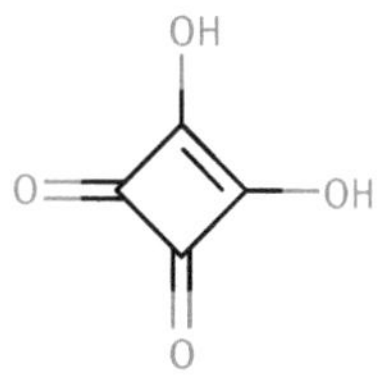

squaric acid

Sulflower {octathia[8]circulene, {1,12:3,4:6,7:9,10-tetraepithiocycloocta[1,2-*c*:3,4-*c*′:5,6-*c*:7,8-*c*′]tetrathiophene}. A portmanteau of sulfur and sunflower.

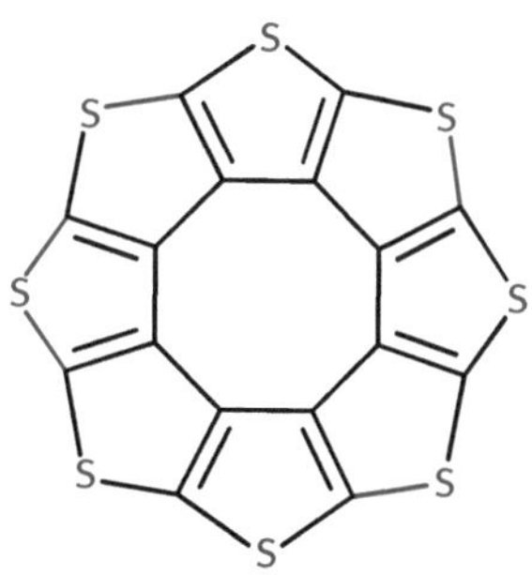

sulflower

Tetrahedrane {tricyclo[1.1.0.0^{2,4}]butane}.

tetrahedrane

Tetraquinane {tetracyclo[7.2.1.0^{4,11}.0^{6,10}]dodecane}. From Latin *quinis*, five apiece. Named as containing four fused pentagons.

[5]Triangulane (tetraspiro[2.0.0.0.2.1.1.1]undecane).

[1.1.1]Triblattane {pentacyclo[6.3.0.0^{2,6}.0^{3,10}.0^{5,9}]undecane}. From German *Blatt*, leaf, sheet.

Triptycene, {10-dihydro-9,10[1′,2′]-benzenoanthracene}, from Greek *triptychos*, threefold, referring to this compound's molecular symmetry.

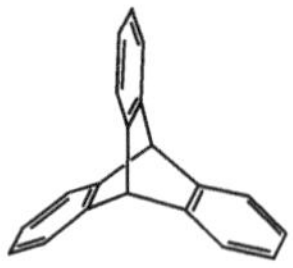

triptycene

Twistane {tricyclo[4.4.0.0^{3,8}]decane}.

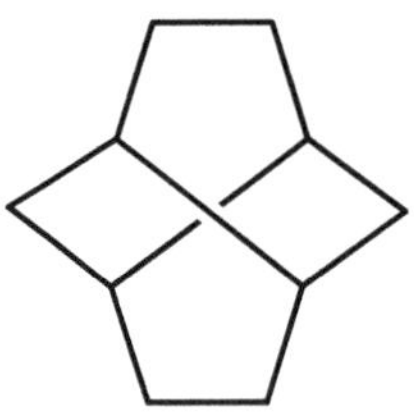

twistane

Ufolane {tricyclo[6.3.1.0^{4,12}]dodecane}. From UFO (unidentified flying object).

Uthrene {1*H*,12*H*-dibenzo[*de,hi*]tetracene-1,12-diyl, a diradical}, named after its shape, roughly like the letter U, and -anthrene.

Zethrene {dibenzo[*de,mn*]tetracene}. Named after its shape, roughly like the letter Z, and -anthrene.

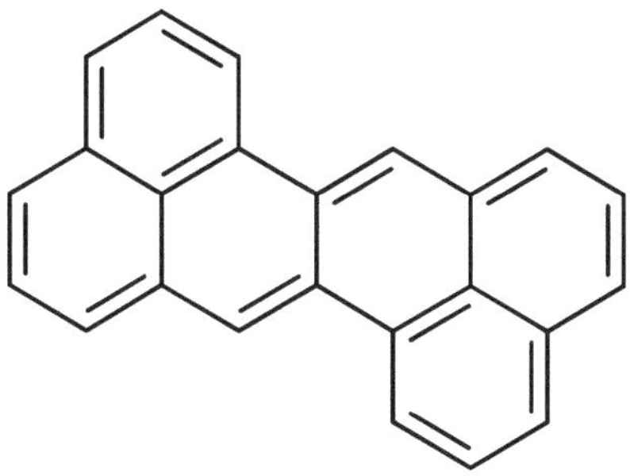

zethrene

2.2.2 Color related names

Contrary to popular belief not all compounds with a color related trivial name are colored. For instance, chrysene, at first conspicuous by its yellow-golden color, was named accordingly, but later shown to be colorless, the observed color having been due to an impurity of yellow tetracene. Neither is pyrrole deep red, as its name might suggest, but rather named after the red color which develops when pine chips, moistened with hydrochloric acid, are exposed to pyrrole vapor.

Alcian blue (a phthalocyanine dye), probably coined by contraction and slight alteration of phthalocyanine.

Amaranth {trisodium 3-hydroxy-4-[(*E*)-2-(4-sulfonaphthalen-1-yl)diazen-1-yl]naphthalene-2,7-disulfonate}, after its color. Amaranth is from Greek *amaranton*, a mythical flower which never fades, from *amarantos*, unfading, from a(n)- and *marainesthai*, to wilt, also influenced by *anthos*, flower.

Anthocyanin (flavylium, 2-phenyl-1-benzopyrylium, a cation), from anth(o)- and cyan(o)-.

Azulene {bicyclo[5.3.0]deca-2,4,6,8,10-pentaene}, named after its blue color, from Spanish *azul*, blue, from Arabic *lazuward*, lapis lazuli.

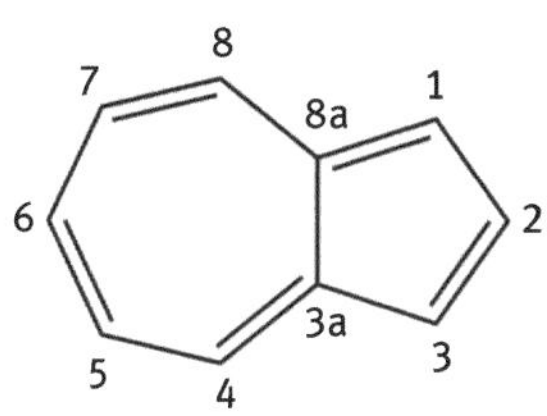

azulene

Berlin blue [diiron potassium hexacyanide], A.k.a. **Milori blue**, after the French dye company A. Milori, **Paris blue, Prussian blue**, and **Turnbull's blue**, named after the dye company Turnbull & Co., Glasgow, UK.

Celestin blue [1-carbamoyl-7-(diethylamino)-3,4-dihydroxyphenoxazin-5-ium chloride], from celest-.

Chalcone [(2*E*)-1,3-diphenylprop-2-en-1-one]. From Greek *chalkos*, copper, referring to such compounds' often yellow color. Also used as class name.

Chamazulene (7-methyl-1,4-dimethylazulene), after chamomile (*Matricaria chamomilla*), and azulene. The specific epithet of the species name is from Greek *chamaimelon*, chamomile, from *chamai*, on the ground, and *melon*, apple, referring to the apple-like smell of this plant.

Chlorogenic acid {(1*S*,3*R*,4*R*,5*R*)-3-{[(2*E*)-3-(3,4-dihydroxyphenyl)prop-2-enoyl] oxy}-1,4,5-trihydroxycyclohexane-1-carboxylic acid}. From Greek ch*loros*, yellow green. This acid is neither green nor does it contain chlorine. The name refers to the green color which develops upon its oxidation.

Chrysene, from Greek *chrysos*, gold, referring to the color of this compound. A misnomer as it turned out later. Rigorously purified chrysene is colorless and the golden-yellow color originally observed was due to the presence of yellow tetracene as an impurity.

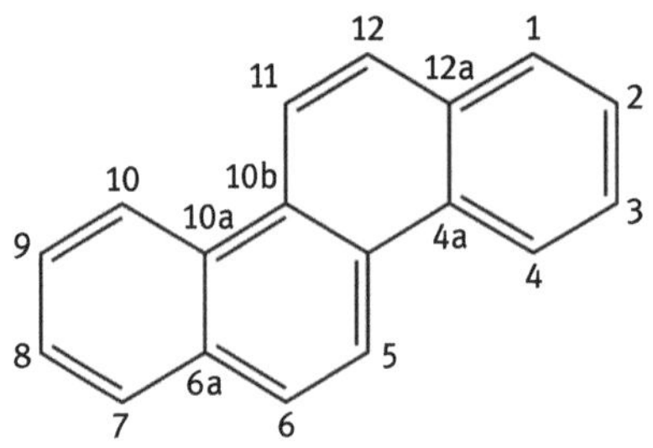

chrysene

Crystal violet {4-[bis[4-(dimethylamino)phenyl]methylidene]-*N,N*-dimethylcyclohexa-2,5-diene-1-iminium chloride}, after its color. A.k.a. **gentian violet**, from gent (io)-, and **methyl violet**.

Croconic acid (4,5-dihydroxycyclopent-4-ene-1,2,3-trione), from Greek *krokos*, saffron, referring to its color.

Cyanide [nitridocarbonate(1−)], from Greek *kyanos*, dark blue, from German *Blausäure*, hydrogen cyanide, literally blue acid, not blue, but named after its preparation by acid treatment of Berlin blue.

Eosin Y [disodium 2-(2,4,5,7-tetrabromo-6-oxido-3-oxo-3*H*-xanthen-9-yl) benzoate)], named after Anna Peters, nicknamed Eos, a friend of its inventor, the German chemist Heinrich Caro (1834–1910), and referring to this dye's pink color, ultimately from Greek *eos*, dawn.

Erythritol [(2*R*,3*S*)-butane-1,2,3,4-tetrol], from erythrin, named after the red color observed upon erythrin's oxidation, from Greek *erythros*, red. Ironically, erythritol is neither red nor does it take part in reactions which generate a red color.

Ferruginol [(4bS,8aS)-4b,8,8-trimethyl-2-(propan-2-yl)-5,6,7,8a,9,10-hexahydrophenanthren-3-ol], from ferruginous, referring to this compound's brownish-yellow color.

Flavan [2-phenylchromane, 2-phenyl-3,4-dihydro-2*H*-1-benzopyran], from Latin *flavus*, light yellow, referring to the colors of its derivatives.

Granaticin {(3a*S*,5*S*,8*R*,9*R*,11*R*,13b*S*,15*R*)-7,8,12,15-tetrahydroxy-5,9-dimethyl-3,3a,5,8,11,13b-hexahydro-8,11-ethanofuro[2,3-*e*]naphtho[2,3-*c*:6,7-*c′*]dipyran-2,6,13 (9*H*)-trione}, from German *Granat*, garnet, referring to this compound's color, cf. Section 13.2.1.

Helianthine {sodium [(*E*)-4-(dimethylamino)phenyl]diazenyl]benzenesulfonate}, after its color, corresponding to that of the sunflower (*Helianthus*). A.k.a. **methyl orange**.

Ilimaquinone {2-hydroxy-5-methoxy-3-[[(1*R*,2*S*,4a*S*,8a*S*)-1,2,4a-trimethyl-5-methylidenedecahydronaphthalen-1-yl]methyl]cyclohexa-2,5-diene-1,4-dione}, after ilima (*Sida fallax*), yellow flowers, from Hawaiian *'ilima*. Ilimaquinone refers to this compound's golden yellow color rather than to its source, the marine sponge *Hippospongia metachroma*.

Leucopterin [2-amino-5,8-dihydro-1*H*-pteridin-4,6,7-trione], from leuc(o)- and pterin, named as white pigment from *Lepidoptera* wings.

Methylene blue [3,7-bis(dimethylamino)phenothiazin-5-ium chloride], from methylene, but without any stated reason.

Methylene green [7-(dimethylamino)-*N*,*N*-dimethyl-4-nitro-3*H*-phenothiazin-3-ium chloride], patterned after methylene blue.

Methylene red [3,6-bis(dimethylamino)thioxanthenium chloride], patterned after methylene blue.

Methyl green {4-[[4-(dimethylamino)phenyl][4-(dimethyliminio)cyclohexa-2,5-dien-1-ylidene]methyl]-*N*,*N*,*N*-trimethylbenzenaminium dichloride}, methyl probably referring to this dye's triphenylmethane core.

Methyl orange {sodium 4-[(*E*)-[4-(dimethylamino)phenyl]diazenyl]benzenesulfonate}, methyl probably referring to the presence of methyl groups in this dye. A.k.a. **helianthine.**

Methyl red {sodium 2-[(*E*)-[4-(dimethylamino)phenyl]diazenyl]benzoic acid}, as methyl orange.

Methyl violet {4-[bis[4-(dimethylamino)phenyl]methylidene]-*N,N*-dimethylcyclohexa-2,5-dien-1-iminium chloride}, as methyl green.

Methyl yellow {*N,N*-dimethyl-4-[(*E*)-phenyldiazenyl]aniline}, as methyl orange. A. k.a. **butter yellow.**

Murexide {ammonium 2,6-dioxo-5-[(2,4,6-trioxotetrahydropyrimidin-5(2H)-ylidene) amino]-1,2,3,6-tetrahydropyrimidin-4-olate}, from M*urex* (murex, sea snails), the source of Tyrian purple, referring to the purple color of aqueous solutions of this compound, also used as a dye. The genus name is ultimately from Greek *myax, myak-*, mussel.

Naples yellow [dilead(II) diantimonate], after the city of Naples, Italy. A.k.a. **jaune d'antimoine** and **bindheimite.**

Neutral red [N^8,N^8,3-trimethylphenazine-2,8-diamine hydrochloride (1/1), referring to this indicator dye's color change at pH 6.8–8.0.

Ponceau (a family of azo dyes), from French *ponceau*, poppy-colored.

Purpurogallin (2,3,4,6-tetrahydroxybenzocyclohepten-5-one), from purpur(o)- and gall, referring to its color and its formation in nutgalls and oak bark.

Pyrrole, from Greek *pyrrhos*, reddish, fiery, referring to the red color which develops when pyrrole vapor acts upon a pine splinter moistened with hydrochloric acid.

Rhodanide [thiocyanate, nitridosulfidocarbonate(1−)], from ^{1}rhod(o)-, named after the red color of this ion's complex with iron(III).

Rhodizonic acid (5,6-dihydroxycyclohex-5-ene-1,2,3,4-tetrone), from Greek *rhodizein*, to color rose red, referring to this compound's dark orange color.

Rhodopin (1,2-dihydro-ψ,ψ-caroten-1-ol), a contraction of ^{1}rhod(o)- and Greek *opsis*, sight, vision, referring its color and function.

Rosalic acid {4-[bis(4-hydroxyphenyl)methylidene]cyclohexa-2,5-dien-1-one}, named after this dye's pink color. A.k.a. **aurin,** from aur(o)-, and **rosolic acid.**

Rubeanic acid [ethanebis(thioamide)], from rubi-, referring to this compound's red color.

Rubicene, named after its red color, from Latin *ruber*, red, madder, and anthracene.

Rufigallol (1,2,3,5,6,7-hexahydroxyanthracene-9,10-dione), from Latin *rufus*, red, gallic acid, and 'ol', referring to this compound's color and its preparation from gallic acid. A.k.a. **rufigallic acid**.

Sulfuretin {2-[(3,4-dihydroxyphenyl)methylidene]-6-hydroxy-1-benzofuran-3(2H)-one}, after the species *Cosmos sulphureus* (sulfur cosmos, yellow cosmos). Here sulfur refers to a color, not to the element. A.k.a. **sulfurein**.

Thymol blue [4,4′-(1,1-dioxido-3*H*-2,1-benzoxathiole-3,3-diyl)bis[6-methyl-2-(propan-2-yl)phenol]], from thymol.

Tropaeolin O {sodium 4-[(2,4-dihydroxyphenyl]diazenyl]benzenesulfonate}, after the color of the plants of the genus *Tropaeolum* (nasturtium). The genus name is from Latin *tropaeum*, trophy. A.k.a. **acid orange 6** and **chrysoine resorcinol**.

Verdigris (copper(II) acetate hydrate), from Old French *vert de Grece*, green of Greece.

Violaxanthin [(3*S*,3′*S*,5*R*,5′*R*,6*S*,6′*S*)-5,6:5′,6″-diepoxy-5,5′,6,6′-tetrahydro-β,β-caroten-3,3′-diol], from violet and xanthin.

Viologen (a class name for 4,4′-bipyridinium compounds), from violet and -gen, referring to the color of the corresponding radical cations.

Violuric acid [5-isonitrosobarbituric acid, 5-(hydroxyimino)-1,3-diazinane-2,4,6-trione], from violet and urea, referring to the color of its nitroso tautomer and its salts.

Xanthic acid (a class name for *O*-alkylcarbonodithioic acids), from xanth(o)-, referring to their salts' usually yellow color.

Xanthine (3,7-dihydropurine-2,6-dione), from xanth(o)-, referring to the yellow residue left after evaporation of xanthine with nitric acid.

Xanthopterin [2-aminopteridine-4(3*H*),6(5*H*)-dione], from xanth(o)- and pterin. Named as yellow pigment of butterfly wings.

Xanthotoxin {9-methoxy-7*H*-furo[3,2-g][1]benzopyran-7-one}, from xanth(o)- and toxin. A.k.a. **ammoidin**, after the species *Ammi majus* (bishop's weed, false bishop's weed, bullwort, lady's lace, false Queen Anne's lace, laceflower). The genus name is from Greek *ammi*, an umbelliferous plant, and **methoxsalen**, a contraction of 8-methoxypsoralen.

2.2.3 Names based on taste or smell

Acrolein (propenal), from Latin *acer, acr-*, sharp, bitter, and *oleum*, oil, referring to the fact that this sharp smelling volatile compound is formed by thermal decomposition of fat.

Acrylic acid (propenoic acid), from Latin *acer, acr-*, sharp, bitter, referring to its smell.

Amarolide (2α,11α-dihydroxypicrasane-1,12,16-trione), from Latin *amarus*, bitter, referring to this lactone's bitter taste.

Anisaldehyde (4-methoxybenzaldehyde), named for its strong aniseed smell.

Anisole (methoxybenzene), from Greek *anison*, anise, dill, referring to its smell.

Bourbonal (3-ethoxy-4-hydroxybenzaldehyde), a synthetic vanilla flavor named after Bourbon vanilla (*Vanilla planifolia*).

Bromine (dibromine), from Greek *bromos*, stench, referring to the obnoxious smell of bromine vapor.

Cacodyl (tetramethyldiarsane), from Greek *kakodes*, evil smelling. Contained in Cadet's liquid, after the French chemist and pharmacist Louis Claude Cadet de Gassicourt (1731–1799). A.k.a. **dicacodyl.**

Cacodylic acid (dimethylarsinous acid), as cacodyl.

Cacodyl oxide (1,1′-oxybis[1,1-dimethylarsane]), as cacodyl.

Cyclamen aldehyde {2-methyl-3-[4-(propan-2-yl)phenyl]propanal}, from cyclamen, referring to this compound's lily-of-the-valley fragrance.

Exaltolide (cyclopentadecanolide, oxacyclopentadecan-2-one), from exaltation, referring to this compound's musk smell.

Geosmin [(4*S*,4a*S*,8a*R*)-4,8a-dimethyloctahydronaphtalen-4a(2*H*)-ol], named as an earthy aroma produced by soil bacteria, from Greek *ge*, earth, land, and *osme*, smell, from *ozein*, to smell.

Glycerol (propane-1,2,3-triol), earlier called glycerin, from Greek *glykys, glyker-*, sweet.

Glycol (ethane-1,2-diol), also used as class name for vicinal diols, as glycerol. A.k.a. **ethylene glycol.**

α-Ionone [(3*E*)-4-(2,6,6-trimethylhex-2-en-1-yl)but-3-en-2-one], from Greek *ion*, the flower violet, referring to the characteristic smell of this compound.

Mustard gas {1-chloro-2-[(2-chloroethyl)sulfanyl]ethane}, referring to its characteristic smell which, however, most likely is due to impurities. This compound is not, as the name suggests, a gas, but a high-boiling liquid. A.k.a. **lost**, an acronym formed by contraction of the names Wilhelm Lommel (1878–1962) and Georg Wilhelm Steinkopf (1879–1949), two German chemists who developed its large-scale synthesis, and **yperite**, a French military code name after the town of Ypres (Ieper), Belgium.

Ozone (trioxygen), from Greek *ozon*, smelling, from *ozein*, to smell, referring to ozone's characteristic unpleasant smell. In vernacular use the word ozone has had the meaning pure and refreshing air, quite the opposite of its chemical meaning.

Picric acid (2,4,6-trinitrophenol), from Greek *pikros*, sharp, bitter, referring to its taste. A.k.a. **carbazotic acid** and **lyddite**, after the town of Lydd, Kent, UK, where this compound was manufactured in WWI.

Raspberry ketone [4-(4-hydroxyphenyl)butan-2-one], named as natural aroma of raspberry (*Rubus idaeus*).

Rhodinol (3,7-dimethyloct-7-en-1-ol), from ^{1}rhod(o)- and with 'ol', referring to this alcohol's flowery scent.

Rose oxide [4-methyl-2-(2-methylprop-1-en-1-yl)tetrahydro-2*H*-pyran], named as natural fragrance of the genus *Rosa*.

Saccharin (1,1-dioxo-1,2-benzothiazol-3-one), referring to the sweet taste of its sodium and calcium salt, respectively, from the adjective saccharine, ultimately from Greek *sakcharon*, sugar, gravel, ultimately from Sanskrit *sarkara*, sugar, gravel, grit.

Sotolone (3-hydroxy-4,5-dimethylfuran-2(5*H*)-one), from Japanese *soto*, raw sugar, and 'ol' and 'one', i.e. named after its taste of raw cane sugar and its structure of an enol lactone. A.k.a. **sotolon**.

Strawberry aldehyde (ethyl 3-methyl-3-phenyloxirane-2-carboxylate). The name of this artificial flavoring agent is not meant to imply that it is an aldehyde, which it is not, but that it possesses both a strawberry and an aldehydic olfactory note.

Strawberry furanone (4-hydroxy-2,5-dimethylfuran-3(2*H*)-one). Named for its olfactory note.

Whisky lactone (5-butyl-4-methyldihydrofuran-2(3*H*)-one), named as ingredient of the aroma of whisky and other alcoholic beverages which have been aged in oak barrels, the ultimate source of this compound. A.k.a. **quercus lactone**, from Latin *quercus*, oak.

Wine lactone [(3*S*,3a*S*,7a*R*)-3,6-dimethyl-3a,4,5,7a-tetrahydro-1-benzofuran-2(3H)-one], named as an important component of the flavor of white wine.

2.2.4 Names based on uses

Anandamide [(5*Z*,8*Z*,11*Z*,14*Z*)-*N*-(2-hydroxyethyl)icosa-5,8,11,14-tetraenamide], from Sanskrit *ananda*, bliss, and amide, referring to this compound's psychopharmacological properties.

Aplasmomycin (an antibiotic), from a(n)-, after the genus *Plasmodium* (protozoans), from Greek *plasma*, form, mold, from *plassein*, to mold, and *-odes*, thing that resembles, as well as -mycin. The name refers to the specific antiprotozoan activity of this antibiotic.

Aqua regia (a mixture of nitric acid and hydrochloric acid), New Latin literally royal water. Named as being able to dissolve gold, the 'king of metals'.

Bacilysin {L-alanyl-3-[(1*R*,2*S*,6*R*)-5-oxo-7-oxabicyclo[4.1.0]heptan-2-yl]-L-alanine}, from bacillus and lysis, referring to this antibiotic's bacteriolytic activity. A.k.a. **tetaine.**

Cardenolide (card-15-enolide), from cardiac and 'olide'.

Cellocidin (but-2-ynediamide), from cello- and -cide, i.e. named as bactericide. A.k.a. **aquamycin**, after the bacterial species *Streptomyces reticuli* var. *aquamyceticus*.

Cervicarcin [3-(2,3-epoxybutanoyl)-1,2,3,4,4a,9a-hexahydro-1,3,4,5,10-pentahydroxy-2-methyl-4a,9a-epoxyanthracen-9(10*H*)-one], with contraction, from cervical cancer, i.e. named as antitumor agent.

Cupferron (ammonium 2-oxo-1-phenylhydrazin-1-olate), from Latin *cuprum*, copper, and *ferrum*, iron, referring to its usefulness as a reagent for copper and iron.

Cuproine (2,2′-biquinoline), from Latin *cuprum*, copper, referring to the usefulness of this reagent for the determination of copper.

Cuprone [(2*E*)-2-(hydroxyimino)-1,2-diphenylethan-1-ol], from Latin *cuprum*, copper, referring to the usefulness of this reagent for the determination of copper, as well as of molybdenum, tungsten, vanadium, and some of the platinum metals.

Doctor solution (sodium plumbite, sodium plumbate(II)), from to doctor, referring to its use in the removal of malodorous sulfur compounds from petroleum distillates.

Ferron (8-hydroxy-7-iodonaphthalene-5-sulfonic acid), from Latin *ferrum*, iron, referring to this compound's usefulness as analytical reagent for iron (and aluminum).

Fissium, from fission and 'ium'. An artificial mixture of the natural elements represented among the products of a nuclear fission.

Goitrin (5-ethenyl-1,3-oxazolidine-2-thione), from goiter, referring to this compound's ability to reduce the production of thyroid hormones. Goiter is ultimately from Latin *guttur*, throat.

Kalignost [sodium tetraphenylborate(1–)], from New Latin *kalium* and diagnostic, referring to this reagent's use for the gravimetric determination of potassium. A.k.a. **kalibor.**

Magneson {4-[(*E*)-2-(4-nitrophenyl)diazenyl]benzene-1,3-diol}, from magnesium, referring to this compound's usefulness for the detection of magnesium (and molybdenum).

Milbemycin (a group of antibiotics), from German *Milbe*, mite, referring to their use as miticides.

Mildiomycin (an antibiotic), from mildew and -mycin, referring to its antimildew activity.

Mirbane oil (nitrobenzene), surreptitiously used as artificial almond aroma in spite of its toxicity. From Middle French *mirobolan, mirobalan*, aromatic nuts used in perfumery, ultimately from Greek *myron*, unguent, perfume, and *balanos*, acorn.

Mirex {1,2,3,4,5,5,6,7,8,9,10,10-dodecachloropentacyclo[5.3.0.$0^{2,6}$.$0^{3,9}$.$0^{4,8}$]decane}, probably derived from pismire. Pismire is from Old Norse, cf. Danish *pissemyre*, pismire, from *pisse*, to urinate, and *myre*, ant, referring to the acrid smell from an anthill. Danish *pisse* is a vulgar word, the polite equivalent being *tisse*. Mirex thus refers to this compound's insecticidal properties.

Narcotine {(3*S*)-6,7-dimethoxy-3-[(5*R*)-4-methoxy-6-methyl-5,6,7,8-tetrahydro[1,3]dioxolo[4,5-*g*]isoquinolin-5-yl]-2-benzofuran-1(3*H*)-one}, from Greek *narkotikos*, benumbing, narcotic, from *narkoun*, to benumb, from *narke*, numbness. A.k.a. **noscapine,** latinized spelling of gnoscopine, from Greek *gnosis*, knowledge, from *gignoskein*, to know, and opium.

Nisin (a polycyclic peptide), a drastic contraction of streptococci group N inhibitory substance.

Nitrin {2-[(phenylhydrazinylidene)methyl]aniline}, named as analytical reagent for nitrite. Cf. nitron.

Nitron [(1,4-diphenyl-1*H*-1,2,4-triazol-4-ium-3-yl)(phenyl)azanide], named as reagent for the quantitative determination of nitrate (and perchlorate). Cf. nitrin.

Phlobaphene (a class name for phenolic natural dyes), from phlor(o)- and Greek *baphe*, dye, from *baptein*, to dip, to dye.

Phlobatannin (oligomeric phenolic natural products), from Greek *phloios, phlor-*, bark, *baphe*, dye, and tannin.

Primocarcin (5-acetamido-4-oxohex-5-enamide), from prim(o)- and carcin(o)-, named as antitumor agent.

Quantum dye ($C_{38}H_{36}EuN_8S_2$), a macrocylic europium chelate, referring to this compound's use as quantitative fluorescent label.

Reductic acid (2,3-dihydroxycyclopent-2-en-1-one), from reduction, referring to the antioxidant action of this compound.

Rongalite (sodium hydroxymethanesulfinate), from French *rongeage*, an etching.

Sapogenin [(25*S*)-spirostan], from saponin, q.v., named as the non-sugar part of a saponin. A.k.a. **neogenin.**

Saponin, a class name for plant glycosides producing a soapy lather, from Latin *sapo, sapon-*, soap.

Substance P (an oligopeptide), from pain, referring to this peptide's analgesic and hyperalgesic properties. A.k.a. **SP**.

Tiron (disodium 4,5-dihydroxybenzene-1,3-disulfonate), a contraction of titanium and iron, referring to the use of this compound for the determination of these metal ions.

Urotropin (1,3,5,7-tetraazaadamantane, 1,3,5,7-tetraazabicyclo[3.3.1.$1^{3,7}$]decane), originally a trademark, referring to this compound's mild antiseptic action on the urinary tract. A.k.a. **hexamethylenetetramine** and **methenamine**.

2.2.5 Names based on chemical or physical properties

Acarbose {4,6-dideoxy-4-[[(1*S*,4*R*,5*S*,6*S*)-4,5,6-trihydroxy-3-(hydroxymethyl)cyclohex-2-en-1-yl]amino]-α-D-glucopyranosyl-(1→4)-α-D-glucopyranosyl-(1→4)-β-D-glucopyranose), named as inhibitor of α-glucosidase.

Allophanic acid (carbamoylcarbamic acid), from Greek *allophanes*, appearing otherwise, referring to the fact that only derivatives of this acid are isolable, not the free acid.

Antipain (a modified polypeptide), with contraction, from anti- and papain, referring to this peptide's inhibition of papain.

L-**Arogenic acid** {1-[(2*S*)-2-amino-2-carboxyethyl]-4-hydroxycyclohexa-2,5-diene-1-carboxylic acid}, from aromatic and -gen, referring to this compound's intermediacy in the biosynthesis of aromatic compounds.

Avidin (a protein), from Latin *avidus*, avid, referring to this protein's avidity for biotin.

Barytes (barium sulfate), from Greek *barytes*, weight, from *barys, baryt-*, heavy. A. k.a. **barite**.

Calcined baryta (barium oxide), from Greek *barys, baryt-*, heavy.

Caustic baryta (barium hydroxide octahydrate), cf. calcined baryta.

Cementite (triiron carbide), from cement, from Latin *caementum*, stone chippings, from *caedere*, to chop.

Charybdotoxin (a polypeptide), from Charybdis, derived from the ancient tale of Odysseus' dangerous passage between the island Scylla and the whirlpool Charybdis, referring to this toxin's action on cellular potassium channels. A.k.a. **CTX**.

Chorismic acid {(3R,4R)-3-[(1-carboxyethen-1-yl)oxy]-4-hydroxycyclohexane-1,5-diene-1-carboxylic acid}, from Greek *chorismos*, separation, from *chorizein*, to separate, referring to chorismic acid's position at a bifurcation of the biosynthetic 'tree' leading to anthranilic acid in one and prephenic acid in the other direction.

Chromotropic acid (4,5-dihydroxynaphthalene-2,7-disulfonic acid), referring to the red coloration which this compound develops in the presence of formaldehyde and certain other compounds.

Cuprene (polyethyne), from cuprum and 'ene', referring to the catalysis of its formation from ethyne by copper and copper oxides.

Dopastine {(2*E*)-*N*-[(2*S*)-2-[hydroxy(nitroso)amino]-3-methylbutan-1-]but-2-enamide}, a contraction of dopamine β-hydroxylase inhibitor.

Erythrin {(2*S*,3*R*)-2,3,4-trihydroxybutyl 4-[(2,4-dihydroxy-6-methylbenzoyl)oxy]-2-hydroxy-6-methylbenzoate}, from Greek *erythros*, red, referring to the red color observed upon this compound's oxidation.

Fasciculin (polypeptides), a class of snake venoms so named because they cause fasciculation of muscle fascicles.

Geissoschizoline (curan-17-ol), from geisso-, schiz(o)-, and indoline, referring to the fact that this compound is formed by hydrolytic cleavage of geissospermine and contains an indoline fragment.

Hygrine {1-[(2*R*)-1-methylpyrrolidin-2-yl]propan-2-one}, from hygr(o)-, referring to the fact that this alkaloid is a liquid at ambient temperature.

Hypoglycin A [3-(2-methylidenecyclopropyl)-L-alanine], from hypoglycinemia, referring to its toxic action.

Indoxyl (1*H*-indol-3-ol), named as synthetic precursor for indigo. Despite the irregular suffix 'yl' it is neither a radical nor a substituent.

Katasulf process, from German *Katalyse*, catalysis, and sulfur.

Luminol (5-amino-2,3-dihydrophthalazine-1,4-dione), from lumi(n)-, referring to the luminescence occurring upon its oxidation.

Lysidine (2-methyl-4,5-dihydro-1*H*-imidazole), from lys(o)- and amidine, referring to the relatively high water solubility of this compound's urate.

Magic acid [antimony pentafluoride—fluorosulfuric acid (1/1), antimony pentafluoride—sulfurofluoridic acid (1/1)], a name referring to this mixture's extreme acidity.

Magic methyl (methyl fluorosulfate, methyl sulfurofluoridate), a name referring to this ester's extreme alkylating power.

Marasmin (a class name), from Greek *marasmos*, a wilting, from *marainein*, to waste away, referring to these compounds'wilting effect on plants.

Muscalure [(9*Z*)-tricos-9-ene], after the genus *Musca* (flies) and lure, referring to this compound's attractant properties.

Oleum (sulfuric acid, sulfur trioxide), from Latin *oleum*, oil. Named as a heavy liquid.

Ormocer, a contraction of organically modified ceramics.

Paktong (Ni,Zn,Cu), from Cantonese *paak t'ung*, paktong, literally white copper.

Parvoline (ethyldimethylpyridine), from Latin *parvus*, small, and 'oline', referring to these compounds' low volatility compared to the isomeric picolines and lutidines.

Pearlite (Fe,Fe_3C), named for this alloy's mother-of-the-pearl appearance under the miscroscope. A.k.a. **sorbite**, after the British microscopist and geologist Henry Clifton Sorby (1826–1908).

Peptide T (an octapeptide), named after its high threonine content.

Piperidine (azinane), from piperine, referring to the fact that it can be obtained from this alkaloid.

Precocene (a class of plant hormones), from precocious metamorphosis, referring to their effect on insect larvae.

Prephenic acid [1-(2-carboxy-2-oxoethyl)-4-hydroxycyclohexa-2,5-diene-1-carboxylic acid]. From phene (benzene), named as biosynthetic precursor of aromatic compounds.

Protostane [(8α,9β,13α,14β)-dammarane], from prot(o)-, sterol, and 'ane', named as biosynthetic precursor of steroids.

Prodlure [(9*Z*,11*E*-tetradeca-9,11-dien-1-yl acetate), probably derived from to prod and lure, named as insect pheromone.

Propylure [(5*E*)-10-propyltrideca-5,9-dien-1-yl acetate], from propyl- and lure, named as insect pheromone.

Sapropterin {(6*R*)-2-amino-6-[(1*R*,2*S*)-1,2-dihydroxypropyl]-5,6,7,8-tetrahydropteridin-4(1*H*)-one}, from sapr(o)- and pterin, named as an enzymatic cofactor. A.k.a. **tetrahydrobiopterin.**

Serotonin (5-hydroxytryptamine), from serum and tonus, referring to its activity as vasoconstrictor. A.k.a. **5-HT**.

Similor (Cu,Zn,Sn), from French *simili*, fake, and *or*, gold, from Latin *aurum*, gold.

Sozolic acid (a mixture of 2-hydroxybenzenesulfonic acid and 4-hydroxybenzenesulfonic acid), from Greek *sozein*, to save, referring to the life-saving antiseptic properties of this preparation.

Tachysterol [(3*S*,6*E*,22*E*)-9,10-secoergosta-5(10),6,8,22-tetraen-3-ol], from tachy-, referring to the fact that this sterol is a liquid at ambient temperature.

Tetracycline [(4*S*,4a*S*,5a*S*,6*S*,12a*S*)-4-(dimethylamino)-3,6,10,12,12a-pentahydroxy-6-methyl-1,11-dioxo-1,4,4a,5,5a,6,11,12a-octahydrotetracene-2-carboxamide], named as tetracyclic antibiotic.

Toxoflavin {1,6-dimethylpyrimido[5,4-*e*][1,2,4]triazine-5,7(1*H*,6H)-trione}, from toxin and Latin *flavus*, light yellow.

Toxopyrimidine {[4-amino-2-methyl-5-(pyrimidin-5-yl)]methanol}, from toxin and pyrimidine.

Tremetone {1-[(2*R*)-2-(prop-1-en-2-yl)-2,3-dihydro-1-benzofuran-5-yl]ethan-1-one, after trembles, a cattle disease caused by tremetone poisoning.

Tremorine [(1,1′-(but-2-yne-1,4-diyl)dipyrrolidine), from tremor, referring to this compound's tremorigenic action.

Trypan blue {3,3′-[(3,3′-dimethyl[1,1′-biphenyl]-4,4′-diyl]di[(*E*)-diazen-2,1-diyl]bis[5-amino-4-hydroxynaphthalene-2,7-disulfonic acid]}, named to indicate this dye's trypanocidal properties.

Tuberin {*N*-[(*E*)-2-(4-methoxyphenyl)ethen-1-yl]formamide}, with contraction, from antitubercular.

Turgorin {3,5-dihydroxy-4-[(6-*O*-sulfo-β-D-glucopyranosyl)oxy]benzoic acid}, also used as a class name, from turgor, referring to its actions in thigmonasty.

Volicitin {N^2-[(9*Z*,12*Z*,15*Z*)-17-hydroxyoctadeca-9,12,15-trienoyl]-D-glutamine}, a contraction of elicitor of volatile biosynthesis.

Yellow baryta (barium chromate), from Greek *barys, baryt-*, heavy.

2.3 Eponyms and demonyms

Eponyms can be subject to Stigler's law of eponymy [1, 2] which states that no scientific discovery is named after its original discoverer.

Adams catalyst (platinum dioxide), after the American chemist Roger Adams (1889–1971).

Adamsite (10-chloro-5,10-dihydrophenarsazine), after the American chemist Roger Adams (1889–1971). A.k.a. **DM.**

Adkins catalyst [(Cr,Cu)O], after the American chemist Homer Burton Adkins (1892–1949).

Ajmaline [(17*R*)-ajmalan-17,21α-diol], after the Muslim Indian physician and Muslim leader Hakim Ajmal Khan (1863–1927).

Aldrin {(1*R*,2*R*,3*R*,6*S*,7*S*,8*S*)-1,8,9,10,11,11-hexachlorocyclo[6.2.1.$1^{3,6}$.$0^{3,7}$]dodeca-4,9-diene}, after the German chemist Kurt Alder (1902–1958), referring to this compound's synthesis by a Diels-Alder reaction.

Andresen's acid (8-hydroxynaphthalene-1,6-disulfonic acid), after the German chemist Momme Andresen (1857–1951).

Angeli's salt (sodium trioxodinitrate), after the Italian chemist Angelo Angeli (1864–1931).

Armstrong's acid (naphthalene-1,5-disulfonic acid), after the British chemist Henry Edward Armstrong (1848–1937).

Arnd's alloy (Cu,Mg), after the German chemist Curt Ulrich Theodor Arnd (1888–1975).

Appel's salt (4,5-dichloro-1,2,3-dithiazol-2-ium chloride), after the German chemist Rolf Appel (1921–2012).

Aten's sulfur (*cyclo*-hexasulfur, hexathiane), after the Dutch chemist Adriaan Hendrik Willem Aten (1908–1979). A.k.a. **ρ-sulfur, ε-sulfur**, and **Engel's sulfur**, after the French chemist Rodolphe Charles Engel (1850–1916).

Austenite [γ-iron, iron(*cF*4)], after the British metallurgist William Chandler Roberts-Austen (1843–1902).

Babbitt metal (Sn.Cu.Pb.Sb), after the American metallurgist Isaac Babbitt (1799–1862).

Baden acid (7-aminonaphthalene-1-sulfonic acid), after the chemical company Badische Anilin- & Soda-Fabrik AG, Ludwigshafen, Germany. A.k.a. **Badische acid.**

Bainite (Fe,Fe_3C), after the American metallurgist Edgar Collins Bain (1891–1988).

Barton esters, a class name, after the British-American chemist Derek Harold Richard Barton (1918–1998).

Bayer's acid (7-hydroxynaphthalene-1-sulfonic acid), after the chemical company Farbenfabriken Bayer AG, Leverkusen, Germany. A.k.a. **croceic acid**, from Latin *croceus*, of saffron, dark red yellow, ultimately from Greek *krokos*, saffron, crocus, of Semitic origin, named as intermediate for synthetic dyes.

Bechgaard salt, a class name, after the Danish chemist Klaus Bechgaard (1945–2017).

Bence Jones protein, a class name, after the British physician Henry Bence Jones (1813–1873).

Bender's salt (potassium *O*-ethylthiocarbonate), after the German chemist Friedrich August Bender (1847–1926).

Benedict's solution (containing sodium carbonate, sodium citrate, and copper(II) sulfate pentahydrate), after the American chemist Stanley Rossiter Benedict (1884–1936).

Berthelot-Thiele reagent (an alkaline solution of phenol and hypochlorite), after the French chemist Marcellin Pierre Eugène Berthelot (1827–1907) and the German chemist Friedrich Karl Johannes Thiele (1865–1918).

Bial's reagent (containing orcinol, hydrochloric acid, and iron(III) chloride), after the German physician Manfred Bial (1869–1908).

Bindschedler's green {the leukobase N^4-[4-(dimethylamino)phenyl]-N^1,N^1-dimethylbenzene-1,4-diamine}, after the chemical company Bindschedler & Busch, Basel, Switzerland.

Bismarck brown Y {(4,4′-[1,3-phenylenedi(*E*)-2,1-diazenediyl]di(benzene1,3-diamine)}, after the German chancellor Otto von Bismarck (1815–1898).

Bredereck's reagent {*N*,*N*,*N*′,*N*′-tetramethyl-1-[(2-methylpropan-2-yl)oxy]methanediamine}, after the German chemist Helmut Bredereck (1904–1981).

Brönner's acid (6-aminonaphthalene-2-sulfonic acid), after the chemical company Farbfabrik vorm. Brönner, Frankfurt am Main, Germany.

Brønsted acid (proton donor) and **Brønsted base** (proton acceptor), after the Danish chemist Johannes Nicolaus Brønsted (1879–1947).

Brown's reagent {bis[(1*S*,2*S*,3*S*,5*R*)-2,6,6-trimethylbicyclo[3.1.1]heptan-3-yl]borane}, after the British-American chemist Herbert Charles Brown (1912–2004).

Bullvalene {tricyclo[3.3.2.0^{2,8}]-deca-3,6,9-triene}, after William "Bull" von Eggers Doering (1917–2011) who had predicted this fluxional compound's rapid degenerate Cope rearrangements.

Bunte salt (*S*-alkylthiosulfuric acid salt), a class name, after the German chemist Hans Bunte (1848–1925).

Burgess reagent {1-methoxy-*N*-[(triethylazaniumyl)sulfonyl]methanimidate}, after the American chemist Edward Meredith Burgess (1934–2018).

Caro's acid [peroxysulfuric acid, sulfuroperoxoic acid, (dioxidanido)hydroxido-dioxidosulfur], after the German chemist Heinrich Caro (1834–1910).

Carother's silk, after the American chemist Wallace Hume Carothers (1896–1937). A.k.a. **nylon**.

Cassius purple (Au,SnO_2), the color being caused by colloidal gold, after the German physician Andreas Cassius (1605–1673).

Chevreul's salt ($Cu_2SO_3{\cdot}CuSO_3{\cdot}2H_2O$), after the French chemist Michel Eugène Chevreul (1786–1889).

Chichibabin's hydrocarbon {[1,1′-biphenyl]-4,4′-diylbis(diphenylmethyl), a diradical}, after the Russian-French chemist Aleksei Yevgenyevich Chichibabin (1871–1945).

Chugaev's reagent [(2*E*,3*E*)-*N*,*N*′-dihydroxybutane-2,3-diimine], after the Russian chemist Lev Aleksandrovich Chugaev (1873–1922). A.k.a. **dimethylglyoxime**.

Clark cell ($Zn/Hg/ZnSO_4$), after the British engineer Josiah Latimer Clark (1822–1898).

Claus benzene (a proposed structure with diagonal bonds between opposite carbon atoms), after the German chemist Adolf Karl Ludwig Claus (1838–1900).

Clerici solution [a solution containing thallium(I) formate and thallium(I) malonate], after the Italian mineralogist and geologist Enrico Clerici (1862–1938).

1,6-Cleve's acid (1-aminonaphthalene-6-sulfonic acid), after the Swedish chemist Per Theodor Cleve (1840–1905).

1,7-Cleve's acid (1-aminonaphthalene-7-sulfonic acid), as 1,6-Cleve's acid.

Collins' reagent [chromium(VI)oxide—pyridine (1/2)], after the American chemist Joseph Charles Collins, Jr. (born 1931).

Collman's reagent (disodium tetracarbonylferrate), after the American chemist James Paddock Collman (born 1932).

Corey aldehyde {(3a*R*,4*R*,5*R*,6aS)-5-hydroxyhexahydro-2-oxo-2*H*-cyclopenta[*b*]furan-4-carbaldehyde}, after the American chemist Elias James Corey (born 1928).

[1]**Corey's reagent** (pyridinium chlorochromate), as Corey aldehyde.

[2]**Corey's reagent** [dimethylsulfoxonium methylide, dimethyl(methylidene)(oxo)-λ^6-sulfane], as Corey aldehyde.

Cori ester (1-*O*-phosphono-α-D-glucopyranose), after the Czech-American biochemists Carl Ferdinand Cori (1896–1984) and Gerty Theresa Cori (1896–1957).

Cornforth reagent (pyridinium dichromate), after the Australian-British chemist John Warcup Cornforth (1917–2013).

Crabtree's catalyst {(*SP*-4)-tris(cyclohexyl)phosphane[(1-2-η:5,6-η)-cycloocta-1,5-diene]pyridineiridium(1+) hexafluoridophosphate (1−)}, after the British-American chemist Robert Howard Crabtree (born 1948).

Dahl's acid (2-aminonaphthalene-5-sulfonic acid), named after the chemical company Dahl & Co., Barmen, Germany. A.k.a. **D acid** and **Dressel acid**, after the German chemist Oskar Dressel (1865–1941).

Dakin's solution (sodium hypochlorite), after the British chemist Henry Drysdale Dakin (1880–1952). A.k.a. **Carrel-Dakin fluid** and **Carrel-Dakin solution**, after the French physician Alexis Carrel (1873–1944) and Dakin.

Daniell's element (Cu,Zn,H_2SO_4), after the British chemist and physicist John Frederic Daniell (1790–1845).

Danishefsky's diene {(1*E*)-1-methoxy-3-[(trimethylsilyl)oxy]buta-1,3-diene], after the American chemist Samuel Joseph Danishefsky (born 1936).

D'Arcet's metal (Bi,Pb,Sn), after the French chemist Jean D'Arcet (1725–1801).

Dess-Martin periodinane [1,1,1-triacetoxy-1λ^5-2-benziodoxol-3(1*H*)-one] after the American chemists Daniel Benjamin Dess (born 1955) and James Cullen Martin (1928–1999).

Devarda's alloy (Cu,Al,Zn), after the Italian chemist Arturo De Varda (Arturo Devarda) (1859–1944).

Dewar benzene (bicyclo[2.2.0]hexa-2,5-diene), after the British chemist James Dewar (1842–1923).

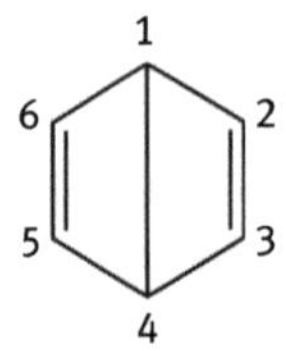

Dewar benzene

Dieldrin {(1a*R*,2*R*,2a*S*,3*S*,6*R*,7*S*,7a*S*)-3,4,5,6,9,9-hexachloro-1a,2,2a,3,6,6a,7,7a-octahydro-2,7:3,6-dimethanonaphtho[2,3-*b*]oxirene}, after the Diels-Alder reaction, q.v.

Diels' hydrocarbon (17-methyl-16,17-dihydro-15*H*-cyclopenta[*a*]phenanthrene), after the German chemist Otto Paul Hermann Diels (1876–1954).

Diesel fuel (a combustible petroleum distillate), after the German scientist and inventor Rudolf Christian Karl Diesel (1858–1913).

Doisynolic acid (1-ethyl-7-hydroxy-2-methyl-3,4,4a,9,10,10a-hexahydro-1*H*-phenanthrene-2-carboxylic acid), after the American biochemist Edward Adelbert Doisy (1893–1986), who synthesized it, and with 'ol'.

Dragendorff's reagent [potassium tetraiodidobismuthate(III)], after the Estonian-German physician and pharmacist Johann Georg Noël Dragendorff (1836–1898).

Duthaler-Hafner reagent {chlorido[[(4*S*,5*S*)-2,2-dimethyl-1,3-dioxolane-4,5-diyl]bis(diphenylmethanolato-κ*O*)(2–)]titanium(1+) cyclopenta-2,4-dienide}, after the Swiss chemists Rudolf O. Duthaler (born 1946) and Andreas Hafner (born 1956).

Ebert-Merz α-acid (naphthalene-2,7-disulfonic acid), after the Swiss chemists R. Ebert and Victor Merz (1839–1904).

Ebert-Merz β-acid (naphthalene-2,6-disulfonic acid), as Ebert-Merz α-acid.

Ehrlich's reagent [4-(dimethylamino)benzaldehyde), after the German physician and bacteriologist Paul Ehrlich (1854–1915).

Ellis carbonylate [sodium hexacarbonyltantalate(1–)—tetrahydrofuran (1/1)], after the American chemist John Emmett Ellis (born 1943).

Ellman's reagent [5,5′-disulfanediylbis(2-nitrobenzoic acid)], after the American chemist George Leon Ellman (1923–2009).

Esbach reagent (picric acid, citric acid), after the French physician Georges Hubert Esbach (1843–1890).

Evans blue {tetrasodium 6,6′-[3,3′-dimethyl[1,1′-biphenyl]-4,4′-diyl]di[(*E*)-diazene-2,1-diyl]bis(4-amino-5-oxidonaphthalene-1,3-disulfonate}, after the American anatomist and embryologist Herbert McLean Evans (1882–1971).

Fehling's reagent [copper(II) sulfate (Fehling solution A), potassium sodium tartrate (Fehling solution B)], after the German chemist Hermann Christian von Fehling (1812–1885).

Feist's acid [(1*S*,2*S*)-3-methylidenecyclopropane-1,2-dicarboxylic acid), after the German chemist Franz Feist (1864–1941).

Fenton's reagent (a solution containing Fe^{2+} and hydrogen peroxide), after the British chemist Henry John Horstman Fenton (1854–1929).

Fetizon's reagent [silver(I) carbonate], after the French chemist Marcel Ernest Paul Fetizon (1926–2015).

Fieser's reagent (chromium trioxide in acetic acid), after the American chemist Louis Fieser (1899–1977).

Fieser's solution (containing potassium hydroxide, sodium dithionite, and sodium 9,10-dioxoanthracene-1-sulfonate), as Fieser's reagent.

Fischer carbene, a class name for carbenes with an electrophilic carbon center, after the German chemist Ernst Otto Fischer (1918–2007).

Fischer's salt [potassium hexanitritocobaltate(III)], after the German chemist Nikolaus Wolfgang Fischer (1782–1850). A.k.a. **aureolin** and **cobalt yellow**.

Folin-Ciocâlteu reagent (phosphomolybdate, phosphotungstate), after the Swedish-American chemist Otto Knut Olof Folin (1867–1934) and the Romanian chemist Vintilă Ciocâlteu (1890–1947). A.k.a. **Folin-Denis reagent** after Folin and the American biochemist and physiologist Willey Glover Denis (1879–1929).

Fowler's solution [containing potassium arsenite, potassium arsenate(III)], after the British physician Thomas Fowler (1736–1801).

Fredericamycin {4′,9,9′-trihydroxy-6′-methoxy-3-[(1*E*,3*E*)-penta-1,3-dien-1-yl]spiro [6,7-dihydro-2*H*-cyclopenta[*g*]isoquinoline-8,2′-cyclopenta[*g*]naphthalene-1,1′,3′,5′,8′-pentone}, after the Frederick Cancer Research Center, Frederick, MD, USA, where this antibiotic was isolated and characterized.

Frémy's salt [potassium nitrosodisulfonate, dipotassium (disulfonatoamino)oxidanyl, oxidobis(trioxidosulfato-κ*S*-nitrate)(•2–), a radical], after the French chemist Edmond Frémy (1814–1894).

Freund's acid (4-aminonaphthalene-2,7-disulfonic acid), after the German chemist Martin Freund (1863–1920).

Freund's adjuvant (antigen emulsified in mineral oil), after the Hungarian-American pathologist Jules Freund (1890–1960).

Friedel's salt [$Ca_2Al(OH)_6(Cl,OH)\cdot 2H_2O$], after the French mineralogist and crystallographer Georges Friedel (1865–1933).

Fröhde's reagent (containing ammonium molybdate and sulfuric acid), after the German pharmacist Johann Karl August Fröhde (born 1831).

Garner's aldehyde [2-methylpropan-2-yl (4*S*)-4-formyl-2,2-dimethyl-1,3-oxazolidine-3-carboxylate], after the American chemist Philip Paul Garner (born 1955).

Gibbs reagent [2,6-dichloro-4-(chloroimino)cyclohexa-2,5-dien-1-one], after the American chemist Harry Drake Gibbs (1872–1934).

Giese salt [ammonium iron(III) hexacyanidoferrate(II)], after the German physicist and veterinarian Werner Giese (born 1936).

Girard reagent P [2-hydrazinyl-2-oxoethan-1-pyridinium chloride (1/1)], after the French chemist André Girard (1901–1968). The P stands for the parent pyridine.

Girard reagent T (2-hydrazino-*N,N,N*-trimethyl-2-oxoethan-1-aminium chloride), as Girard reagent P. The T. stands for the parent trimethylamine.

Glauber's salt (sodium sulfate decahydrate), after the Dutch-German chemist and physician Johann Rudolph Glauber (1644–1688). A.k.a. **sal mirabilis, sal mirabile,** New Latin, miraculous salt, and **mirabilite.**

Glover acid (78–80% sulfuric acid), after the British chemist John Glover (1817–1902).

Goldschmidt's radical [DPPH(•), 2,2-diphenyl-1-picrylhydrazinyl, 2,2-diphenyl-1-(2,4,6-trinitrophenyl)hydrazinyl], after the German chemist Stefan Goldschmidt (1889–1971).

Gold's reagent {[[[(dimethylamino)methylidene]amino]methylidene]dimethylammonium chloride}, after the German chemist Heinrich Gold (born 1912).

Gomberg's radical (triphenylmethyl), after the Ukrainian-American chemist Moses Gomberg (1866–1947).

Good's buffers (twenty zwitterionic buffers), after the Canadian-American botanist and plant pathologist Norman E. Good (1917–1992).

Graham's salt (sodium metaphosphate), after the British chemist Thomas Graham (1805–1867). A.k.a. **Kurrol's salt**, after the Estonian chemist Julius Ludwig Kurrol (born 1865), and **Maddrell's salt**, after the British chemist Robert Maddrell (1819–1900).

Grubbs' first generation catalyst [dichlorido(phenylmethylidene)bis(tricyclohexylphosphane)ruthenium], after the American chemist Robert Howard Grubbs (born 1942).

Grubbs' second generation catalyst {dichlorido[1,3-bis(2,4,6-trimethylphenylimidazolidin-2-ylidene](phenylmethylidene)(tricyclohexylphosphane)ruthenium}, as Grubbs' first generation catalyst.

Guignet's green [chromium(III) oxide dihydrate], after the French industrialist Charles-Ernest Guignet (1829–1906). A.k.a. **Mittler's green** and **viridian**, from Latin *viridis*, green.

Günzburg reagent (a mixture of benzene-1,3,5-triol and 4-hydroxy-3-methoxybenzaldehyde), after the German physician Alfred Günzburg (1861–1945).

Guthrie's alloy (Bi,Sn,Pb,Cd), after the American physician and chemist Samuel Guthrie, Jr. (1782–1848).

Gutzeit test, after the German chemist Heinrich Wilhelm Theodor Gutzeit (1845–1888).

Hamycin (a polyene antibiotic), from the initials of the drug company Hindustan Antibiotics Limited, Pune, India, which developed this drug, and -mycin. A.k.a. **primamycin**.

Hanuš reagent [iodine(I) chloride, chloroiodane], after the Czech chemist Josef Hanuš (1872–1955). A.k.a. **Wijs reagent**.

Harden-Young ester (1,6-di-*O*-phosphono-β-D-fructofuranose), after the British biochemist Arthur Harden (1865–1940) and the Australian biochemist William John Young (1878–1942).

Harper's alloy (Bi,Pb,Sn), after the American chemist H. W. Harper (1859–1943).

Hayem's solution [isotonic mercury(II) chloride, sodium sulfate, and sodium chloride], after the French physician Georges Hayem (1841–1933).

Hemi-Dewar biphenyl {2-phenylbicyclo[2.2.0]hexa-2,5-diene}, from hemi- and Dewar benzene, after the British chemist James Dewar (1842–1923).

Heusler compound, a class name for certain alloys, after the German engineer and chemist Carl Ludwig David Friedrich Heusler (1866–1947).

Heyns catalyst (Pt,C), after the German chemist Kurt Fritz Heyns (1908–2005).

Hittorf phosphorus (polymeric phosphorus), after the German chemist and physicist Johann Wilhelm Hittorf (1824–1914). A.k.a. **monoclinic phosphorus** and **violet phosphorus**.

Homberg's alloy (Pb,Bi,Sn), after the German physician and chemist Wilhelm Homberg (1652–1715).

Hopcalite (MnO_2,CuO,Co_2O_3,Ag_2O), coined as a trademark by contraction of Johns Hopkins University and University of California.

Hoveyda-Grubbs first generation catalyst {dichlorido[[[2-(propan-2-yl)oxy]phenyl]methylidene](trichcylohexylphosphane)ruthenium}, after the American chemists Amir H. Hoveyda (born 1959) and Robert Howard Grubbs (born 1942).

Hoveyda-Grubbs second generation catalyst {dichlorido[1,3-bis(2,4,6-trimethylphenyl)imidazolidin-2-ylidene][[2-[(propan-2-yl)oxy]phenyl]methylidene]ruthenium}, as Hoveyda-Grubbs first generation catalyst. Simultaneously and independently discovered by the German chemist Siegfried Blechert (born 1946).

H oxime (a group of oximes useful in the treatment of anticholinesterase poisoning), after the German chemist Ilse Hagedorn (1921–2005).

Hünig's base [*N*-ethyl-*N*-(propan-2-yl)propan-2-amine], after the German chemist Siegfried Helmut Hünig (born 1921).

Jacobsen's catalyst {chlorido[(1*S*,2*S*)-cyclohexane-2,2′-diyl]bis[nitrilo-(*E*)-methylidene] bis[4,6-bis(2-methylpropan-2-yl)phenolato-κ*O*(2−)]manganese}, after the American chemist Eric Niels Jacobsen (born 1960).

Jones reagent (containing chromium trioxide and sulfuric acid), after the British chemist Ewart Ray Herbert Jones (1911–2002).

Jorissen's reagent [containing vanadic acid (trihydroxidooxidovanadium) and sulfuric acid], after the Belgian chemist Armand Jules Joseph Jorissen (1853–1920).

Kaempferol [3,5,7-trihydroxy-2-(4-hydroxyphenyl)-4*H*-1-benzopyran-4-one], after the German physician, naturalist, and explorer Engelbert Kämpfer (1651–1716).

Kahane's reagent [magnesium uranyl acetate, dioxidouranium(2+) magnesium acetate (1/1/4)], after the French biochemist Ernest Kahane (1903–1996). A.k.a. **Blanchetière's reagent**, after the French chemist Alexandre Marie Blanchetière (1875–1934).

Kalle's acid (1-aminonaphthalene-2,7-disulfonic acid), after the German chemist Wilhelm Ferdinand Kalle (1870–1954).

Karl Fischer reagent (a solution of iodine, sulfur dioxide, and imidazole in ethanol), after the German chemist Karl Fischer (1901–1958).

Karstedt's catalyst [platinum(0) alkene complexes], after the American chemist Bruce David Karstedt (1939–2011).

Kekulene {15,23:16,22-dimethenobenzeno[6,7]phenanthro[3,2-*a*]naphtho[2,3-*o*]pentaphene}, after the German chemist Friedrich August Kekulé von Stradonitz (1829–1896). A.k.a. **superbenzene**.

Keller's reagent (a mixture of nitric acid, hydrochloric acid, and hydrofluoric acid), after the American chemist Fred Keller (born ca. 1898). A.k.a. **Dix-Keller**

reagent, after the American chemists Edgar Hutton Dix, Jr. (1892–1963) and Fred Keller.

Kemp's triacid (1,3,5-trimethylcyclohexane-1,3,5-tricarboxylic acid), after the American chemist Daniel Schaeffer Kemp (born 1936).

Kiliani's reagent (a solution of chromic acid in sulfuric acid), after the German chemist Heinrich Kiliani (1855–1945).

Klein's solution (containing cadmium borotungstate), after the German mineralogist J. F. C. Klein (1842–1907).

Koch's acid (8-aminonaphthalene-1,3,6-trisulfonic acid), after the American chemist Julius Arnold Koch (1864–1932).

Koelsch radical {9-[(9*H*-fluoren-9-ylidene)phenylmethyl]fluoren-9-yl}, after the American chemist Charles Frederick Koelsch (1907–1999).

Kohler's ketone [3,3-diphenyl-1-(2,4,6-trimethylphenyl)propan-1-one], after the American chemist Elmer Peter Kohler (1865–1938).

König's salt [5-aminopenta-2,4-dien-1-iminium chloride (1/1)], after the German chemist Walter Georg König (1878–1964).

Koser's reagent {hydroxy[[(4-methylphenyl)sulfonyl]oxy]phenyl-λ^3-iodane}, after the American chemist Gerald Franklin Koser (born 1942). A.k.a. **HTIB**.

Kurrol's salt (sodium metaphosphate; potassium metaphosphate), after the Estonian chemist Julius Ludwig Kurrol (born 1865). A.k.a. **Graham's salt**, after the British chemist Thomas Graham (1805–1867) and **Maddrell's salt,** after the British chemist Robert Maddrell (1819–1900).

Ladenburg benzene (tetracyclo[2.2.0.0^{2,6}.0^{3,5}]hexane), after the German chemist Albert Ladenburg (1842–1911). A.k.a. **prismane, triprismane**, and **[3]prismane.**

Lalancette's reagent [sodium tetrahydridoborate(I), sulfur], after the French chemist J.-M. Lalancette (born 1934).

Lappaconitine (20-ethyl-8,9-dihydroxy-1,14α,16β-trimethoxyaconitan-3-yl 2-acetamidobenzoate), after the Lapp people, a.k.a. Sami, and the genus *Aconitum* (monkshoods), cf. aconitane.

Laurell buffer [containing sodium barbital (sodium diethylbarbiturate) and calcium lactate], after the Swedish biochemist Carl-Bertil Hugo Laurell (1919–2001).

Laurent's acid (5-aminonaphthalene-1-sulfonic acid), after the Canadian chemist Auguste Laurent (1807–1853). A.k.a. **L acid** and **purpurin acid.**

Lauth's violet (3,7-diaminophenothiazin-5-ium, a cation), after the French chemist Charles Lauth (1836–1913). A.k.a. **thionine.**

Lawesson's reagent [2,4-bis(4-methoxyphenyl)-1,3,2,4-dithiadiphosphetane-2,4-dithione], not invented, but developed and promoted by the Swedish-Danish chemist Sven-Olov Lawesson (1926–1985).

Lazier catalyst (copper chromite, dichromium dicopper pentaoxide), after the American chemist Wilbur Arthur Lazier (1900–1976). A.k.a. **Adkins catalyst**, after the American chemist Homer Burton Adkins (1892–1949).

Lemieux-Johnson reagent (a mixture of sodium periodate and osmium tetraoxide), after the Canadian chemist Raymond Urgel Lemieux (1920–2000) and the American chemist William Summer Johnson (1914–1995).

Lemieux-von Rudloff reagent (a mixture of sodium periodate and potassium permanganate), after Lemieux and the German-Canadian chemist Ernst Max von Rudloff (born 1923).

Leuchs anhydride (a class name for certain 1,3-oxazole-2,5-diones), after the German chemist Hermann Leuchs (1879–1945).

Lewis acid (lone-electron pair acceptor) and **Lewis base** (lone-electron pair donor), after the American chemist Gilbert Newton Lewis (1875–1946).

Lewisite [(2-chloroethen-1-yl)arsonous dichloride], after the American chemist and soldier Winford Lee Lewis (1878–1943).

Lichtenberg alloy (Bi,Sn,Pb), after the German physicist Georg Christoph Lichtenberg (1742–1799).

Liebermann reagent (containing potassium nitrite and sulfuric acid), after the Hungarian chemist Leo Liebermann (1852–1926).

Lifschitz salts [nickel(II) salts with diamino ligands], after the Dutch chemist Israel Lifschitz (1888–1953).

Lindane [γ-hexachlorocyclohexane, (1*r*,2*R*,3*S*,4*r*,5*R*,6*S*)-1,2,3,4,5,6-hexachlorocyclohexane], after the Dutch chemist Teunis van der Linden (1884–1965). A.k.a. **BCH** and **HCH.**

Lindlar catalyst (Pd, partially poisoned with Pb or S), after the British-Swiss chemist Herbert Wilson Lindlar (1909–2009).

Luff-Schoorl solution [containing sodium carbonate, citric acid, and copper(II) sulfate], after the German chemist Gustav Luff (deceased 1925) and the Dutch chemist Nicholaas Schoorl (1872–1942).

Lunge's reagent [*N*-(1-naphthyl)ethane-1,2-diamine], after the German-Swiss chemist Georg Lunge (1839–1923). A.k.a. **Griess' reagent**, after the German-British industrial chemist Peter Griess (1829–1888).

Maddrell's salt (sodium metaphosphate), after the British chemist Robert Maddrell (1819–1900). A.k.a. **Graham's salt,** after the British chemist Thomas Graham (1805–1867) and **Kurrol's salt**, after the Estonian chemist Julius Ludwig Kurrol (born 1865).

Magnus' green salt [tetraammineplatinum(II) tetrachloridoplatinate(II)], a polymorph of Magnus's pink salt, after the German chemist Heinrich Gustav Magnus (1802–1870).

Magnus's pink salt [tetraammineplatinum(II) tetrachloridoplatinate(II)], a polymorph of Magnus's green salt, q.v.

Marquis reagent (containing formaldehyde and sulfuric acid), after the Estonian pharmacologist Eduard Marquis (1871–1944).

Marshall's acid (peroxydisulfuric acid), after the British chemist Hugh Marshall (1868–1913).

Martensite (Fe,C), after the German metallurgist Adolf Martens (1850–1914).

Martius yellow (2,4-dinitronaphthalen-1-ol), after the German chemist Carl Alexander von Martius (1838–1920).

Meerwein salt (triethyloxonium tetrafluoridoborate), after the German chemist Hans Leberecht Meerwein (1879–1965).

Meisenheimer complex (a class name), after the German chemist Jakob Meisenheimer (1876–1934). A.k.a. **Jackson-Meisenheimer complex**, after the American chemist Charles Loring Jackson (1845–1935) and Meisenheimer.

Meldola blue {*N*,*N*-dimethyl-9*H*-benzo[*a*]phenoxazin-9-iminium chloride}, after the British chemist Ralphael Meldola (1849–1915). A.k.a. **Basic Blue 6.**

Meldrum's acid (2,2-dimethyl-1,3-dioxane-4,6-dione), after the American chemist Norman Andrew Meldrum (1876–1934).

Michler's base [4,4′-methylenebis(*N*,*N*-dimethylaniline)], after the German chemist Wilhelm Traugott Michler (1846–1889).

Michler's ketone {bis[4-(dimethylamino)phenyl]methanone)}, as Michler's base.

Midland's reagent [9-(pinan-3-yl)-9-borabicyclo[3.3.1]nonane], after the American chemist Michael Mark Midland (born 1946).

Millon's base (mercury hydroxide nitride), after the French physician and chemist Auguste Nicolas Eugène Millon (1812–1867).

Mitis green {tetracopper(2+) diacetate hexakis[(dioxidoarsenate(1−)]}, after the Austrian industrialist Ignaz Edler von Mitis (1771–1842). A.k.a. **Paris green** and **Schweinfurt green.**

Mohr's salt [diammonium iron(II) sulfate hexahydrate], after the German chemist Carl Friedrich Mohr (1806–1879).

Mond metal (Ni,Cu,Mn), after the German-British inventor and industrialist Ludwig Mond (1839–1909).

Mond nickel (Ni), as Mond metal.

Monel (Ni,Cu), after the American industrialist and philanthropist Ambrose Monell (1873–1921).

Monellin (a polypeptide), named for the Monell Chemical Senses Center, Philadelphia, PA, USA where this natural compound was isolated and characterized. This research center was in turn named after the American industrialist and philanthropist Ambrose Monell (1873–1921).

Moore's ketene [3,3-dimethyl-2-(oxomethylidene)butanenitrile], after the American chemist Harold Wesley Moore (born 1936).

Mosher's acid (3,3,3-trifluoro-2-methoxy-2-phenylpropanoic acid), after the American chemist Harry Stone Mosher (1915–2001).

^{1}**Mukaiyama's reagent** (2-chloro-1-methylpyridinium iodide), after the Japanese chemist Teruaki Mukaiyama (1927–2018).

^{2}**Mukaiyama's reagent** (2-fluoro-1-methylpyridinium 4-methylbenzenesulfonate), cf. ^{1}Mukaiyama's reagent.

Muntz metal (Cu,Zn,Fe), after the British industrialist George Frederick Muntz (1794–1857).

Muthmann's liquid (1,1,2,2-tetrabromoethane), after the German chemist and crystallographer Friedrich Wilhelm Muthmann (1861–1913).

Muthmann's sulfur I (α-*cyclo*-octasulfur, octathiocane), as Muthmann's liquid. A.k.a. **orthorhombic sulfur.**

Muthmann's sulfur II (β-*cyclo*-octasulfur, octathiocane), as Muthmann's liquid. A.k.a. **monoclinic sulfur.**

Muthmann's sulfur III (γ-*cyclo*-octasulfur, octathiocane), as Muthmann's liquid. A.k.a. **nacreous sulfur, Gernez's sulfur**, after the French chemist Dèsiré Gernez

(1834–1903), and **rosickyite**, a mineral named after the Czech mineralogist Vojtěch Rosický (1880–1942).

Nazarov's reagent {ethyl 3-oxopent-4-enoate}, after the Russian chemist Ivan Nikolayevich Nazarov (1906–1957).

Negishi reagent [dibutylbis(cyclopentadienyl)zirconium], after the Japanese-Americam chemist Ei-ichi Negishi (born 1935).

Nessler's reagent [potassium tetraiodidomercurate(2–)], after the German chemist Julius Neßler (1827–1905).

Neuberg ester (6-*O*-phosphono-β-D-fructofuranose), after the German biochemist Carl Neuberg (1877–1956). A.k.a. **Embden** ester, after the German chemist Gustav Embden (1874–1933)

Noyori reagent [(*S*)-BINAP/(*S*)-diamine-Ru], after the Japanese chemist Ryoji Noyori (born 1938).

Newton's metal (Bi,Pb,Sn) after the British physicist Isaac Newton (1642–1727).

Olah's reagent [pyridine hydrofluoride (1/1)], after the Hungarian-American chemist George Andrew Olah (Olâh György) (1927–2017).

Park nucleotide (an oligopeptide derivative), after the American biochemist James Theodore Park (born 1922).

Pearlman's catalyst [palladium(II) hydroxide on carbon], after the American chemist William M. Pearlman (1919–2007).

Pelletierine {1-[(2*R*)-piperidin-2-yl]propan-2-one}, after the French chemist Pierre Joseph Pelletier (1788–1842).

Peyrone's salt [(*SP*-4-2)-diamminedichloridoplatinum(II)], after the Italian chemist Michele Peyrone (1813–1883). A.k.a. by the INN **cisplatin.**

Pfitzner-Moffatt reagent (dimethyl sulfoxide, dicyclohexylcarbodiimide), after the American chemist K. E. Pfitzner and the Canadian-American chemist John Gilbert Moffatt (born 1930).

Piloty's acid (*N*-hydroxybenzenesulfonamide), after the German chemist Oskar Piloty (1866–1915).

Pinchbeck (Cu,Zn), after the British watchmaker Christopher Pinchbeck (1670–1732).

Pinner's salt (1-iminio-1-alkoxyalkane chloride, a class name), after the German chemist Adolf Pinner (1842–1909).

Piria's acid (4-aminonaphthalene-1-sulfonic acid), after the Italian chemist Raffaele Piria (1815–1865), A.k.a. **napththionic acid.**

Pirkle's alcohol [1-(anthracen-9-yl)-2,2,2-trifluoroethanol], after the American chemist William Howard Pirkle (1934–2018).

Queen's metal (Sn,Sb,Pb.Bi), an alloy with an originally secret composition and only used in the British Royal Household.

Raney's alloy (Ni,Al), after the American mechanical engineer Murray Raney (1885–1966). Cf. Urushibara nickel.

Raney nickel (Ni,H), as Raney's alloy.

Ringer solution (containing sodium chloride, potassium chloride, calcium chloride, and sodium hydrogencarbonate), after the British physician Sydney Ringer (1835–1910).

Rose's metal (Bi,Pb,Sn), after the German pharmacist and chemist Valentin Rose the Elder (1736–1771).

Roussin's black salt (potassium heptanitrosyl-μ^3-thiotetraferrate), after the French chemist and pharmacist François-Zacharie Roussin (1827–1894).

Sarin (propan-2-yl methylphosphonofluoridate), probably coined as an acronym of the names of the German chemists Gerhard Schrader (1903–1990), Otto Ambros (1901–1990), Gerhard Ritter (1902–after 1977), and Hans-Jürgen von der Linde (born 1900). A.k.a. **GB** and **trilon 46.**

^{1}**Schlenk-Brauns hydrocarbon** ($C_{32}H_{24}$, a diradical), after the German chemists Wilhelm Schlenk (1879–1943) and M. Brauns.

^{2}**Schlenk-Brauns hydrocarbon** ($C_{38}H_{28}$, a diradical), cf. ^{1}Schlenk-Brauns hydrocarbon.

Schlippe's salt [sodium tetrasulfidoantimonate(V)], after the German-Russian pharmacist and chemist Karl Friedrich von Schlippe (1799–1869).

Schradan {*N*-bis(dimethylamino)phosphoryloxy-(dimethylamino)phosphoryl]-*N*-methylmethanamine}, after the German chemist Gerhard Schrader (1903–1990).

Schrock carbene, a class name for carbenes with a nucleophilic carbon center, after the American chemist Richard Royce Schrock (born 1945).

Schrock's molybdenum catalyst ($C_{30}H_{35}F_{12}MoNO_2$), as Schrock carbene.

Schwartz's reagent (chloridobis(η^5-cyclopentadienyl)hydridozirconium), after the American chemist Jeffrey Schwartz (born 1945).

Schweizer's reagent [tetraamminediaquacopper(II) dihydroxide], commonly misspelled Schweitzer's reagent, after the Swiss chemist Matthias Eduard Schweizer (1818–1860). A.k.a. **cuoxam**, a contraction of cuprum, ox(o)-, and ammonia.

Shvo catalyst (RN 104439-27-2). after the Israeli chemist Youval Shvo (1930–2007).

Sidot's blende (ZnS), after the French chemist Théodore Sidot who first reported this material's phosphorescence in 1866.

Sobrerol (*p*-menth-1-ene-6,8-diol), after the Italian chemist Ascanio Sobrero (1812–1888) who determined this terpenoid's structure.

Sonnenschein's reagent [phosphomolybdic acid, molybdophosphoric acid, tetrahydrogen(phosphododecamolybdate)], after the German forensic chemist Franz Leopold Sonnenschein (1819–1879).

Sorbite (Fe,Fe_3C), after the British geologist and metallurgist Henry Clifton Sorby (1826–1908). A.k.a. **pearlite**.

Sorel cement ($MgCl_2,MgO$), after the French inventor Stanislas Sorel (1802–1871).

Sørensen buffer (a solution containing potassium dihydrogenphosphate and disodium hydrogenphosphate), after the Danish chemist Søren Peder Lauritz Sørensen (1868–1939).

Spiegelman's Monster (a self-reproducing RNA chain of 218 nucleotides), after the American molecular biologist Sol Spiegelman (1914–1983).

Staudinger's ketene (diphenylethenone), after the German chemist Hermann Staudinger (1881–1965).

Tobias acid (2-aminonaphthalene-1-sulfonic acid), after the German chemist Georg Tobias (deceased 1909).

Tuftsin [L-threonyl-L-lysyl-L-prolyl-N^5-(diaminomethylidene)-L-ornithine, after Tufts University, Boston, MA, USA, in turn named after the American businessman and philantropist Charles Tufts (1781–1876).

Tutton salt [$M^IM^{II}(SO_4)_2{\cdot}6H_2O$], after the British chemist Alfred Edwin Howard Tutton (1864–1938).

Tyrode's solution (containing sodium chloride, potassium chloride, magnesium chloride, sodium hydrogencarbonate, sodium dihydrogenphosphate, and glucose), after the French-American pharmacologist Maurice Vejux Tyrode (1878–1930).

Urushibara nickel (Ni,Zn), after the Japanese chemist Yoshiyuki Urushibara (1901–1972). Cf. Raney alloy.

Vaska's complex [*trans*-carbonylchloridobis(triphenylphosphane)iridium(I)], after the Estonian-American chemist Lauri Vaska (1925–2015).

Vauquelin's salt [tetraamminepalladium(II) tetrachloridopalladate(II)}, after the French chemist Nicolas Louis Vauquelin (1783–1829).

Vedejs' reagent [dioxidooxidomolybdenum—hexamethylphosphoric triamide—pyridine (1/1/1)], after the Latvian-American chemist Edwin Vedejs (1941–2017).

Viehe's salt [*N*-(dichloromethylene)-*N*-methylmethanaminium chloride], after the German-Belgian chemist Heinz Günther Viehe (1929–2010). A.k.a. **Vilsmeier reagent**, after the German chemist Anton Vilsmeier (1894–1962).

Vleminckx' solution (containing calcium polysulfides), after the Belgian physician Jean François Vleminckx (1800–1876).

Vortmann's sulfate {μ-dioxidobis[pentaamminecobalt(III)] sulfate}, after the German chemist Georg Vortmann (1854–1932).

Wackenroder solution (containing polythionic acids $H_2S_nO_6$, n = 3–20), after the German chemist Heinrich Wilhelm Ferdinand Wackenroder (1798–1854).

Warfarin [4-hydroxy-3-(3-oxo-1-phenylbutyl)-2*H*-1-benzopyran-2-one], with contraction, after WARF (Wisconsin Alumni Research Foundation) and coumarin.

West's solution (containing tetraphosphorus octasulfide and diiodomethane), after the American mineralogist Cutler Delong West (1903–1993).

Wieland-Gumlich aldehyde [(19*E*)-18-hydroxycur-19-en-17-al], after the German chemists Heinrich Otto Wieland (1877–1957) and Walter Gumlich.

Wieland-Miescher ketone [(8a*S*)-8a-methyl-3,4,8,8a-tetrahydronaphthalene-1,6(2*H*,7*H*)-dione], after the Swiss chemist Peter Wieland (born 1920) and the Swiss chemist Karl Miescher (1892–1974).

Wijs' reagent (a solution of iodine monochloride in acetic acid), after the Dutch chemist J. J. A. Wijs (1864–1942).

Wilkinson's catalyst [(*SP*-4)-chloridotris(triphenylphosphane)rhodium(I)], after the British chemist Geoffrey Wilkinson (1921–1996).

Winterstein's acid [(3*R*)-3-(dimethylamino)-3-phenylpropanoic acid], after the German-Swiss chemist Ernst Winterstein (1865–1949).

Wittig reagent, such as methylidenetriphenyl-λ^5-phosphane, a class name for phosphanium ylides, after the German chemist Georg Wittig (1897–1987).

Woollins' reagent (2,4-diphenyl-1,3,2,4-diselenadiphosphetane 2,4-diselenide), after the British chemist John Derek Woollins (born 1955).

Wood's metal (Bi,Pb,Sn,Cd), after the American dentist and inventor Barnabas Wood (1819–1875). A.k.a. **Lipowitz's alloy**, after the Polish-German-Austrian pharmacist, chemist, and industrialist Friedrich Julius Alexander Lipowitz (Fryderyk Juliusz Alexander Lipowitz) (1810/1811–1873), **Bendalloy**, and **Cerrobend**.

Woodward's reagent K [3-(2-ethyl-1,2-oxazol-2-ium-5-yl)benzenesulfonate], after the American chemist Robert Burns Woodward (1917–1979).

Woodward's reagent L [5-methyl-2-(2-methylpropan-2-yl)-1,2-oxazol-2-ium perchlorate], as Woodward's reagent K.

Wurster's blue (N^1,N^1,N^4,N^4-tetramethylbenzene-1,4-diamine), after the German chemist Casimir Wurster (1854–1913). Contrary to popular belief this compound is colorless. It has been named after the intensely blue radical cation which is formed upon its oxidation and which also, and more appropriately, has been called Wurster's blue. A.k.a. **TMPD.**

Wurster's red (N^1,N^1-dimethylbenzene-1,4-diamine), as Wurster's blue.

Zeise's salt [potassium trichlorido(η^2-ethene)platinate(II)], after the Danish chemist William Christopher Zeise (1789–1847).

2.4 Toponyms

Aklavin {methyl (1*R*,2*R*,4*S*)-2-ethyl-2,5,7-trihydroxy-6,11-dioxo-4-[[2,3,6-trideoxy-3-(dimethylamino)-α-L-*lyxo*-hexopyranosyl]oxy]-1,2,3,4,6,11-hexahydrotetracene-1-carboxylate}, after its first source, fungi in a soil sample from Aklavik, NWT, Canada. A.k.a. **aclacinomycin T.**

Brazilin {(6a*S*,11b*R*)-7,11b-dihydroindeno[2,1-*c*][1]benzopyran-3,6a,9,10(6*H*)-tetrol}, from brazilwood (*Caesalpina echinata*), ultimately from Brazil. The toponym Brazil is from Portuguese *Terra do Brasil*, land of the brazilwood (*Paubrasilia echinata*). Brazilwood, from Portuguese *pau-brasil*, from *pau*, wood, stick, and *brasa*, ember, was named after its orange-red color

Britannia metal (Sn,Sb,Cu,Bi,Zn), from Latin *Britannia*, Britain, developed by the British industrialist James Vickers (deceased 1809).

Chicago sky blue 6B {tetrasodium 6,6′-[(3,3′-dimethoxy[1,1′-biphenyl]-4,4′-diyl)bis(diazene-2,1-diyl)]bis(4-amino-5-hydroxynaphthalene-1,3-disulfonate}, after the city of Chicago, IL, USA. A.k.a. **Direct Blue 1** and **Pontamine Sky Blue.**

Ethiops iron (triiron tetraoxide), from Greek *Aithiops*, Ethiopian, negro, from *aithein*, to kindle, to blaze, and *ops*, face, and iron.

Indigo, ultimately from Greek *indikon pharmakon*, i.e. blue dye from India.

Jesaconitine [8-acetoxy-20-ethyl-3α,13,15α-trihydroxy-1α,6α,16β-trimethoxy-4-(methoxymethyl)aconitan-14α-yl 4-methoxybenzoate], from Jeso, the pre-1869 name of the Japanese island of Hokkaido, referring to the habitat of the jesaconitine-producing *Aconitum* (monkshood) species, cf. aconitane.

Kalkitoxin {(2*R*)-*N*,2-dimethyl-*N*-[(3*S*,5*S*,6*R*)-7-[(4*R*)-4-ethenyl-4,5-dihydro-1,3-thiazol-2-yl]-3,5,6-trimethylheptyl]butanamide}, after Playa Kalki, Curaçao, the habitat of the source organism, the blue-green alga *Lyngbya majuscula*.

Manoalide {(5*R*)-5-hydroxy-4-[(2*R*,6*R*)-6-hydroxy-5-[(3*E*)-4-methyl-6-(2,6,6-trimethylcyclohex-1-en-1-yl)hex-3-en-1-yl]-3,6-dihydro-2*H*-pyran-2-yl]furan-2(5H)-one}, after Manoa Valley, Oahu, HI, USA and with 'olide'.

α-Muurolene (1β-cadina-4,9-diene), after its discovery in specimens of *Pinus silvestris*, (Scotch pine, European redwood) collected in the vicinity of the small town of Muurola, Rovaniemi, Finland.

^{1}**Norge saltpeter** (calcium nitrate), from Norwegian *Norge*, Norway. A.k.a. **Norway saltpeter**. Cf. ^{2}**Norge saltpeter**.

^{2}**Norge saltpeter** (ammonium nitrate), as ^{1}Norge saltpeter.

Nybomycin {8-(hydroxymethyl)-6,11-dimethyl-4*H*-[1,3]oxazolo[5,4,3-*ij*]pyrido[3,2-*g*]quinoline-4,10(11*H*)-dione}, with contraction, from New York Botanical Garden and -mycin.

Nystatin (a polyene macrolide), with contraction, from New York State and statin.

Pimaricin (a macrocylic natural fungicide), with contraction, after Pietermaritzburg, Natal, South Africa, the habitat of the bacterial species *Streptomyces natalensis*, which produces pimaricin. A.k.a. **natamycin.**

Rochelle salt [potassium sodium (2*R*,3*R*)-2,3-dihydroxybutanedioate tetrahydrate], named after the town of La Rochelle, Nouvelle Aquitaine, France. A.k.a. **Seignette salt**, after the French apothecary Pierre Seignette (1660–1719) of La Rochelle.

Rose bengal {4,5,6,7-tetrachloro-3′,6′-dihydroxy-2′,4′,5′,7′-tetraiodo-3*H*-spiro[2-benzofuran-1,9′-xanthen]-3-one}. A pink colored dye named after Bengal, India and Pakistan. A.k.a. **Rose Bengal.**

Thebaine (3,6-dimethoxy-17-methyl-6,7,8,14-tetradehydro-4,5α-epoxymorphinan), from thebaic, of Thebes, Egypt. By the 18th century it had acquired the meaning pertaining to opium which then came to Europe from Egypt. Thebaine is a constituent of opium.

Thebenine {6-methoxy-3-[2-(methylamino)ethyl]phenanthrene-1,5-diol}, as hebaine.

Wyerone {methyl (2*Z*)-3-[5-[3-(3-ethyloxiran-2-yl)prop-2-ynoyl]furan-2-yl]prop-2-enoate}, after Wye College, University of London, Ashford, Kent, UK, where this compound was isolated from *Vicia faba* (broad bean). The name of the college is after the village of Wye, a toponym from Old English *weoh*, idol, shrine, possibly referring to worship there by pre-Christian Angles.

2.5 Miscellaneous trivial names

Acmonital (Fe,Cr,Si,Mg,C,S,P), a contraction of Italian *Acciaio Monetale Italiano*, Italian monetary steel.

Alite (Ca_3SiO_5), named after the letter A, i.e. as the first member of a series of synthetic minerals, cf. belite and celite. A.k.a. **hatrurite**, after its type locality, the Hatrurim Formation, Negev, Israel.

Alizarin (1,2-dihydroxyanthracene-9,10-dione), ultimately from Arabic *al asarah*, the juice, referring to this compound's occurrence in plant juice.

Anatabine [(2*R*)-1,2,3,6-tetrahydro-2,3′-bipyridine}, from anabasine and German *Tabak*, tobacco.

Belite (Ca_2SiO_4), named after the letter B, i.e. as the second member of a series of synthetic minerals, cf. alite and celite. A.k.a. **larnite**, after its type locality Scawl Hill, Larne, Northern Ireland, UK.

Benzidine ([1,1′-biphenyl]-4,4′-diamine), from 'benz(o)' and 'idine'.

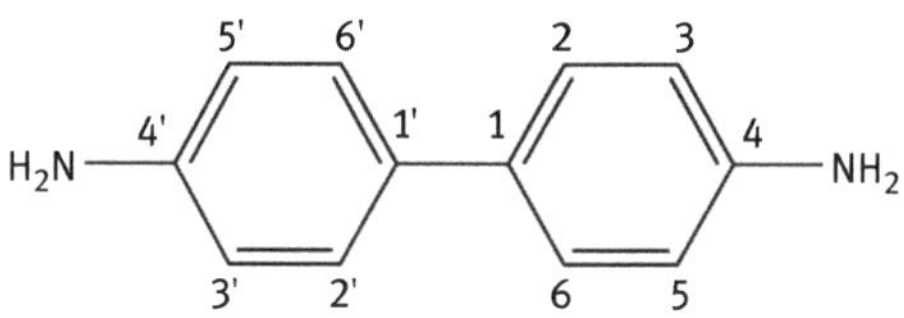

benzidine

Benzil (diphenylethandione), an arbitrary name derived from benzoin.

Biopterin [2-amino-6-(1,2-dihydroxypropyl)-1*H*-pteridin-4-one], from bio- and pterin, named as enzyme cofactor. A.k.a. **primapterin**, from Latin *primus*, the first, and pterin.

Brass (Cu,Zn), of uncertain etymology. Cf. Middle Low German *bras*, metal, and Old Frisian *bres*, copper.

Bronze (Cu,Sn), either from Persian *biring*, bronze, or from Latin *aes brundusinum*, metal from Brundisium, now Brindisi, Italy.

Celite [($Ca_2(Al,Fe)_2O_5$), named after the letter C, i.e. as the third member of a series of synthetic minerals, cf. alite and belite. A.k.a. **brownmillerite**, after the American chemist Lorrin Thomas Brownmiller (1902–1990).

Cerotin (hexacosan-1-ol), from cerotic acid. A.k.a. **ceryl alcohol.**

Chrysaniline [5-(4-aminophenyl)acridin-2-amine], from chrys(o)- and aniline.

Colcothar (diiron trioxide), ultimately from Arabic *qulqutar*, probably derived from Greek *chalkanthos*, from *chalkos*, copper, and *anthos*, flower.

Copperas [iron(II) sulfate heptahydrate], from Medieval Latin *aqua cuprosa*, copperish water. This name referred originally to solutions of copper(II) sulfate, later known as 'blue copperas' while the corresponding iron(II) compound was called 'green copperas'. The epithet green was subsequently lost.

Dicyclopentadiene {tricyclo[5.2.1.$0^{2,6}$]deca-3,8-diene}. of cyclopenta-1,3-diene. While this name may look systematic at first glance, it is for all practical purposes trivial.

Diketene (4-methylideneoxetan-2-one), of ketene. Cf. dicyclopentadiene.

Diphosgene (trichloromethyl carbonochloridate), of phosgenee. Cf. dicyclopentadiene.

Dithionic acid [bis(hydroxidodioxidosulfur)(*S*—*S*), $H_2S_2O_6$], from Greek *theion*, sulfur.

Freon (a class of halocarbons such as dichlorodifluoromethane), from freezing and (halogenated) hydrocarbon, referring to their use as refrigerants.

Fusel oil [amyl alcohol (pentan-1-ol) and isomers], from German *Fusel*, poor-quality liquor.

Gasoline (a volatile flammable liquid hydrocarbon mixture), from gas and Latin *oleum*, oil. A.k.a. **petrol**, from French *essence de pétrole*, essence of petroleum.

Heroin [17-methyl-7,8-didehydro-4,5α-epoxymorphinan-3,6α-diyl diacetate], originally coined as a trademark, from Greek *heros*, hero, referring to the strength of this compound as antitussive remedy for tuberculosis patients. A.k.a. **diamorphine**, a contraction of diacetylmorphine.

Halomon [(3*S*,6*R*)-6-bromo-3-(bromomethyl)-2,3,7-trichloro-7-methyloct-1-ene], a contraction of halomethyloctene or halomonoterpene, contrary to common practice not named after its source, *Portieria hornemannii* (a red alga).

Hystazine (2,3-dihydroxyanthracene-9,10-dione), from Greek *hysteros*, later, and alizarin, named as a by-product of an alizarin synthesis.

Kerosene (a flammable hydrocarbon oil), from Greek *keros*, wax. A.k.a. **kerosine**.

Litmus (a coloring matter from lichens), ultimately from Old Norse *litmosi*, herbs used in dyeing, from *litr*, color, and *mosi*, moss. Interestingly, the German equivalent *Lackmus* has a different etymology, being derived from Indo-European *leg*, to drip, and German *Mus*, mush.

Misch metal (an alloy of rare earth metals), from German *Mischmetall*, mixed metal, from *mischen*, to mix.

Myronic acid {1-*S*-[(1*E*)-*N*-(sulfooxy)but-3-enimidoyl]-1-thio-β-D-glucopyranose}, from Greek *myron*, unguent, perfume.

Naphtha (a liquid hydrocarbon mixture), from Greek *naphtha*, naphtha, ultimately from Persian *neft*, naphtha.

^{1}**Natron** (sodium hydrogencarbonate), ultimately derived from Ancient Egyptian *ntry*, niter.

^{2}**Natron** (sodium carbonate decahydrate), as ^{1}natron.

Nitroprusside [pentacyanidonitrosylferrate(2–), an anion], from nitr(o)- and, ultimately, Prussian blue. A misnomer since it is a nitroso compound and does not contain nitro groups.

Pantoic acid [(2*R*)-2,4-dihydroxy-3,3-dimethylbutanoic acid], from Greek *pas, pan-*, all, every, referring to the widespread occurrence of this acid in food.

Pantothenic acid {3-[(2*R*)-2,4-dihydroxy-3,3-dimethylbutanamido]propanoic acid}, from Greek *pantothen*, from everywhere,referring to the widespread occurrence of this acid in food.

Paraprotein (a class name), named as abnormal serum globulins with unique physical and electrophoretic characteristics.

Petroleum (an oily flammable bituminous liquid), from petr(o)- and Latin *oleum*, oil.

Pewter (Sn,Cu,Sb,Bi,Ag), from Vulgar Latin *piltrum*, pewter, of uncertain origin.

Phosgene (carbonyl dichloride), from Greek *phos, phot-*, light, referring to the photochemical formation of this compound from carbon monoxide and chlorine.

Rifamycin (a class of macrocyclic antibiotics), a variation of rifomycin.

Rifomycin (a class of macrocyclic antibiotics), arbitrarily named after the classical French gangster movie *Rififi*, from French slang *rififi*, heist.

Sulfurol [2-(4-methylthiazol-5-yl)ethanol], named as sulfur-containing alcohol.

Superhydride [lithium triethyl(hydrido)borate(1–)], named as superior hydride donor compared to lithium tetrahydridoborate.

Supermesityl [2,4,6-tris(2-methylpropan-2-yl)phenyl], named for this substituent's superior bulk compared to mesityl (2,4,6-trimethylphenyl).

Supersilyl [tris(trimethylsilyl)silyl], named for this substituent's superior bulk compared to trimethylsilyl.

Suprasterol II [(3*S*,6*R*,7*S*,22*E*)-6,8:7,10-dicyclo-9,10-secoergosta-5(10),22-dien-3-ol}, so named for the prolonged irradiation needed to produce this compound from calciferol.

Tartrazine {trisodium 5-hydroxy-1-(4-sulfonatophenyl)-4-[(*E*)-(4-sulfonatophenyl) diazenyl]-1*H*-pyrazole-3-carboxylate}, from tartaric acid and hydrazine, referring to the synthesis of the intermediate pyrazole derivative. A.k.a. **Yellow No. 5.**

Tombac (Cu,Zn), from Malay *tembaga*, copper, ultimately from Sanskrit *tamraka*, copper.

Triphosgene [bis(trichloromethyl) carbonate], of phosgene. Cf. diketene.

Uranediol (17α-methyl-*D*-homo-5α-androstane-3β,17α-diol), a contraction of urine, androstane, and 'diol'.

Uvitic acid (5-methylbenzene-1,3-dicarboxylic acid), from Latin *uva*, grape, referring to the possible formation of this acid from tartaric acid.

Uvitonic acid (6-methylpyridine-2,4-dicarboxylic acid), named as pyridine analog of the benzene derivative uvitic acid.

2.6 Names for isomers of known compounds

It has been customary to create names for such compounds by adding the prefix 'iso', from Greek *isos*, equal, such as here:

Nicotinic acid (pyridine-3-carboxylic acid), from nicotine, referring to its formation by oxidation of nicotine. A.k.a. **niacin.**

Isonicotinic acid (pyridine-4-carboxylic acid).

When more than one isomer had to be named, additional prefixes came into play, i.e. 'allo', from Greek *allos*, other, different, 'anteiso', from Latin *ante*, before (in place or time), in front of, against, and 'iso', 'neo', from Greek *neos*, new, 'para', from Greek *para*,beside, similar, near, beyond, irregular, a modification of, and 'pseudo', ultimately from Greek *pseudein*, to lie, to cheat, to falsify. Examples are:

Cinnamic acid [(2*E*)-3-phenylprop-2-enoic acid], after cinnamon, ultimately from Hebrew *qinnamon*, cinnamon.

Allocinnamic acid [(2*Z*)-3-phenylprop-2-enoic acid].

Cumene [propan-2-yl)benzene], after the genus *Cuminum* (cumin). The genus name is from Latin *cumenum*, Roman caraway.

Isocumene (propylbenzene), named as isomer of cumene.

Pseudocumene (1,2,4-trimethylbenzene), named as isomer of cumene.

Pentadecanoic acid.

Isopentadecanoic acid (13-methyltetradecanoic acid), named as isomer of pentadecanoic acid.

Anteisopentadecanoic acid (12-methyltetradecanoic acid), named as isomer of isopentadecanoic acid.

Isomers of the pentyl group, also called amyl, ultimately from Greek *amylon*, starch, literally not ground (starch was obtained from unground grain), have been named as follows:

***n*-Pentyl** (pentyl, pentan-1-yl).

***sec*-Pentyl** (pentan-2-yl).

***tert*-Pentyl** (2-methylbutan-2-yl).

3-Pentyl (pentan-3-yl).

Isopentyl (3-methylbutyl, 3-methylbutan-1-yl).

***sec*-Isopentyl** (3-methylbutan-2-yl).

Neopentyl (2,2-dimethylpropyl, 2,2-dimethylpropan-1-yl).

Active pentyl (2-methylbutyl, 2-methylbutan-1-yl), from active amyl alcohol [(2*S*)-2-methylbutan-1-ol].

Isocetane (2,2,4,4,6,8,8-heptamethylnonane) is a name which was given to one particular isomer of the many isomers of cetane (hexadecane).

^{1}**Isooctane** (2-methylheptane) is an arbitrary name for one particular of the many isomers of octane.

^{2}**Isooctane** (2,2,4-trimethylpentane) is an equally arbitrary name given to another of the many isomers of octane. Engine fuel's knocking performance is rated by the octane number, defined as the percentage of ^{2}isooctane in a test fuel composed of heptane and ^{2}isooctane. Octane number is, of course, a monumental misnomer, if taken literally, since any mixture of heptane and octane would have an octane number of practically zero.

In the case of inositol, where nine stereoisomeric cyclitols had to be named, this was accomplished with italicized prefixes:

Inositol (cyclohexane-1,2,3,4,5,6-hexol), named as a sugar alcohol, from Greek *is, inos-*, fiber, muscle, sinew, tendon.

allo-Inositol [(1R,2R,3S,4R,5S,6S)-cyclohexane-1,2,3,4,5,6-hexol], the stereodescriptor allo- is from Greek allos, other.

D-*chiro*-Inositol [(1R,2*R*,3*S*,4*S*,5*S*,6*S*)-cyclohexane-1,2,3,4,5,6-hexol], the stereodescriptor *chiro-* is from Greek *cheir, chiro-*, hand. A.k.a. **quercitol**, from Latin *quercus*, oak.

L-*chiro*-Inositol [(1*R*,2*R*,3*R*,4*R*,5*S*,6*S*)-cyclohexane-1,2,3,4,5,6-hexol]. Cf. the D-isomer.

***cis*-Inositol** [(1*s*,2*s*,3*s*,4*s*,5*s*,6*s*)-cyclohexane-1,2,3,4,5,6-hexol], the stereodescriptor *cis-* is from Latin *cis*, on this side.

***epi*-Inositol** [(1*R*,2*R*,3*r*,4*S*,5*S*,6*s*)-cyclohexane-1,2,3,4,5,6-hexol], the stereodescriptor *epi-* is from Greek *epi*, beside,on, upon, after.

***muco*-Inositol** [(1*R*,2*S*,3*r*,4*R*,5*S*,6*r*)-cyclohexane-1,2,3,4,5,6-hexol], the stereodescriptor *muco-* is from Latin *mucus*, mucus.

***myo*-Inositol** [(1*R*,2*S*,3*r*,4*R*,5*S*,6*s*)-cyclohexane-1,2,3,4,5,6-hexol], the stereodescriptor *myo-* is from Greek *mys, myo-*, muscle. A.k.a. **inositol.**

***neo*-Inositol** [(1*R*,2*R*,3*S*,4*S*,5*S*,6*S*)-cyclohexane-1,2,3,4,5,6-hexol], the stereodescriptor *neo-* is from Greek *neos, neo-*, new.

***scyllo*-Inositol** [(1*r*,2*r*,3*r*,4*r*,5*r*,6*r*)-cyclohexane-1,2,3,4,5,6-hexol], the stereodescriptor *scyllo-* is after the genus *Scyllium* (catsharks), later renamed *Scyliorhinus*, ultimately from Greek *skylion*, dogfish. A.k.a. **scyllitol, cocositol**, after the genus *Cocos* (coconut tree), the genus name being from Portuguese *coco*, grimace, referring to the nut with its three holes and thus appearance of a face, and **quercinitol.**

However, historically, a group of stereoisomers could also receive independent names:

Tartaric acid (2,3-dihydroxybutanedioic acid), named after tartar, ultimately from Persian *durd* (lees). A.k.a. **threaric acid**, from threose.

D-Tartaric acid [(2*S*,3*S*)-2,3-dihydroxybutanedioic acid]. A.k.a. **D-threaric acid**, from D-threose.

L-Tartaric acid [(2*R*,3*R*)-2,3-dihydroxybutanedioic acid]. A.k.a. **L-threaric acid**, from L-threose.

Racemic acid [DL-tartaric acid, (2*RS*,3*RS*-dihydroxybutanedioic acid), *rac*-(2*R*,3*R*-dihydroxybutanedioic acid], from Latin *racemus*, cluster of grapes. A.k.a. **erythraric acid**, from erythrose, **uvic acid**, from Latin *uva*, grape, and **vinic acid**, from Latin *vinum*, wine.

Mesotartaric Acid [(2*R*,3*S*)-2,3-dihydroxybutanedioic acid], 'meso', from Greek *mesos*, middle, referring to its achiral nature. A.k.a. **erythraric acid**, from erythrose.

Another way to name isomers of known compounds was by anagramming the name of the known compound. Examples are:

Xylose (*xylo*-pentose), from xyl(o)-.

Lyxose (*lyxo*-pentose), anagrammatically derived from xylose.

However, anagrams have also been used to name related, but not isomeric, compounds from the same source. Examples are:

Meconic acid (3-hydroxy-4-oxo-4*H*-pyran-2,6-dicarboxylic acid), from Greek *mekon*, poppy.

Comenic acid (5-hydroxy-4-oxo-4*H*-pyran-2-carboxylic acid), anagrammatically derived from meconic acid.

Comanic acid (4-oxo-4*H*-pyran-2-carboxylic acid), anagrammatically derived from meconic acid.

2.7 Semitrivial names

These older names operate with prefixes and suffixes by and large like modern systematic nomenclature, but lack its rigor such as in the rational choice of parent hydrides.

The obsolete infix 'ol', used to modify trivial names to show the presence of a carbon-carbon triple bond, as in propionic acid/propiolic acid, corresponds to 'yne' used in current systematic names.

One should note the highly arbitrary use of 'meta', 'ortho', and 'para' in some of the following semitrivial names. In trivial acid names the prefix 'ortho' stands for maximum hydration while 'meta' appears in names of acids which are formally derived from an 'ortho' acid by loss of water.

Acetone (propanone), earlier called pyroacetic acid, named as derivative of acetic acid. The suffix 'one' was chosen arbitrarily.

Acetophenone (1-phenylethan-1-one), from phene (benzene). Contrary to common logic both the prefix and the suffix of this name indicate the presence of the keto group. The same is true of the names benzophenone, butyrophenone, and propiophenone. The systematic name of acetophenone involves an interesting dilemma. The choice of parent was ethanone or ethanal. The first was chosen over the other in spite of the fact that there is no parent compound called ethanone. In the second case 1-phenylethanal would have been a misnomer since the substitution converts the parent aldehyde to a ketone. The same dilemma appears in the systematic names of benzophenone, butyrophenone, and propiophenone. A.k.a. **hypnone**, referring to its sleep-inducing properties.

Acesulfame [6-methyl-2,2-dioxo-1,2,3-oxathiazin-4(3*H*)-one], with contraction, from acetoacetic ester and sulfamate.

Acetin (2,3-dihydroxypropyl acetate), with contraction from German *Acetylglycerin*, acetylgycerol. A.k.a. **monoacetin.**

Aceturic acid (*N*-acetylglycine), from acetic acid and urine, patterned after hippuric acid.

Aconic acid (5-oxo-4,5-dihydrofuran-3-carboxylic acid), with contraction, from aconitic acid (prop-1-ene-1,2,3-tricarboxylic acid), from which it can be formed.

L-Alanosine, a contraction of 3-[hydroxy(nitroso)amino]-L-alanine.

Allene (propa-1,2-diene), named as formed from the allyl radical by loss of a hydrogen atom.

Alloxan [pyrimidine-2,4,5,6(1*H*,3*H*)-tetrone], with contraction, from allantoin and oxalic acid.

Alloxanthin [(3*R*,3′*R*)-7,7′,8,8′-tetradehydro-β,β-carotene-3,3′-diol], named as carotenoid. Cf. alloxanthine and alloxantin.

Alloxanthine {1*H*-pyrazolo[3,4-*d*]pyrimidine-4,6(5*H*,7*H*)-dione}, named as isomer of xanthine. Cf. alloxanthin and alloxantin.

Alloxantin [5,5′-dihydroxy-5,5′-bipyrimidine-2,2′,4,4′,6,6′(1*H*,1′*H*,3*H*,3′*H*,5*H*,5′*H*)-hexone], from alloxan. Cf. alloxanthin and alloxanthine.

Amidogen (azanyl, a radical), from amide and -gen, patterned after halogen.

Amylene (pentene), ultimately a back formation from amyl alcohol.

Amsonic acid [2,2′-(ethene-1,2-diyl)bis(5-aminobenzenesulfonic acid)], a contraction of 4,4′-diamino-2,2′-stilbenedisulfonic acid.

Anabsinthin {(1*S*,2*R*,5*S*,8*S*,9*S*,12*S*,13*R*,14*S*,!5*R*,!7*S*,19*S*,22*S*,23*S*,26*S*,27*R*)-12-hydroxy-3,8,12,17,19,23-hexamethyl-6,18,25-trioxaoctacyclo[13.11.1.0^{1,17}.0^{2,14}.0^{4,13}.0^{5,9}.0^{19,27}.0^{22,26}]heptacos-3-ene-7,24-dione}, with contraction, named as analog of absinthin.

Anapterin {2-amino-7-[(1*S*,2*R*)-dihydroxypropyl]-1*H*-pteridin-4-one}, from analog and pterin, named as derivative of pterin.

Aniline mustard [*N*,*N*-bis(2-chloroethyl)aniline], from aniline and mustard gas {1-chloro-2-[(2-chloroethyl)sulfanyl]ethane}.

Anisidine (methoxyaniline), named as amino derivative of anisole (methoxybenzene).

Anisomycin {(2R,3S,4S)-4-hydroxy-2-[(4-methoxyphenyl)methyl]pyrrolidin-3-yl acetate}, from anisole and -mycin.

Anthragallol (1,2,3-trihydroxyanthracene-9,10-dione), from anthracene and gallic acid, referring to the vicinal position of the three hydroxy groups. A.k.a. **anthragallic acid.**

Anthranthene {naphtho[7,8,1,2,3-*nopqr*]tetraphene}, with contraction, from anthracene.

Anthranilic acid (2-aminobenzoic acid), from 'anthra' in the sense of carbon and aniline, referring to the formal formation of its core structure by addition of one carbon atom to aniline.

Atrolactic acid (2-hydroxy-2-phenylpropanoic acid), from atropine and lactic acid.

Atropic acid [(2*E*)-2-phenylprop-2-enoic acid], from atropine.

Azaserine, a contraction of *O*-(diazoacetyl)-L-serine.

Azide [trinitride(1–)], from French *azote*, nitrogen.

Benzilic acid (hydroxydiphenylacetic acid), from benzil, referring to its formation by rearrangement of benzil.

Benzophenone (diphenylmethanone), from phene (benzene). Cf. acetophenone.

Besyl- (benzenesulfonyl-), a contraction of the systematic name.

Bialaphos {L-alanyl-L-alanylphosphinothricin, 4-[hydroxy(methyl)phosphinoyl]-l-homoalanyl-L-alanyl-L-alanine}, named as containing two alanine residues and a phosphinic acid group.

Bicine [*N*,*N*-bis (2-hydroxyethyl)glycine], a contraction of the systematic name.

Bicozamycin {(1*S*,6*R*)-6-hydroxy-5-methylidene-1-[(1*S*,2*S*)-1,2,3-trihydroxy-2-methylpropyl]-2-oxa-7,9-diazabicyclo[4.2.2]decane-8,10-dione}, a contraction of the earlier names **bicyclomycin**, referring to this antibiotic's bicyclic structure, and **aizumycin**, after the bacterial species name *Streptomyces aizunensis*. The specific epithet is a toponym after the Aizu area, Fukushima Prefecture, Japan.

Biguanide (imidodicarbonimidic diamide), from guanidine, referring to its composition, i.e. two condensed guanidine units.

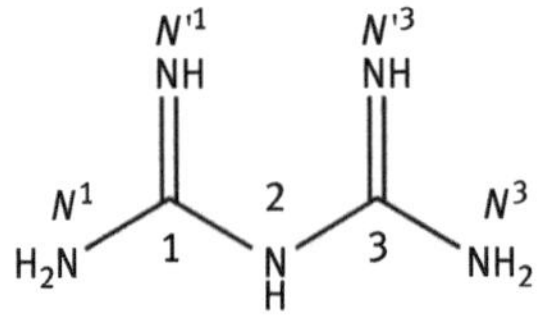

biguanide

Biuret (dicarbonimidic diamide), from urea, referring to its composition, i.e. two condensed urea units.

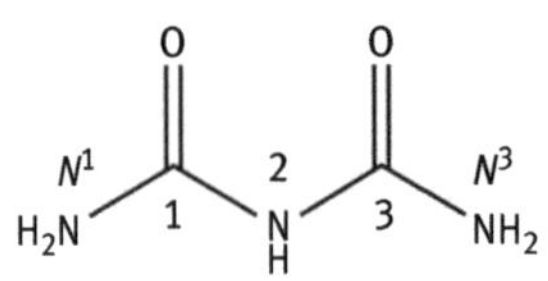

biuret

Boric acid (boranetriol, trihydroxidoboron), from boron.

Borinic acid (boranol, dihydridohydroxidoboron), as boric acid.

Boronic acid (boranediol, hydridodioxidoboron), as boric acid.

Bromal (tribromoethanal), a contraction of tribromoacetaldehyde.

Bromoform (tribromomethane), named as formic acid where both the hydroxyl group and the oxo group have been replaced with bromine.

Brosyl- (4-bromobenzenesulfonyl-), a contraction of the systematic name.

Butenolide (but-2-en-1,4-olide, furan-2(5*H*)-one), from but-2-enoic acid and 'olide'.

Butyrolactone (γ-butyrolactone, butan-1,4-olide, oxolan-2-one), from butyric acid and lactone. A.k.a. as **butanolide**.

Butyrophenone (1-phenylbutan-1-one), from butyric acid and phene (benzene). Cf. acetophenone.

Camphoric acid (1,2,2-trimethylcyclopentane-1,3-dicarboxylic acid), from camphor.

Camsyl- (camphor-10-sulfonyl-), a contraction of the semisystematic name.

ε-Caprolactam (6-hexanelactam, azepan-2-one), named as lactam of ε-aminocaproic acid (6-aminohexanoic acid).

Caprylene (oct-1-ene), ultimately a back formation from caprylic acid.

Carbanil (phenyl isocyanate), from carbon and anil.

Carbanilic acid (phenylcarbamic acid), from carbanil.

[1]**Carbazide** (hydrazinecarbohydrazide), a contraction of carbohydrazide.

[2]**Carbazide** (carbonyl diazide), a contraction of carbonyl azide.

α-Carboline (9*H*-pyrido[2,3-*b*]indole), a contraction of carbazole and quinoline.

β-Carboline (9*H*-pyrido[3,4-*b*]indole), named as isomer of α-carboline.

γ-Carboline (5*H*-pyrido[4,3-*b*]indole), as β-carboline.

δ-Carboline (9*H*-pyrido[3,2-*b*]indole), as β-carboline.

Carbanilide (*N*,*N*′-diphenylurea), from carbanilic acid.

Carbapenem {(5*R*)-1-azabicyclo[3.2.0]hept-2-en-7-one}, named as penem analog where the sulfur atom has been replaced with carbon.

Carbazic acid (hydrazinecarboxylic acid), from carbon and French *azote*, nitrogen.

Carbic anhydride {(1*R*,2*S*,6*R*,7*S*)-4-oxatricyclo[5.2.1.0^{2,6}]dec-8-ene-3,5-dione}, uncharacteristically from carbon. A.k.a. **nadic anhydride**, a contraction of norbornenedicarboxylic anhydride.

Carbostyril [quinolin-2(1*H*)-one], in an arbitrary manner from carbon and styrene.

Carbyl sulfate (1,3,2,4-dioxadithiane 2,2,4,4-tetraoxide), from carbon and sulfate. A.k.a. **ethionic anhydride**, from ethionic acid.

Carone {3,7,7-trimethylbicyclo[4.1.0]heptan-2-one}, by variation of carvone.

Cathyl- (ethoxycarbonyl-), a contraction of carboethoxyl.

Ceramide (sphingolipids), with contraction, from cerebroside and amide.

Cerebroside (sphingolipids), from cerebrose (galactose) and ‘ide’.

Chanoclavine {[(8*E*)-6-methyl-8,9-didehydro-6,7-secoergolin-8-yl]methanol}, with latinization, from chain and clavi(c)-, referring to the opened D ring relative to the parent ergoline. A.k.a. **secaclavine.**

Chloral (trichloroethanal), a contraction of trichloroacetaldehyde.

Chloral hydrate (2,2,2-trichloroethane-1,1-diol), named as the product of the addition of water to chloral. Not a crystal hydrate in the modern sense.

α-Chloralose {1,2-*O*-[(1*R*)-2,2,2-trichloroethane-1,1-diyl]-α-D-glucofuranose}, a contraction of the systematic name.

Chloramphenicol {2,2-dichloro-*N*-[(1*R*,2*R*)-1,3-dihydroxy-1-(4-nitrophenyl)propan-2-yl]acetamide}, a contraction of chloro-, amine, phenyl, nitro-, and glycol.

Chloranil (2,3,5,6-tetrachlorocyclohexa-2,5-diene-1,4-dione), from chlorine and aniline, referring to this compound's synthesis from aniline.

Chloranilic acid (2,5-dichloro-3,6-dihydroxycyclohexa-2,5-diene-1,4-dione), named as a product of the hydrolysis of chloranil.

Chlordane {1,3,4,7,8,9,10,10-octachlorotricyclo[5.2.1.0^{2,6}]dec-8-ene}, named as chlorine-substituted indane.

Chlorendic anhydride {(1*R*,2*S*,6*R*,7*S*)-1,7,8,9,10,10-hexachloro-4-oxatricyclo[5.2.1.0^{2,6}]dec-8-en-3,5-dione}, a contraction of the systematic name 1,4,5,6,7,7-hexachloro-*endo*-norborn-5-ene-2,3-dicarboxylic anhydride.

Chloroform (trichloromethane), named as formic acid where both the hydroxyl group and the oxo group have been replaced with chlorine.

Chlorohydrin (3-chloropropan-1,2-diol), named as the product of the addition of hypochlorous acid to an alkene.

Chloropicrin (trichloronitromethane), from ^{2}chlor(o)- and picric acid, referring to a synthesis of this compound by exhaustive chlorination of picric acid (2,4,6-trinitrophenol).

Chloroprene (2-chlorobuta-1,3-diene), from ^{2}chlor(o)- and isoprene, in the sense that the methyl group of isoprene has been exchanged for a chlorine atom.

Closyl- (4-chlorobenzenesulfonyl-), a contraction of the systematic name.

Cholamine [(2-aminoethyl)trimethylammonium, a cation], a contraction of choline and amine.

Citicoline {5′-*O*-[hydroxy[[[2-(trimethylazaniumyl)ethoxy]phosphinato]oxy]phosphoryl]cytidine}, a contraction of cytidine diphosphate choline ester. A.k.a. **CDP-choline**.

Citraconic acid [(2*Z*)-2-methylbut-2-enedioic acid], from citrus and aconic acid.

Citraurin (8′-apozeaxanthinal), from citrus and aur(o)-.

Cobaltosic oxide (tricobalt tetraoxide), a contraction of cobaltoso-cobaltic oxide, cobalto standing for Co^{2+} and cobalti for Co^{3+}.

Colamine (2-aminoethan-1-ol), a contraction of alcohol and amine. A.k.a. **ethanolamine**.

Copellidine (trimethylpiperidine), from collidine and piperidine.

Corphin (3,7,8,12,13,17-hexahydro-2*H*,21*H*-porphine), from corrin and porphine, referring to the fact that corphin combines the structural elements of corrin and porphyrine.

Cotarnine {4-methoxy-6-methyl-5,6,7,8-tetrahydro[1,3]dioxolo[4,5-*g*]isoquinolin-5-ol}, anagrammatically derived from narcotine.

Cotinine [(5*S*)-1-methyl-5-(pyridin-3-yl)pyrrolidin-2-one], anagrammatically derived from nicotine.

Coumalic acid (2-oxo-2*H*-pyran-5-carboxylic acid), from coumarin and malic acid.

Coumestrol {3,9-dihydroxy-6*H*-benzofuro[3,2-*c*][1]benzopyran-6-one}, from coumarin and estr(o)-.

Creatinine (2-amino-1-methyl-5*H*-imidazol-4-one), derived from creatine.

***m*-Cresotinic acid** (2-hydroxy-4-methylbenzoic acid), derived from *m*-cresol (3-methylphenol). A.k.a. ***m*-cresotic acid** and **4-methylsalicylic acid**.

Cromoglycic acid {5,5′-[(2,hydroxy-1,3-propanediyl)bis(oxy)]bis[4-oxo-4*H*-1-benzopyran-2-carboxylic acid}, with contraction, from chromene and glycerol. A.k.a. **cromoglicic acid**.

Cryptopyrrole (3-ethyl-2,4-dimethyl-1*H*-pyrrole), from crypt(o)- and pyrrole, referring to the fact that this compound is only formed upon drastic reduction of heme.

Cyacetazide (2-cyanoacetohydrazide), a contraction of the systematic name.

Cyamelide (1,3,5-trioxane-2,4,6-triimine), from cyan(o)- and melam. A.k.a. **insoluble cyanuric acid**.

Cyanamide (carbononitridic amide), from cyan(o)- and amide.

Cyanic acid (carbononitridic acid), from cyan(o)-.

Cyanidin [2-(3,4-dihydroxyphenyl)-3,5,7-trihydroxy-1-benzopyrylium, a cation], from cyan(o)-.

Cyanidine (1,3,5-triazine), from cyan(o)-.

Cyanin (cyanidin 3,5-*O*-diglucoside), from cyan(o)-. Not to be confused with cyanine.

Cyanine {1-(3-methylbutyl)-4-[(*Z*)-[1-(3-methylbutylquinolin)-4(1*H*)-ylidene]methyl] quinolinium, a cation}, from cyan(o)-. Not to be confused with cyanin.

Cyanoform (methanetricarbonitrile), named as formic acid where both the hydroxyl group and the oxo group have been replaced with cyano groups.

[1]**Cyanogen** [azanylidynemethyl, nitridocarbon(•), a radical], from cyan(o)-, patterned after halogen. A.k.a. **cyanyl**.

[2]**Cyanogen** (ethanedinitrile), from cyan(o)-, patterned after halogen. A.k.a. **dicyan, dicyanogen**, and **oxalonitrile.**

Cyclamic acid (cyclohexylsulfamic acid), a contraction of the systematic name.

Cyclamine, a class name, a contraction of cyclic amine.

Cyclane, a class name, a contraction of cycloalkane.

Cyclanone, a class name, a contraction of cycloalkanone.

Cyclene, a class name, a contraction of cycloalkene.

Cyclitol, a class name, from cycl(o)- and 'itol'.

Cyclovalone {2,6-bis[(4-hydroxy-3-methoxyphenyl)methylidene]cyclohexan-1-one}, a contraction of 2,6-divanillydenecyclohexanone.

Cystamine (2,2′-disulfanediyldiethan-1-amine), a contraction of cysteine and amine.

Cystathionine {*S*-[(2*R*)-2-amino-2-carboxyethyl]-L-homocysteine}, a contraction of cysteine and methionine, referring to its intermediacy in the conversion of methionine to cysteine.

Cysteamine (2-aminoethane-1-thiol), from cysteine and amine.

Dactinomycin (an antibiotic), an anagrammatic contraction of actinomycin D.

Dalapone (2,2-dichloropropanoic acid), an anagrammatic contraction of α,α-dichloropropionic acid.

Damsyl- [4-(dimethylamino)benzenesulfonyl-], a contraction of the systematic name.

Dansyl- [5-(dimethylamino)naphthalene-1-sulfonyl-], a contraction of the systematic name.

Deanol [2-(diethylamino)ethan-1-ol], a contraction of the systematic name.

Decalin (decahydronaphthalene), a contraction of German *Decahydronaphthalin*, decahydronaphthalene.

Dehydroacetic acid [3-acetyl-6-methyl-2*H*-pyran-2,4(3*H*)-dione], referring to the fact that its empirical formula corresponds to the loss of one mole of water from one mole of acetic acid.

Deoxybenzoin (1,2-diphenylethanone), from deoxy- and benzoin.

Desaurin {dimethyl (2*Z*,2′*Z*)-[1,3-dithietane-2,4-diylidenebis[[(2*E*,3*Z*)-butane-3,2-diylidene-(1*Z*,2*E*)-hydrazine-2,1-diylidene]]bis(phenylacetate)}, also a class name, probably a contraction of German *Desoxybenzoin*, deoxybenzoin, a typical starting material for desaurins, and aur(o)-, referring to the yellow color of many desaurins.

1,2-Diacetin (3-hydroxypropane-1,2-diyl diacetate), a contraction of German *1,2-Diacetylglycerin*, 1,2-diacetylglycerol.

1,3-Diacetin (2-hydroxypropane-1,3-diyl diacetate), a contraction of German *1,3-Diacetylglycerin*, 1,3-diacetylglycerol.

Diacetone alcohol (4-hydroxy-4-methylpentan-2-one), named as a formal dimer of acetone.

Diacetyl (butane-2,3-dione), from di- and acetyl.

Diacetylene (buta-1,3-diyne), named as an alkadiyne.

Dicoumarol {3,3′-methylenebis[4-hydroxy-2*H*-1-benzopyran-2-one]}, named as a compound containing two coumarol units.

Didanosine (2′,3′-dideoxy-D-inosine), a contraction and alteration of dideoxyinosine.

Digallic acid {3,4-dihydroxy-5-[(3,4,5-trihydroxybenzoyl)oxy]benzoic acid}, from di- and gallic acid.

Diglyme [1-methoxy-2-(2-methoxyethoxy)ethane], a contraction of diethylene glycol dimethyl ether.

Digol (2,2′-oxydiethanol), a contraction of diethylene glycol.

Diguanide (imidodicarbonimidic diamide), from di- and guanidine. A.k.a. **biguanide**.

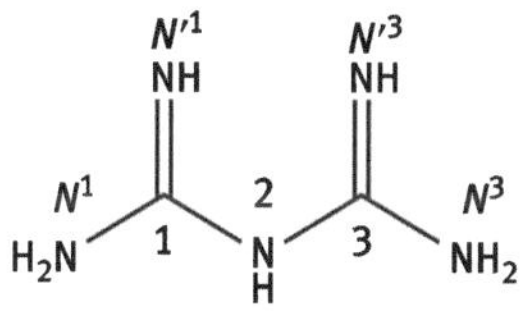

diguanide

Diketopiperazine (piperazine-2,5-dione), from di-, keto-, and piperazine.

Dilactide (3,6-dimethyl-1,4-dioxane-2,5-dione), from di- and lactide. A.k.a. **lactide**.

Dimedone (5,5-dimethylcyclohexane-1,3-dione), a contraction of the systematic name.

Dimercaprol (2,3-disulfanylpropan-1-ol), a contraction of 2,3-dimercaptopropan-1-ol. A.k.a. **BAL** and **British antilewisite**.

Dimethylglyoxime [(2*E*,3*E*)-*N*,*N*′-dihydroxybutane-2,3-diimine], from di-, methyl-, glyoxal, and oxime.

Dimsyl [(methylsulfinyl)methanide, an anion], a contraction of dimethyl sulfoxide anion.

1,2-Diolein [3-hydroxypropane-1,2-diyl bis[(9*Z*)-octadec-9-enoate)], a contraction of German *1,2-Dioleylglycerin*, 1,2-dioleylglycerol.

1,3-Diolein [2-hydroxypropane-1,3-diyl bis[(9*Z*)-octadec-9-enoate)], a contraction of German *1,3-Dioleylglycerin*, 1,3-dioleylglycerol.

Dioxin, a vernacular class name coined by contraction of polychlorodibenzo[1,4] dioxin.

1,2-Dipalmitin (3-hydroxypropane-1,2-diyl dihexadecanoate), a contraction of German *1,2-Dipalmitoylglycerin*, 1,2-dipalmitoylglycerol.

1,3-Dipalmitin (2-hydroxypropane-1,3-diyl dihexadecanoate), a contraction of German *1,3-Dipalmitoylglycerin*, 1,3-dipalmitoylglycerol.

Dipentene [1-methyl-4-(prop-1-en-2-yl)cyclohex-1-ene], named as a formal dimer of pentene. A.k.a. **cajeputene**, after the cajeput tree (*Melaleuca cajuputi*), from Malay *kayu putih*, literally white tree.

Diphenic acid {[1,1′-biphenyl]-2,2′-dicarboxylic acid}, from diphenyl (1,1′-biphenyl).

Diphenolic acid [4,4-bis(4-hydroxyphenyl)pentanoic acid], from phenol.

Diphenyl (1,1′-biphenyl), from phenyl.

Diphos [1,2-bis(diphenylphosphino)ethane], a contraction of the systematic name.

Dipicolinic acid (pyridine-2,6-dicarboxylic acid), from picolinic acid.

Dipicrylamine [2,4,6-trinitro-*N*-(2,4,6-trinitrophenyl)aniline], from picric acid (2,4,6-trinitrophenol). A.k.a. **hexyl.**

Diprenyl- (3,7-dimethyloct-2,6-dien-1-yl-), from di- and prenyl (3-methylbut-2-en-1-yl), from isoprene (2-methylbuta-1,3-diene). A.k.a. **geranyl,** from geraniol [(2*E*)-3,7-dimethyloct-2,6-dien-1-ol].

Dipropargyl (hexa-1,5-diyne), from di- and propargyl (prop-2-yn-1-yl), named as alkadiyne.

Dipyrrin {2-[2*H*-pyrrol-2-ylidene)methyl]-1*H*-pyrrole}, from di- and pyrrole.

1,2-Distearin (3-hydroxypropane-1,2-diyl dioctadecanoate), a contraction of German *1,2-Distearoylglycerin*, 1,2-distearoylglycerol.

1,3-Distearin (2-hydroxypropane-1,3-diyl dioctadecanoate), a contraction of German *1,3-Distearoylglycerin*, 1,3-distearoylglycerol.

Dithionic acid [bis(hydroxidodioxidosulfur)(S—*S*)], from di- and thi(o)-.

Dithionous acid [bis(hydroxidooxidosulfur)(S—*S*)], from di- and thi(o)-.

Dithizone [(*E*)-*N*′,2-diphenyldiazenecarbothiohydrazide}, a contraction of diphenyl thiocarbazone.

Divicine (2,4-diamino-1,4-dihydropyrimidine-5,6-dione), named as degradation product of vicine,

Divinyl (buta-1,3-diene), from di- and vinyl, named as alkadiene.

Duplodithio (3,3,6,6-tetramethyl-1,2,4,5-tetrathiane). This name is a loner with its unparalleled use of the odd numeral 'duplo', from Latin *duplum*, a double. It describes this compound as a dimer of acetone in which each oxygen atom has been replaced with two sulfur atoms.

Dypnone [(2*E*)-1,3-diphenylbut-2-en-1-one], a contraction of dihypnone, named as condensation product of hypnone (acetophenone, 1-phenylethan-1-one).

Edesyl- (ethane-1,2-disulfonyl-), a contraction of the systematic name.

Ellagic acid {2,3,7,8-tetrahydroxy[1]benzopyrano[5,4,3-*cde*][1]benzopyran-5,10-dione}, from French *galle*, gall, spelled backwards, an anagram of gallic acid.

Embonic acid, 4,4′-methylenebis(3-hydroxynaphthalene-2-carboxylic acid), an approximate contraction of the systematic name. A.k.a. **pamoic acid**, probably derived from pamaquine [N^1,N^1-diethyl-N^4-(6-methoxyquinolin-8-yl)pentane-1,4-diamine], the INN of an antimalarial which was the first liquid basic drug to be pharmaceutically formulated as its solid pamoate.

Endrin {(1*R*,2*S*,3*R*,6*S*,7*R*,8*S*,9*S*,11*R*)-3,4,5,6,13,13-hexachloro-10-oxapentacyclo[6.3.1.$1^{3,6}$.$0^{2,7}$.$0^{9,11}$]tridec-4-ene}, a contraction of *endo*-dieldrin.

Epichlorohydrin [2-(chloromethyl)oxirane], named as the epoxy compound which is formed by loss of hydrogen chloride from chlorohydrin (3-chloropropane-1,2-diol). A.k.a. **glycidyl chloride.**

Epithilone (a class of metabolites produced by bacteria such as *Sorangium cellulosum*), from epoxide, thiazole, and 'one'.

Eritadenine [(2*R*,3*R*)-4-(6-amino-9*H*-purin-9-yl)-2,3-dihydroxybutanoic acid], with contraction, from erythronic acid and adenine.

Erythorbic acid {(5*R*)-5-[(1*R*)-1,2-dihydroxyethyl]-3,4-dihydroxyfuran-2(5*H*)-one}, with contraction, from erythrose and ascorbic acid.

Esyl- (ethanesulfonyl-), a contraction of the systematic name.

Ethacrynic acid {[2,3-dichloro-4-(2-methylidenebutanoyl)phenoxy]acetic acid}, probably from ethane, acrylic acid, phene, and acetic acid.

Ethamphetamine (*N*-ethyl-1-phenylpropan-2-amine), a contraction of *N*-ethylamphetamine.

Ethionic acid [2-(sulfooxy)ethane-1-sulfonic acid], with contraction, from ethane and sulfonic acid.

Ethionine (*S*-ethyl-L-homocysteine), a contraction of the systematic name.

Etidronic acid (1-hydroxyethane-1,1-diphosphonic acid), a contraction of the unsystematic name ethane-1-hydroxy-1,1-diphosphonic acid.

Etryptamine [1-(1*H*-indol-3-yl)butan-2-amine), a contraction of α-ethyltryptamine.

Fluoral (trifluoroethanal), a contraction of trifluoroacetaldehyde.

Fluoran {3*H*-spiro[2-benzofuran-1,9′-xanthen]-3-one}, named as parent compound of fluorescein. Not to be confused with fluorane (HF).

Fluoroform (trifluoromethane), named as formic acid where both the hydroxyl group and the oxo group have been replaced with fluorine.

Formicin [*N*-(hydroymethyl)acetamide], named as the product of the addition of acetamide to formaldehyde.

Gallein {3′,4′,5′,6′-tetrahydroxy-3*H*-spiro[2-benzofuran-1,9′-xanthen]-3-one}, a contraction of pyrogallolphthalein.

Glycide [(oxiran-2-yl)methanol], from glycerol (propane-1,2,3-triol). A.k.a. **glycidol.**

Glycidic acid (oxirane-2-carboxylic acid), as glycide. A.k.a. **epihydrinic acid.**

Glycolide (1,4-dioxane-2,5-dione), from glycolic acid (hydroxyacetic acid) and 'olide'.

Glyme (1,2-dimethoxyethane), a contraction of glycol dimethyl ether.

Glyoxal (ethanedial), named as aldehyde derived from glycol (ethane-1,2-diol).

Glyoxylic acid (oxoacetic acid), named as acid derived from glyoxal (ethanedial).

Graphene (two-dimensional graphite), from graphite.

Graphene oxide (two-dimensional graphite oxide), from graphene.

Graphite oxide, from graphite. A.k.a. **graphitic acid** and **graphitic oxide**.

Guanidine (iminomethanediamine), after guanine, from which its was first obtained by degradation.

Halon (haloalkane), a contraction of halohydrocarbon.

Halothane (2-bromo-2-chloro-1,1,1-trifluoroethane). a contraction of haloethane.

Hemimellitic acid (benzene-1,2,3-tricarboxylic acid), from hemi- and mellitic acid, referring to the fact that this acid contains half as many carboxyl groups as mellitic acid.

Hemimellitene (1,2,3-trimethylbenzene), back formation from hemimellitic acid, named as its formal reduction product.

Histopine [*N*-(1-carboxyethyl)-L-histidine], with contraction, from histidine and opine.

Holophosphoric acid (λ^5-phosphanepentol, pentahydroxidophosphorus), from hol (o)- and phosphoric acid.

Homosotolone [5-ethyl-3-hydroxy-3-methylfuran-2(5*H*)-one], named as homolog of sotolone. A.k.a. **abhexone** and **maple furanone**.

Hopantenic acid {4-[[(2*R*)-2,4-dihydroxy-3,3-dimethylbutanoyl]amino]butanoic acid}, a contraction of homopantothenic acid.

Hotrienol [(5*E*)-3,7-dimethylocta-1,5,7-trien-3-ol], by and large a contraction of the systematic name.

Hydantoin (imidazolidine-2,4-dione), named, with contraction, as obtained by hydrogenation of allantoin.

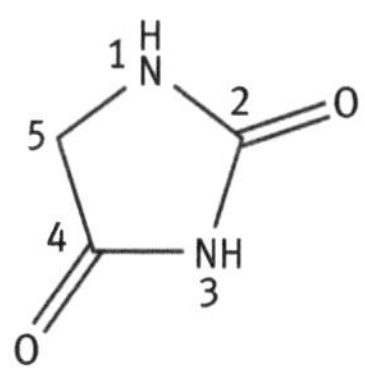

hydantoin

Hydantoic acid (*N*-carbamoylglycine), from hydantoin.

Hydracrylic acid (3-hydroxypropanoic acid), named as the formal product of the addition of water to acrylic acid (prop-2-enoic acid).

Hydrazoic acid [hydrogen azide, hydrogen trinitride(1−), hydrido-1κ*H*-trinitrogen (2*N*–*N*), triaza-1,2-dien-2-ium-1-ide], from hydrogen and French *azote*, nitrogen. Ak. a. **azoimide**.

Hydrindantin {2,2′-dihydroxy-1*H*,1′*H*-[2,2′-biindene]-1,1′,3,3′(2*H*,2′*H*)-tetrone}, a contraction of the systematic name.

Hydrocarbostyril [3,4-dihydroquinolin-2(1*H*)-one], from [2]hydr(o)- and carbostyril.

Hydrocinnamic acid (3-phenylpropanoic acid), from ^{2}hydr(o)- and cinnamic acid.

Hydroquinone (benzene-1,4-diol), named as reduction product of *p*-quinone (cyclohexa-2,5-diene-1,4-dione).

Hyprolose (a polysaccharide), a contraction of 2-hydroxypropyl cellulose.

Hypromellose (a polysaccharide), a contraction of 2-hydroxypropyl methyl cellulose.

Hypusine {[N^6-(2*S*)-4-amino-2-hydroxybutyl]-L-lysine}. a contraction of hydroxyputrescine and lysine.

Indican (1*H*-indol-3-yl-β-D-glucopyranoside), from indigo and 'an'.

Iodoform (triiodomethane), named as formic acid where both the hydroxyl group and the oxo group have been replaced with iodine.

Iridosmine (Ir,Os,Rh,Pt), a contraction of iridium and osmium. A.k.a. **iridosmium**.

Isatoic acid [2-(carboxyamino)benzoic acid], from isatin, q.v.

Isatoic anhydride [2*H*-3,1-benzoxazine-2,4(1*H*)-dione], from isatoic acid.

Isethionic acid (2-hydroxyethane-1-sulfonic acid), improperly from 'iso' and ethionic acid.

Isopentane (2-methylbutane) and **neopentane** (2,2-dimethylpropane); the decision which of the two isomers should receive the 'iso' and which the 'neo' label was purely arbitrary.

Isophorone (3,5,5-trimethylcyclohex-2-en-1-one), from 'iso' and phorone.

Isophthalic acid (benzene-1,3-dicarboxylic acid), named as isomer of phthalic acid.

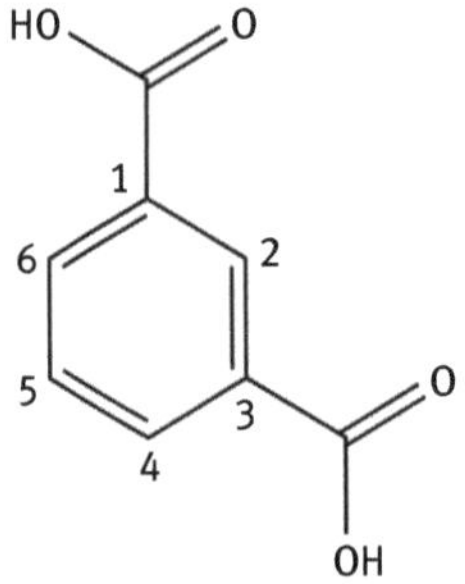

isophthalic acid

Isuretine (*N′*-hydroxyimidoformamide), from 'iso' and urea, i.e. named as isomer of urea. A.k.a. **formamidoxime** and **isuret**.

Itaconic acid (2-methylidenebutanedioic acid), an anagram of aconitic acid.

Ketene (ethenone), also used as class name, from ketone and 'ene'.

Ketipic acid (3,4-dioxohexanedioic acid), a contraction of 3,4-diketoadipic acid.

Kyanmethin (2,6-dimethylpyrimidin-4-amine), from cyan(o)- and a contraction of the systematic name.

Kyaphenine (2,4,6-triphenyl-1,3,5-triazine), from cyan(o)- and a contraction of the systematic name, i.e. named as formal trimer of benzonitrile. A.k.a. **cyaphenine**.

Laurocapram (*N*-laurylcaprolactam, 1-dodecylazepan-2-one), a contraction of laurylcaprolactam.

Levodopa (3-hydroxy-L-tyrosine), from lev(o)- and dopa (3-hydroxytyrosine).

Liothyronine (3,3′,5-triiodo-L-thyronine), from L-, ²iod(o)-, and thyronine.

Lupetidine (dimethylpiperidine), from lutidine and piperidine.

Lutidine (dimethylpyridine), a contracted anagram of toluidine, referring to their isomerism.

Lysopine {N^2-[(1*R*)-1-carboxyethyl]-L-lysine}, a contraction of lysine and opine.

Magnalium (Al,Mg), a contraction of magnesium and aluminum.

Maleuric acid [(2*Z*)-4-(carbamoylamino)-4-oxobut-2-enoic acid], a contraction of maleic acid and monoureide.

Mecrylate (methyl 2-cyanoprop-2-enoate), a contraction of methyl 2-cyanoacrylate.

Mellose, a contraction of methyl cellulose.

Menadiol, a contraction of 2-methylnaphthalene-1,4-diol.

Menadione, a contraction of 2-methylnaphthalene-1,4-dione.

Menaquinone {2-methyl-3-[(2*E*,6*E*,10*E*,14*E*)-3,7,11,15,19-pentamethylicosa-2,6,10,14,18-pentaen-1-yl]naphthalene-1,4-dione}, a contraction of (3-substituted) 2-methyl-1,4-naphthoquinone.

Mesaconic acid [(2*E*)-2-methylbut-2-enedioic acid], most likely named as an isomer of citraconic acid by contraction of mesocitraconic acid.

Mesidine (2,4,6-trimethylaniline), from mesitylene and 'idine'.

Mesoperiodic acid (trihydroxidodioxidoiodine, H_3IO_5), from meso- and periodic acid, referring to this acid's intermediate position between orthoperiodic acid and metaperiodic acid.

Mesoxalic acid (oxopropanedioic acid), from meso- and oxalic acid, named as analog of oxalic acid.

Mesyl- (methanesulfonyl-), a contraction of the systematic name.

^{1}**Metaarsenic acid** (arsorous acid, H_3AsO_3), named as tricoordinated arsenic compound.

^{2}**Metaarsenic acid** (diarsenic acid, arsonooxyarsonic acid, $H_4As_2O_7$), irregular name for a tetracoordinated arsenic compound.

^{1}**Metaboric acid** (hydroxidooxidoboron, HBO_2), named as dicoordinated boron compound.

^{2}**Metaboric acid**, this name has also been used as trivial name for the polymer $[HBO_2]_n$.

^{1}**Metaformaldehyde** (1,3,5-trioxane), named as trimer of formaldehyde.

^{2}**Metaformaldehyde** (polyoxymethylene), named as polymeric formaldehyde.

^{1}**Metagallic acid** (hydroxidooxidogallium, $HGaO_2$), named as dicoordinated gallium compound. Metagallic acid has also been used as trivial name for the polymer $[HGaO_2]_n$.

^{2}**Metagallic acid** (3,4-dihydroxy-5-[(3,4,5-trihydroxybenxoyl)oxy]benzoic acid), from 'meta' and gallic acid.

Metaldehyde (2,4,6,8-tetramethyl-1,3,5,7-tetraoxocane), a contraction of metaacetaldehyde, i.e. named as tetramer of acetaldehyde.

Metaphosphoric acid (hydroxidodioxidophosphorus, HPO_3), named as tricoordinated phosphorus compound.

Metasilicic acid (oxosilanediol, H_2SiO_3), named as tricoordinated silicon compound.

Metatelluric acid (telluric acid, H_2TeO_4), named as tetracoordinated tellurium compound.

Metavanadic acid (hydroxidodioxidovanadium, HVO_3), named as tricoordinated vanadium compound.

Metanilic acid (3-aminobenzenesulfonic acid), a contraction of *m*-anilinesulfonic acid.

Metaperiodic acid (hydroxidotrioxidoiodine, HIO_4), from meta- and periodic acid, named as tetracoordinated iodine compound.

Metasilicic acid (^{1}silicic acid, dihydroxy(oxo)silane, dihydroxidooxidosilicon), named as tricoordinated silicon compound.

Methacrylic acid (2-methylprop-2-enoic acid), a contraction of methylacrylic acid.

Methadone [6-(dimethylamino)-4,4-diphenylheptan-3-one], a drastic contraction of the systematic name.

Methamphetamine (*N*-methyl-1-phenylpropan-2-amine), a contraction of *N*-methylamphetamine.

Methionol [3-(methylsulfanyl)propan-1-ol], a contraction of methylthiopropanol.

Mevaldic acid (3-hydroxy-3-methyl-5-oxopentanoic acid), from mevalonic acid and aldehyde.

Mevalonic acid (3,5-dihydroxy-3-methylpentanoic acid), a modified contraction of dihydroxymethylvalerolactone. Related compounds are labeled as **mevinoid.**

Monacetin (2,3-dihydroxypropyl acetate), a contraction of German *Monoacetylglycerin*, monoacetylglycerol.

Monactin (a macrotetrolide complex), a contraction of monohomononactin.

Mucochloric acid [(2*Z*)-2,3-dichloro-4-oxobut-2-enoic acid], with contraction, from pyromucic acid and ^{2}chlor(o)-.

Muconic acid (hexa-2,4-dienedioic acid), from mucic acid, q.v., and patterned after itaconic acid.

Nalidixic acid (1-ethyl-7-methyl-4-oxo-1,8-naphthyridine-3-carboxylic acid), probably from naphthyridine and carboxylic acid.

Napadisilate (naphthalene-1,5-disulfonate), a contraction of the systematic name.

Napalm (gelled gasoline), a contraction of naphthenic acid and palmitic acid.

Naphthalic acid (naphthalene-1,8-dicarboxylic acid), from naphthalene, patterned after phthalic acid.

Naphthene (cycloalkanes), a class name, from naphtha and 'ene', referring to the fact that their molecular formulas are the same as those of the corresponding alkenes.

Naphthenic acid (cycloalkanecarboxylic acids), from naphthene.

Naphthionic acid (4-aminonaphthalene-1-sulfonic acid), from naphthalene and 'thio'.

Naphthol AS (3-hydroxy-*N*-phenylnaphthalene-2-carboxamide), from German *Naphtol AS*, from obsolete German *Naphtol*, naphthol, and *Amid einer Säure*, amide of an acid.

Napsyl- (2-naphthalenesulfonyl-), a contraction of the systematic name.

Neamine [2-deoxy-4-*O*-(2,6-diamino-2,6-dideoxy-α-D-glucopyranosyl)-D-streptamine], with contraction, from neomycin and amine.

Necic acid [(2*R*,3*R*,5*Z*)-5-ethylidene-2-hydroxy-2,3-dimethylhexanedioic acid), from necine alkaloids, q.v.

Necine alkaloids, a class name, after the genus *Senecio* (ragworts, groundsels), from Latin *senecio, senecion-*, groundsel, from *senex, senec-*, old man, referring to the gray hair on the seeds of these plants.

Neohexane (2,2-dimethylbutane), named as homolog of neopentane.

Neophyl- (2-methyl-2-phenylpropan-1-yl-), from neopentane and phenyl-.

Ninhydrin [2,2-dihydroxy-1*H*-indene-1,3-(2*H*)-dione], a contraction of triketohydrindene hydrate.

Nitroglycerol (propane-1,2,3-triyl trinitrate), earlier called nitroglycerin. From German *Nitroglycerin*. Not a nitro compound, but an ester of nitric acid.

Nipecotic acid (piperidine-3-carboxylic acid), a contraction of nicotinic acid and piperidine.

Nitinol (Ni,Ti), from Ni and Ti, and the initials of Naval Ordnance Laboratory.

Nitramine (*N*-methyl-*N*,2,4,6-tetranitrobenzenamine), a contraction of the systematic name. A.k.a. **tetryl**, from tetr(a)-, referring to the presence of four nitro groups in this explosive, and patterned after trotyl.

Nitroform (trinitromethane), named as formic acid where both the hydroxyl group and the oxo group have been replaced with nitro groups.

Nitroprusside [pentacyanidonitrosoferrate(2−)], from ^{1}nitrosyl and ^{1}prusside. Not a nitro compound as the name might suggest.

^{1}**Nitrosyl** [oxidonitrogen (1+), a cation], from New Latin *nitrosus*, pertaining to nitrous acid, from nitrogen.

^{2}**Nitrosyl-** (oxoazanyl-), from New Latin *nitrosus*, pertaining to nitrous acid, from nitrogen.

Nomilin {(5*S*,5a*R*,5b*R*,7a*S*,8*S*,10a*S*,11a*R*,11b*R*,13a*R*)-5-(acetyloxy)-8-(furan-3-yl)decahydro-1,1,5a,7a,11b-pentamethyloxireno[4′,5′]pyrano[4′,3′:5,6]naphth[2,1-*c*]oxepin-3,10,12(1*H*,4*H*,10a*H*)-trione}, an anagram of limonin, q.v.

Nonoxynol (a class name), a contraction of nonylphenoxypolyethoxyethanol.

Nopinic acid {2-hydroxy-6,6-dimethylbicyclo[3.1.1]heptane-2-carboxylic acid}, an anagram of the name of its isomer, pinonic acid, q.v.

Norpseudoephedrine [(1*S*,2*S*)-2-amino-1-phenylpropan-1-ol], named from nor- and pseudoephedrine, q.v. A.k.a. **cathine**, after khat (*Catha edulis*). Khat is from Arabic *qat*, khat.

Octalin (1,2,3,4,5,6,7,8-octahydronaphthalene), a contraction of German *Octahydronaphthalin*, octahydronaphthalene.

Octogen (1,3,5,7-tetranitro-1,3,5,7-tetrazocane), with contraction, from oct(a)- and nitrogen, referring to this compound's eight nitrogen atoms. A.k.a. **HMX**.

Onocerane I [(1*S*,2*R*,4a*S*,8a*R*,1′*S*,2′*R*,4a′*S*,8a′*R*)-1,1′-(ethane-1,2-diyl)bis(2,5,5,8a-tetramethyldecahydronaphthalene], from onocerin, q.v., and with 'ane', cf. sterane.

Opine (a class name), a contraction of octopine-type compound.

Ornaline [*N*-(4-amino-1-carboxybutyl)glutamic acid], a contraction of ornithine and nopaline.

Ornipressin (an oligopeptide), acontraction of ornithine and vasopressin.

Orthanilic acid (2-aminobenzenesulfonic acid), a contraction of *o*-anilinesulfonic acid.

Orthoarsenic acid (arsenic acid, H_3AsO_4), named as hydrated arsenic acid and tetracoordinated arsenic compound.

Orthoacetic acid (ethane-1,1,1-triol), named as hydrated ethanoic acid.

Orthoboric acid (boric acid, H_3BO_3), named as hydrated boric acid and tricoordinated boron compound.

Orthocarbonic acid (methanetetrol), named as hydrated carbonic acid.

Orthocarboxylic acid (alkane-1,1,1-triol), named as hydrated carboxylic acid.

Orthoformic acid (methanetriol), named as hydrated formic acid (methanoic acid).

Orthonitric acid (nitroric acid, trihydroxidooxidonitrogen, H_3NO_4), named as hydrated nitric acid and tetracoordinated nitrogen compound.

Orthonitrous acid (azanetriol, trihydroxidonitrogen, H_3NO_3), named as hydrated nitrous acid and tricoordinated nitrogen compound.

Orthoperiodic acid (pentahydroxidooxidoiodine, H_5IO_6), from ortho- and periodic acid, named as dihydrated periodic acid and hexacoordinated iodine compound. A. k.a. **paraperiodic acid**.

Orthophosphoric acid (phosphoric acid, H_3PO_4), named as tetracoordinated phosphorus compound.

Orthosilicic acid ([2]silicic acid, tetrahydroxysilane, silanetetrol, H_4SiO_4), named as hydrated metasilicic acid and tetracoordinated silicon compound. A.k.a. **monosilicic acid.**

Orthosomycin (a class of antibiotics), named as containing orthoester functions, and from -mycin.

Orthotelluric acid [hexahydroxidotellurium, $Te(OH)_6$], named as hexacoordinated tellurium compound.

Orthovanadic acid (vanadic acid, trihydroxidooxidovanadium, H_3VO_4), named as tetracoordinated vanadium compound.

Oxalacetic acid (oxobutanedioic acid), named as condensation product of oxalic and acetic acid.

Oxazole (1,3-oxazole) and **isoxazole** (1,2-oxazole), by elision of thia-oxa-ole; the decision which of the isomers should receive the 'iso' label was purely arbitrary.

Oxenin [(8*E*)-10-hydroxy-11,12-didehydro-7,10-dihydroretinol], a drastic contraction of the systematic name. A.k.a. **oxenine.**

Oxonic acid (4,6-dioxo-1,4,5,6-tetrahydro-1,3,5-triazine-2-carboxylic acid), a drastic contraction of the systematic name.

Oxymorphone (3,14-dihydroxy-17-methyl-4,5α-epoxymorphinan-6-one), a contraction of the systematic name.

Palmidrol [*N*-(2-hydroxyethylhexadecanamide)], a contraction of palmitic acid, amide, hydroxy-, and 'ol'.

Pantetheine {(2*R*)-2,4-dihydroxy-3,3-dimethyl-*N*-[3-oxo-3-[(2-sulfanylethyl)amino] propyl]butanamide}, from pantothenic acid, patterned after cysteine.

Pantolactone [3-hydroxy-4,4-dimethyldihydrofuran-2(3*H*)-one], from pantoic acid and lactone.

Paraben (4-hydroxybenzoic acid), a contraction of *p*-hydroxybenzoic acid. Also used as a class name.

Paraldehyde (2,4,6-trimethyl-1,3,5-trioxane), named as trimeric isomer of acetaldehyde.

Paraformaldehyde (polyoxymethylene), named as polymeric isomer of formaldehyde.

Paramolybdate (heptamolybdate, $Mo_7O_{24}{}^{6-}$), named as oligomer of molybdate.

Paraperiodic acid (pentahydroxidooxidoiodine, H_5IO_6), from para- and periodic acid, named as dihydrated metaperiodic acid and hexacoordinated iodine compound. A.k.a. **orthoperiodic acid.**

Parasorbic acid (6-methyl-5,6-dihydro-2*H*-pyran-2-one), named as isomer of sorbic acid. Patterned after ascorbic acid.

Paratungstate (dodecatungstate, $H_2W_{12}P_{42}{}^{10-}$), named as oligomer of tungstate.

Penaldic acid (*N*-formyl-3-oxoalanine), from penicillin and aldehyde.

D-**Penicillamine** (3-sulfanyl-D-valine), from penicillin and amine.

Penicillanic acid {(2*S*,5*R*)-3,3-dimethyl-7-oxo-4-thia-1-azabicyclo[3.2.0]heptane-2-carboxylic acid}, from penicillin, 'ane', and acid.

Penicillenic acid {*N*-[(5-oxooxazol-4(5*H*)-ylidene)methyl]-3-sulfanylvaline}, from penicillin, 'ene', and acid.

Penicillic acid (3-methoxy-5-methyl-4-oxohexa-2,5-dienoic acid), from penicillin and acid.

Pentaerythritol [2,2-bis(hydroxymethyl)propan-1,3-diol], named as five-carbon compound with features similar to erythritol.

Pentagastrin, from pentapeptide and gastrin, i.e. named as pentapeptide with gastrin-like activity.

Pentahomoserine (5-hydroxynorvaline), from penta- and homoserine with penta- referring to the locant in the systematic name.

1-Pentol (3-methylpent-2-en-4-yn-1-ol), a contraction of the systematic name.

Perxenic acid (tetrahydroxidodioxidoxenon, H_4XeO_6), from per- and xenon, named as hexacoordinated xenon compound.

Phenacetin [*N*-(4-ethoxyphenyl)acetamide), from phenyl- and acet(o)-.

Phenacyl- (2-oxo-2-phenylethyl-), an anagrammatic contraction of acetophenone (1-phenylethan-1-one), and 'yl'.

Phenetidine (ethoxyaniline), from phenetole and 'idine'.

Phenetole (ethoxybenzene), by contraction of phene (benzene), ethyl, and 'ole'.

Phenethyl- (2-phenylethyl-), a contraction of the systematic name.

Phenobarbital [phenylethylbarbituric acid, 5-ethyl-5-phenylpyrimidine-2,4,6 (1*H*,3*H*,5*H*)-trione], from phen(o)- and barbituric acid. A.k.a. **phenobarbitone** and **luminal**, with Latin *lumen, lumin-*, light, irregularly used in the sense of phen(o)-.

Phentermine [*N*-(2-methylpropan-2-yl)aniline], a contraction of phenyl-*tert*-butylamine.

Phenytoin (5,5-diphenylimidazolidin-2,4-dione), a contraction of diphenylhydantoin.

Phostamic acid (2-hydroxy-1,2-azaphospholidine 2-oxide), from phosphorus and lactam.

Phostonic acid (2-hydroxy-1,2-oxaphospholidine 2-oxide), from phosphorus and lactone.

Phosvitin (a phosphoprotein), from phosphorus and Latin *vitellus*, egg yolk.

Phthalane (1,3-dihydro-2-benzofuran), named as reduction product of phthalic anhydride.

Phthalic acid (benzene-1,2-dicarboxylic acid), from naphthalene, referring to its formation by oxidation of naphthalene.

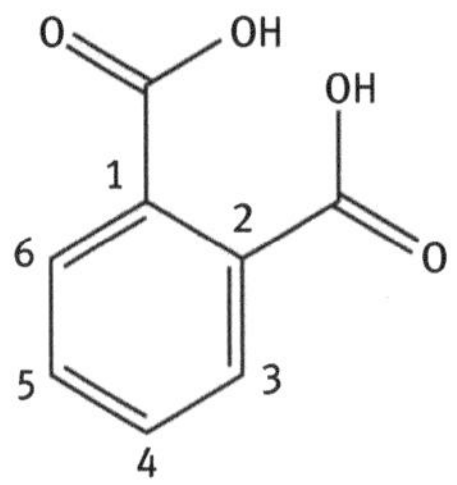

phthalic acid

Phthalide [2-benzofuran-1(3*H*)-one], from phthalic acid and 'olide', named as phthalic acid derived lactone.

Picloram (2-amino-3,5,6-trichloropyridine-2-carboxylic acid), an anagrammatic contraction of 4-amino-3,5,6-trichloropicolinic acid.

Picolinic acid (pyridine-2-carboxylic acid), named as oxidation product of α-picoline (2-methylpyridine).

Picramide (2,4,6-trinitroaniline), named as amide of picric acid.

Picrolonic acid [5-methyl-4-nitro-2-(4-nitrophenyl)-2,4-dihydro-3*H*-pyrazol-3-one], from picr(o)- and pyrazolone.

Pinacolin (3,3-dimethylbutan-2-one), a variation of pinacol, referring to the fact that pinacolin is a rearrangement product of pinacol. A.k.a. **pinacolone.**

Pipecolic acid (piperidine-2-carboxylic acid), from pipecoline.

2-Pipecoline (2-methylpiperidine), a contraction of piperidine and picoline.

Pipsyl- (4-iodobenzenesulfonyl-), a contraction of *p*-iodobenzenesulfonyl-.

Pivalic acid (2,2-dimethylpropanoic acid), with contraction, from pinacolin, referring to the fact that this acid can be obtained by oxidation of pinacolin, and from valeric acid, to show that pivalic acid is an isomer of valeric acid.

Pixyl- (9-phenyl-9*H*-xanthen-9-yl-), a contraction of the systematic name.

Prasterone (3β-hydroxyandrost-5-en-17-one), a contraction of dehydroepiandrosterone. A.k.a. **DHEA.**

Prednisolone (11β,17,21-trihydroxypregna-1,4-diene-3,20-dione), an anagrammatic contraction of the systematic name.

Prednisone (17,21-dihydroxypregna-1,4-diene-3,11,20-trione), an anagrammatic contraction of the systematic name.

Prehnitene (1,2,3,4-tetramethylbenzene), from prehnitic acid, q.v.

Prenyl- (3-methylbut-2-en-1-yl-), back formation from isoprene.

Propiophenone (1-phenylpropan-1-one), from phene (benzene). Cf. acetophenone.

Protopine {7-methyl-6,8,9,16-tetrahydrobis[1,3]benzodioxolo[4,5-*c*:5′,6′-*g*]azecin-15 (7*H*)-one}, from prot(o)- and opium.

Pseudoephedrine [(1*S*,2*S*)-2-(methylamino)1-phenylpropan-1-ol], named as isomer of ephedrine, q.v.

Pseudouridine 5-(β-D-ribofuranosyl)uracil], named as isomer of uridine.

Psicofuranine [9-(β-D-psicofuranosyl)-9*H*-purin-6-amine), from psicofuranose.

Pteroic acid {4-[[(2-amino-4-oxo-1,4-dihydropteridin-6-yl)methyl]amino]benzoic acid}, from pteridine.

Pulvinic acid [3-hydroxy-5-oxo-4-phenylfuran-2(5*H*)-ylidene)(phenyl)acetic acid], an anagram of vulpinic acid. A.k.a. **pulvic acid.**

Pyrindine (1*H*-cyclopenta[b]pyridine), from pyridine and indene.

Pyrogallol (benzene-1,2,3-triol), from ‘pyro’ and gallic acid, referring to its formation by decarboxylation of gallic acid.

Pyromellitic acid (benzene-1,2,4,5-tetracarboxylic acid), from pyr(o)- and mellitic acid, referring to its formation by pyrolysis of mellitic acid.

Pyromucic acid (furan-2-carboxylic acid), named as formed by heating of mucic acid.

Pyrophosphoric acid (diphosphoric acid, $H_4P_2O_7$), named as anhydride obtained by heating of phosphoric acid.

Pyrosulfuric acid (disulfuric acid, $H_2S_2O_7$), named as anhydride obtained by heating of sulfuric acid.

Pyruvic acid (2-oxopropanoic acid), from pyr(o)- and uvic acid, referring to its formation by pyrolysis of, inter alia, tartaric acid.

Quinaldine (2-methylquinoline), from quinoline and aldehyde, named as a quinolin derivative prepared from aniline and acetaldehyde.

Quinhydrone [quinone—hydroquinone (1/1), cyclohexa-2,5-diene-1,4-dione—benzene-1,4-diol (1/1)], a contraction of quinone and hydroquinone.

Quinitol (cyclohexane-1,4-diol), with contraction, from hydroquinone and 'itol', named as hydrogenated hydroquinone (benzene-1,4-diol).

Radio-iodine (^{131}I), from radio- and iodine.

Radio-phosphorus (^{32}P), from radio- and phosphorus.

Radio-sodium (^{24}Na), from radio- and sodium.

Resacetophenone [1-(2,4-dihydroxyphenyl)ethan-1-one], a contraction of resorcine and acetophenone.

Resveratrol {5-[(*E*)-2-(4-hydroxyphenyl)ethen-1-yl]benzene-1,3-diol}, a contraction of resorcine and veratrol.

Rhodanine (2-sulfanylidene-1,3-thiazolidin-4-one), named after ammonium rhodanide (ammonium thiocyanate), from which it can be prepared by reaction with chloroacetic acid. A.k.a. **rhodanic acid** and **rhodaninic acid.**

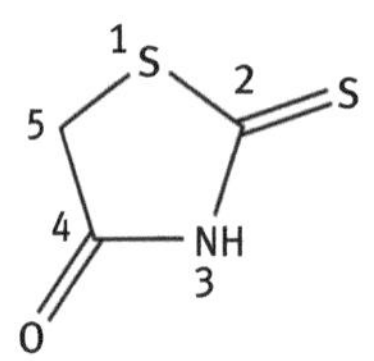

rhodanine

α-Ribazole [5,6-dimethyl-1-α-D-ribofuranosyl-1*H*-benzimidazole], a contraction of ribose and benzimidazole.

Saccharopine {*N*-[(5*S*)-5-amino-5-carboxypentan-1-yl]-L-glutamic acid}, after the fungal genus *Saccharomyces*, referring to this compound's role as a lysine precursor in yeast metabolism, and from octopine, referring to this compound's structure, two amino acids (here glutamine and lysine) with a shared amino group, as first observed in octopine. The genus name is from sacchar(o)- and -myces.

Sebacil (cyclodecane-1,2-dione), from sebacic acid, patterned after benzil.

Sebacoin (2-hydroxycyclodecan-1-one), from sebacic acid, patterned after benzoin.

Semibullvalene [(4a*s*,4b*s*)-2a,2b,4a,4b-tetrahydrocyclopropa[*cd*]pentalene], from semi- and bullvalene.

Semicarbazide (hydrazinecarboxamide), with contraction, from 'semi', carbamide, and hydrazide.

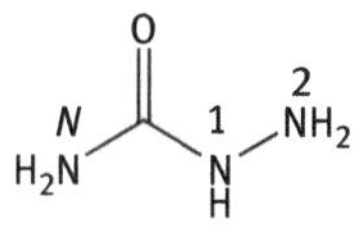

semicarbazide

Semidine (*N*-phenylbenzene-1,4-diamine), from semi- and benzidine.

Sialon (Al,N,O,Si), a contraction of silicon aluminum oxynitride.

Sorbide (a class name), named as anhydride of sorbitol.

Sphingosine [(2*S*,3*R*,4*E*)-2-aminooctadec-4-ene-1,3-diol], from sphingolipid, q.v.

Succimer (2,3-disulfanylbutanedioic acid), an anagrammatic contraction of dimercaptosuccinic acid.

Sulfamic acid (sulfuramidic acid), from sulfuric acid and amide.

Sulfamide (sulfuric diamide), from sulfuric acid and amide.

Sulfanilic acid (4-aminobenzenesulfonic acid), a contraction of *p*-anilinesulfonic acid.

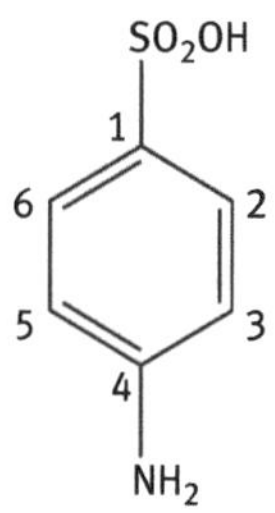

sulfanilic acid

Sulfanuric chloride (1,3,5-trichloro-1λ^6,3λ^6,5λ^6-1,3,5,2,4,6-trithiatriazine 1,3,5-trioxide), named as sulfur analog of cyanuric chloride.

Sulfene [(dioxido-λ^6-sulfanylidene)methane], also used as a class name, named as sulfur analog of ketene.

Sulfine [(oxido-λ^4-sulfanylidene)methane], also used as a class name, named as λ^4 analog of sulfene.

Sulfolane (thiolane 1,1-dioxide), from sulfone and thiolane.

Sulforaphane [1-isothiocyanato-4-(methylsulfinyl)butane], from sulforaphene.

Sulfuryl chloride (sulfuryl dichloride, dichloridodioxidosulfur), from sulfuric acid.

Tacrine (1,2,3,4-tetrahydroacridin-9-amine), a contraction of the systematic name.

Tartronic acid (2-hydroxypropanedioic acid), a contraction of tartaric acid and malonic acid.

Terephthalic acid (benzene-1,4-dicarboxylic acid), from phthalic acid and terpene, named as isomer of phthalic acid isolated from turpentine.

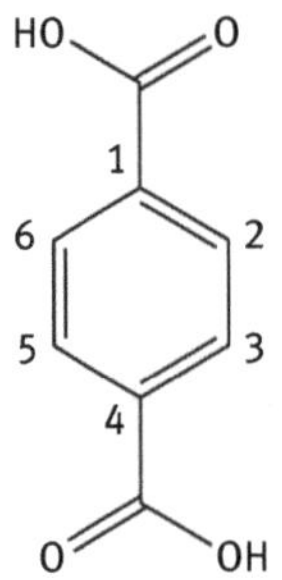

terephthalic acid

Terpenylic acid [(2,2-dimethyl-5-oxotetrahydrofuran-3-yl)acetic acid], from terpene. A.k.a. **terpenolic acid**.

Terprenin {2^2,2^5-dimethoxy-3^4-[(3-methylbut-2-en-1-yl)oxy]-[1^1,2^1:2^4,3^1-terphenyl]-1^4,2^3,3^4-triol}, from terphenyl and prenyl-.

Tetraglyme (2,5,8,11,14-pentaoxapentadecane), from tetra- and glyme.

Tetralin (1,2,3,4-tetrahydronaphthalene), a contraction of German *Tetrahydronaphthalin*, tetrahydronaphthalene.

Tetralol (1,2,3,4-tetrahydronaphthalen-2-ol), a contraction of the systematic name.

Tetrolic acid (but-2-ynoic acid), named as four-carbon acid with a carbon-carbon triple bond, patterned after stearolic acid.

Tetroquinone (2,3,5,6-tetrahydroxycyclohexa-2,5-diene-1,4-dione), a contraction of 2,3,5,6-tetrahydroxy-*p*-quinone.

Thalidomide [2-(2,6-dioxopiperidin-3-yl)-1*H*-isoindole-1,3(2*H*)-dione], a contraction of 3-(phthalimido)glutarimide.

Thebenidine (benzo[*lmn*]phenanthridine), a name referring to this compound's formation upon zinc dust destillation of thebenine, q.v.

2-Thenoic acid (thiophene-2-carboxylic acid), from thiophene, patterned after benzoic acid.

Thenyl- [(thiophen-2-yl)methyl-], from thiophene, patterned after benzyl-.

Thetin [(dimethylsulfonio)acetate], from thi(o)- and betaine. A.k.a. **sulfobetaine**.

Thexyl- (2,3-dimethylbutan-2-yl-), a contraction of *tert*-hexyl-.

Thialdine (2,4,6-trimethyl-1,3,5-dithiazinane), from 'thio' and aldehyde, i.e. named as a dithioacetal of acetaldehyde.

Thianaphthene (1-benzothiophene), from thiophene and naphthalene. A.k.a. **thiophthene**.

Thiarubrine A [3-(hex-5-en-1,3-diyn-1-yl)-6-(prop-1-yn-1-yl)-1,2-dithiine], from thi(a)- and Latin *ruber*, rød.

Thiazole (1,3-thiazole) and **isothiazole** (1,2-thiazole), by elision of thia-aza-ole; the decision which of the two isomers should receive the 'iso' label was purely arbitrary.

Thienyl (thiophenyl), from thiophene, patterned after phenyl.

Thiochrome {2-(2,7-dimethyl-5*H*-pyrimido[4,5-*d*][1,3]thiazolo[3,2-*a*]pyrimidin-8-yl)ethanol}, from thiamine and -chrome.

Thioindigo [(2*E*)-2-(3-oxo-1-benzothiophen-2(3*H*)-ylidene)-1-benzothiophen-3(2*H*)-one], from indigo, referring to the replacement of indigo's two imino groups with sulfur.

Thiolutin {*N*-[4-methyl-5-oxo-4,5-dihydro[1,2]dithiolo[4,3-*b*]pyrrol-6-yl]acetamide}, from thi(o)- and Latin *luteus*, yellow, referring to this compound's sulfur content and color.

Thionalide [*N*-(2-naphthyl)-2-sulfanylacetamide], a contraction of the obsolete name thioglycolic β-aminonaphthalide.

Thionyl chloride (sulfurous dichloride, dichloridooxidosulfur), from thi(o)- and 'one'.

Thiophenine (thiophenamine), a contraction of the systematic name, patterned after aniline.

Thiosinamine [*N*-allylthiourea, *N*-(prop-2-en-1-yl)thiourea], from thi(o)- and Latin *sinapis*, mustard, referring to its preparation from allyl mustard oil. This compound is not an amine as the name suggests.

Thiotolene (methylthiophene), from thi(o)- and toluene.

Thiotropocin (8-sulfanylcyclopenta[*c*][1,2]oxathiole-3,4-dione), from 'thio', tropane, and 'ocine'.

Thioxene (dimethylthiophene), from thi(o)- and xylene.

Thiuram disulfide (2-dithioperoxy-1,3-dithiodicarbonic diamide), thiuram probably by contraction of thiourea and amide.

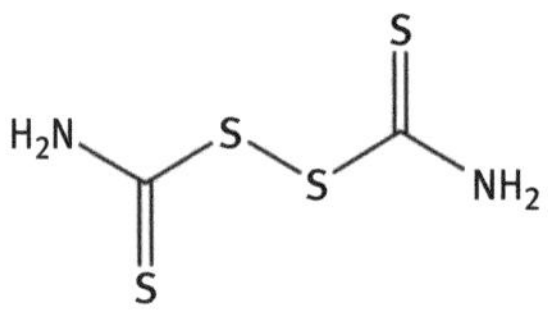

thiuram disulfide

Thiuram monosulfide (1,2,3-trithiodicarbonic diamide), with irregular use of 'mono', thiuram probably by contraction of thiourea and amide.

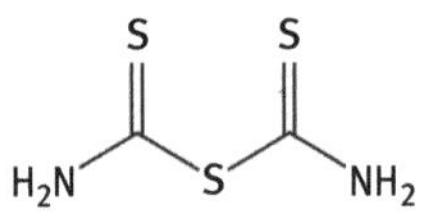

thiuram monosulfide

Thymolphthalein {3,3-bis[4-hydroxy-2-methyl-5-(propan-2-yl)phenyl]-2-benzofuran-1(3*H*)-one}, from thymol and phenolphthalein.

Tolane (1,1′-ethyne-1,2-diyldibenzene), derived from toluene, but without indication of the presence of two toluene units, and with the irregular ending 'ane'. A.k.a. **tolan.**

Tolidine {3,3′-dimethyl-[1,1′-biphenyl]-4,4′-diamine}, from toluene, patterned after benzidine.

Tolyl- (methylphenyl-), from toluene. A.k.a. **toluyl.**

Toluoyl- (methylbenzoyl-), from toluic acid.

Tosyl- (4-methylbenzenesulfonyl-), a contraction of *p*-toluenesulfonyl-.

Triacetin (tri-*O*-acetylglycerol, propane-1,2,3-triyl triacetate), with contraction, from German *Triacetylglycerin*, triacetylglycerol. Cf. stearin.

Tricarballylic acid (propane-1,2,3-tricarboxylic acid), named as tricarboxylic acid with the same carbon skeleton as the allyl radical.

Tricine (*N*-[tris(hydroxymethyl)methyl]glycine), a contraction of the systematic name.

Trientine [N^1,N^2-bis(2-aminoethyl)ethane-1,2-diamine], a contraction of triethylenetetramine.

Triflic acid (trifluoromethanesulfonic acid), a contraction of the systematic name.

Triflinic acid (trifluoromethanesulfinic acid), as triflic acid.

Triflyl- (trifluoromethanesulfonyl-), as triflic acid.

Triglyme (2,5,8,11-tetraoxadodecane), from tri- and glyme, q.v.

Trimellitic acid (benzene-1,2,4-tricarboxylic acid), named as tribasic acid obtained from mellitic acid.

Trimesic acid (benzene-1,3,5-tricarboxylic acid), named as tribasic acid obtained from mesitylene.

Trinactin (a macrotetrolide), a contraction of trihomononactin.

Triprenyl- [(2*E*,6*E*)-3,7,11-trimethyldodeca-2,6,10-trien-1-yl-}, from prenyl. A.k.a. **farnesyl-**, from farnesol, q.v.

Tris [*N*,*N*,*N*-tris(hydroxymethyl)methanamine], a contraction of the systematic name. A.k.a. **trometamol** and **THAM**.

Trityl- (triphenylmethyl-), a contraction of the systematic name.

Triuret (tricarbonodiimidic diamide), from urea, referring to its being composed of three fused urea units.

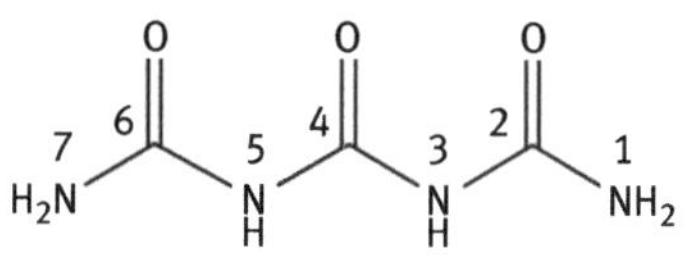

triuret

Tropic acid (3-hydroxy-2-phenylpropanoic acid), from atropine, from which it can be obtained by hydrolysis.

Tropilidene (cyclohepta-1,3,5-triene), ultimately from atropine.

Tropine {(tropan-3-ol, (3-*endo*)-8-methyl-8-azabicyclo[3,2,1]octan-3-ol}, ultimately from atropine.

Tropyl- (cyclohepta-2,4,6-trien-1-yl-), with contraction, from tropilidene and 'yl'.

Tropylium (cyclohepta-2,4,6-trien-1-ium), from tropyl-.

Trotyl (1-methyl-2,4,6-trinitrobenzene), a contraction of 2,4,6-trinitrotoluene. The suffix 'yl' is unsystematic. A.k.a. **TNT**.

Troxerutin {3′,4′,7-tris(2-hydroxyethyl)rutin}, a contraction of tris(hydroxyethyl) rutin.

Tryptazan {3-(2*H*-indazol-3-yl)alanine], from tryptophan and 'aza'.

Tryptophol [2-(1*H*-indol-3-yl)ethan-1-ol], from tryptophan and 'ol'.

Tuaminoheptane (heptan-2-amine), from 2-aminoheptane.

Uramil [5-aminopyrimidine-2,4,6(1*H*,3*H*,5*H*)-trione], from urea and amide.

Urazole (1,2,4-triazolidine-2,5-dione), from urea, 'aza', and 'ole'.

Urethane (ethyl carbamate), from urea and ethane.

Urochloralic acid [6-*O*-(2,2,2-trichloroethyl) β-D-glucopyranosiduronic acid), from uronic acid and chloral (trichloroethanal), although not a derivative of chloral, but of 2,2,2-trichloroethanol.

Uvitic acid (5-methylbenzene-1,3-dicarboxylic acid), from pyruvic acid, from which it can be made.

Vanilmandelic acid [hydroxy(4-hydroxy-3-methoxyphenyl)acetic acid], from vanillin and mandelic acid.

Viquidil {3-[(3*R*,4*R*)-3-ethenylpiperidin-4-yl]-1-(6-methoxyquinolin-4-yl)propan-1-one}, a contraction of vinyl- and quinoline.

Xanthydrol (9*H*-xanthen-9-ol), a contraction of xanthene and -hydrol, i.e. named as secondary alcohol with a xanthene skeleton.

2.8 Ambiguous names

No, traumatic acid will not traumatize the unsuspecting consumer of legumes. However, it does promote the healing of wounds sustained by a plant by stimulating cell division. Traumatic is from Greek *trauma*, wound, from *titroskein*, to wound.

And no, the closest relation periodic acid has with periodicity is that iodine does indeed occupy a place in the Periodic Table. Correspondingly the adjective part of this name is pronounced [**pur**-ahy-**od**-ik] and not [peer-ee-**od**-ik].

And again no, moronic acid is not a euphemism for idiotic acid, but formed upon oxidation of morolic acid, a secondary metabolite of the South American mora tree (*Mora excelsa*). And yes, a Google search for idiotic acid does generate a fair number of hits, though none concerning a well-defined chemical compound.

The homophones fluorine (F_2) and fluorene ($C_{13}H_{10}$) have received their names for different reasons. The element fluorine was named after the mineral fluorite (CaF_2) whose name in turn refers to its use as flux in iron smelting. The name fluorene, on the other hand, refers to its violet fluorescence. Both flux and fluorescence originate ultimately from Latin *fluere*, to flow.

The element chlorine (Cl_2) was named after its green color while chlorin (2,3-dihydroporphyrin) has its name as the core structure of the green pigment chlorophyll.

Confusion is the name of the game when homophones are created. The alkaloid serpentine [16-(methoxycarbonyl)-19α-methyl-3,4,5,6,16,17-hexadehydro-18-oxayohimban-4-ium-1-ide] was first found in Indian snakeroot (*Rauwolfia serpentina*), the specific epithet stemming from Latin *serpens, serpent-*, snake, literally the creeping, from *serpere*, to creep. The same name was given to the mineral serpentine [$Mg_3Si_2O_5(OH)_4$] with reference to its serpent-like colors and patterns or to the ancient belief that this mineral protects its bearer from snake bites.

The French bean (*Phaseolus vulgaris*) contains, among others, three, chemically unrelated, secondary metabolites all of which have been named phaseolin.

The inorganic sodium metagallate (gallium sodium dioxide), has an organic namesake, i.e. sodium metagallate (sodium 3,4-dihydroxy-5-[(3,4,5-trihydroxybenzoyl)oxy]benzoate).

The chemical warfare agent lewisite is unrelated to the mineral lewisite nor do the two eponyms refer to the same person. And no, warfarin is not a chemical warfare agent, but a rodenticide, cf. Section 2.3. Dioptase is not an enzyme, nor is glutathione a thioketone. The former is a mineral and the latter a tripeptide without any thione group.

Deuteroporphyrin [3,3′-(3,7,12,17-tetramethylprophyrin-2,18-diyl)dipropanoic acid], from Greek *deuteros*, the second, is not a deuterated porphyrin, but named as 'second-tier' bacterial degradation product of protoporphyrin.

And again no, osmium and osmosis are not etymological congeners. The former name is ultimately derived from Greek *ozein*, to smell, and the latter from Greek *othein*, to push.

Copperas is mainly an iron(II) salt though it may also contain copper.

2.9 Prefixes, suffixes, and adjectives commonly used in chemical names

2.9.1 Examples of general prefixes and adjectives

'allo', as in allosteric, from Greek *allos*, other, different.

'anthr(a)', as in anthracene, from Greek *anthrax, anthrak-*, coal.

'apo', as in apoenzyme, from Greek *apo*, off, away from.

'ep(i)', as in epoxycyclohexane, epithiocyclohexane (1,2-sulfanocyclohxane), and epiminocyclohexane (1,2-azanocyclohexane), from Greek *epi*, beside, on, upon, after.

'geminal' (*gem*-), as in *gem*-diol [$R^1R^2C(OH)_2$], from Latin *gemini*, twins, denoting 1,1-disubstitution.

'hemi', as in hemiacetal, from Greek *hemisys*, half.

'holo', as in holoenzyme, fromGreek *holos*, complete.

'homo', as in homoserine, from Greek *homos*, same.

'hypo', as in hypochlorous acid, from Greek *hypo*, under, below, from below.

$^{1'}$**in'** (ig-, il-, im-, ir-), as in infinite, from Latin *in*-, non-.

$^{2'}$**in'** (il-, im-, ir-), as in insertion, from Latin *in*, into, on.

'iso', as in isobutane (2-methylpropane), from Greek, *isos*, equal, denoting isomerism.

$^{1'}$**meso'**, as in mesoperiodic acid ($H_6I_2O_{10}$), from Greek *mesos*, middle. In this case mesoperiodic acid has an intermediate position, with regard to hydration, between orthoperiodic acid (H_5IO_6) and metaperiodic acid (HIO_4).

$^{2'}$**meso'**, as in mesotartaric acid, indicating achirality in spite of the presence of stereogenic centers.

'meta' (*m*-), as in metaphosphoric acid [$(HPO_3)_n$], from Greek *meta*, between, with, after, often denoting 1,3-substitution of benzene, as in *m*-xylene (1,3-dimethylbenzene), but also used in other ways.

'mono', as in carbon monoxide, from Greek *monos*, alone.

'neo', as in neopentane (2,2-dimethylpropane), from Greek, *neos*, new, often denoting isomerism.

'ortho' (*o*-), as in orthosilicic acid (H_4SiO_4), from Greek *orthos*, straight, right, true, often denoting 1,2-substitution of benzene, such as in *o*-xylene (1,2-dimethylbenzene), but also used in other ways.

'para' (*p*-), as in parasorbic acid (6-methyl-5,6-dihydro-2*H*-pyran-2-one), from Greek *para*, beside, similar, near, beyond, irregular, a modification of, often denoting 1,4-disubstitution of benzene, such as in *p*-xylene (1,4-dimethylbenzene), but also used in other ways. In the case of parasorbic acid isomerism with sorbic acid [(2*E*,4*E*)-hexa-2,4-dienoic acid] is denoted.

'per', as in periodic acid (hydroxidotrioxidoiodine, HIO_4) and perfluorooctanoic acid (2,2,3,3,4,4,5,5,6,6,7,7,8,8,8-pentadecafluorooctanoic acid), from Latin *per*, through, by means of, thoroughly. In the first example 'per' is used to show that the central atom is present in its highest possible valence, in the second example exhaustive substitution is indicated, obviating the inclusion of a long series of locants in the name.

'peri', as in *peri*-substitution, from Greek *peri*, around, about, from *peran*, to pass through, often referring to 1,8-disubstitution of naphthalene and analogous aromatic ring systems.

'proto', from Greek *protos*, the first.

'pseudo', from Greek *pseudein*, to lie, to cheat, to falsify.

'pyr(o)', as in pyrosulfuric acid (disulfuric acid) or pyruvic acid (2-oxopropanoic acid), referring to the formation of a compound by strong heating of a precursor.

'semi', as on semiquinone, from Latin *semis, semi-*, half.

'sesqui', as in sesquiterpene, from Latin *sesqui*, one and a half, a contracted form of *semisque*, one and a half.

'sester', as in sesterterpene, from Latin *sestertius*, two and a half, also a coin worth two and a half as.

'sub', as in carbon suboxide (propa-1,2-diene-1,3-dione), from Latin *sub*, under.

'super', as in sodium superoxide [sodium dioxide(•1−)], from Latin *super*, above, over.

'supra', as in suprafacial, from Latin *supra*, above over.

'vicinal' (*vic*-), as in *vic*-diol, from Latin *vicinalis*, neighborly, denoting 1,2-substitution.

2.9.2 Examples of specific prefixes, suffixes, and adjectives

'ace', as in acenaphthylene, from acetic acid.

'aceto', as in acetoacetic acid, from acetic acid

'anhydr(o)', as in anhydroglucose, from anhydrous.

'antara', as in antarafacial, from Sanskrit *antara*, the other.

'anthra', as in anthraazepine, from anthracene.

'anthrene', as in phenanthrene, from anthracene.

'aphene', as in tetraphene, from phenanthrene.

'az(a)', as in triazole, corresponding to replacement of carbon ny nitrogen, cf. az(o)-.

'azi', as in azibenzil (1-diazo-1,2-diphenylethan-1-one), cf. az(o)-.

'az(o)', as in azobenzene (diphenyldiazene), from French *azote*, nitrogen. *Azote* is from a(n)- and Greek *zoein*, to live, referring to nitrogen's inability to support life.

$^{1\prime}$**benz(o)'**, as in 1-benzofuran, from benzene.

$^{2\prime}$**benz(o)'**, as in benzophenone, from benzoic acid.

'hapto', as in hapticity, from Greek *haptein*, to fasten.

'idine', as in piperidine, from pyridine.

'il(e)', as in benzil, ultimately from Latin *-ilis*, suffix denoting function or relationship.

'ket(o)', as in ketose, from ketone.

'naphth(o)', as in naphthacene, from naphthalene.

'olide', as in macrolide, from glycolide.

'phen(o)', as in phenothiazine, from phene (benzene).

References

[1] https://en.wikipedia.org/wiki/Stigler%27s_law_of_eponymy (retrieved August 18, 2018).
[2] https://en.wikipedia.org/wiki/List_of_examples_of_Stigler%27s_law (retrieved August 18, 2018).
[3] https://english.stackexchange.com/questions/238004/where-on-earth-is-penguin-from (retrieved November 20, 2018).

3 Rudimentary systematic nomenclature

The desire to name new compounds by structure rather than by source or properties developed as soon as structural relations between different compounds were recognized. However, some early attempts were based on incomplete or erroneous conclusions.

Acetic acid (ethanoic acid), from Latin *acetum*, vinegar, from *acer*, sharp.

Acetaldehyde (ethanal), from acetic acid.

Acetoacetic acid (3-oxobutanoic acid), named as acetyl substituted acetic acid.

Acetoin (3-hydroxybutan-2-one), patterned after benzoin.

Acetol (1-hydroxypropan-2-one), with contraction, from and alcohol.

Acetone (propan-2-one), named as a condensation product of acetic acid, obtained by heating a metal acetate, and provided with the arbitrary suffix 'one'.

Acetonitrile (ethanenitrile), from acetic acid and nitrile.

^{1}**Acetyl-** (ethanoyl-), named as the acyl part of acetic acid.

^{2}**Acetyl-** (vinyl-, ethenyl-), an obsolete irregular name for the vinyl radical.

Acetylene (ethyne), named as the alkyne formed by loss of one hydrogen atom from the vinyl radical, cf. ^{2}acetyl-.

Acrindoline (7*H*-indolo[3,2-*c*]acridine), from acridine and indole.

Adiponitrile (hexanedinitrile), from adipic acid.

Alcohol, a class name, cf. Section 3.5.

Aldehyde, a class name, contraction of New Latin *alcohol dehydrogenatum*, referring to the formation of aldehydes by dehydrogenation of primary alcohols.

Aliphatic compound, from Greek *aleiphar, aleiphat-*, fat, from *aleiphein*, to smear, referring to the occurrence of these compounds in fats.

Alkane, back formation from alkyl.

Alkene, back formation from alkyl.

Alkyl-, from alkali, referring to a perceived analogy between alkyl radicals and alkali metals, such as the formation of methyl chloride (chloromethane) vs. sodium chloride (NaCl).

Alkyne, back formation from alkyl.

https://doi.org/10.1515/9783110612714-003

Allene (propa-1,2-diene), a contraction of allylene.

Allyl- (prop-2-en-1-yl-), referring to the first discovery of allyl substituted compounds in the genus *Allium* (garlic), from Latin *allium*, garlic.

Allylene (now obsolete, propa-1,2-diene), named as the alkene formed by the loss of one hydrogen atom from the allyl radical.

Amine, a class name, formed by contraction of ammonia and with the suffix -ine. The different spelling of amine and ammine is intentional to avoid confusion. Amine is also the name of ammonia in the role of parent compound. Thus, there is no single compound called amine.

Ammine (ammonia as complex ligand), formed by contraction of ammonia and the suffix -ine. The different spelling of amine and ammine is intentional to avoid confusion.

Ammonia (nitrogen hydride, nitrogen trihydride, azane), from the adjective ammoniac, from Latin *sal ammoniacum*, salt of Ammon, named after the Ancient Egyptian god Amon, literally the hidden one. There has always been consensus that ammonia should be called by its common name and that the systematic synonyms should only serve for the construction of composite names. Cf. amine.

Amyl- (pentyl-, pentan-1-yl-), a name derived from the first part of the radicofunctional name amyl alcohol.

Amyl alcohol (pentan-1-ol), named as a product of the fermentation of starch, from Greek *amylon*, starch, literally not ground; starch was obtained from unground grain.

Amylene (pent-1-ene), named as the alkene formed by the loss of one hydrogen atom from the amyl radical.

Aniline, from Arabic *al-nil*, indigo plant, ultimately from Sanskrit *nila*, dark blue, referring to the fact that aniline (which is colorless) was first obtained by degradation of the blue dye indigo.

Aqua (water, hydrogen oxide, dihydrogen oxide, oxidane), from Latin *aqua*, water, only used for water in the role as complex ligand.

Aromatic compound, ultimately from Greek *aromatikos*, fragrant. Originally used of both aromatic compounds in the modern sense and of odoriferous alkenes such as terpenes, cf. Sections 3.4. and 3.5.

Benzal chloride [(dichloromethyl)benzene], from benzaldehyde.

Benzaldehyde (benzaldehyde, benzenecarbaldehyde), from benzoic acid.

Benzoic acid (benzoic acid, benzenecarboxylic acid), from German *Benzoesäure*, from *Benzoeharz*, gum benzoin, from French *benzoin*, an odoriferous resin of the tree

Styrax benzoin (gum benjamin tree, onycha, Sumatra benzoin tree), ultimately from Arabic *luban jawi*, Javanese incense. The Arabic name of the resin is a misnomer, since it did not come from Java, but from Sumatra.

Benzoin (2-hydroxy-1,2-diphenylethan-1-one), named as a major constituent of gum benzoin.

Benzol (benzene), earlier also called phene. from French *benzoin*, cf. benzoic acid.

Benzonitrile (benzenecarbonitrile), from benzoic acid and nitrile.

Benzotrichloride [(trichloromethyl)benzene], from benzoic acid.

Benzyl- (phenylmethyl-), derived from the first part of the radicofunctional name benzyl alcohol.

Benzyl alcohol (phenylmethanol), named as reduction product of benzaldehyde.

Butane (butane), back formation from butyl, with the isomer **isobutane** (2-methylpropane).

Butyl- (butan-1-yl-), derived from the first part of the radicofunctional name butyl alcohol.

Butyl alcohol (pentan-1-ol), ultimately from butyric acid, with the isomers ***sec*-butyl alcohol** (butan-2-ol), **isobutyl alcohol** (2-methylpropan-1-ol), and ***tert*-butyl alcohol** (2-methylpropan-2-ol).

[1]**Butylene** (butene), back formation from butyl.

[2]**Butylene-** (butane-1,4-diyl-), back formation from butyl.

Butyraldehyde (butanal), named as the aldehyde corresponding to butyric acid, with the isomer **isobutyraldehyde** (2-methylpropanal).

Butyric acid (butanoic acid), from Greek *butyron*, butter, from *bous*, cattle, and *tyros*, cheese, with the isomer **isobutyric acid** (2-methylpropanoic acid).

Ester, a class name, formed by contraction of obsolete German *Essigäther* (ethyl acetate, ethyl ethanoate). Earlier ether was used in a broader sense, i.e. as a class name for volatile liquids.

Ether, a class name, ultimately from Greek *aither*, ether, from *aithein*, to kindle, to blaze. In Greek mythology the sphere in which the gods lived, i.e. the upper atmosphere. In early science ether was a hypothetical, essentially massless or nearly massless, medium everywhere in space, but also used in a broader sense, i.e. as a class name for volatile liquids. Also used in the sense diethyl ether (1,1′-oxydiethane).

Ethyl- (ethyl-), from ether, after, the Greek mythological personification of the upper sky, space, and heaven, and Greek *hyle*, matter.

Ethyl alcohol (ethanol), named as an ethyl radical joined to a hydroxyl group.

[1]**Ethylene** (ethene), named as the alkene formed by the loss of one hydrogen atom from the ethyl radical.

[2]**Ethylene-** (ethane-1,2-diyl-), back formation from ethyl.

Formonitrile (methanenitrile, hydrogen cyanide), from formic acid and nitrile.

Glyceric acid (2,3-dihydroxypropanoic acid), from glycerin.

Glycerin (glycerol, propane-1,2,3-triol), from Greek *glykys, glyker-*, sweet, after its taste.

[1]**Glyceryl-** (2,3-dihydroxypropyl-), back formation from glycerin.

[2]**Glyceryl-** (propane-1,2,3-triyl), back formation from glycerin.

Glycidic acid [(oxiran-2-yl)acetic acid], from glycidol.

Glycidol [(oxiran-2-yl)methanol], from glycerin. A.k.a. **glycide.**

Glycol, a class name, from glycerin and 'ol'.

Hydrazine (diazane), from hydrogen and French *azote*, nitrogen, i.e. named as hydrogenated dinitrogen (diazyne).

Hydrazobenzene (1,2-diphenylhydrazine), from hydrazine and benzene.

Hydroxylamine (azanol). The irregular letter l probably inserted for euphonic reasons.

Ketone (ketone), a class name, from.

Malononitrile (propanedinitrile), from malonic acid and nitrile.

Methane (methane), back formation from methyl.

Methyl- (methyl-), back formation from methylene.

Methyl alcohol (methanol), named as a methyl radical joined to a hydroxyl group.

[1]**Methylene-** (methanediyl-), cf. Section 3.1.

[2]**Methylene-** (methylidene-), cf. Section 3.1.

Olefiant gas (ethene), from French *gaz oléfiant*, oil making gas. Cf. olefin.

Olefin (alkene), a contraction of olefiant gas, referring to the fact that heavy liquid products, i.e. 1,2-dihaloethanes, are formed by addition of halogens to ethene.

Perchloromethyl mercaptan [trichloromethanesulfenyl chloride, trichloro(chlorosulfanyl)methane], named as a derivative of methyl mercaptan (methanethiol) where all four hydrogen atoms have been replaced by chlorine. An obvious misnomer, since without a sulfanyl group it does not qualify as a mercaptan (thiol). This misnomer has occasionally been misinterpreted as describing the elusive **trichloromethanethiol** (CCl_3SH).

Phene (now obsolete, benzene), from Greek *phainein*, to show, referring to this compound's occurrence in illuminating gas.

Phenol (benzenol), named as hydroxy substituted benzene, cf. phene. A.k.a. **carbolic acid**.

Phenyl- (phenyl-), named as the radical formed by loss of a hydrogen atom from benzene, cf. phene.

Phenylene- (benzenediyl-), back formation from phenyl.

Phthalonitrile (benzene-1,2-dicarbonitrile), from phthalic acid and nitrile.

Propane (propane), ultimately from propionic acid.

Propionaldehyde (propanal), from propionic acid.

Propionic acid (propanoic acid), with contraction derived from Greek *proteon pion*, the first fat, referring to the fact that propionic acid is to be regarded as the first member of the homologous series of fatty acids judged by the soapy feel of its salts.

Propiononitrile (propanenitrile), from propionic acid and nitrile.

Propyl- (propan-1-yl-), derived from the first part of the radicofunctional name propyl alcohol.

Propyl alcohol (propan-1-ol), ultimately from propionic acid, with the isomer **isopropyl alcohol** (propan-2-ol).

[1]**Propylene** (propene), back formation from propyl.

[2]**Propylene-** (propane-1,3-diyl-), back formation from propyl.

Succinonitrile (butanedinitrile), from succinic acid (butanedioic acid) and nitrile.

Toluene (methylbenzene), earlier called toluol, named after its source, tolu balsam, obtained from the tropical tree *Myroxylon balsamum*, and named after the town of Santiago de Tolú, Colombia.

Toluidine (methylaniline), from toluene and 'idine', patterned after pyridine.

Vinyl- ([2]acetyl-, ethenyl-), from New Latin *spiritus vini*, wine spirit, i.e. ethanol. A name proposed by the German chemist Hermann Kolbe (1808–1884) to show this radical's relationship with the ethyl radical, i.e. containing two hydrogen atoms less.

Water (hydrogen oxide, dihydrogen oxide, oxidane). There has always been consensus that water should be called by its common name and that the systematic synonyms should only serve for the construction of composite names. Cf. aqua.

Xylene (dimethylbenzene), earlier called xylol, from Greek *xylon*, wood, referring to these compounds' occurrence in wood tar.

Xylenol (dimethylphenol), from xylene.

Xylidine (dimethylaniline), from xylene and 'idine, patterned after pyridine.

3.1 Methylene hydrate, a productive monstrosity

While the formation of methanol by pyrolysis of wood had already been used by ancient Egyptian embalmers in the preparation of embalming fluids it was only in 1834 that the two French chemists Jean-Baptiste Dumas (1800–1884) and Eugène-Melchior Peligot (1811–1890) prepared and examined pure methanol. At that time the atomic weights of the elements were still disputed and this led them to formulate the structure of this alcohol as 'methylene hydrate', in modern terms CH_3OH. Their creation of the name 'methylene' for this 'radical' can be traced to a linguistic mishap which has not always been fully recognized. The intention was to call this radical 'wood spirit'. As the spirit part quite logically Greek *methy*, wine, alcoholic liquid, was chosen. In the search for the Greek word for wood they found and used *hyle* which, however, means wood in the sense of woods, forest, while the proper Greek equivalent of wood (material) is *xylon* (which has been the source of other chemical names). Once the name methylene hydrate had been coined and adopted this linguistic monstrosity begot the names methyl alcohol (once its exact structure had become clear), methanol, methyl, and methane, all important constituents of modern systematic nomenclature [1]. It is interesting to note that methyl hydrate, a corruption of obsolete methylene hydrate, is still in use as a commercial label for methanol.

It should be noted that the above-mentioned Greek word *hyle* has two discrete meanings, forest and matter. This explains that the etymologies of 'methyl' (with Greek *hyle* in the sense of forest, but mistaken for wood) and of 'ethyl' (with Greek *hyle* in the sense of matter, combined with ether) are by no means analogous.

3.2 Benzoic acid, an unsystematic cornerstone of systematic nomenclature

One of the first sources of benzene derivatives was gum benzoin, a resin of the styrax tree (*Styrax benzoin*), ultimately from Arabic *lubān jāwi*, literally frankincense of Java, a misnomer since gum benzoin came from Sumatra. Dry distillation of gum benzoin gave benzoic acid (benzenecarboxylic acid, cf. Chapter 3) which was named accordingly. Benzoic acid could then be converted to benzaldehyde (benzenecarbaldehyde, cf. Chapter 3), benzyl alcohol (phenylmethanol, cf. Chapter 3), and benzene (cf. Chapter 3). Thus, the parent compound benzene received a trivial name without any implication of structure, but this name has been retained as component of the systematic nomenclature as has the fusion prefix 'benzo'. On the other hand, parallel with these developments, benzene was also named phene (cf. Chapter 3) which led to names like phenyl (benzenyl) and the fusion prefix 'pheno'.

Also the systematic nomenclature's suffix 'oic acid', for names like pentanoic acid, has its roots in benzoic acid and thus gum benzoin.

A plethora of trivial names has developed from this source, such as

Benzal, as in benzal chloride [(dichloromethyl)benzene], named as the product of the replacement of benzaldehyde's oxygen atom with two chlorine atoms.

Benzaldehyde (benzenecarbaldehyde).

Benzhydrol (diphenylmethanol), from hydrol (secondary alcohol).

Benzil (1,2-diphenylethan-1,2-dione).

Benzilic acid (2,2-diphenyl-2-hydroxyacetic acid).

^{1}**Benz(o)-**, as in 2-benzoxepine.

^{2}**Benz(o)-**, as in benzotrichloride [(trichloromethyl)benzene].

Benzoin (2-hydroxy-1,2-diphenylethan-1-one).

Benzophenone (diphenylmethanone).

Benzoquinone (cyclohexadienedione).

Benzoyl-.

Benzyl- (phenylmethyl-).

3.3 Early notions of aliphaticity

Many of the first characterized organic compounds came from fats and were therefore called aliphatic, from Greek *aleiphar*, fat, from *aleiphein*, to smear. The observation that alkanes were by and large unreactive led to their naming as paraffins, from Latin *parum affinis*, little connected to, little related to. Alkenes, on the other hand, readily reacted with, say, halogens, to form heavy oily products and were therefore assigned the class name olefin, a contraction of olefiant gas. Upon realization that alkyl radicals could survive reactions unchanged, just as alkali metal ions do, the class name alkyl, and hence the names alkane, alkene, and alkyne were coined.

3.4 Early notions of aromaticity

For obvious reasons prior to the discovery of the electron in the late 19th century chemists were unable to pin down reliable connections between a substance's structure and its reactivity. When the German chemist August Wilhelm Hofmann (1818–1892) in 1855 coined the term 'aromatic' for a group of compounds comprising both alkenes, such as terpenes, and arenes [2], the justification was on one hand the pleasant smell of e.g. terpenes and, on the other hand, the common high degree of unsaturation characteristic of all these hydrocarbons. It was the German chemist and physicist Erich Armand Arthur Joseph Hückel (1896–1980) who in 1931 formulated the Hückel rule which predicts that mancude [3] planar ring compounds will be aromatic (in the contemporary sense) if the number n of π-electrons satisfies the condition $4n + 2$ and anti-aromatic when there are $4n$ π-electrons. Thus, the unfortunate early lack of discrimination is perpetuated in contemporary terminology, i.e. the chemical terminus technicus 'aromatic' has been dissociated from its common language meaning 'pleasant smelling' and the systematic names of the only distantly related alkenes and arenes still share the ending 'ene'. More about this in Section 4.3.5.

3.5 The 'ol' ambiguity

Oleum, from Latin *oleum*, oil, is the trivial name of fuming sulfuric acid, a viscous solution of sulfur trioxide in sulfuric acid. Since the amount of dissolved sulfur trioxide is traditionally given as sulfuric acid equivalents such solutions can even be labeled as 103% sulfuric acid, etc.

From the same source stem many, now obsolete, trivial names ending in 'ol' such as benzol (benzene) and furfurol (furan-2-carbaldehyde), in order to characterize the compounds in question as oils or, more precisely, relatively high-boiling liquids.

The word alcohol is of Arabic origin. *Al-kuhl*, also *al-kuhul* (derived from Arabic *kahala*, to stain, to paint) is the name of kohl, powdered stibnite (Sb_2S_3), used for cosmetic purposes since antiquity. The alchemistic notion that rigorous purification of an impure substance would lead to its essence or spirit in the shape of a dry powder led later, via 'alcohol of wine' to the modern meaning of first ethanol and then, from 1850, alcohols in general and hence to the introduction of 'ol' as a suffix for systematic names of alcohols. The name alcohol is also the ultimate source of the systematic class names alkane, alkene, alkyne, alkyl, etc. cf. Chapter 4. Thus, both halves of the word alcohol have been scavenged for nomenclature purposes. This is, as far as etymology is concerned, ironic since the *al* part of *al-kuhl* is nothing else but the Arabic definite article the.

The creation of the systematic suffix 'ol' for alcohols led to a conflict with preexisting names ending in 'ol' because of these compounds' low volatility. This conflict could in many cases be resolved by appropriate name changes. Thus, benzol, toluol, and xylol became benzene, toluene, and xylene, while furfurol, having been shown to be an aldehyde, was relabeled as furfural. It should be noted that in German, *Benzol, Toluol,* and *Xylol* are still valid names.

A third ambiguity raised its ugly head when the Hantzsch-Widman nomenclature (H-W) for heterocycles, first presented in German, claimed the apparently arbitrary suffix 'ol' to denote five-membered rings. In the English version of the H-W rules this 'ol' became 'ole', helping to keep confusion in check.

3.6 The mesityl radical

An early name for acetone (propan-2-one) was mesite, from Greek *mesites*, mediator. This name should show that acetone was less volatile than ether, but more volatile than alcohol. Although it could be formed from acetic acid, for instance by heating of lead acetate, it was not closely related to acetic acid.

When a condensation product of acetone was obtained it was called mesityl oxide in the belief that it had been formed by loss of water from two molecules of the enol form of acetone to give the corresponding ether. Mesityl oxide was later shown to be 4-methylpent-3-en-2-one, but the original name is still in use. Interestingly, further condensation of mesityl oxide with acetone leads, as one could expect, to 2,6-dimethylhepta-2,5-dien-4-one, but this compound had already received the unrelated trivial name phorone, because it was first obtained by dry distillation of calcium camphorate.

The Irish chemist Robert John Kane (1809–1890) found that acetone upon treatment with concentrated sulfuric acid yielded a hydrocarbon which was called mesitylene in the belief that it was the mesityl radical which had lost one hydrogen atom, i.e. propyne. Later it was shown that mesitylene is the corresponding trimer, i.e. 1,3,5-trimethylbenzene, but the common name mesitylene remained in use.

The substituent 1,3,5-trimethylphenyl, derived from mesitylene and called mesityl for brevity, is used to add bulk and thus stability to otherwise labile molecular systems. This mesityl group is, of course, a far cry from the mesityl group implied in the obsolete name mesityl oxide.

3.7 Valency woes

While a name like sodium chloride (NaCl) is unambiguous, copper chloride is not. Here the valency of the metal needs to be specified and this was achieved by coining the names cuprous chloride [CuCl, copper(I) chloride] and cupric chloride [$CuCl_2$, copper(II) chloride]. Moving from copper to iron we note that we could encounter e.g. ferrous oxide [FeO, iron(II) oxide] and ferric oxide [Fe_2O_3, iron(III) oxide, diiron trioxide], but also mixed valency compounds such as ferrous ferric oxide [Fe_3O_4, iron(II) diiron(III) oxide, triiron tetraoxide].

Similar problems are seen with non-metals. With chlorine in its 'standard valency', whatever that might be, we talk about chloric acid ($HClO_3$, valency 5), the lower valency analogs being chlorous acid ($HClO_2$, valency 3) and hypochlorous acid (HClO, valency 1). What about valency 7, as befits a group 17 element? Yes, indeed, we also encounter perchloric acid ($HClO_4$).

On the other hand, persulfuric acid (H_2SO_5) does not contain sulfur in a higher valency than it has in sulfuric acid (H_2SO_4), but, as a derivative of hydrogen peroxide (H_2O_2, dioxidane), contains an oxygen-oxygen bond.

These few examples show the shortcomings of such a naming system. The exact meaning of the prefixes and suffixes used in this rudimentary naming system has to be defined and memorized for every single element. Neither does a name of this type reflect the stoichiometry of a compound in any straightforward manner.

Confusion is almost inevitable when closed-shell molecules are named without distinction alongside radicals. Thus, the five oxides of sodium have received rather unsystematic traditional names:

Sodium oxide (Na_2O) [disodium oxide, sodium oxide(2−)].

Sodium monoxide (NaO) [sodium oxide(•1−)].

Sodium superoxide (NaO_2) [sodium dioxide(•1−)].

Sodium peroxide (Na_2O_2) [disodium peroxide, sodium dioxide(2−)].

Sodium ozonide (NaO_3) [sodium trioxide(•1−)].

3.8 The derivatization game

In the early days of chemistry solid products were much easier to handle than liquids. They could be reliably purified by recrystallization, characterized by their melting point, their crystal habit, and by combustion analysis. Liquids and oils, and especially water-soluble substances, were much more difficult to purify and to characterize.

For this reason it became a standard procedure to convert the latter to solid derivatives with all the virtues mentioned above. Typically aldehydes and ketones were converted to oximes or hydrazones and monosaccharides to osazones.

Nomenclaturewise these products were not considered as 'regular' compounds to be named as such, but as 'derivatives'. This led to names like benzaldoxime (benzaldehyde oxime), acetone hydrazone, or glucose osazone which did not do justice to the structure of these compounds, but rather involved a description of their formation. These traditional names are still widely used and more so than the modern systematic names.

***anti*-Benzaldoxime** [(*Z*)-*N*-hydroxy-1-phenylmethan-1-imine].

***syn*-Benzaldoxime** [(*E*)-*N*-hydroxy-1-phenylmethan-1-imine].

Contrary to the modern *E,Z*-convention (which is based on the positions of the highest CIP prioritized groups) the old *anti,syn*-convention for aldoximes was based on the position of the hydroxyl group relative to the aldehydic hydrogen atom, i.e. it had the opposite sense.

Formaldehyde (2,4-dinitrophenyl)hydrazone [1-(2,4-dinitrophenyl)-2-methylidenehydrazine].

Acetone hydrazone [(propan-2-ylidene)hydrazine).

Acetone azine [di(propan-2-ylidene)hydrazine].

D-Glucose osazone [(2*R*,3*S*,4*R*,5*Z*,6*E*)-5,6-bis(phenylhydrazinylidene)hexane-1,2,3,4-tetrol].

One should note that a name like D-glucose osazone neither reflects the presence of phenyl groups nor the redox reactions in the course of this derivatization of D-glucose.

3.9 Hydrogenation of unsaturated cyclic compounds

Prior to the current systematic nomenclature several methods were employed to name partially or fully hydrogenated unsaturated cyclic compounds. Thus, naphthaline (naphthalene) could be hydrogenated to tetraline (1,2,3,4-tetrahydronaphthalene), octaline (1,2,3,4,5,6,7,8-octahydronaphthalene), and decaline (tetrahydronaphthalene), the position of the added hydrogen atoms in the partially hydrogenated compounds just being understood since it followed from the relative ease of hydrogenation.

In the case of, for instance, pyrrole suffixes were used. The two partially hydrogenated compounds were called Δ^2-pyrroline (2,3-dihydro-1*H*-pyrrole) and Δ^3-pyrroline (2,5-dihydro-1*H*-pyrrole), respectively. The Δ descriptors showed the position of the remaining double bond. Finally, the fully hydrogenated compound was called pyrrolidine (pyrrolidine, tetrahydropyrrole).

3.10 Superstition in chemistry: Salvarsan and neosalvarsan

Salvarsan (arsphenamine, from arsenic, phenol, and amine) and neosalvarsan (neoarsphenamine) were designed as antisyphilitic agents and are among the very first synthetic drugs. Based on their empirical formulas and in the absence of solid structural information they were presented as derivatives of arsenobenzene (diphenyldiarsene), in analogy with the well-known azobenzene (diphenyldiazene), the parent of the important azo dyes.

However, as early as 1932 the double bond rule [4] stated that compounds with double or triple bonds between heavy atoms are intrinsically unstable. This is best illustrated by a comparison of elemental nitrogen with elemental (white) phosphorus. The former employs a nitrogen-nitrogen triple bond to form stable N_2 molecules. The latter is limited to phosphorus-phosphorus single bonds which leads to tetrahedral P_4 molecules as the smallest possible stable structure. In 1960 a crystallographic study [5] showed that the compound formerly regarded as arsenobenzene was in fact the corresponding trimer, i.e. 1,2,3,4,5,6-hexaphenylcyclohexaarsane.

Thus, since 1932 and especially from 1960, any formulation of salvarsan and neosalvarsan as arsenobenzene derivatives could be nothing else but in bad faith. However, as late as 2001 the venerable Merck Index stuck to its arsenobenzene structures [6] and even in 2018 Wikipedia [7] and ChemSpider [8] still misrepresented the structure of neosalvarsan. In its salvarsan entry [9] Wikipedia, however, acknowledges that salvarsan is a mixture of cyclooligoarsanes.

In an interesting twist, the double bond rule has subsequently been compromised by the observation that compounds with double or triple bonds between heavy atoms such as diarsenes (R^1–As=As–R^2) can indeed be prepared and isolated if these bonds are protected by massive steric hindrance [10].

References and comments

[1] https://en.wikipedia.org/wiki/MethanolHistory (retrieved August 2, 2018).

[2] https://en.wikipedia.org/wiki/Aromaticity (retrieved August 2, 2018).

[3] mancude = maximum number of noncumulative double bonds.

[4] https://en.wikipedia.org/wiki/Double_bond_rule (retrieved September 18, 2018).

[5] K. Hedberg, E. W. Hughes, J. Waser, *Acta Cryst.* **14**, 369–374 (1960); https://doi.org/10.1107/S0365110X61001261.

[6] The Merck Index, Merck & Co., Inc, Whitehouse Station, NJ, USA, 2001, entries 820 and 6474.

[7] https://en.wikipedia.org/wiki/Neosalvarsan (retrieved September 18, 2018).

[8] http://www.chemspider.com/Chemical-Structure.9586.html?rid=46947be9-1fac-45c1-aa7c-2272864b827bHYPERLINK"http://www.chemspider.com/Chemical-Structure.9586.html?rid=46947be9-1fac-45c1-aa7c-2272864b827b&page_num=0"&HYPERLINK"http://www.chemspider.com/Chemical-Structure.9586.html?rid=46947be9-1fac-45c1-aa7c-2272864b827b&page_num=0"page_num=0 (retrieved September 18, 2018).

[9] [https://en.wikipedia.org/wiki/Arsphenamine (retrieved September 18, 2018).

[10] A. H. Cowley, J. G. Lasch, N. C. Norman, and M. Pakulski, *J. Am. Chem. Soc.* **105**, 5506–5507 (1983); DOI: 10.1021/ja00354a069.

4 The IUPAC systematic nomenclature

With its recommendations the International Union of Pure and Applied Chemistry (IUPAC) [1], founded in 1919, standardizes chemical nomenclature in general [2] and has, inter alia, issued recommendations for the nomenclature of inorganic chemistry in the shape of the 'Red Book' [3] and of organic chemistry as the 'Blue Book' [4]. The 2013 issue of the 'Blue Book' is beset by a sheer overwhelming number of misprints and errors and IUPAC has initiated a project to provide a list of errata to the IUPAC 'Blue Book', an improved subject index, some corrected, improved, or revised sections, and some additional sections to deal with omissions from the current edition [5]. A dedicated IUPAC Nomenclature webpage provides a wealth of information including many structural formulas with the appropriate locants shown [6].

By and large IUPAC nomenclature operates with three naming systems, i.e. compositional, substitutive, and additive names. Even in simple cases, such as PF_3, this gives rise to three valid IUPAC names, i.e. phosphorus trifluoride, trifluorophosphane, and trifluoridophosphorus.

4.1 Retained names

Contrary to popular belief even basically systematic nomenclature cannot avoid the use of trivial names in the shape of 'retained names'. Trivial names have been retained for mainly two reasons: names which are deeply rooted in common language and names whose fully systematic equivalent would be unwieldy and tedious.

4.1.1 Parent hydrides

The current IUPAC nomenclature is the result of a tug-of-war between tradition, systematic rigor, and convenience. Already IUPAC's naming of the alkanes breaks with the otherwise fundamental rule that parent hydride names of non-metallic elements should be composed of the modified element name and 'ane' which, in the case of carbon, gives the names carbane, dicarbane, tricarbane, etc., is broken in deference to tradition: the first four alkanes, methane, ethane, propane, and butane, retain their well-established trivial names, and the following pentane, hexane, etc. lack the otherwise mandatory element name, also here following a strong and ingrown tradition. Thus, a very basic IUPAC rule has never been extended to apply to the many millions of organic compounds.

https://doi.org/10.1515/9783110612714-004

An interesting conflict arose in the naming of the parent hydride AlH_3. The straightforward name according to the general rule would have been aluminane. However, this name was already assigned to the saturated six-membered ring compound ($C_5H_{11}Al$) with five carbon atoms and one aluminum atom. The alternative alane was discarded because it would lead to the substituent name alanyl for H_2Al–, a name already given to the acyl group corresponding to the amino acid alanine. Therefore the only remaining option, alumane, was adopted.

Similar considerations complicated the naming of the the parent hydride InH_3. Here the default would have been indane, a retained name earlier assigned to a bicyclic hydrocarbon (cf. Section 4.1.2.). This dilemma was resolved by calling indium hydride indigane. The latter name is not as far-fetched as it looks at first glance, since the name of the element indium was derived from the color indigo.

Here follow samples of retained names:

Ammonia (azane)
Butane
Ethane
Hydrazine (diazane)
Hydroxylamine (azanol), not a parent hydride *stricto sensu*. Hydroxylamine in IUPAC organic nomenclature, azanol in IUPAC inorganic nomenclature.
Methane
Pentane, hexane, etc.
Propane

4.1.2 Ring compounds

Also here retained names dominate:

Aceanthrylene, named as anthracene with a *peri*-fused five-membered carbon ring.

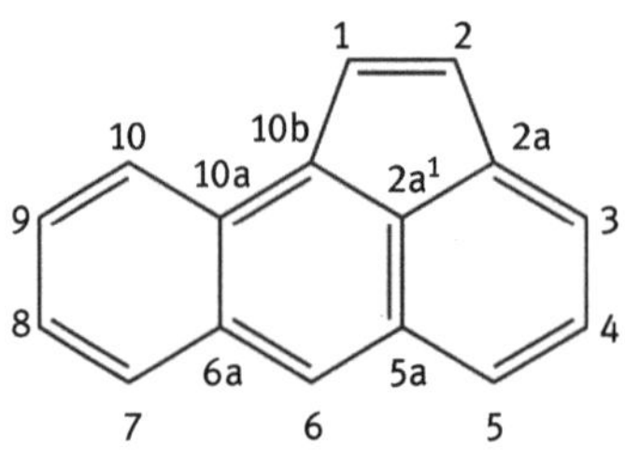

aceanthrylene

Acenaphthylene {cyclopenta[*de*]naphthalene}, named as naphthalene with a *peri*-fused five-membered carbon ring.

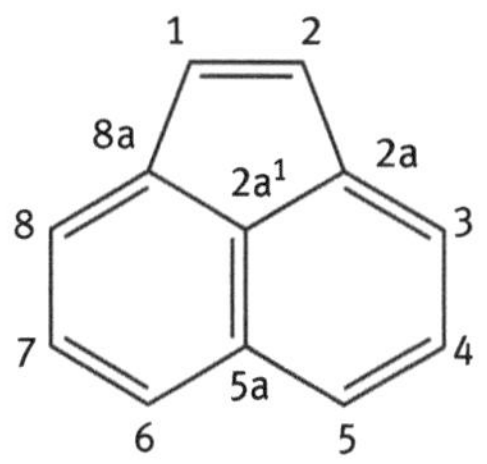

acenaphthylene

Acephenanthrylene {benz[*e*]acenaphthylene}, named as phenanthrene with a *peri*-fused five-membered carbon ring.

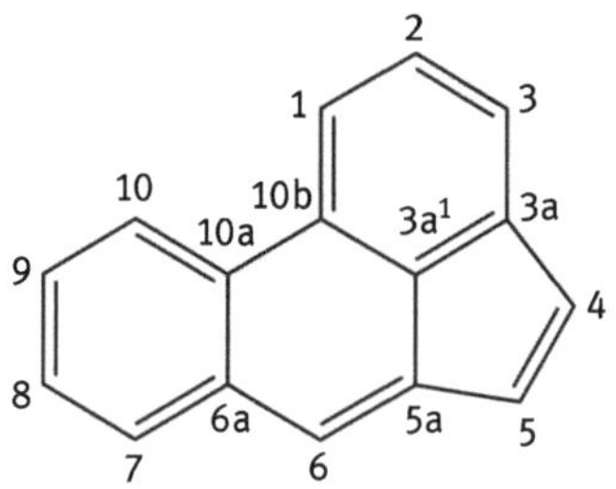

acephenanthrylene

Acridine {dibenzo[*d,e*]pyridine}, from acrid, referring to this compound's smell.

Anthracene {tricyclo[8.4.0.0^{3,8}]tetradeca-1,3,5,7,9,11,13-heptaene}, named after its first source, coal, from Greek *anthrax, anthrak-*, coal. Earlier called **paranaphthalene**.

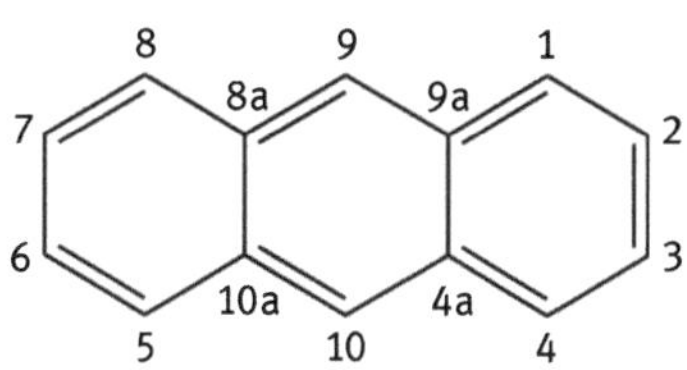

anthracene

Azulene {bicyclo[5.3.0]deca-2,,4,6,8,10-pentaene}, named after its blue color, from Spanish *azul*, blue.

Bacteriochlorin (2,3,12,13-tetrahydroporphyrin), from chlorin.

Benzene, from French *benzoin*, cf. benzoic acid. Earlier **benzol** and **phene**.

Biphenylene (dibenzocyclobutadiene), named as two fused *o*-phenylene units.

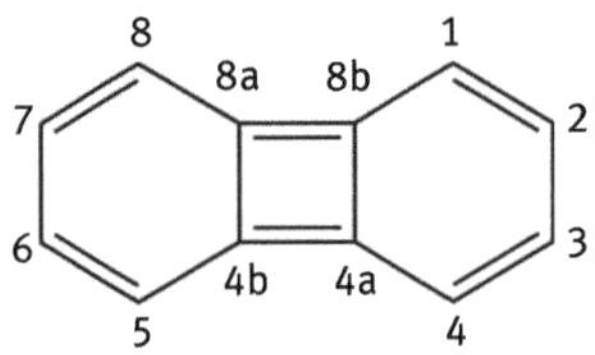

biphenylene

Butalene (bicyclo[2.2.0]buta-1,2,5-triene), named as two fused cyclobutadiene rings.

Carbazole (dibenzopyrrole), named as a five-membered nitrogen-containing ring with two fused benzene rings.

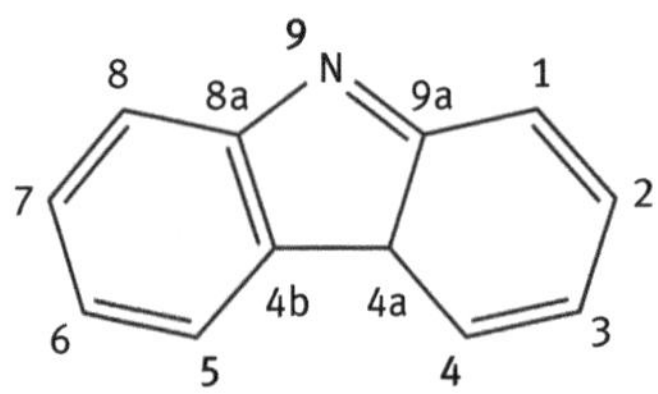

4a*H*-carbazole

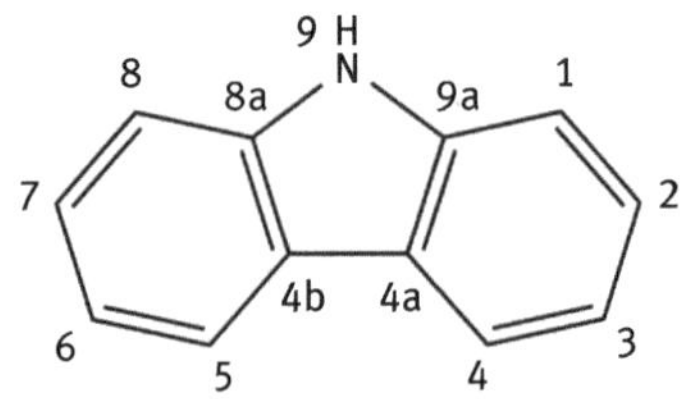

9*H*-carbazole

Chlorin (2,3-dihydroporphyrin), named as the core structure of chlorophyll.

Chromene (1-benzopyran), ultimately from Greek *chroma*, color, referring to the occurrence of this ring system in colored natural products.

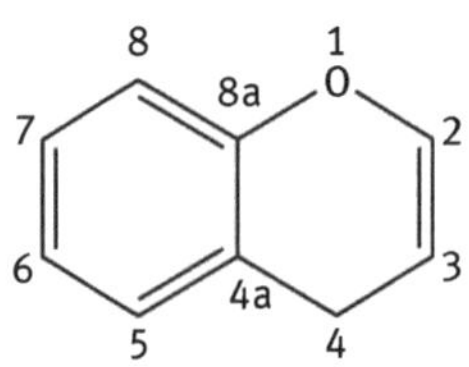

4*H*-1-benzopyran

Chrysene {benzo[*a*]phenanthrene}, from Greek *chrysos*, gold, referring to the color of this compound. A misnomer as it turned out later. Rigorously purified chrysene is colorless and the golden-yellow color originally observed was due to the presence of yellow tetracene as an impurity.

Cinnoline {benzo[*c*]pyridazine}, a name created by variation of German *Chinolin*, quinoline.

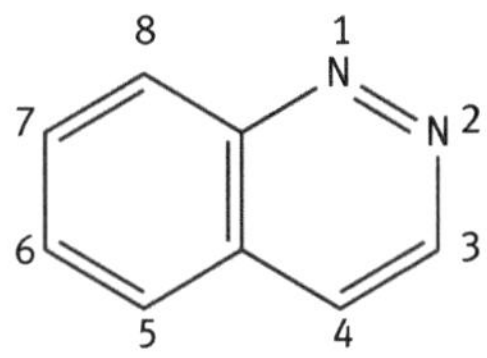

cinnoline

Coronene {dibenzo[*ghi,pqr*]perylene}, from Latin *corona*, crown, referring to the shape of the molecule.

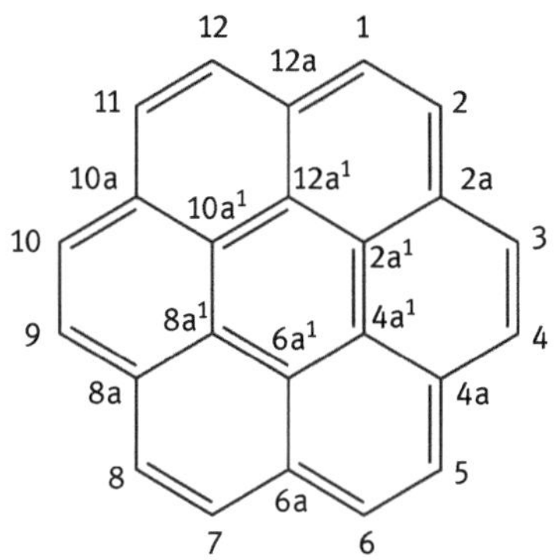

coronene

Corrin, earlier corphyrin, named as the core of vitamin B_{12}.

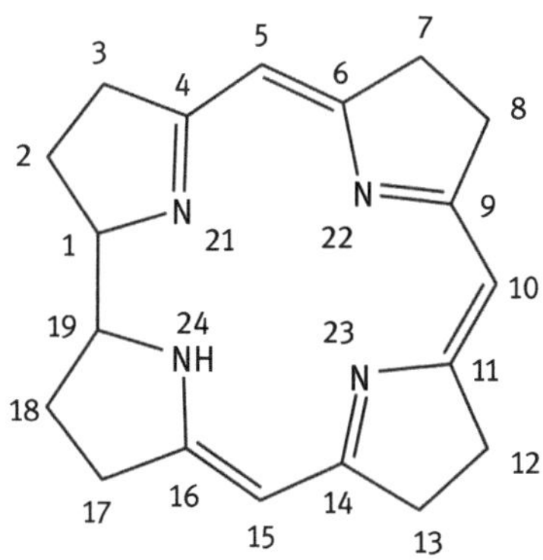

corrin

Ferrocene [bis(η^5-cyclopentadienyl)iron, bis(η^5-cyclopentadienido)iron], provisionally coined as ferrozene, from ferro- and benzene, referring to the benzene-like character of its cyclopentadiene rings, but then changed to ferrocene to make it look and sound more attractive.

Fluoranthene {benzo[*j*,*k*]fluorene}, a portmanteau of fluorescence and anthracene.

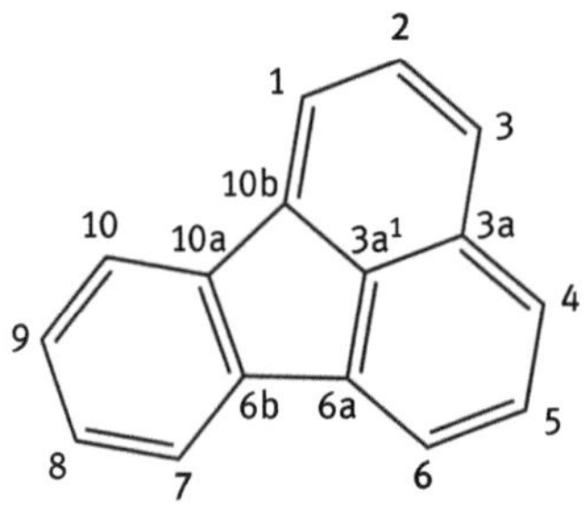

fluoranthene

Fluorene (9*H*-fluorene), from fluorescence.

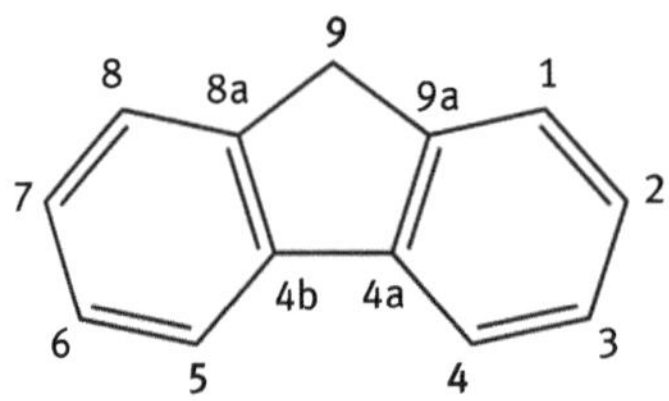

9*H*-fluorene

Fulvalene [1,1′-bis(cyclopenta-2.4-dien-1-ylidene], from fulvene and naphthalene, referring to this compound's isomerism with naphthalene.

Fulvene (5-methylidenecyclopenta-1,3-diene), from Latin *fulvus*, tawny, referring to this compound's reddish-yellow color.

Furan, coined by contraction of furfuran which in turn was named as the parent ring system of furfurol (furfural, furan-2-carbaldehyde), ultimately from Latin *furfur*, bran. The German chemist Heinrich Limpricht (1827–1909), who was the first to prepare furan, called it tetraphenol, i.e. as four-carbon analog of phenol.

Heptalene {bicyclo[5.5.0]dodeca-2,5,6,8,10,12-hexaene}, named as two fused cycloheptatriene rings.

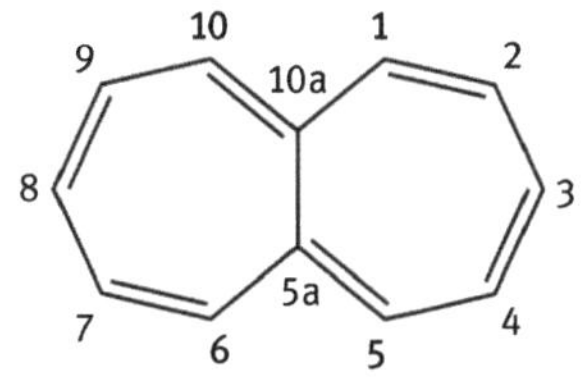

heptalene

Hexahelicene {benzo[*c*]naphtho[1,2-*g*]phenanthrene}, named as helix-shaped molecule with six fused benzene rings. Higher homologs are named in the same manner (heptahelicene etc.).

Imidazole (1*H*-imidazole, 1,3-diazole), named as a five-membered ring with two nitrogen atoms.

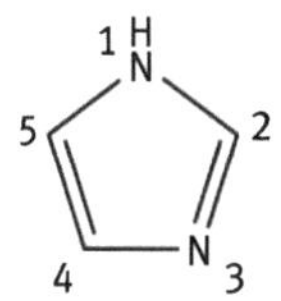

1*H*-imidazole

Imidazolidine (1,3-diazolidine), named as fully hydrogenated imidazole.

***as*-Indacene** {cyclopenta[*h*]indene}, named by contraction of indene and anthracene; *as*- stands for asymmetric.

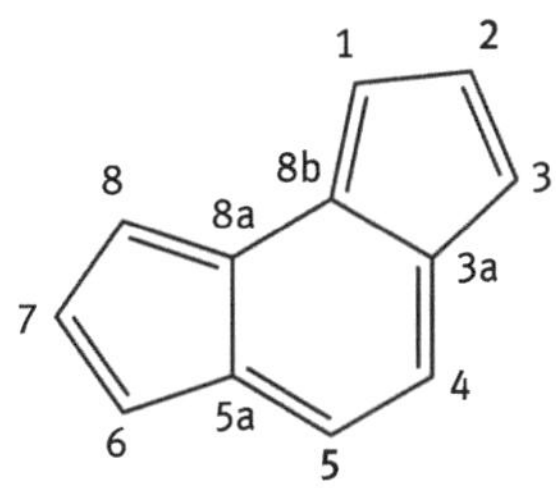

as-indacene

***s*-Indacene** {cyclopenta[*g*]indene}, named by contraction of indene and anthracene; *s*- stands for symmetric.

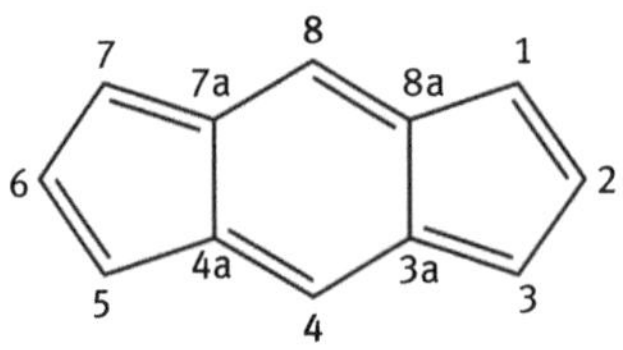

s-Indacene

Indane (2,3-dihydro-1*H*-indene), named as partially saturated derivative of indene.

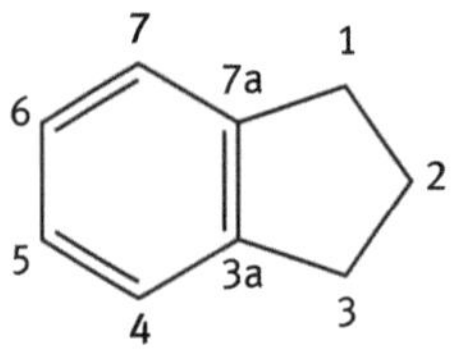

indane

Indazole {benzo[*b*]pyrazole}, named as derivative of indole with an additional nitrogen ring atom.

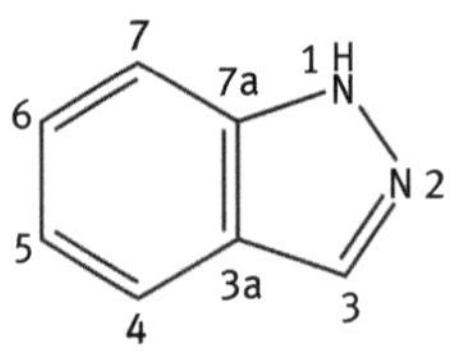

1*H*-indazole

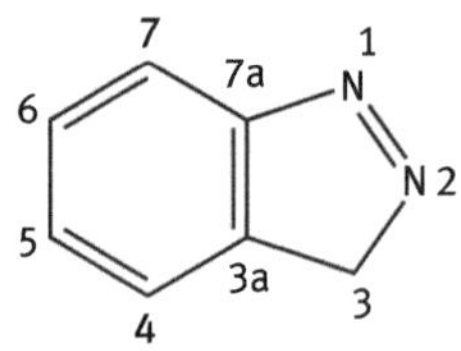

3*H*-indazole

Indene (1*H*-indene, benzocyclopentene), named as all-carbon analog of indole.

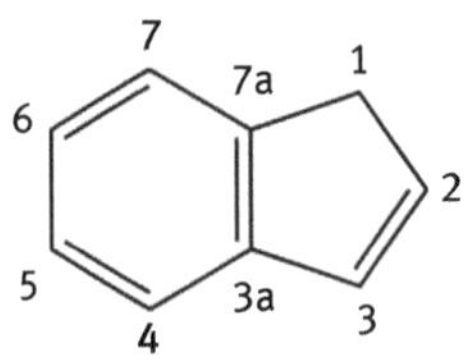

1*H*-indene

Indole {1*H*-indole, benzo[*b*]pyrrole}, named as a degradation product of indigo.

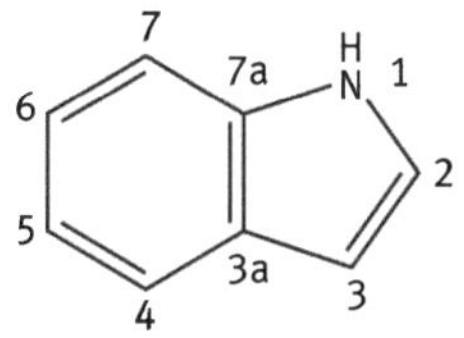

1*H*-indole

Indolizine {pyrrolo[1,2-*a*]pyridine}, named as five-membered analog of quinolizine.

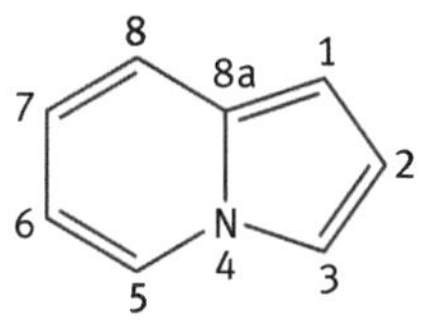

indolizine

Isobacteriochlorin (2,3,7,8-tetrahydroporphyrin), from chlorin.

Isochromene (2-benzopyran), from iso- and chromene.

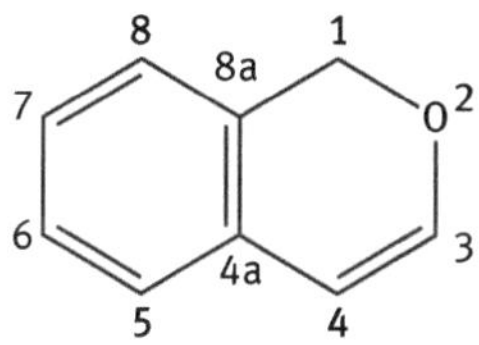

1*H*-2-benzopyran

Isoindole {2*H*-isoindole, benzo[*c*]pyrrole}, from indole.

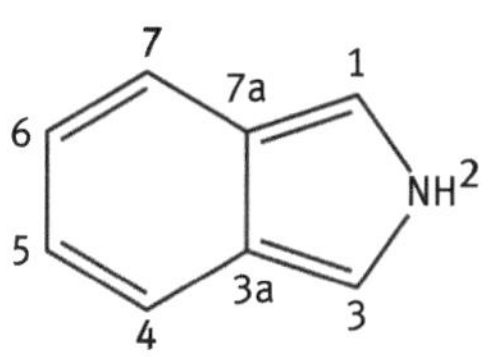

2*H*-isoindole

Isoquinoline {benzo[*c*]pyridine}, from quinoline.

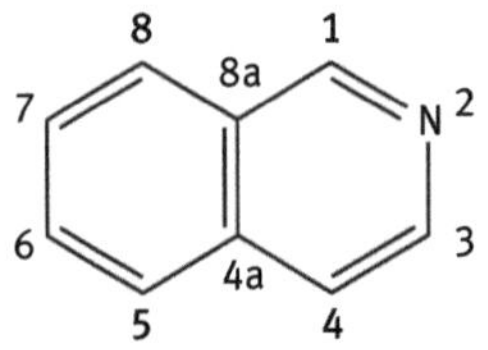

isoquinoline

Morpholine (tetrahydro-1,4-oxazine), a contraction of morphine and quinoline, referring to an erroneous belief that morphine contains a morpholine ring as part of its structure.

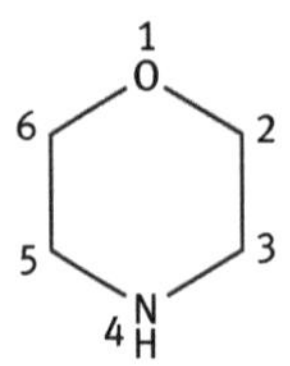

morpholine

Naphthalene, named after its occurrence in naphtha, ultimately from Persian *neft*, naphtha. Earlier called **naphthaline**.

Naphthyridine {1,8-naphthyridine, pyrido[2,3-*b*]pyridine}, named by contraction of naphthalene and pyridine.

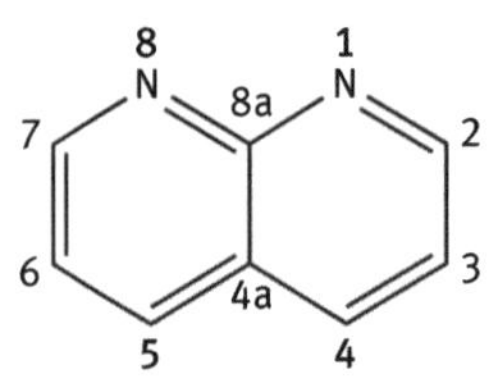

1.8-naphthyridine

Octalene {bicyclo[6.6.0]tetradeca-1(8),2,4,6,9,11,13-hexaene}, named as two fused cyclooctatetraene rings.

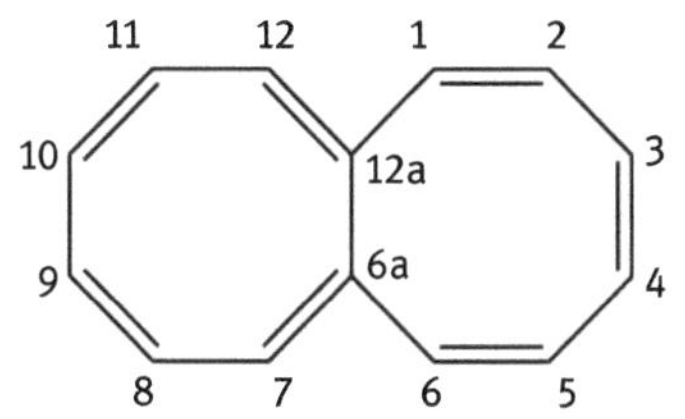

octalene

Ovalene {phenanthro[3,4,5,6-*qrabc*]coronene}, named after the oval shape of this molecule.

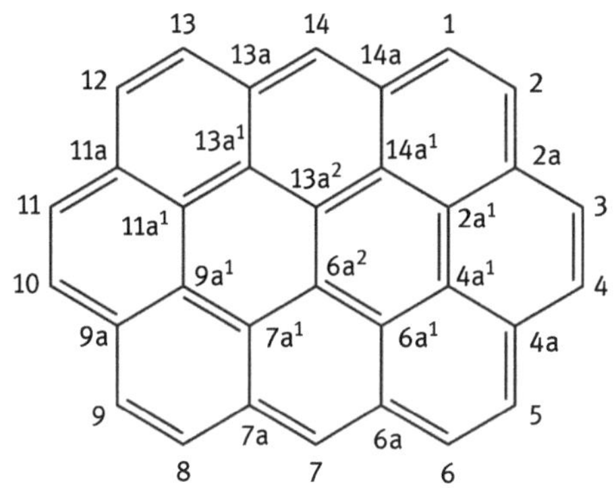

ovalene

Oxanthrene (dibenzo[1,4]dioxin), named as oxygen analog of anthracene.

1,2-Oxazolidine, named as fully hydrogenated 1,2-oxazole.

1,3-Oxazolidine, named as fully hydrogenated 1,3-oxazole.

Pentalene {bicyclo[3.3.0]octa-1,3,5,7-tetraene}, named as two fused cyclopentadiene rings.

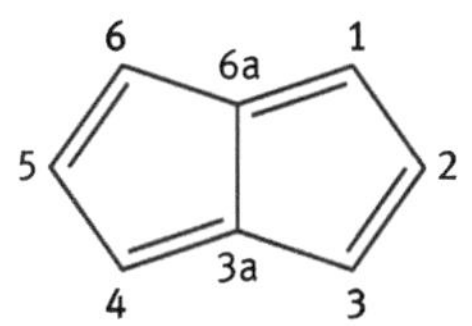

pentalene

Perimidine {1*H*-perimidine, 1*H*-benzo[*de*]quinazoline}, named as a *peri*-substituted naphthalene containing an amidine group.

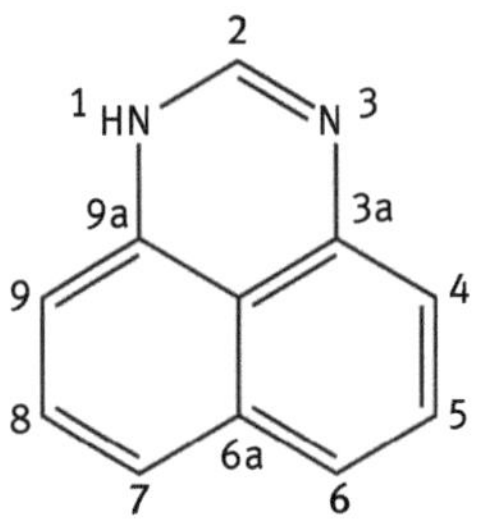

1*H*-perimidine

Perylene {dibenz[*de*,*kl*]anthracene}, a contraction of *peri*-dinaphthylene. A.k.a. ***peri*-dinaphthalene**.

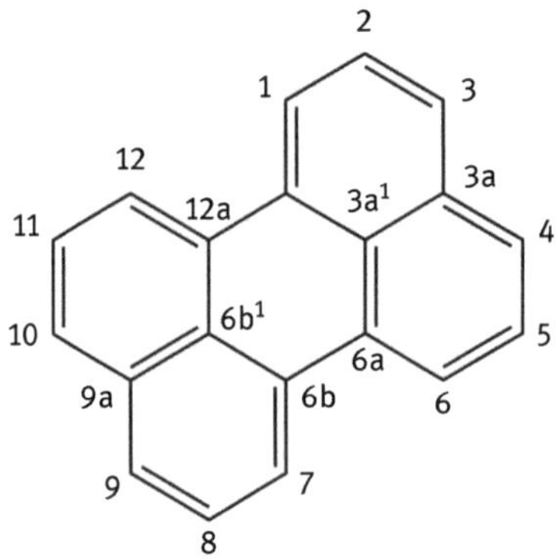

perylene

Phenalene (1*H*-phenalene, 1*H*-benzonaphthene), a contraction of phen(o)- and naphthalene. A.k.a. ***peri*-naphthene**.

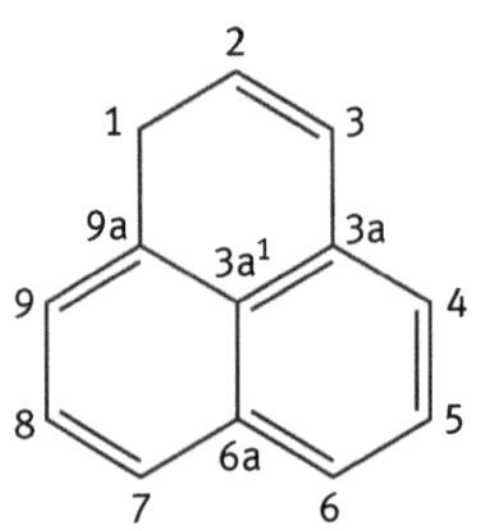

1*H*-phenalene

phenazine, from phene and anthra-, named as isomer of anthracene.

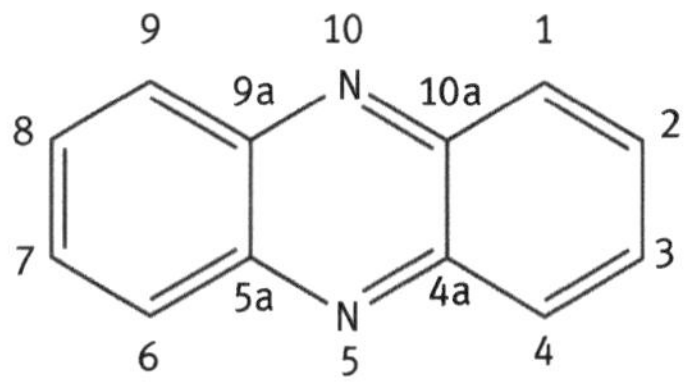

phenazine

Phenanthrene, from phene and anthra-, named as isomer of anthracene.

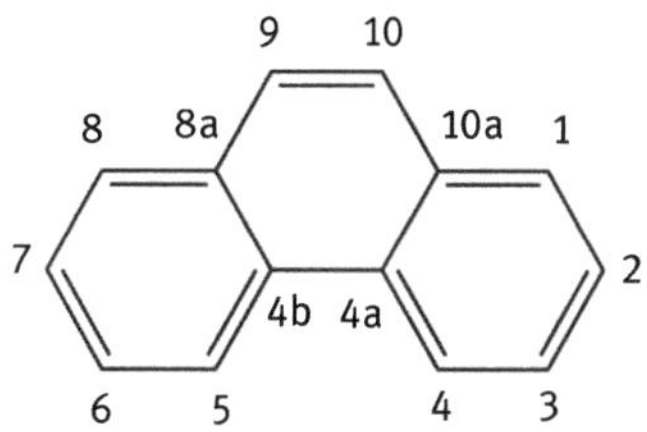

phenanthrene

Phenanthridine {benzo[c]quinoline}, named as a hybrid of phenanthrene and pyridine.

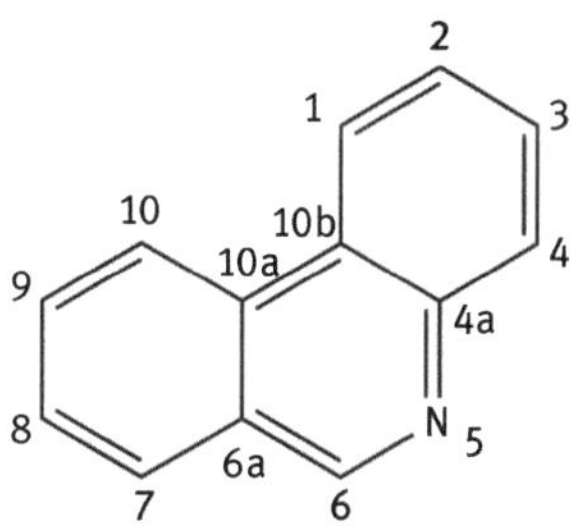

phenanthridine

Phenanthroline (1,10-phenanthroline, 4,5-diazaphenanthrene), named as hybrid of phenanthrene and quinoline.

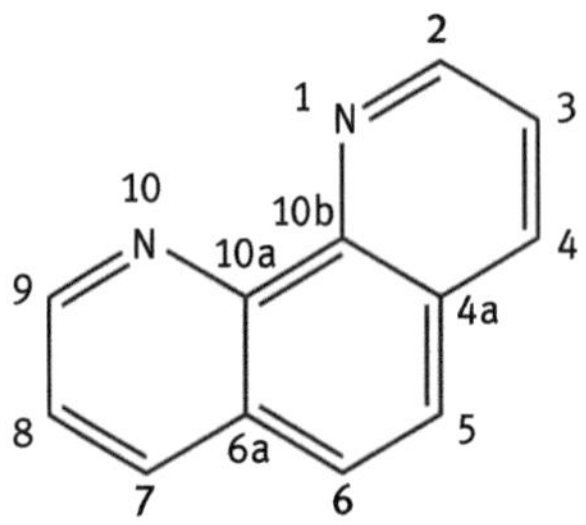

phenanthroline

Phenothiazine [10*H*-phenothiazine, 10*H*-dibenzo[*b*,*e*][1,4]thiazine}, named as thiazine with two fused benzene rings.

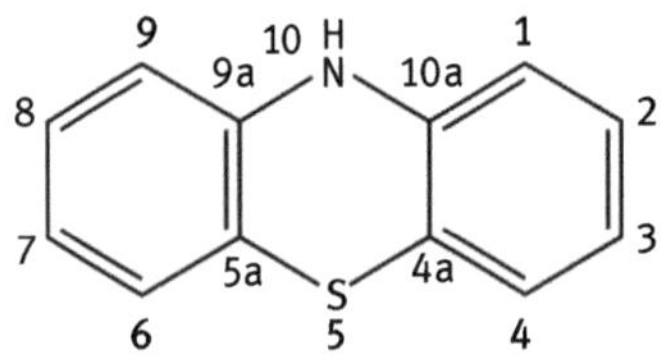

10*H*-phenothiazine

Phthalazine {benzo[*d*]pyridazine, 2,3-diazanaphthalene}, named as the azine of phthalaldehyde (benzene-1,2-dicarbaldehyde).

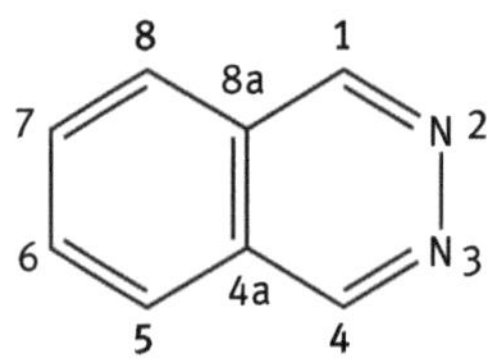

phthalazine

Phthalocyanine {2,11,20,29,37,38,38,40-octazanonacyclo[29.6.1.$1^{3,1}$0.$1^{12,19}$.$1^{21,28}$.$0^{4,9}$.$0^{13,18}$.$0^{22,27}$.$0^{31,36}$]tetraconta-1,3,5,7,9,11,13,15,17,19(39),20,22,24,26,28,30(37,31,33,35-nonadecaene}, from phthalic acid and cyanine, a class of dyes.

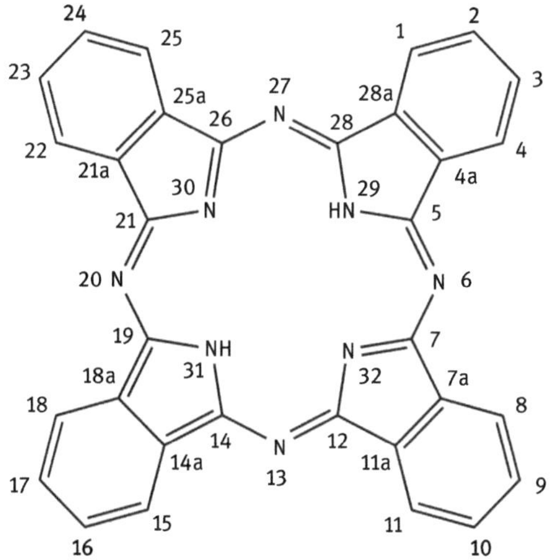

phthalocyanine

Picene {dibenzo[*a*,*i*]phenanthrene}, named after its first source, pitch, from Latin *pix, pic-*, pitch.

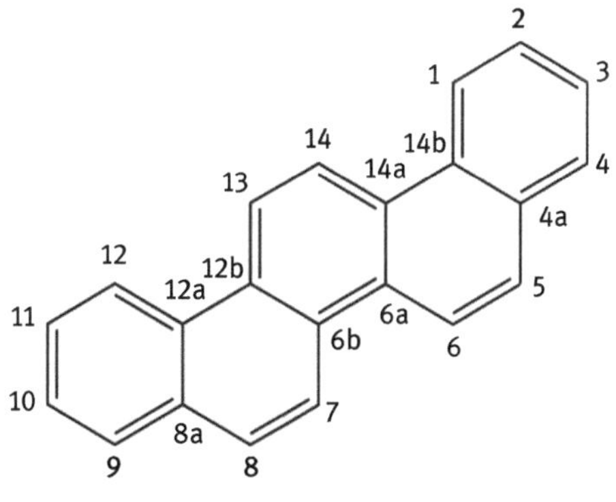

picene

Piperazine (hexahydropyrazine), named as derivative of piperidine with an additional nitrogen atom.

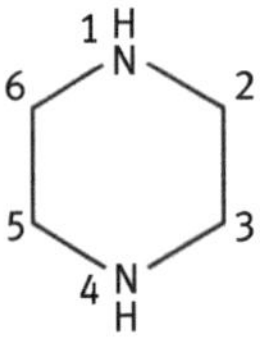

piperazine

Piperidine (hexahydropyridine), named as degradation product of the alkaloid piperine, found in black pepper (*Piper nigrum*). The genus name is from Latin *piper*, pepper.

Porphyrin, {21,22,23,24-tetraazapentacyclo[16.2.1.1^{3,6}.1^{8,11}.11^{3,16}]tetracosa-1(21),2,4,6,8(23),9,11,13,15,17,19-undecaene}. From Greek *porphyros*, purple, referring to this heterocycle's deep red color. Earlier **porphine**.

Propalene {bicyclo[1.1.0]buta-1,3-diene}, named as two fused cyclopropene rings.

Pteridine {pyrimido[4,5-*b*]pyrazine, 1,3,5,8-tetraazanaphthalene}, from Greek *pteron*, feather, wing, referring to the fact that this ring system is part of natural products found in butterfly wings.

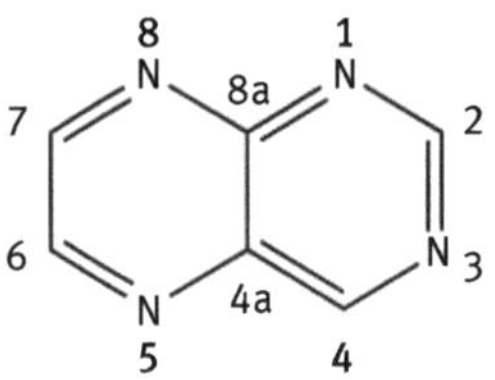

pteridine

Purine {6*H*-imidazo[4,5-*d*]pyrimidine}, from Latin *purus*, pure, and New Latin *acidum uricum*, uric acid, referring to this ring system as the core structure of uric acid and related natural products. Here, pure is to be understood as devoid of peripheral groups.

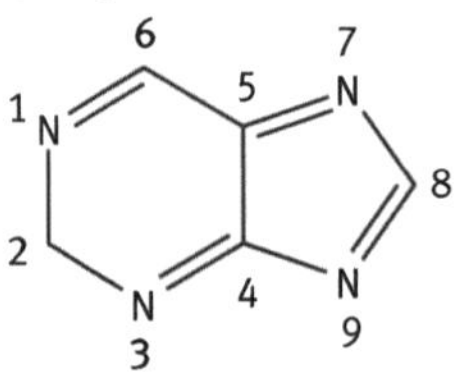

2*H*-purine

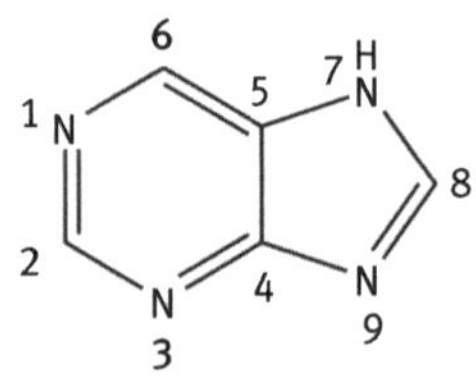

7*H*-purine

Pyran, named as parent structure of γ-pyrone (4*H*-pyran-4-one). Pyrone is a contraction of earlier pyrocomane, so named because it was first obtained by heating of comanic acid (4-oxo-4*H*-pyran-2-carboxylic acid).

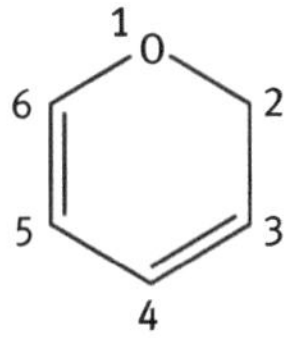

2*H*-pyran

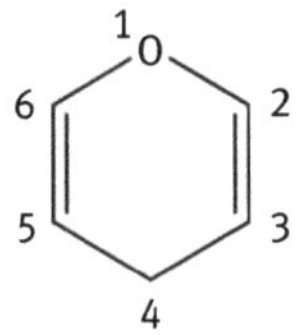

4*H*-pyran

Pyranthrene {tribenzo[*ghi,m,pqr*]picene, benzo[*c*]naphtho[3,2,1,8,7-*rstuv*]pentaphene}, from pyr(o)- and anthracene.

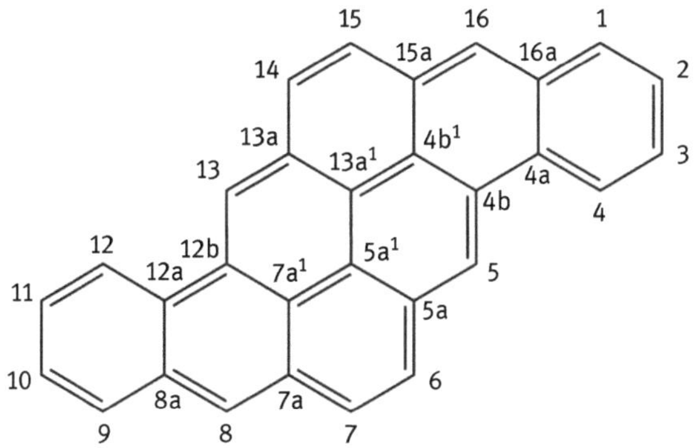

pyranthrene

Pyrazine (1,4-diazine), named as an anlog of pyridine with one additional nitrogen atom in the ring.

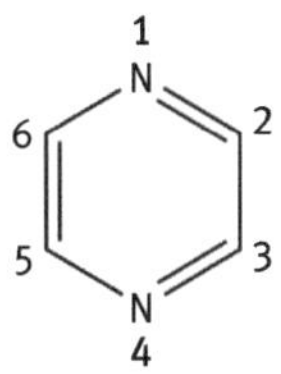

pyrazine

Pyrazole (1,2-diazole), named as an anlog of pyrrole with one additional nitrogen atom in the ring.

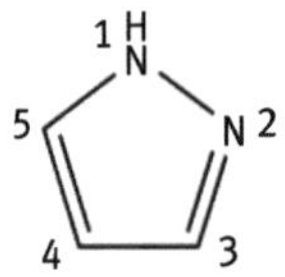

1*H*-pyrazole

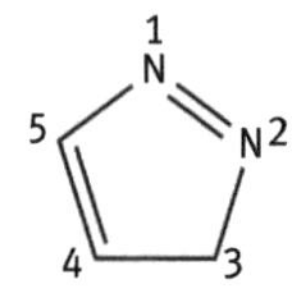

3*H*-pyrazole

Pyrazolidine (tetrahydro-1,2-diazole), named as fully hydrogenated pyrazole.

Pyrene {benzo[*def*]phenanthrene}, from Greek *pyr*, fire, referring to this compound's formation upon dry distillation of coal.

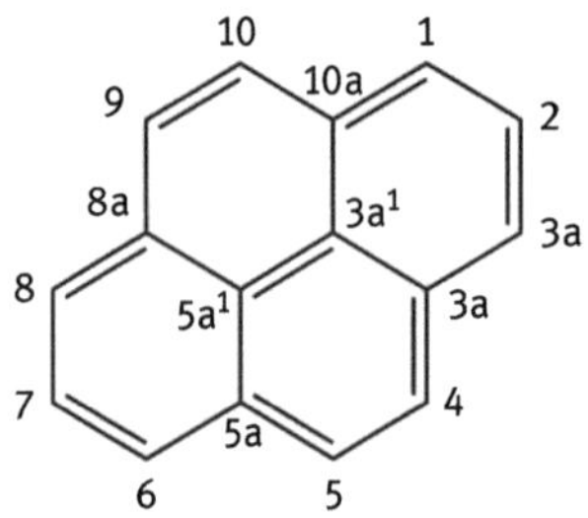

pyrene

Pyridazine (1,2-diazine), named by contraction of pyridine and hydrazine.

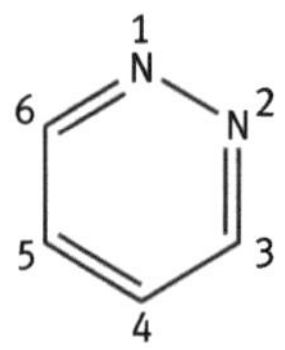

pyridazine

Pyridine, after pyrrole; named in the mistaken belief that it was a pyrrole derivative.

Pyrimidine (1,3-diazine), named as derivative of pyridine with on additional nitrogen atom in the ring.

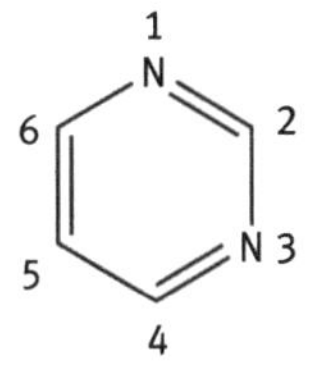

pyrimidine

Pyrrole, from Greek *pyrrhos*, reddish, fiery, referring to the red color which develops when pyrrole vapor acts upon a pine splinter moistened with hydrochloric acid. A.k.a. **divinylenimine**.

Pyrrolidine (tetrahydropyrrole), named as fully hydrogenated pyrrole.

Pyrrolizine (1*H*-pyrrolizine), from pyrrol; the suffix 'izine'is arbitrary.

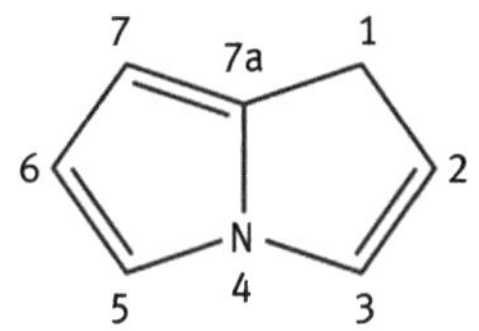

1*H*-pyrrolizine

Quinazoline {benzo[*a*]pyrimidine, 1,3-diazanaphthalene}, named as analog of quinoline with an additional nitrogen atom.

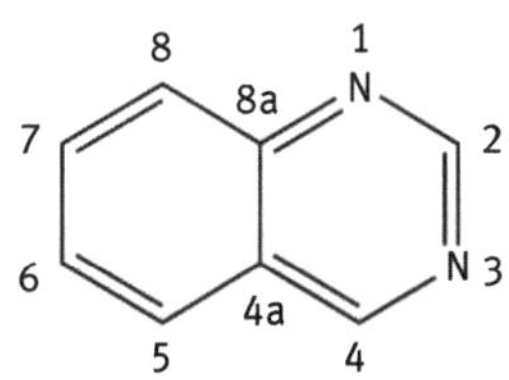

quinazoline

Quinoline {benzo[*b*]pyridine, 1-azanaphthalene}, named as degradation product of the alkaloid quinine.

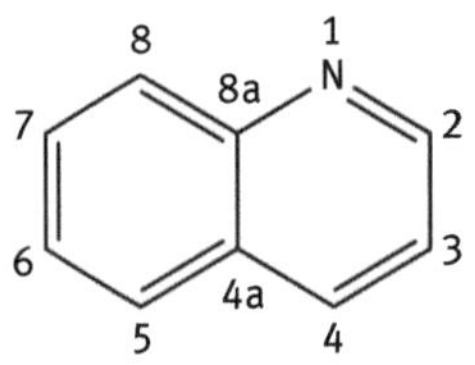

quinoline

Quinolizine (4*H*-quinolizine), named as analog of quinoline.

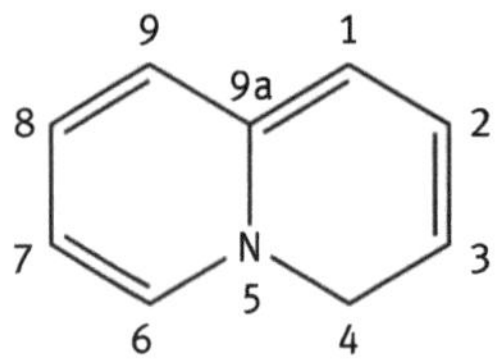

4*H*-quinolizine

Quinoxaline {benzo[*a*]pyrazine}, named as analog of quinoline and as derivative of glyoxal (ethanedial).

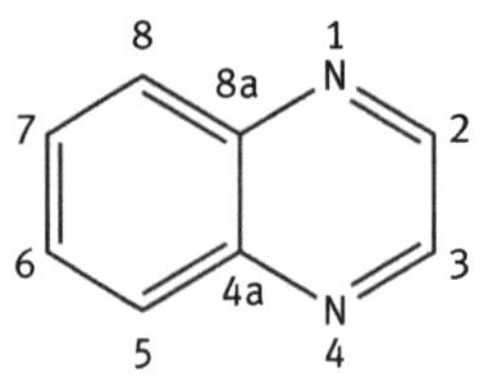

quinoxaline

Quinuclidine {1-azabicyclo[2.2.2]octane}, from Latin *nucleus*, kernel, nut, named as structural part of alkaloids.

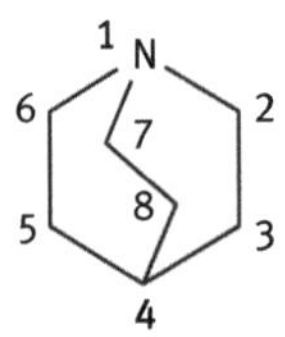

quinuclidine

Rubicene, named after its red color, from rubi-, and anthracene.

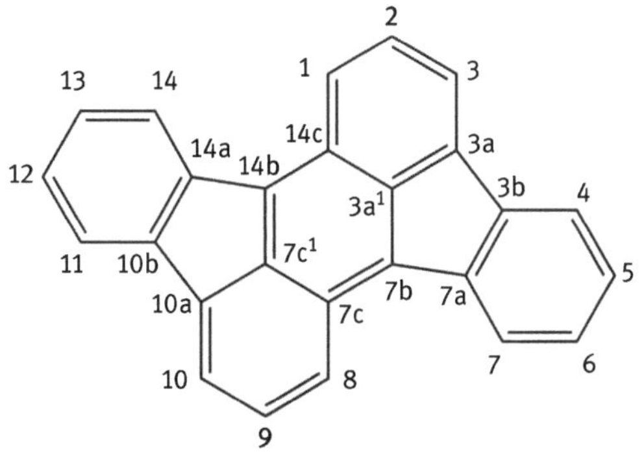

rubicene

Sapphyrin {(6*Z*,11*Z*,15*Z*,20*Z*)-25,26,27,28,29-pentaazahexacyclo[20.2.1.1^{2,5}.1^{7,10}.1^{12,15}.1^{17,29}]nonacosa-1(24),2,4,6,8,10(28),11,13,15,17(26),18,20,22-tridecaene}, named after its blue color, from sapphire.

Selenophene, named as selenium analog of thiophene.

Tellurophene, named as tellurium anlog of thiophene.

Tetracene {benz[*b*]anthracene}, named as four-ring homolog of anthracene. Higher homologs are named in the same manner (pentacene etc.). Earlier called **naphthacene**.

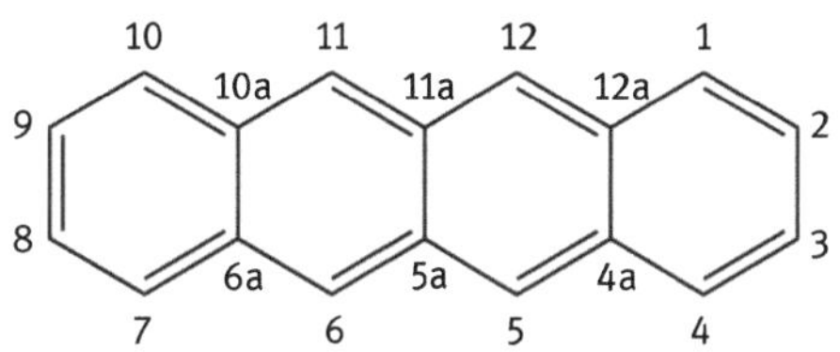

tetracene

Tetraphene {benz[*a*]anthracene}, named as four fused benzene rings.

Thianthrene (dibenzo-1,4-dithiine, 9,10-dithiaanthracene), named as sulfur-containing analog of anthracene.

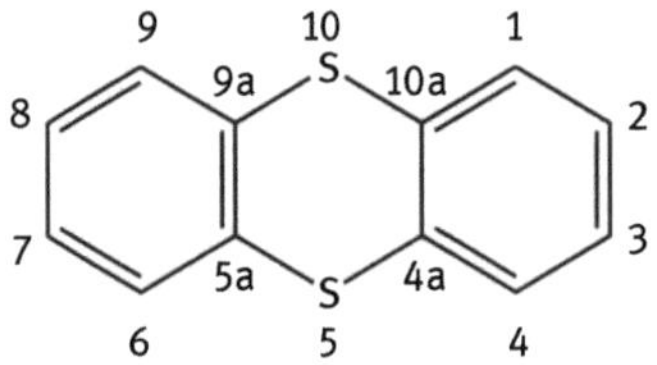

thianthrene

Thiophene, from Greek *theion*, sulfur, literally the divine, and phene (benzene), from Greek *phainein*, to show.

Trinaphthylene {dinaphth[2,3-*a*:2′,3′-*c*]anthracene, naphtho[2,3-*h*]pentaphene}, named as three fused 2,3-naphthylene units.

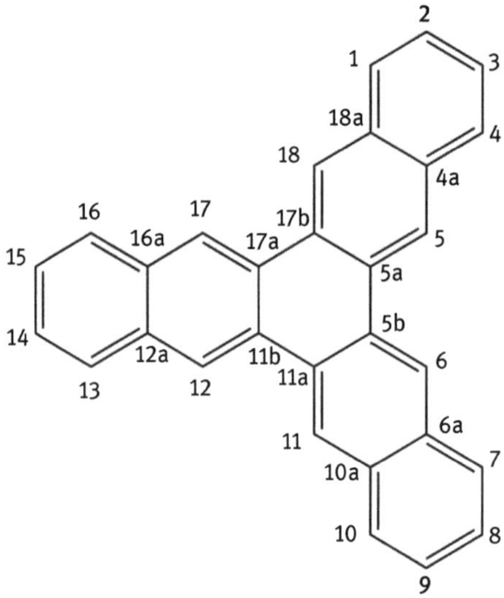

trinaphthylene

Triphenylene {benzo[*l*]phenanthrene}, named as three fused *o*-phenylene units. A.k.a. **isochrysene.**

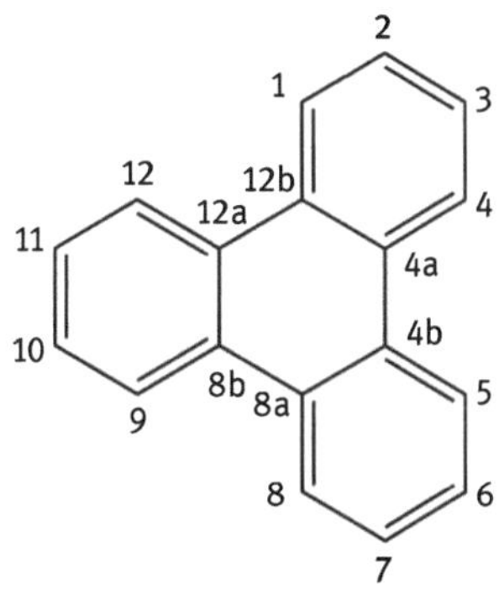

triphenylene

Xanthene {9*H*-xanthene, dibenzo[*a*,*e*]pyran}, from Greek *xanthos*, yellow, referring to this compound's yellow color.

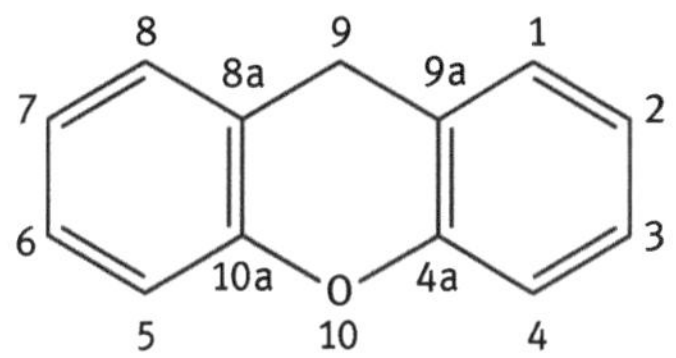

9*H*-xanthene

4.1.3 Natural products

IUPAC nomenclature operates with special procedures for the naming of, especially chiral, compounds such as alkaloids (cf. Section 4.1.3.1.), amino acids and peptides (cf., Chapter 10), carbohydrates (cf. Chapter 9), carotenes (cf. Chapter 8), corrinoids, cyclitols, lipids (cf. Chapter 7), nucleosides, nucleotides, steroids, terpenes (cf. Chapter 8), and tetrapyrroles, based on a large portfolio of stereoparents, each provided with a specific array of locants.

4.1.3.1 Alkaloid stereoparent hydrides

Aconitane {(1*R*,2*R*,3*R*,5*R*,8*S*,9*R*,13*R*,17*R*)-11-azahexacyclo[7.7.2.$1^{2,5}$.$0^{1,10}$.$0^{3,8}$.$0^{13,17}$]nonadecane}, after the genus *Aconitum* (aconite, monkshoods, wolf's bane, women's bane, devil's helmet, queen of poisons, blue rocket). Suggestions to derive the genus name from Greek *akonitos*, without dust, one referring to these plants' habitat on rocky ground, and another, in terms of able to prevail in a fight without effort, i.e. invincible, bear the hallmark of folk etymology. Another possibility is a derivation from Greek *akon*, dart, javelin, referring to the practice of poisoning the tips of weapons with the poisonous juice of such a plant. Most likely the Greek name of the plant is from a Semitic source.

Ajmalan {(1*S*,9*R*,10*S*,12*S*,13*S*,16*S*)-13-ethyl-8-methyl-8,15-diazahexacyclo[14.2.1.$0^{1,9}$.$0^{2,7}$.$0^{10,15}$.$0^{12,17}$]nonadeca-2,4,6-triene}, after the alkaloid ajmaline, named after the Indian physician and Muslim leader Hakim Ajmal Khan (1863–1927).

Akuammilan {(2*R*,3*E*,7a*R*,12b*S*,13*R*)-3-ethylidene-13-methyl-1,2,4,6,7,12b-hexahydro-2*H*-2,7a-methanoindolo[2,3-*a*]quinolizine}, from Twi *akuamma*, owala tree (*Pentaclethra africana*).

Alstophyllan {(4a*R*,6*S*,13*S*,13a*R*)-4-ethyl-7,14-dimethyl-1,4a,5,6,7,12,13,13a-octahydro-6,13-iminopyrano[3′,4′:5,6]cycloocta[1,2-*b*]indole}, with contraction, after the species *Alstonia macrophylla* (batino, devil tree). The genus name is after the British botanist Charles Alston (1685–1760).

Aporphine {6-methyl-5,6,6a,7-tetrahydro-4*H*-dibenzo[*de*,*g*]quinoline}, a contraction of apomorphine, so named as the product of the formal loss of one molecule of water from morphine.

Aspidofractinine {(9*R*,21*S*)-2,12-diazahexacyclo[14.2.2.$1^{9,12}$.$0^{1,8}$.$0^{3,8}$.$0^{16,21}$]henicosa-3,5,7-triene}, with contraction, after the species *Aspidosperma refractum* (oleander), a.k.a. *Aspidosperma pyrifolium*. The genus name is from Greek *aspis, aspid-*, shield, and *sperma*, seed. The former specific epithet is from Latin *refractus*, broken up, open.

Aspidospermidine {(1*R*,9*R*,12*R*,19*R*)-12-ethyl-8,16-diazapentacyclo[10.6.1.$0^{1,9}$.$0^{2,7}$.$0^{16,19}$]nonadeca-2,4,6-triene}, cf. aspidofractinine.

Atidane {(1*S*,2*R*,4*S*,5*S*,7*R*,10*R*,11*R*)-5-methyl-13-azapentacyclo[9.3.3.$2^{4,7}$.$0^{1,10}$.$0^{2,7}$]nonadecane}, named after the atis plant (*Aconitum heterophyllum*), from Hindi *atis*, atis plant.

Atisine {(1*S*,2*R*,4*S*,6*R*,7*S*,10*R*,11*R*,17*S*)-11-methyl-5-methylidene-16-oxa-13-azahexacyclo[9.6.3.$2^{4,7}$.$0^{1,10}$.$0^{2,7}$.$0^{13,17}$]docosan-6-ol}, cf. atidane.

Atropine (tropan-3α-yl 3-hydroxy-2-phenylpropanoate), after the species *Atropa belladonna* (belladonna, deadly nightshade). The genus name is from Greek *Atropos*, literally she who may not be turned aside, one of the Three Fates and cutter of the thread of life, referring to the toxicity of these plants. A.k.a. **duboisine**, after the genus *Duboisia* (corkwood tree), named after the French agronomist and writer

Louis Du Bois (1773–1855). A.k.a. **daturine**, after the genus *Datura* (jimsonweed); this genus name is from Hindi *dhatura*, jimsonweed. A.k.a. **hyoscyamine**, after the genus *Hyoscyamus* (henbane); this genus name is from Greek *hyoskyamos*, henbane, from *hys*, *hyo-*, pig, and *kyamos*, bean.

Berbaman {(1*S*,14*R*)-7,23-dioxa-15,30-diazaheptacyclo[22.6.2.$2^{3,6}$.$1^{8,12}$.$1^{14,18}$.$0^{27,31}$.$0^{22,33}$]hexatriaconta-3(36),4,6(35),8,10,12(34),18(33),19,21,24(32),25,27(31)-dodecaene}, after the genus *Berberis* (barberry). The genus name is ultimately from Arabic *barbaris*, barberry.

Berbine {5,8,13,13a-tetrahydro-6*H*-isoquinolino[3,2-*a*]isoquinoline}, cf. berbaman.

Cephalotaxine {11b*S*,12*S*,14a*R*)-13-methoxy-2,3,5,6,11b,12-hexahydro-1*H*-[1,3]dioxolo[4′,5′:4,5]benzo[1,2-*d*]cyclopenta[b]pyrrolo[1,2-*a*]azepin-12-ol}, after the genus *Cephalotaxus* (plum yew). The genus name is from Greek *kephale*, head, and Latin *taxus*, yew tree.

Cerin {2α-hydroxyfriedelan-3-one, (2*R*,4*R*,4a*S*,6a*S*,6b*R*,8a*R*,12a*R*,12b*S*,14a*S*,14b*S*)-2-hydroxy-4,4a,6b,8a,11,11,12b,14a-octamethylicosahydropicen-3(2*H*)-one}, from Greek *keros*, wax, referring to its erroneous identification as a wax after its isolation from cork.

Cevane {(1*R*,2*S*,6*S*,9*S*,10*R*,11*S*,14*S*,15*S*,18*S*,23*S*,24*S*)-6,10,23-trimethyl-4-azahexacyclo[12.11.0.$0^{2,11}$.$0^{3,9}$.$0^{15,24}$.$0^{18,23}$]pentacosane}, from Spanish *cebadilla*, sabadilla (*Schoenocaulon officinale*), diminutive of *cebada*, barley, ultimately from Latin *cibus*, food, meal.

Chelidonine {(5b*R*,8*S*,12b*S*)-13-methyl-5b,6,7,12b,13,14-hexahydro[1,3]dioxolo[4,5]benzo[1,2-*c*][1,3]dioxolo[4,5-*i*]phenanthridin-6-ol}, after the genus *Chelidonium* (celandine). The genus name is from Greek *chelidonios*, of the swallow, from *chelidon*, swallow. The genus name reflects the ancient belief that these plants' season paralleled the migration of swallows.

Cinchonan {4-[[(2*R*,4*S*,5*R*)-5-ethenyl-1-azabicyclo[2.2.2]octan-2-yl]methyl]quinoline}, after the genus *Cinchona* (tropical trees). The genus name is after Doña Francisca Henriquez de Ribera, vicereine of Peru (1569–1647).

Conanine {(1*R*,2*S*,5*S*,6*S*,9*R*,12*S*,13*S*)-6,7,13-trimethyl-7-azapentacyclo[10.9.0.$0^{2,9}$.$0^{5,9}$.$0^{13,18}$]icosane}, after the conessi tree, a.k.a. coral swirl and tellicherry bark (*Wrightia antidysenterica*), of unknown etymology.

Corynan {(2*S*,3*R*,12b*S*)-2,3-diethyl-1,2,3,4,6,7,12,12b-octahydroindolo[2,3-*a*]quinolizine}, after the genus *Corynanthe* (flowering plants). The genus name is from Greek *koryne*, club, and *anthos*, flower.

Corynoxan {(3*R*,6′*S*,7′*S*,8′a*S*)-6′,7′-diethylspiro[1,2-dihydroindole-3,1′-3,5,6,7,8,8a-hexahydro-2*H*-indolizine]}, cf. corynan.

Crinan {(1*R*,13*R*)-5,7-dioxa-12-azapentacyclo[10.5.2.$0^{1,13}$.$0^{2,10}$.$0^{4,8}$]nonadeca-2,4(8),9-triene}, after the genus *Crinum* (swamplilies). The genus name is from Greek *krinon*, lily.

Curan {(3a*S*,5*S*,6*R*,6a*S*,11b*S*,12*S*)-12-ethyl-6-methyl-1,2,3a,4,5,6,6a,7-octahydro-3,5-ethano-3*H*-pyrrolo[2,3-*d*]carbazole}, after the poison curare, from Tupi *kurari*, he to whom it comes, falls.

Daphnane {(3*S*,10*R*,13*S*,14*R*)-1-methyl-14-(propan-2-yl)-2-propyl-12-azapentacyclo[8.6.0.$0^{2,13}$.$0^{3,7}$.$0^{7,12}$]hexadecane}, after the genus *Daphne* (shrubs). The genus name is from Greek *daphne*, laurel.

Dendrobane {(1*S*,4*S*,7*S*,8*R*,11*R*,12*R*,13*S*)-2,12-dimethyl-13-(propan-2-yl)-10-oxa-2-azatetracyclo[5.4.1.$1^{8,11}$.$0^{4,12}$]tridecane}, after the genus *Dendrobium* (orchids). The genus name is from Greek *dendron*, tree, and *bios*, life.

Eburnamenine {(15*R*,19*R*)-15-ethyl-1,11-diazapentacyclo[9.6.2.$0^{2,7}$.$0^{8,18}$.$0^{15,19}$]nonadeca-2,4,6,8(18),16-pentaene}, after the species *Hunteria eburnea* (hunteria root). The specific epithet is from Latin *eburneus*, of ivory. The genus name is after the British physician and botanist William Hunter (1755–1812).

Emetan {(2*S*,3*R*,11b*S*)-3-ethyl-2-[[(1*R*)-1,2,3,4-tetrahydroisoquinolin-1-yl]methyl]-2,3,4,6,7,11b-hexahydro-1*H*-benzo[*a*]quinolizine}, after emetin, from Greek *emetos*, vomiting, referring to emetin's emetic property.

Ergoline {(6a*R*,10a*R*)-4,6,6a,7,8,9,10,10a-octahydroindolo[4,3-*fg*]quinoline}, from ergot (*Claviceps purpurea*), from French *ergot* (cock's spur).

Ergotaman {(1S,2*S*,4*R*)-*N*-[[(6a*R*,9*S*)-7-methyl-6,6a,8,9-tetrahydro-4*H*-indolo[4,3-*fg*]quinolin-9-yl]methyl]-3-oxa-6,9-diazatricyclo[7.3.0.$0^{2,6}$]dodecan-4-amine}, cf. ergoline.

Erythrinan {(4a*S*,13b*S*)-2,3,4,4a,5,6,8,9-octahydro-1*H*-indolo[7a,1-*a*]isoquinoline}, after the genus *Erythrina* (coral tree, flame tree). The genus name is from Greek *erythros*, red.

Evonimine (a macrolide), after the genus *Euonymus* (spindle tree, burning-bush, strawberry bush, wahoo, wintercreeper). The genus name is from Greek *euonymos*, having an auspicious name, from *eu*, well, and *onoma*, name.

Evonine {[(1*S*,3*R*13*S*,14*S*,17*S*,18*R*,20*S*,21*S*,23*R*,24*R*)-18,19,21,24-tetra(acetyloxy)-25-hydroxy-3,13,14,25-tetramethyl-6,15,22-trioxo-2,5,16-trioxa-11-azapentacyclo[15.7.1.$0^{1,20}$.$0^{3,23}$.$0^{7,12}$]pentacosa-7(12),8,10-trien-20-yl]methyl acetate}, cf. evonimine.

Formosanan {(3*S*,4′a*S*,5′a*S*,10′aR)-spiro[1,2-dihydroindole-3,6′-1,4a,5,5a,7,8,10,10a-octahydropyrano[3,4-*f*]indolizine}, after the species *Uncaria formosana* (a tropical woody vine). The specific epithet is from New Latin *formosanus*, from the island of Formosa, now called Taiwan.

Galanthamine {(1*S*,12*S*,14*R*)-9-methoxy-4-methyl-11-oxa-4-azatetracyclo[8.6.1.$0^{1,12}$.$0^{6,17}$]heptadeca-6(17),7,9,15-tetraen-14-ol}, after the plant genus *Galanthus* (snowdrops). The genus name is from Greek *gala*, milk, and *anthos*, flower.

Galanthan {(1*S*,12*S*,16*R*)-9-azatetracyclo[7.6.1.$0^{2,7}$.$0^{12,16}$]hexadeca-2,4,6-triene}, cf. galanthamine.

Hasubanan {(1*R*,10*S*)-17-azatetracyclo[8.4.3.$0^{1,10}$.$0^{2,7}$]heptadeca-2,4,6-triene}, from Japanese *hasu-no-ha-kazura*, snake vine (*Stephania japonica*).

Hetisan {(1*R*,5*R*,11*R*,14*R*,16*S*,17*R*,18*R*)-5-methyl-12-methylidene-7-azaheptacyclo[9.6.2.$0^{1,8}$.$0^{5,17}$.$0^{7,16}$.$0^{9,14}$.$0^{14,18}$]nonadecane}, from hetisine, an alkaloid named by contraction of heteroatisine.

Ibogamine {(1*R*,15*R*,17*S*,18*S*)-17-ethyl-3,13-diazapentacyclo[13.3.1.$0^{2,10}$.$0^{4,9}$.$0^{13,18}$]nonadeca-2(10),4,6,8-tetraene}, after iboga (*Tabernanthe iboga*). Iboga is a native African name for this tree.

Kopsan {(1*R*,4*R*,12*R*,13*S*,18*R*)-5,14-diazaheptacyclo[12.5.3.$0^{1,13}$.$0^{4,12}$.$0^{4,18}$.$0^{6,11}$.$0^{12,16}$]docosa-6,8,10-triene}, after the plant genus *Kopsia* (vinca). The genus name is after the Danish botanist Jan Kops (1765–1849).

Lunarine {(18*E*)-14-oxa-3,21,25-triazatetracyclo[24.2.1.$0^{7,15}$.$0^{8,13}$]nonacosa-7,9,12,18,27-pentaene-4,11,23-trione}, after the genus *Lunaria* (honesty). The genus name is from Latin *lunarius*, moon-like.

Lycopodane {(1*S*,2*R*,10*S*,13*S*)-6-azatetracyclo[8.6.0.$0^{1,6}$.$0^{2,13}$]hexadecane}, after the genus *Lycopodium* (ground pines, creeping cedar). The genus name is from Greek *lykos*, wolf, and *podion*, diminutive of *pous*, foot.

Lycorenan {(5a*R*,11b*S*,11c*S*)-1,2,3,5,5a,7,1b,11c-octahydro[2]benzopyrano[3,4-*g*]indole}, after the genus *Lycoris* (spider lily, hurricane lily, cluster amaryllis). The genus name is from Greek *lykoros*, twilight.

Lythran {(1*S*,13*Z*,17*S*,19*S*)-16-oxa-24-azapentacyclo[15.7.1.$1^{8,12}$.$0^{1,7}$.$0^{19,24}$]hexacosa-2,4,6,8,10,12(26),13-heptaene}, after the genus *Lythrum* (loosestrife). The genus name is from Greek *lythron*, gore, referring to the color of the flowers.

Lythranidine {23-methoxy-25-azatetracyclo[18.3.1.$1^{2,6}$.$1^{11,15}$]hexacosa-1(23),2,4,6(26),20(24),21-hexaene-3,9,17-triol}, as lythran.

Matridine {(1*R*,2*R*,9*S*,17*S*)-7,13-diazatetracyclo[7.7.1.$0^{2,7}$.$0^{13,17}$]heptadecane}, after the genus *Matricaria* (mayweed, chamomile). The genus name is from Latin *matricaria*, feverfew (*Tanacetum parthenium*), from *matrix*, womb, referring to the ancient use of *Matricaria chamomilla* for treating uterus diseases.

Morphinan {(4a*R*)-1,2,4,8,10,10aα-hexahydro-2*H*-10α,4aα-[iminoethano]phenanthrene}, from morphine, named after the Greek god of dreams Morpheus, literally the maker of shapes, from Greek *morphe*, form, shape, referring to this compound's analgesic and sedative properties.

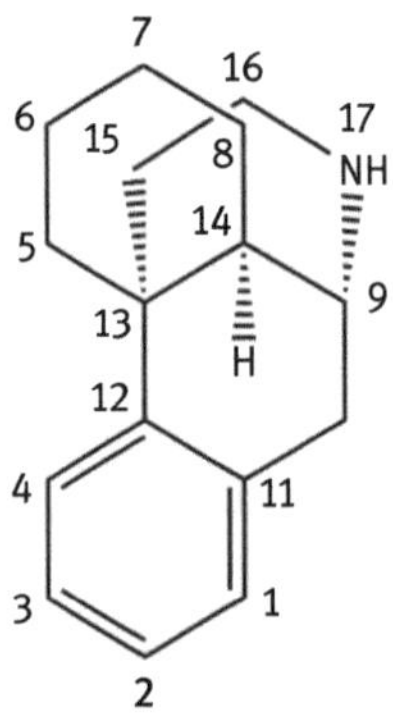

morphinan

Nupharidine {(1*R*,4*S*,5*R*,7*S*,9a*S*)-4-(furan-3-yl)-1,7-dimethyl-5-oxido-2,3,4,6,7,8,9,9a-octahydro-1*H*-quinolizin-5-ium, a cation}, after the genus *Nuphar* (water-lily, pond-lily, alligator-bonnet, bonnet-lily, spatterdock). The genus name is ultimately from Sanskrit *nilotpala*, blue lotus flower, from *nila*, dark blue, and *utpala*, nenuphar blossom.

Ormosanine {(1*R*,2*R*,7*S*,9*S*,10*R*)-1-[(2*R*)-piperidin-2-yl]-3,15-diazatetracyclo[7.7.1.$0^{2,7}$.$0^{10,15}$]heptadecane}, after the genus *Ormosia* (horse-eye beans, ormosias). The genus name is from Greek *hormos*, chain, necklace.

18-Oxayohimban {(1*S*,15*R*,20*R*)-17-oxa-3,13-diazapentacyclo[11.8.0.$0^{2,10}$.$0^{4,9}$.$0^{15,20}$]henicosa-2(10),4,6,8-tetraene}, after the species yohimbe (*Pausinystalia yohimbe*). The plant name is from a native language of Cameroon.

Oxyacanthan {(1*R*,14*S*)-8,23-dioxa-15,30-diazaheptacyclo[22.6.2.$2^{9,12}$.$1^{3,7}$.$1^{14,18}$.$0^{27,31}$.$0^{22,33}$]hexatriaconta-3(36),4,6,9(35),10,12(34),18(33),19,21,24(32),25,27(31)-dodecaene}, after the species *Crataegus oxyacantha* (hawthorn). The specific epithet is from Greek *oxys*, sharp, and *akantha*, thorn, spine.

Pancracine {(1*S*,13*S*,15*S*,16*S*)-5,7-dioxa-12-azapentacyclo[10.6.1.$0^{2,10}$.$0^{4,8}$.$0^{13,18}$]nonadeca-2,4(8),9,17-tetraene-15,16-diol}, after the genus *Pancratium* (daffodils, narcissus). The genus name is from Greek *pankrates*, all-powerful.

Rheadan {(1*R*,11*R*)-19-oxa-10-azatetracyclo[9.8.0.$0^{2,7}$.$0^{12,17}$]nonadeca-2,4,6,12,14,16-hexaene}, after the species *Papaver rhoeas* (common poppy, corn poppy, corn rose, field poppy, Flanders poppy, red poppy). The specific epithet is from Greek *rhoias* (corn poppy).

Rodiasine {(1*R*,14*S*)-9,20,21,25-tetramethoxy-15,30-dimethyl-23-oxa-15,30-diazaheptacyclo[22.6.2.1^{3,7}.1^{8,12}.1^{14,18}.0^{27,31}.0^{22,33}]pentatriaconta-3(35),4,6,8,10,12(34),18,20,22(33),24,26,31-dodecaen-6-ol}, after the species *Chlorocardium rodiei* (cogwood, demerara greenheart, ispingo moena, sipiri, bebeeru, bibiru). The specific epithet is after the 19th century Royal Navy surgeon Hugh Rodie, who first described the medicinal uses of this plant.

Samandarin {(1*R*,2*S*,3*S*,6*R*,8*S*,10*S*,11*S*,14*R*,16*S*)-2,6-dimethyl-19-oxa-17-azapentacyclo[14.2.1.0^{2,14}.0^{3,11}.0^{6,10}]nonadecan-8-ol}, with contraction, after the genus *Salamander* (amphibians). The genus name is from Greek *salamandra*, lizard, probably of oriental origin. A.k.a. **samandarine**.

Santonin {(3*S*,3a*S*,5a*S*,9b*S*)-3,5a,9-trimethyl-3a,4,5,9b-tetrahydro-3*H*-benzo[*g*][1]benzofuran-2,8-dione}, from Latin *herba santonica*, Levant wormseed (*Artemisia pauciflora*). The Latin name is from *herba*, herb and *santonicus*, of the Santones, an ancient Celtic people.

Sarpagan {(6*S*,9*E*)-9-ethylidene-11-methyl-5,6α,8,9,10α,11β,11aβ,12-octahydro-6,10-methanoindolo[3,2-*b*]quinolizine}, after Hindi *sarpagandha*, Indian snakeroot (*Rauwolfia serpentina*).

Senecionan {(1*R*,4*Z*,6*R*,7*S*,17*R*)-4-ethylidene-6,7-dimethyl-2,9-dioxa-14-azatricyclo[9.5.1.0^{14,17}]heptadec-11-ene}, after the genus *Senecio* (ragworts, groundsel). The genus name is from Latin *senecio, senecion-*, old man, groundsel, from *senex, senec-*, old man, referring to the gray hair on the seeds of these plants.

Solanidine {(1*S*,2*S*,7*S*,10*R*,11*S*,14*S*,15*R*,16*S*,17*R*,20*S*,23*S*)-10,14,16,20-tetramethyl-22-azahexacyclo[12.10.0.0^{2,11}.0^{5,10}.0^{15,23}.0^{17,22}]tetracos-4-en-7-ol}, after the genus *Solanum* (nightshades). The genus name is from Latin *solanum*, nightshade. The Latin plant name *solanum* is of obscure etymology.

Sparteine {(1*S*,2*R*,9*S*,10*S*)-7,15-diazatetracyclo[7.7.1.0^{2,7}.0^{10,15}]heptadecane}, after the genus *Spartium* (Spanish broom). The genus name is from Latin *spartum*, the herb broom.

Spirosolane {(1*R*,2*S*,4*S*,7*S*,8*R*,9*S*,12*S*,13*S*)-5′,7,9,13-tetramethylspiro[5-oxapentacyclo[10.8.0.0^{2,9}.0^{4,8}.0^{13,18}]icosane-6,2′-piperidine]}, named as spiro compound; cf. solanidine. A.k.a. **tomatanine.**

Strychnidine {4a*R*,5a*S*,8a*R*,13a*S*,15a*S*,15b*R*)-2,4a,5,5a,7,8,13a,14,15,15a,15b,16-dodecahydro-4,6-methanoindolo[3,2,1-*ij*]oxepino[2,3,4-*de*]pyrrolo[2,3-*h*]quinoline}, after the genus *Strychnos* (trees, vines). The genus name is from Greek *strychnos*, nightshade, from *strychnos*, acrid, bitter.

Tazettine {18-methoxy-15-methyl-5,7,12-trioxa-15-azapentacyclo[11.7.0.0^{1,16}.0^{2,10}.0^{4,8}]icosa-2,4(8),9,19-tetraen-13-ol}, after the species *Narcissus tazetta* (paperwhite, bunch-flowered narcissus, bunch-flowered daffodil, Chinese sacred lily, cream narcissus, joss flower, polyanthus narcissus). The specific epithet is from Italian *tazzetta*, diminutive of *tazza*, cup. A.k.a. **sekisanine** and **ungernine**.

Trichothecane {(1*S*,2*R*,7*R*,9*R*,12*S*)-1,2,5,12-tetramethyl-8-oxatricyclo[7.2.1.$0^{2,7}$]dodecane}, after the fungal genus *Trichothecium*. The genus name is from trich(o)- and Greek *thekion*, diminutive of *theke*, case.

Tropane {(1*S*,5*R*)-8-methyl-8-azabicyclo[3.2.1]octane}, after atropine which in turn was named after the genus *Atropa* (nightshades). The genus name is from Greek *Atropos*, literally she who may not be turned aside, one of the Three Fates and cutter of the thread of life, referring to the toxicity of these plants.

Tubocuraran {(1*S*,16*R*)-7,23-dioxa-15,30-diazaheptacyclo[22.6.2.$2^{3,6}$.$1^{8,12}$.$1^{18,22}$.$0^{27,31}$.$0^{16,34}$]hexatriaconta-3(36),4,6(35),8,10,12(34),18(33),19,21,24(32),25,27(31)-dodecaene}, from tubocurare, i.e. curare shipped in sections of hollow bamboo.

Tubulosan {(2*S*,3*R*,11b*S*)-3-ethyl-2-[[(1*R*)-2,3,4,9-tetrahydro-1*H*-pyrido[3,4-*b*]indol-1-yl]methyl]-2,3,4,6,7,11b-hexahydro-1*H*-benzo[*a*]quinolizine}, after the species *Pogonopus tubulosus* (a tropical tree). The specific epithet is from Latin *tubulosus*, resembling or having the form of a tube.

Veratraman {(2*R*,5*S*)-2-[(1*S*)-1-[(6a*S*,6b*S*,9*S*,11a*S*,11b*R*)-10,11b-dimethyl-1,2,3,4,6,6a,7,8,9,11,11a-dodecahydrobenzo[*a*]fluoren-9-yl]ethyl]-5-methylpiperidine}, after the genus *Veratrum* (false hellebore, corn lily). The genus name is from Latin *veratrum*, hellebore, from *verus*, true, referring to the ancient belief that subsequent sneezing confirms the truth of a statement.

Vincaleukoblastine {methyl (3a*R*,3a^1*R*,4*R*,5*S*,5a*R*,10b*R*)-4-acetoxy-3a-ethyl-9-[(3*S*,5*S*,7*S*,9*S*)-5-ethyl-5-hydroxy-9-(methoxycarbonyl)-2,4,5,6,7,8,9,10-octahydro-1*H*-3,7-methano[1]azacycloundecino[5,4-*b*]indol-9-yl]-5-hydroxy-8-methoxy-6-methyl-3a,3a1,4,5,6,11,12-octahydro-1*H*-indolizino[8,1-*cd*]carbazole-5-carboxylate}, cf. vincane.

Vincane {(15*S*,19*S*)-15-ethyl-1,11-iazapentacyclo[9.6.2.$0^{2,7}$.$0^{8,18}$.$0^{15,19}$]nonadeca-2,4,6,8(18)-tetraene}, after the genus *Vinca* (periwinkle). The genus name is from Latin *pervinca*, periwinkle, perhaps from *pervincire*, to entwine, to bind.

Vobasan {(1*R*,14*R*,15*E*,18*S*)-15-ethylidene-17,18-dimethyl-10,17-diazatetracyclo[12.3.1.$0^{3,11}$.$0^{4,9}$]octadeca-3(11),4,6,8-tetraene}, after the genus *Voacanga* (voacanga), from Malagasy, and from base.

Vobtusine {methyl (1*R*,1′*R*,7′*S*,11′*R*,12*R*,13′*R*,16*S*,17*S*,22*R*,24′*R*,25′*S*)-24′-hydroxy-19′-methoxyspiro[15-oxa-8,19-diazahexacyclo[10.9.1.$0^{1,9}$.$0^{2,7}$.$0^{12,16}$.$0^{19,22}$]docosa-2,4,6,9-tetraene-17,15′-8-oxa-4,17-diazaheptacyclo[11.10.1.$1^{1,4}$.$0^{7,11}$.$0^{17,24}$.$0^{18,23}$.$0^{11,25}$]pentacosa-18(23),19,21-triene-10-carboxylate}, with contraction, after the species *Voacanga thouarsii* var. *obtusa*. The subspecific epithet is from Latin *obtusus*, dulled, from ob- and *tundere*, to beat.

Yohimban {(4a*R*,13b*S*,14a*S*)-1,3,11,12,14,15,16,17,18,19,20,21-dodecahydroindolo[2′,3′:3,4]pyrido[1,2-*b*]isoqinoline}, cf. 18-oxayohimban.

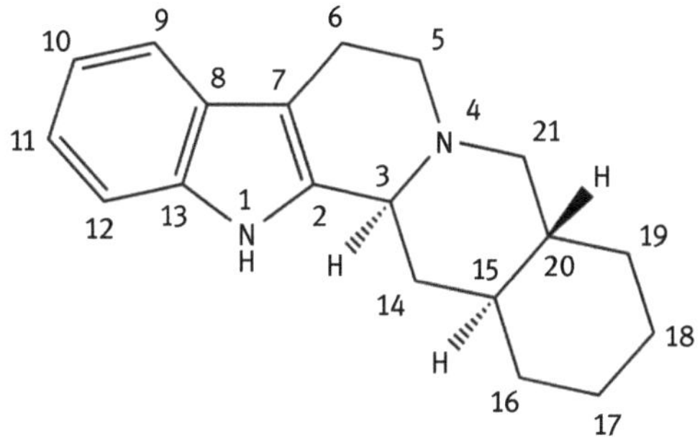

yohimban

4.1.3.2 Steroid stereoparent hydrides

Androstane {5α-androstane, (5*R*,8*S*,9*S*,10*S*,13*S*,14*S*)-10,13-dimethyltetradecahydro-1*H*-cyclopenta[*a*]phenanthrene}, from Greek *aner, andro*-, male person, stamen, and sterol. A.k.a. **etioallocholane**. Testane is the corresponding 5β-isomer.

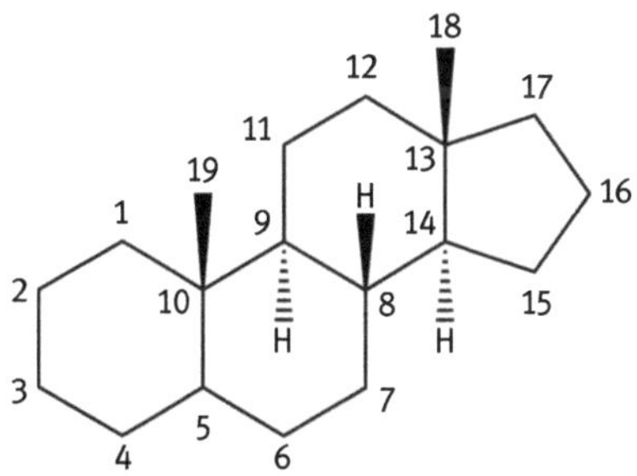

5α-androstane

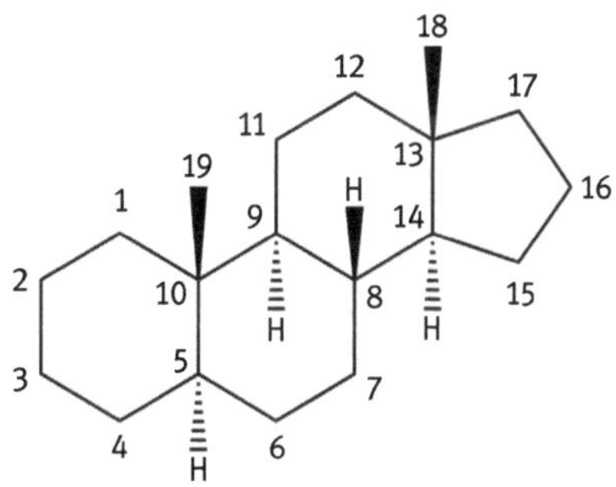

5β-androstane

Bufanolide {(5*R*)-5-[(8*R*,9*S*,10*S*,13*S*,14*R*,17*R*)-10,13-dimethyltetradecahydro-1*H*-cyclopenta[*a*]phenanthren-17-yl]oxan-3-one}, after the genus *Bufo* (toads), from Latin *bufo, bufon*-, toad, and ‘olide’.

Campestane {(8*R*,9*S*,10*S*,13*R*,14*S*,17*R*)-17-[(2*R*,5*R*)-5,6-dimethylheptan-2-yl]-10,13-dimethyltetradecahydro-1*H*-cyclopenta[*a*]phenanthrene}, after the species *Brassica campestris* (field mustard), from Latin *campester*, of the field, and sterol.

Cardanolide {(4*R*)-4-[(8*R*,9*S*,10*S*,13*S*,14*R*,17*R*)-10,13-dimethyltetradecahydro-1*H*-cyclopenta[*a*]phenanthren-17-yl]oxolan-2-one}, a contraction of cardiac and 'olide', cf. bufanolide.

Cholane {5β-cholane, (5*S*,8*R*,9*S*,10*S*,13*R*,14*S*,17*R*)-19,13-dimethyl-17-[(2*R*)-pentan-2-yl]tetradecahydro-1*H*-cyclopenta[*a*]phenanthrene}, from Greek *chole*, bile. **Allocholane** is the corresponding 5α-isomer.

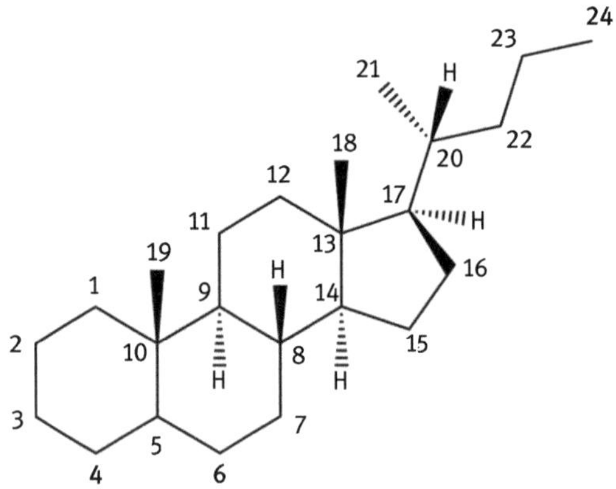

5α-cholane

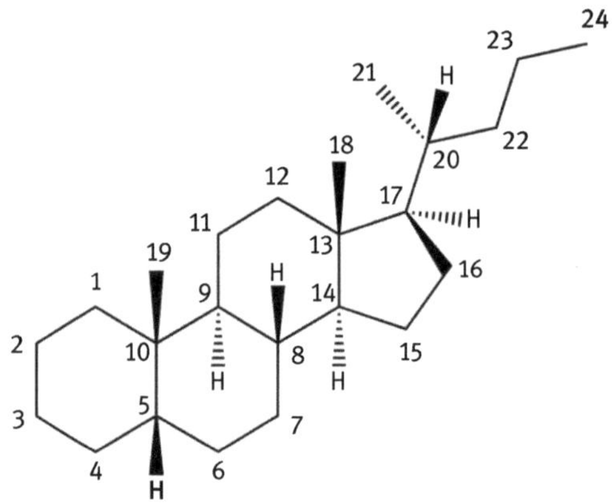

5β-cholane

Cholestane {(8*R*,9*S*,10*S*,13*R*,14*S*,17*R*)-10,13-dimethyl-17-[(2*R*)-6-methylheptan-2-yl] tetradecahydro-1*H*-cyclopenta[*a*]phenanthrene}, from Greek *chole*, bile, and sterol.

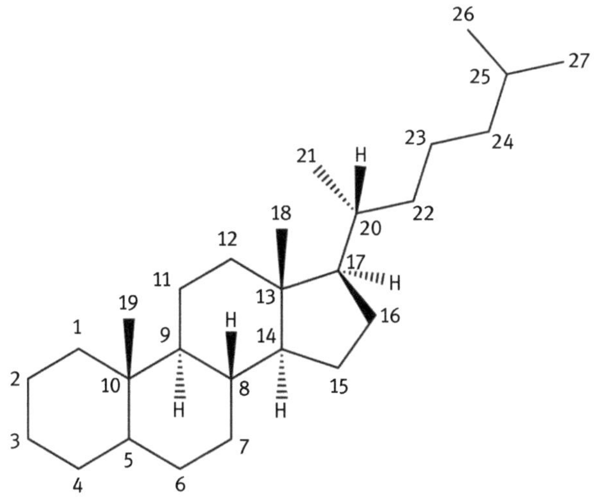

5α-cholestane

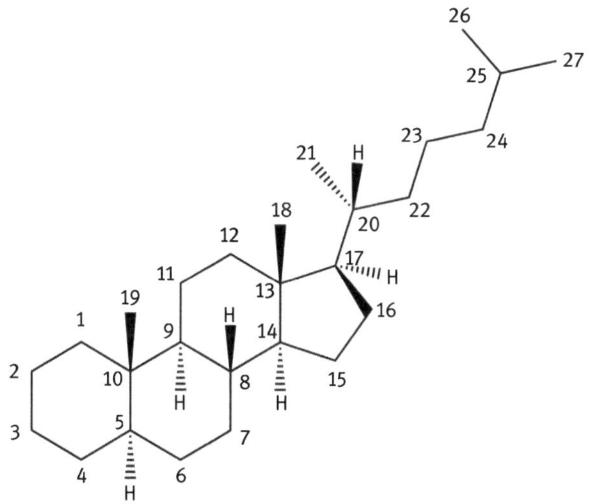

5β-cholestane

Ergostane {(8*R*,9*S*,10*S*,13*R*,14*S*,17*R*)-17-[(2*R*,5*S*)-5,6-dimethylheptan-2-yl]tetradeca-hydro-1*H*-cyclopenta[*a*]phenanthrene}, from ergot and sterol.

Estrane {(8*R*,9*R*,10*S*,13*S*,14*S*)-13-methylhexadecahydrocyclopenta[*a*]phenanthrene}, from estrus, ultimately from Greek *oistros*, gadfly, frenzy.

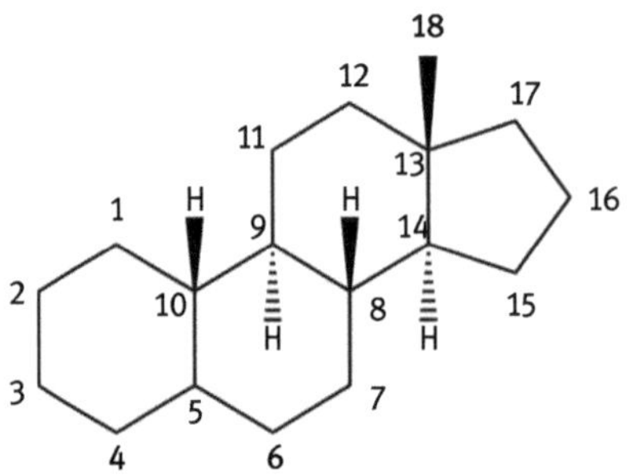

5α-estrane

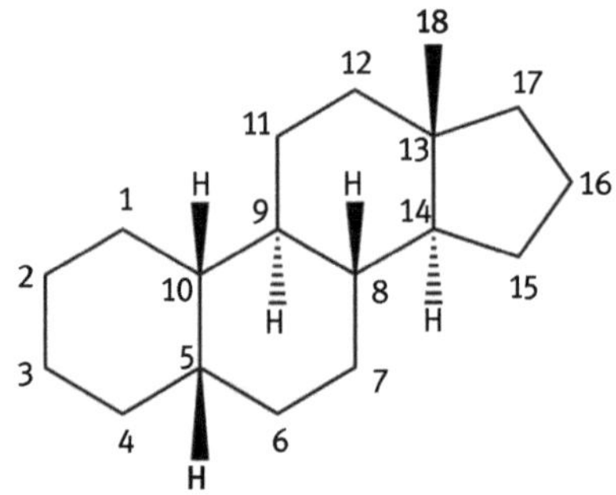

5β-estrane

Furostan {(1*R*,2*S*,4*S*,7*S*,8*R*,9*S*,12*S*,13*S*)-7,9,13-trimethyl-6-(3-methylbutyl)-5-oxapentacyclo[10.8.0.$0^{2,9}$.$0^{4,8}$.$0^{13,18}$]icosane}, from furan and sterol.

Gonane {(8*S*,9*R*,10*S*,13*S*,14*R*)-hexadecahydro-1*H*-cyclopenta[*a*]phenanthrene}, from gonad, ultimately from New Latin *gonas, gonad-*, gonad, ultimately from Greek *gone*, child, offspring. A.k.a. **18-norestrane.**

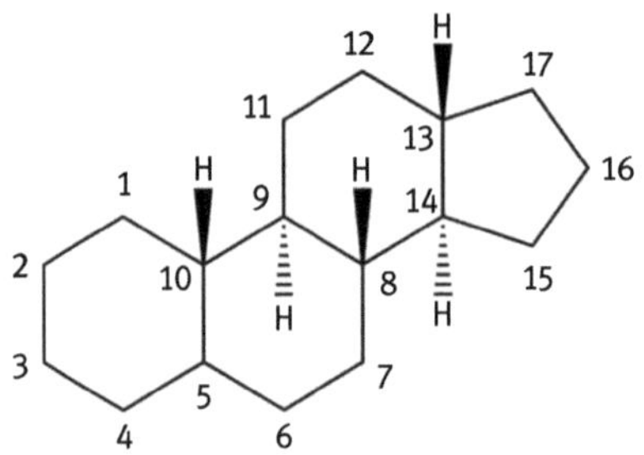

5α-gonane

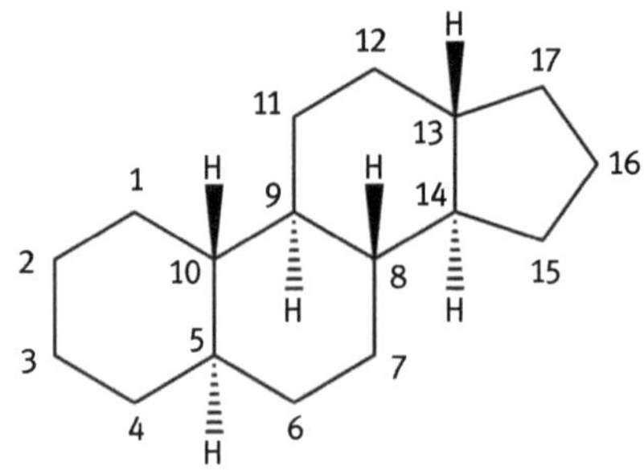

5β-gonane

Gorgostane {(8*R*,9*S*,10*S*,13*R*,14*S*,17*R*)-10,13-dimethyl-17-[(1*S*)-1-[(1*R*,2*R*)-2-methyl-2-[(2*R*)-3-methylbutan-2-yl]cyclopropan-1-yl]ethyl]tetradecahydro-1*H*-cyclopenta[*a*]phenanthrene}, after the order *Gorgonacea* (gorgonians), later renamed *Alcyonacea*, from Latin *gorgonia*, coral, ultimately from Greek *gorgos*, dreadful.

Poriferastane {(8*R*,9*S*,10*S*,13*R*,14*S*,17*R*)-17-[(2*R*,5*S*)-5-ethyl-6-methylheptan-2-yl]-10,13-dimethyltetradecahydro-1*H*-cyclopenta[*a*]phenanthrene}, after the phylum *Porifera* (sponges), from Latin *poriferus*, pore-carrying, and sterol.

Pregnane {5β-pregnane, (5*S*,8*S*,9*S*,10*S*,13*R*,14*S*,17*S*)-17-ethyl-10,13-dimethyltetradecahydro-1*H*-cyclopenta[*a*]phenanthrene}, from pregnant. **Allopregnane** is the corresponding 5α-isomer.

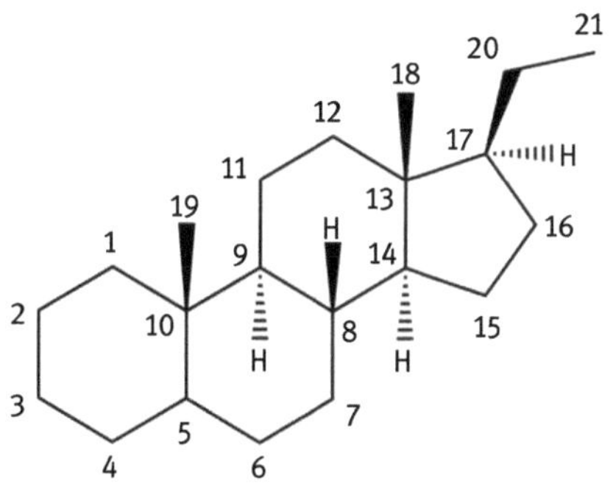

5α-pregnane

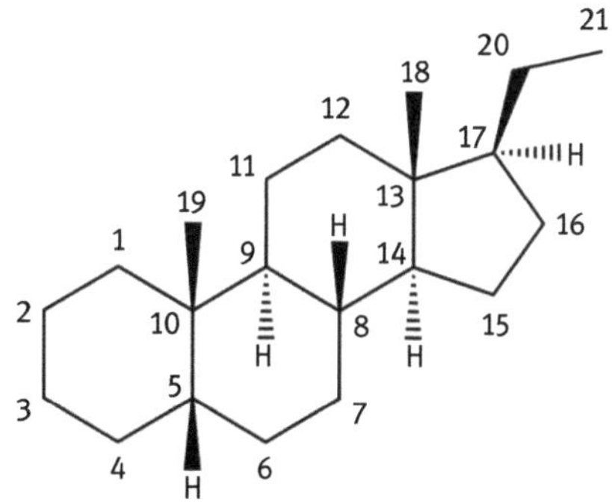

5β-pregnane

Spirostan {(1*R*,2*S*,4*S*,6*R*,7*S*,8*R*,9*S*,12*S*,13*S*)-5′,7,9,13-tetramethylspiro[5-oxapentacyclo[10.8.0.0^{2,9}.0^{4,8}.0^{13,18}]icosane-6,2′-oxane]}, from 'spiro', from Greek *speira*, coil, twist, indicating two rings with a shared atom, and sterol.

Stigmastane {(8*R*,9*S*,10*S*,13*R*,14*S*,17*R*)-17-[(2*R*,5*R*)-5-ethyl-6-methylheptan-2-yl]-10,13-dimethyltetradecahydro-1*H*-cyclopenta[*a*]phenanthrene}, after the genus *Physostigma* (vines), from Greek *physa*, bladder, and *stigma*, mark, tattoo, referring to the large hood covering the stigma, and sterol.

4.1.3.3 Terpene stereoparent hydrides

Abietane {(2*S*,4a*S*,4b*R*,8a*S*,10a*S*)-4b,8,8-trimethyl-2-(propan-2-yl)dodecahydrophenanthrene}, after the genus *Abies* (firs), from Latin *abies, abiet-*, fir.

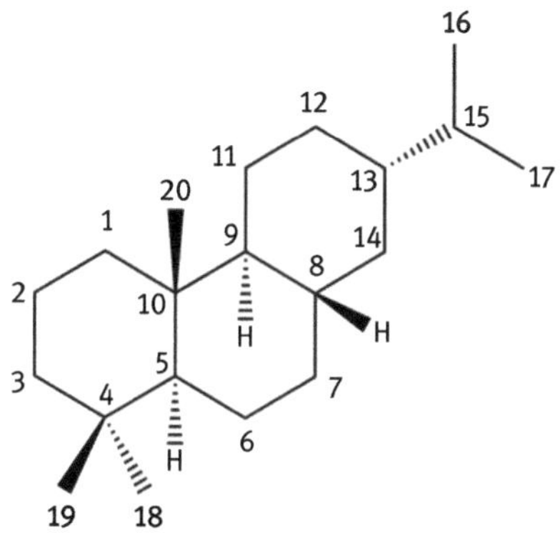

abietane

Ambrosane [(3a*S*,5*R*,8*S*,8a*S*)-3a,8-dimethyl-5-(propan-2-yl)decahydroazulene], after the genus *Ambrosia* (ragweeds), from ambrosia, the Greek gods' immortalizing food, from Greek *ambrotos*, immortal. A.k.a. **pseudoguaiane.**

Aristolane [(1a*R*,3a*R*,7*R*,7a*S*,7b*S*)-1,1,7,7a-tetramethyloctahydro-1a*H*-cyclopropa[*a*] naphthalene], after the genus *Aristolochia* (birthwort, pipevine, Dutchman's pipe), from Greek *aristos*, best, and *locheia*, childbirth. A.k.a. **ferulane.**

Atisane {(1*S*,4*S*,9*S*,10*S*,12*R*,13*R*)-5,5,9,13-tetramethyltetracyclo[10.2.2.$0^{1,10}$.$0^{4,9}$]hexadecane}, from Hindi *atis*, atis plant (*Aconitum heterophyllum*).

Beyerane {(1*S*,4*R*,9*R*,10*R*,13*R*)-5,5,9,13-tetramethyltetracyclo[11.2.1.$0^{1,10}$.$0^{4,9}$]hexadecane}, after the genus *Beyeria* (pinkwood, ticky wallaby bush). The genus name is after the Dutch municipal official Adriaan de Beijer (1773–1843).

Bisabolane [1-methyl-4-(6-methylheptan-2-yl)cyclohexane], after the gum resin bisabol, obtained from the genus *Commiphora* (myrrh). Bisabol is from Wolof *bisap u ala*, bisabol.

Bornane {1,7,7-trimethylbicyclo[2.2.1]heptane}, named after borneol (Borneo camphor), i.e. after the island of Borneo, Indonesia. A.k.a. **bornylane** and **camphane.**

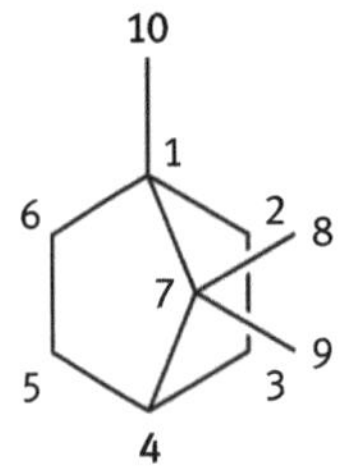

bornane

Cadinane [(1*S*,4*S*,4a*S*,6*S*,8a*S*)-1,6-dimethyl-4-(propan-2-yl)decahydronaphthalene], after cade oil (juniper tar oil), from Medieval Latin *catanus*, cedar juniper (*Juniperus oxycedrus*).

Carane {*cis*-carane, 3,7,7-trimethylbicyclo[3.1.0]heptane}, after the genus *Carum* (caraway), ultimately from Greek *karon*, caraway.

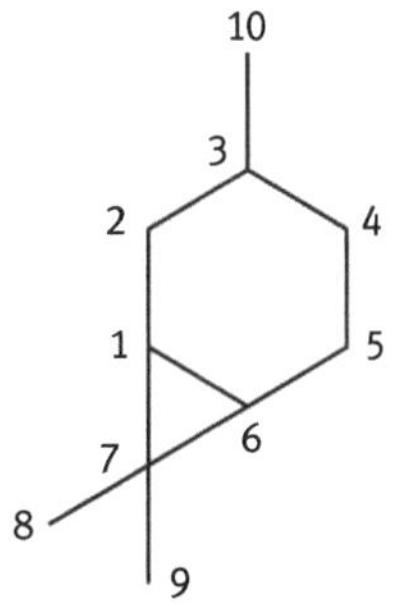

carane

Carotene {α-carotene, 1,2,2-trimethyl-2-[(1*E*,3*E*,5*E*,7*E*,9*E*,11*E*,13*E*,15*E*,17*E*)-3,7,12,16-tetramethyl-18-[(1*R*)-2,6,6-trimethylcyclohex-3-en-1-yl]octadeca-1,3,5,7,9,11,13,15,17-nonaen-1-yl]cyclohex-1-ene}, from Latin *carota*, carot (*Daucus carota*).

Caryophyllane {2,6,10,10-tetramethylbicyclo[7.20]undecane}, after the genus *Caryphyllus* (clove), from Greek *karyophyllon*, clove tree, from *karion*, grain, seed, kernel, and *phyllon*, leaf.

Cedrane {α-cedrane, (1*S*,2*R*,5*S*,7*S*,8*R*)-2,6,6,8-trimethyltricyclo[5.3.1.$0^{1,5}$]undecane}, after the genus *Cedrus* (cedar), from Greek *kedros*, cedar.

Dammarane {(5*S*,8*R*,9*R*,10*S*,13*R*,14*R*,17*R*)-4,4,8,10,14-pentamethyl-17-[(2*R*)-6-methylheptan-2-yl]dodecahydro-1*H*-cyclopenta[*a*]phenanthrene}, after dammar gum, obtained from the genera *Agathis*, *Shorea*, or *Hopea*, from Malay *damar*, dammar.

Drimane (1,1,4a,5,6-pentamethyldecahydronaphthalene), after the genus *Drimys* (woody evergreen flowering plants), from Greek *drimys*, sharp, acrid.

Eremophilane [(1*S*,4a*R*,7*R*,8a*R*)-1,8a-dimethyl-7-(propan-2-yl)decahydronaphthalene], after the genus *Eremophila* (emu bush, poverty bush), from Greek *eremos*, lone, solitary, and *philos*, loving.

Eudesmane [(1*R*,4a*R*,7*R*,8a*S*)-1,4a-dimethyl-7-(propan-2-yl)decahydronaphthalene], after the subgenus *Eudesmia* (eucalyptus), from Greek *eu*, easily, fine, and *desmos*, band, bond.

Fenchane {1,3,3-trimethylbicyclo[2.2.1]heptane}, from German *Fenchel*, fennel, ultimately from Latin *foeniculum*, fennel, diminutive of *foenum*, hay.

Gammacerane [γ-cerane, (4a*S*,6a*R*,6b*R*,8a*S*,12a*S*,12b*R*,14a*R*,14b*S*)-4,4,6a,6b,9,9,12a,14b-octamethyldocosahydropicene]. It has been impossible to trace the exact origin of this name. It appears in the scientific literature for the first time in 1957 where Allard and Ourisson [7] suggested to IUPAC that gammacerane should be assigned the role of a fundamental stereoparent structure, but without any further information about the compound itself and the rationale for its name. The name could have been created upon the recognition of gammacerane as the core structure of a collection of triterpenes to be named systematically, but even then such an unusual name ought to have been presented with an accompanying explanation. Any further mention of gammacerane in the literature, the first appearing about a decade later, deals with gammacerane found in sediments, and playing a role in the 'fingerprinting' of crude oil, with its name taken for granted. The fact that gammacerane is not a member of a related series consisting of α-cerane, β-cerane, γ-cerane, etc., and that the trivial name cerane was already earlier assigned to the unrelated normal alkane hexacosane shrouds this obscure name in even deeper mystery. There is also no convincing relationship between the octamethylperhydropicene derivative cerin, q.v., and the octamethylperhydropicene gammacerane since diagenetic exhaustive hydrogenation of cerin would be expected to lead to the octamethylperhydropicene friedelane, rather than its isomer gammacerane unless one assumes a radical shuffling of methyl groups in the course of this process.

Germacrane [(1*R*,4*s*,7*S*)-1,7-dimethyl-4-(propan-2-yl)cyclododecane], with contraction, after the species *Geranium macrorhizum* (bigroot geranium, Bulgarian geranium, rock crane's-bill), ultimately from Greek *geranion*, geranium, from *geranos*, crane, and *makros*, long, large.

Gibbane {(1*R*,3*S*,12*S*)-tetracyclo[10.2.1.0^{1,9}.0^{3,8}]pentadecane}, after the fungal genus *Gibberella*, from Latin *giberella*, diminutive of *gibber*, hump on the back.

Grayanotoxane {(1*R*,4*S*,8*S*,9*R*,10*S*,13*R*,14*S*)-5,5,9,14-tetramethyltetracyclo[11.2.1.0^{1,10}.0^{4,8}]hexadecane}, from grayanotoxin, after the Japanese shrub *hana-hirinoki* (*Leucothoe grayana*) and toxin. The specific epithet is after the American botanist Asa Gray (1810–1888).

Guaiane [(1*S*,3a*S*,4*S*,7*R*,8a*S*)-1,4-dimethyl-7-(propan-2-yl)decahydroazulene], after the genus *Guaiacum* (guaiac tree, lignum-vitae), from Taino *guayacan*, guaiac.

Himachalane {(4a*S*,9a*S*)-2,5,9,9-tetramethyldecahydro-1*H*-benzocycloheptene}, after the Himalayan deodar (*Cedrus deodara*).

Hopane {(3*R*,3a*S*,5a*R*,7a*S*,11a*S*,11b*R*,13a*R*,13b*S*)-5a,5b,8,8,11a,13b-hexamethyl-3-(propan-2-yl)hexadecahydrocyclopenta[*a*]chrysene}, after the genus *Hopea* (tropical trees), named after the British physician and botanist John Hope (1725–1786). A.k.a. ***a'*-neogammacerane**.

Humulane (1,1,4,8-tetramethylcycloundecane), after the genus *Humulus* (hops), a name of Germanic origin, cf. Old Norse *humli*, hop.

Kaurane {(1*R*,4*R*,9*R*,10*R*,13*R*,14*S*)-5,5,9,14-tetramethyltetracyclo[11.2.1.$0^{1,10}$.$0^{4,9}$]hexadecane}, after the New Zealand kauri pine (*Agathis australis*), from Maori *kawri*, kauri pine.

***ent*-Kaurane** {(1*S*,4*S*,9*S*,10*S*,13*S*,14*R*)-5,5,9,14-tetramethyltetracyclo[11.2.1.$0^{1,10}$.$0^{4,9}$]hexadecane}, named as enantiomer of kaurane, q.v.

Labdane {(1*S*,2*S*,4a*S*,8a*R*)-2,5,5,8a-tetramethyl-1-[(3*R*)-3-methylpentan-1-yl]octahydronaphthalene}, after the labdanum resin from the species *Cistus ladanifer* and *Cistus creticus* (rock roses), ultimately from Greek *ledon*, rock rose.

Lanostane {(5*S*,8*R*,9*S*,10*R*,13*R*,14*S*,17*R*)-4,4,19,13,14-pentamethyl-17-[(2*R*)-6-methylheptan-2-yl]dodecahydro-1*H*-cyclopenta[*a*]phenanthrene}, from Latin *lana*, wool, and sterol.

Linamarin {2-[(β-D-glucopyranosyl)oxy]-2-methylpropanenitrile}, after the genus *Linum* (flax) and Latin *amarus*, bitter. The genus name is from Latin *linum*, flax.

Linarin {7-[[6-*O*-(6-deoxy-α-L-mannopyranosyl)-β-D-glucopyranosyl]oxy]-5-hydroxy-2-(4-methoxyphenyl)-4*H*-1-benzopyran-4-one}, after the genus *Linaria* (toadflax). The genus name is from Latin *linarius*, pertaining to flax, from *linum*, flax.

Lupane {(1*S*,3a*R*,5a*R*,5b*R*,7a*S*,11a*S*,11b*R*,13a*R*,13b*R*)-3a,5a,5b,8,8,11a-hexamethyl-1-(propan-2-yl)hexadecahydrocyclopenta[*a*]chrysene}, after the species *Humulus lupulus* (hop), from Latin *lupulus*, diminutive of *lupus*, hop.

***p*-Menthane** [1-methyl-4-(propan-2-yl)cyclohexane], after the genus *Mentha* (mint plant), ultimately from Greek *mintha*, mint plant.

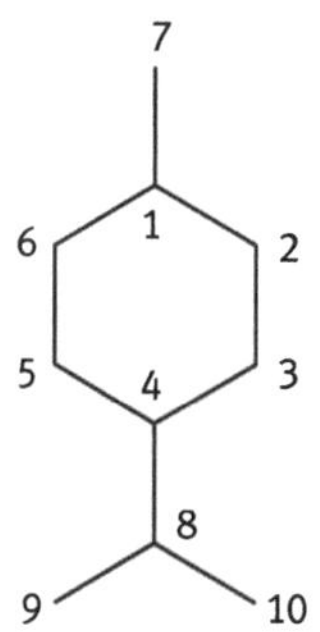

p-menthane

Oleanane {(4a*S*,6a*R*,6b*R*,8a*R*,12a*S*,14a*R*,14b*S*)-4,4,6a,6b,8a,11,11,14b-octamethylhexadecahydropicene}, after oleander (*Nerium*), obscurely akin to Medieval Latin *laurandrum*, oleander, perhaps a conflation of Latin *laurus*, laurel, and *rhododendron.*

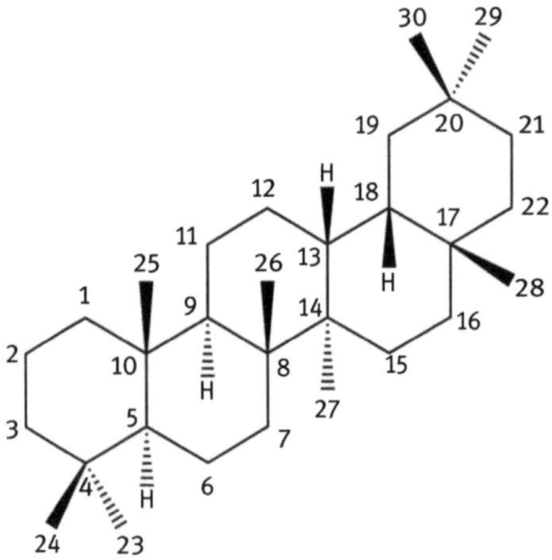

oleanane

Ophiobolane {(1*R*,3*R*,4*S*,7*R*,8*S*,11*S*,12*R*)-1,4,8-trimethyl-12-[(2*S*)-6-methylheptan-2-yl]tricyclo[9.3.0.0^{3,7}]tetradecane}, after the fungal genus *Ophiobolus*, from Greek *ophis*, snake, and *bolos*, a throwing, a casting, from *ballein*, to throw, to cast.

Picrasane {(1*R*,2*S*,6*R*,7*S*,9*R*,13*S*,14*S*,17*S*)-2,6,14,17-tetramethyl-10-oxatetracyclo[7.7.1 .0^{2,7}.0^{13,17}]heptadecane}, after the genus *Picrasma* (bitterwood), from Greek *pikrasmos*, bitterness.

Pimarane [(2*R*,4a*S*,4b*R*,8a*S*,10a*S*)-2-ethyl-2,4b,8,8-tetramethyldecahydrophenanthrene], with contraction, after the species *Pinus maritima* (cluster pine), from Latin *pinus*, pine, and *maritimus*, of the sea, from *mare*, sea.

Pinane {2,6,6-trimethylbicyclo[3.1.1]heptane}, after the genus *Pinus* (pines), cf. pimarane.

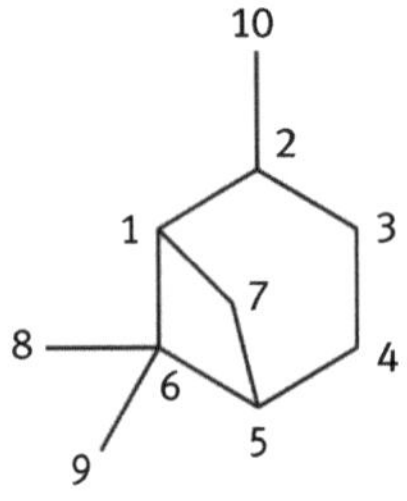

pinane

Podocarpane [(4a*R*,4b*S*,8a*R*,10a*S*)-1,1,4a-trimethyldecahydrophenanthrene], after the genus *Podocarpus* (yellowwood), from Greek *pous*, foot, and *karpos*, fruit.

Prostane [(1*S*,2S)-1-heptyl-2-octylcyclopentane], after prostaglandin, from prostate gland, from Greek *prostates*, prostate gland, from *proistanai*, set in front, and Latin *glans, gland-*, acorn.

Protostane {5α-protostane, (5*S*,8*S*,9*S*,10*S*,13*S*,14*S*,17*R*)-4,4,8,10,14-pentamethyl-17-[(2*R*)-6-methylheptan-2-yl]dodecahydro-1*H*-cyclopenta[*a*]phenanthrene}, from Greek *protos*, the first, and sterol, named as biosynthetic precursor of many steroids.

Retinal [(2*E*,4*E*,6*E*,8*E*)-3,7-dimethyl-9-(2,6,6-trimethylcyclohex-1-en-1-yl)nona-2,4,6,8-tetraenal], from retina, from Greek *rete*, net, and 'al', suffix for aldehydes. A.k.a. **vitamin A aldehyde** and **axerophthal**, from a(n)- and xerophthalmia, referring to this compound's activity against xerophthalmia.

Rosane [(2*R*,4a*R*,4b*R*,8a*R*,10a*R*)-2,4a,8,8-tetramethyldecahydrophenanthrene], after the genus *Rosa* (roses), from Latin *rosa*, rose.

Taxane [(4*R*,4a*R*,6*S*,9*R*,10*S*,12a*R*)-4,9,12a,13,13-pentamethyltetradecahydro-6,10-methanobenzocyclodecene], after the genus *Taxus* (yews), from Latin *taxus*, yew.

Trichothecane {(2*R*,7*R*,9*S*,12*S*)-1,2,5,12-tetramethyl-8-oxatricyclo[7.2.1.0^{2,7}]dodecane}, after the fungal genus *Trichothecium*. The genus name is from trich(o)- and -thecium.

Thromboxane [(2*R*,3*S*)-3-heptyl-2-octyloxane], from Greek *thrombos*, lump, blood clot. A.k.a. **TX**.

Thujane {4-methyl-1-(propan-2-yl)bicyclo[3.1.0]hexane}, after the genus *Thuja* (thuja), ultimately from Greek *thyia*, a kind of cedar. A.k.a. **sabinane**.

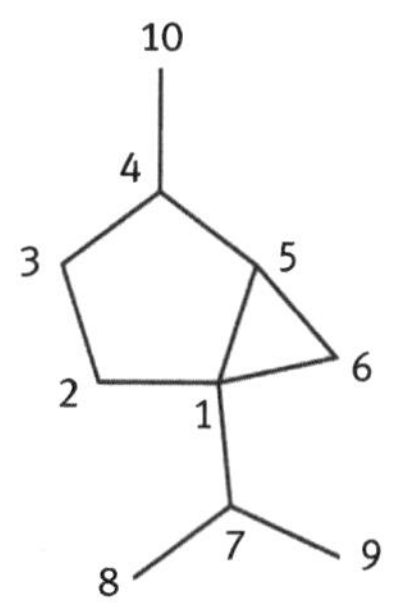

thujane

Trichothecane {(2*R*,7*R*,9*S*,12*S*)-1,2,5,12-tetramethyl-8-oxatricyclo[7.2.1.$0^{2,7}$]dodecane}, after the fungal genus *Trichothecium*, from Greek *thrix*, *tricho-*, hair, and *thekion*, diminutive of *theke*, case, chest.

Ursane [(1*S*,2*R*,4a*R*,6a*R*,6b*R*,8a*S*,12a*S*,14a*R*,14b*S*)-1,2,4a,6a,6b,9,9,12a-octamethylhexadecahydropicene], after the species *Arctostaphylos uva-ursi* (bearberry), from Latin *ursus*, bear.

4.1.3.4 Other stereoparent hydrides

Bilane {2-[(1*H*-pyrrol-2-yl)methyl]-5-[[[5-(1*H*-pyrrol-2-yl)methyl]-1*H*-pyrrol-2-yl]methyl]-1*H*-pyrrole}, from Latin *bilis*, bile.

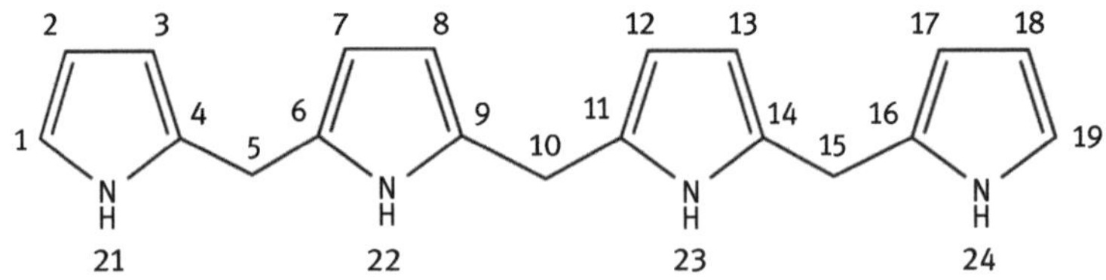

bilane

21*H*-Bilin {2-[(pyrrol-2-ylidene)methyl]-5-[[[5-(pyrrol-2-ylidene)methyl]-1H-pyrrol-2-yl]methylidene]pyrrole}, as bilan.

Cepham {(6*R*)-5-thia-1-azabicyclo[4.2.0]octan-8-one}, with contraction, from cephalosporanic acid and lactam. Cephalosporanic acid is after the fungal genus *Cephalosporium*, from Greek *kephale*, head, and *speirein*, to sow, to strew.

Flavan (2-phenyl-3,4-dihydro-2*H*-1-benzopyran), from Latin *flavus*, light yellow.

Isoflavan (3-phenyl-3,4-dihydro-2*H*-1-benzopyran), cf. flavan.

Lignan (polyphenols), from Latin *lignum*, wood.

Neoflavan (4-phenyl-3,4-dihydro-2*H*-1-benzopyran), cf. flavan.

Neolignan (polyphenols), named as isomer of lignan.

Penam {(5*R*)-4-thia-1-azabicyclo[3.2.0]heptan-7-one}, from penicillin and lactam. Penicillin (a class of antibiotics) is after the fungal genus *Penicillium*, ultimately from Latin *penicillus*, diminutive of *penis*, tail, male member.

Penem {(5*R*)-4-thia-1-azabicyclo[3.2.0]hept-2-en-7-one}, from penam and 'ene'.

Porphyrin (21,22-dihydroprophyrin), from Greek *porphyros*, purple.

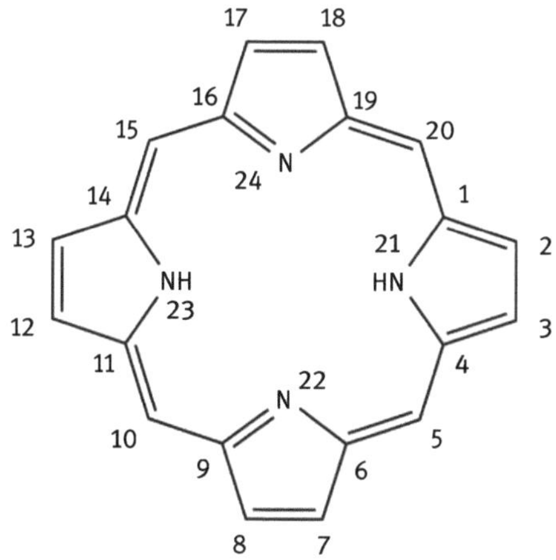

porphyrin

4.1.4 Modification of retained names

Apart from the usual possibility of substitution and functionalization, retained names can also be modified with the prefixes 'de(s)', 'dide(s)', 'tride(s)', etc., (from Latin *de*, away from, in the sense of removal of a characteristic group), 'nor', 'dinor', 'trinor', etc., (derived from normal, in the sense of reduced to its core structure by removal of a peripheral carbon atom), and 'homo', 'dihomo', 'trihomo', etc., (by contraction of homologous, in the sense of extension of a carbon skeleton by one carbon atom). Examples are:

2-Deoxy-D-ribose, also called **2-deoxy-D-arabinose** [(3*S*,4*R*)-3,4,5-trihydroxypentanal].

1,2-Didehydrobenzene (earlier called benzyne).

63-Desulfatohirudin, also called 63-desulfohirudin. The name hirudin is after the source of this sulfated polypeptide, the species *Hirudo medicinalis* (European medicinal leech), from Latin *hirudo, hirudin-*, leech.

Homoserine (2-amino-4-hydroxybutanoic acid).

Dihomo-γ-linoleic acid [(8*Z*,11*Z*,14*Z*)-icosa-8,11,14-trienoic acid].

Norbornane {bicyclo[2.2.1]heptane}. The name bornane is ultimately from Borneo camphor.

Dinoroleic acid (hexadec-7-enoic acid).

Another operation is the closure or opening of a ring in the parent compound. The corresponding prefixes are 'cyclo' and 'seco' (from Latin *secare*, to cut). Examples are:

D-Cycloserine [(4*R*)-4-amino-1,2-oxazolidin-3-one].

10,11-Secoergoline (3-{(2*S*)-[(piperidin-2-yl)methyl]}-1*H*-indole). The name ergoline is ultimately from ergot (*Claviceps purpurea*).

Still another operation, marked 'abeo', is the movement of a single bond in a parent structure. from Latin *abeo*, I go away, from Latin *ire*, to go.

Podocarpane. The name podocarpane is from the genus name *Podocarpus* (yellowwood), from Greek *pous, pod-*, foot, and *karpos*, fruit.

(3α*H*)-5(4→3)-Abeo-podocarpane.

4.2 Systematic names

While relying on a large portfolio of retained trivial names IUPAC nomenclature allows for rational modification of both these and fully systematic names.

4.2.1 Numerals

While the choice and use of numerals, both as locants and as multiplicative prefixes, ought to be a trivial task the original concept to use Greek numerals has been strongly modified by the addition of Latin and Greek-Latin hybrid numerals. The driving force for this has been widespread unease with the Greek numerals *hen* (one) and *ennea* (nine). The numeral eicosa (20) is still used by CAS while IUPAC has changed it to icosa. Also, in the absence of a Classical Latin word for zero a New Latin word, *nil*, had to be created. The examples of Table 4.1. illustrate the current use of numerals (a different set of numerals has been created for the naming of undiscovered elements, cf. Section 6.1.).

Second-order multipliers (for composite name parts) are to be used when the name element to be multiplied contains first-order multipliers such as in 1,3-bis(2,2-dichloroethyl)cyclopentane.

When the unit to be multiplied does not contain numerals, the multiplicators can be derived from the Latin adverbial numerals beyond *semel* (once), i.e. *bis* (twice), *ter* (thrice), *quater* (four times), *quinquies, quinquiens*, (five times), as in 1,1′-biphenyl (earlier diphenyl), $1^1,2^1:2^4,3^1$-terphenyl, $1^1,2^1:2^4,3^1:3^4,4^1$-quaterphenyl,

Table 4.1: Multiplicative prefixes and composite multiplicative prefixes used in systematic nomenclature.

Number	**Multiplicative prefix** (first-order multiplier)	**Multiplicative prefix for composite name parts** (second-order multiplier)
1	**mono** (from Greek *monos*, alone, instead of *hen*, one)	–
2	**di** (from Greek *dyo, di-*, two)	**bis** (from Latin)
3	**tri** (from Greek *treis, tri-*, three)	**tris** (from Greek)
4	**tetra** (from Greek *tettares, tetra-*)	**tetrakis** (from Greek)
5	**penta** (from Greek *pente, penta-*, five)	**pentakis** (from Greek)
6	**hexa** (from Greek *hex, hexa-*, six)	**hexakis** (from Greek)
7	**hepta** (from Greek *hepta*, seven)	**heptakis** (from Greek)
8	**octa** (from Greek *okto, okta-*, eight)	**octakis** (Greek) (from Greek *oktakis*, eight times)
9	**nona** (from Latin *novem, nona-*, nine, instead of Greek *ennea*)	**nonakis** (from Latin/Greek)
10	**deca** (from Greek *deka*, ten)	**decakis** (from Greek *dekakis*, ten times) etc.
11	**undeca** (from Latin/Greek, instead of earlier hendeca, hen- is still used inhigher numbers, i.e. from21 onward)	
12	**dodeca** (from Greek *dodeka*, twelve)	
13	**trideca** (from Greek *treis kai deka, trideca-*, thirteen)	
19	**nonadeca** (Latin/Greek)	
20	**icosa**[1] (from Greek *eikosi, eikosa-, twenty*)	
21	**henicosa** (from Greek)	
22	**docosa** (from Greek)	
30	**triaconta** (from Greek)	
31	**hentriaconta** (from Greek)	
90	**nonaconta** (from Latin/Greek)	

[1] as in IUPAC nomenclature, eicosa in CAS nomenclature

quinquephenyl, etc. However, some other IUPAC names, like triphenylene and tetraphenylene, have uncharacteristically been coined with Greek cardinal numbers while at the same time biphenylene is explicitly preferred to diphenylene. Another discrepancy can be seen in IUPAC names like 2,2′-bipyridine which is preferred to 2,2′-bipyridyl. It should also be noted that while biphenyl does indeed consist of two phenyl groups the names of the higher homologs terphenyl etc., although convenient to use, are strictly speaking misnomers since they also contain phenylene groups in addition to their two terminal phenyl groups. This imperfect homology among the oligophenyls also shows in the different construction of the locants in 1,1′-biphenyl vs. $1^1,2^1:2^4,3^1$-terphenyl and the following members of the series.

In most cases it is unnecessary to explicitly state the number one, but it does appear in cases where it is required to avoid ambiguity such as in carbon monoxide (CO).

It should be noted that commonly used name constructions like hemihydrate or sesquihydrate are unsound since they suggest addition compounds with fractions of water molecules.

The use of these numerals requires elision rules which vary from compound class to compound class, thus tetra plus 'one' is tetrone while tetra plus 'oxide' is tetraoxide.

Numbers in names and formulas are written with arabic numerals except for oxidation numbers which are shown with Roman numerals, supplemented with 0 (zero).

4.2.1.1 Obsolete locant systems

Older rules gave names where Greek letters were used as locants for aliphatic compounds and simple heterocycles as well as monosubstituted naphthalenes. Special terms like *ortho-*, *meta-* and *para-* for 1,2-, 1,3-, and 1,4-disubstituted benzenes, respectively, and vicinal (*vic-*, from Latin *vicinalis*, neighborly, from *vicus*, village, farm, town quarter), asymmetric (*as-*), and symmetric (*sym-*) for 1,2,3-, 1,2,4-, and 1,3,5-trisubstituted benzenes, respectively. The locant α (lower-case Greek alpha) was placed on the first position which was substitutable. The current rules place the locant 1 on the business end of the parent structure, in the case of heterocycles on the highest prioritized heteroatom.

Even in the current numbering system for fused ring systems only hydrogen-carrying carbon atoms are numbered at first as shown for naphthalene. This is sufficient as long as only substitution products need to be named. If the two interior carbon atoms require numbering they receive the locants 4a and 8a. In the case of more extensive ring systems even locants such as 6b, 6c, etc. can appear.

Before the current rules for the numbering of the atoms of parent ring systems were adopted such numbering was often at the discretion of the discoverer of the compound in question. This could lead to highly irrational results as can be seen in the mind-boggling numbering of the purine skeleton which is neither clockwise nor anticlockwise and contrary to the general rule of numbering the exterior and the interior atoms of a fused ring system separately. This maverick numbering of purine, created by the German chemist (Hermann) Emil (Louis) Fischer (1852–1919), is still official IUPAC numbering.

The Greek-letter locants became confusing when applied to long-chain fatty acids. Few chemists were able to memorize the whole Greek alphabet, and, of course, there were not enough letters to label very long-chain compounds.

In certain class names the position of unsaturation relative to the far end of the molecule was focused upon. Here the locant ω (lower-case Greek omega) was placed on the ultimate carbon, and the numbering went backwards with the locant (ω–1) (omega minus one) on the penultimate carbon, etc. The ω convention is only used in class names like (ω–3)-unsaturated fatty acids, not in the naming of individual compounds, cf. Chapter 7.

Here are a few examples of obsolete names with their current equivalent in parentheses: α-naphthylamine (naphthalen-1-amine), β-naphthoic acid (2-naphthoic acid, naphthalene-2-carboxylic acid), α,α′-dimethylthiophene (2,5-dimethylthiophene), γ-hydroxycaproic acid (4-hydroxyhexanoic acid).

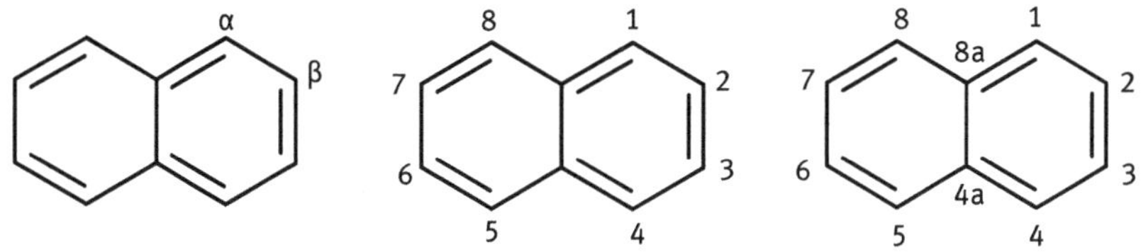

naphthalene

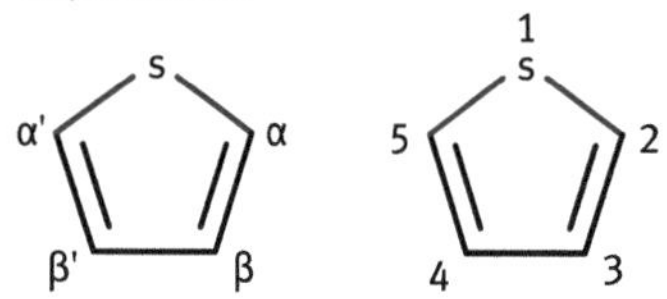

thiophene

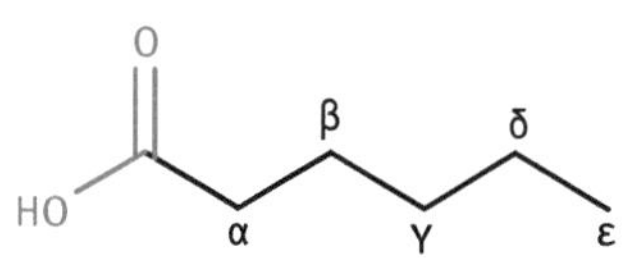

caproic acid

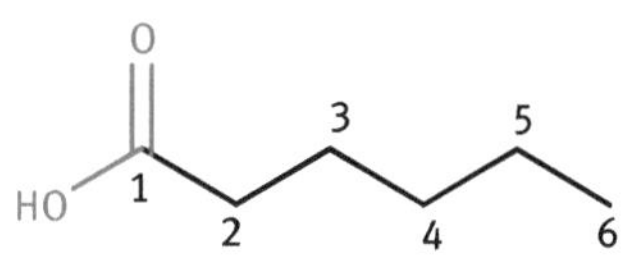

hexanoic acid

4.2.2 Stereodescriptors

Over the years not only relative stereochemistry, but also absolute stereochemistry needed to be shown in systematic names. Below is a list of important stereodescriptors:

(+)-[1], as in (+)-lactic acid [(2*R*)-2-hydroxypropanoic acid], dextrorotatory.

(–)-[1], as in (–)-camphor {(1*S*,4*S*)-1,7,7-trimethylbicyclo[2.2.1]heptan-2-one}, levorotatory.

d-[1,2], as in d-glucose (D-glucose), dextrorotatory, from Latin *dexter*, right.

l-[1,2], as in l-fructose (D-fructose), levorotatory, from Latin *laevus*, left.

D-[3], as in D-glyceraldehyde [(2*R*)-2,3-dihydroxypropanal], written with lower-case small caps, from Latin *dexter,* right.

L-[3], as in L-alanine [(2S)-2-aminopropanoic acid], written with lower-case small caps, from Latin *laevus*, left.

(+)-[1], as in (+)-lactic acid [(2*R*)-2-hydroxypropanoic acid], dextrorotatory.

(*R*)-[4], **(R*)-**[3], as in (1*R**,3*R**,5*R**)-1-[(2S)-butan-2-yloxy-3-chloro-5-nitrocyclohexane, from Latin *rectus*, right.

(Re)-[4], describing prochirality.

(*r*)-[4], for pseudoasymmetry.

(*S*)-[4], **(S*)-**[3], from Latin *sinister*, left.

(Si)-[4], describing prochirality.

(*s*)-[4], for pseudoasymmetry.

***rac*-**[4], as in *rac*-(1*R*,2*S*)-2-chlorocyclopentane-1-carboxylic acid, from racemate, from racemic acid, from Latin *racemus*, cluster of grapes.

***rel*-**[4], as in *rel*-(2*R*,3*R*)-3-bromobutan-2-ol, relative.

ent-[4], as in *ent*-kauran, for an enantiomer of a natural product, short for *enantio*.

'*ambo*'[4], as in 2-*ambo*-(2*R*,4′*R*,8′*R*)-α-tocopherol, from Latin *ambo*, both, referring to an unequal mixture of two epimers.

'α'[4], as in α-D-glucopyranose, in this example defining the stereochemistry at C-1.

'β'[4], as in cholest-5-en-3β-ol, in this example defining the stereochemistry at C-3.

'(*M*)'[5], as in (*M*)-hexahelicene, from minus in the sense of anticlockwise.

'(*m*)'[5], for pseudoasymmetry.

'(*P*)'[5], as in (*P*)-hexahelicene, from plus in the sense of clockwise.

'(*p*)'[5], for pseudoasymmetry.

'(*A*)'[5], from anticlockwise.

'(*C*)'[5], from clockwise.

'(Δ)'[5], from Latin *dexter*, right.

(+)-[1], as in (+)-lactic acid [(2*R*)-2-hydroxypropanoic acid], dextrorotatory.

'(Λ)'[5], from Latin *laevus*, left.

'anti'[2,6], as in *anti*-benzaldoxime [(*Z*)-*N*-hydroxy-1-phenylmethanimine], from Greek *anti*-, opposite.

'syn'[2,6], as in *syn*-benzaldoxime [(*E*)-*N*-hydroxy-1-phenylmethanimine], from Greek *syn*-, with, together with, by means of, at the same time.

'cis'[6], as in *cis*-stilbene [(1*Z*)-1,2-diphenylethene], from Latin *cis*, on this side. The name stilbene is from Greek, *stilbein*, to glitter, referring to the appearance of this substance's crystals.

'trans'[6], as in *trans*-stilbene [(1*E*)-1,2-diphenylethene], Latin *trans*, on the other side.

'(*E*)'[6,7], as in (2*E*)-but-2-ene, from German *entgegen*, opposite.

'(*Z*)'[6,7], as in (2*Z*)-prop-2-enoic acid, from German *zusammen*, together.

'endo', as in {(2-*endo*)-bromo-(7-*anti*)-fluorobicyclo[2.2.1]heptane}.

'exo', as in {(5-*exo*)-bromo-(5-*endo*, 7-*anti*)-dimethylbicyclo[2.2.1]hept-2-ene.

[1]relative enantiomeric configuration
[2]obsolete
[3]absolute enantiomeric configuration, based on dextrorotatory glyceraldehyde, designated as D-glyceraldehyde,
[4]absolute enantiomeric configuration, based on the Cahn-Ingold-Prelog (CIP) priority rules, tetrahedral geometry
[5]absolute configuration, for geometries other than tetrahedral
[6]geometric configuration
[7]geometric isomerism, based on the CIP priority rules

In addition, geometry can be shown by italicized prefixes like '*antiprismo*', '*asym*' (asymmetric), '*catena*' (from Latin *catena*, chain), '*dodecahedro*', '*fac*' (facial), '*hexahedro*' (from hexahedron i.e. cube), '*hexaprismo*', '*icosahedro*', '*mer*' (meridional), '*octahedro*', '*pentaprismo*', '*quadro*' (square), '*sym*' (symmetric), '*tetrahedro*', '*triangulo*', and '*triprismo*'. The structural prefixes '*arachno*' (from Greek *arachnion*, spider's web, from *arachne*, *arachno*-, spider), '*closo*' (incorrectly derived from Greek *klobos*, cage), '*conjuncto*' (from Latin *coniunctus*, joined, from *coniungere*, to join), '*hypho*' (from Greek *hyphe*, net), '*klado*' (from Greek *klados*, branch), and '*nido*' (from Latin *nidus*, nest) are employed in the naming of oligoboranes.

4.2.3 Endings

The most important endings are

'an', as in dextran, a contraction of dextrose anhydride.

'ane', as in methane, derived from the Latin suffix -*anus*, denoting a relationship.

'ene', as in ethene, derived from Greek -*ene*, suffix for female descendant.

'ium', as in bohrium, derived from Latin -*ium*, suffix for adjectives in the neutral gender.

'yne', as in ethyne, an arbitrary ending patterned after -ine.

4.2.4 Suffixes

Over time IUPAC has fine-tuned its system of suffixes. For example, 'ide' denotes an anion created by the loss of a proton while 'uide' corresponds to an anion formed by addition of a hydride ion.

Also the suffixes for cationic groups discern between 'ylium' for cations created by the loss of a hydride ion and 'ium' for cations formed by addition of a hydron.

IUPAC's terminology can at times be confusing. Thus the 'ium' in cation names is a suffix while the 'ium' in element names is an ending.

'acid', as in benzoic acid, from Latin *acidum*, acid.

'al', as in ethanal, from aldehyde.

'aric acid', as in glucaric acid, from tartaric acid.

'ate', as in ethanolate, from Latin -*atus*, suffix denoting function.

'carbaldehyde', as in cyclohexanecarbaldehyde, from carbon and aldehyde.

'carboxylic acid', as in cyclohexanecarboxylic acid, from carbon.

'diyl', as in ethane-1,2-diyl, from 'di' and 'yl'.

'ide', as in methanide, from oxide, ultimately from French *acide*, acid.

'ium', as in methylium, derived from Latin -*ium*, suffix for adjectives in the neutral gender.

'oic acid', as in pentanoic acid, from benzoic acid.

'ol', as in ethanol, from alcohol.

'one', as in propanone, from ketone.

'thial', as in pentanethial, from 'thio' and 'al'.

'thiol', as in methanethiol, from 'hio' and 'ol'.

'thione', as in cyclohexanethione, from 'thio' and 'one'.

'uide', as in tetramethylboranuide, an arbitrary variation of 'ide'.

'yl', as in ethyl, from Greek *hyle*, matter, referring to organic chemistry's radicals as 'ultimate matter', first used in benzoyl.

'ylidene', as in methylidene, from 'yl', 'ide', and 'ene'.

4.2.5 Prefixes

'az(a)', as in triazole, indicating replacement of carbon by nitrogen, from French *azote*, nitrogen.

'azoxy', as in azoxybenzene (diphenyldiazene 1-oxide), from az(o)- and oxygen.

'carboxy', as in 4-carboxybenzenesulfonic acid, from carboxylic acid.

'cyclo', as in cyclohexane and *cyclo*-octasulfur, from Greek *kyklos*, ring, circle, cycle, wheel.

'diazo', as in diazomethane, from di- and az(o)-.

'formyl', as in 2-formylbenzoic acid, from formic acid.

'homo', as in homocysteine, from homologous, i.e. expanded with one carbon atom.

'hydroxy', as in 2-hydroxybenzoic acid, from hydrogen and oxygen.

'hydroxyl', as in hydroxylamine, from hydrogen and oxygen.

'nor', as in noradrenaline, from normal, in the sense of reduced to its core structure by removal of one peripheral carbon atom.

'oxa', as in oxayohimban, indicating replacement of carbon by oxygen, from oxygen.

'oxo' (oxidanylidene, O=), as in 2-oxopropanedioic acid, from oxygen.

'oxy'(oxidanediyl, –O–), as in 1,1′-oxydibenzene, from oxygen.

'sulfanediyl' (–S–), as in 2,2′-sulfanediyldiacetic acid, from sulfane.

'sulfanyl', as in 2-sulfanylbenzoic acid, from sulfane.

'sulfanylidene' (S=), as in 2-sulfanylidenepropanedial, from sulfane and 'ylidene'.

4.2.6 Seniorities

A useful systematic name must be built upon no more than one parent hydride and no more than one characteristic group. This requires both parent hydrides and characteristic groups to be ordered by senority, the most senior parent hydride becomes the core of the name while all other parent hydrides are relegated to the role of substituents. Likewise, the most senior characteristic group is allowed as such while less senior characteristic groups become substituents. Seniority sequences have been established in a more or less arbitrary manner and belong to the basic features of IUPAC and CAS nomenclature.

A few examples:

Hydroxypentanoic acid, i.e. carboxylic acid more senior than alcohol.

2-Aminoethanol, i.e. alcohol more senior than amine.

Methylbenzene, i.e. benzene more senior than methane.

2-Phenylpyridine, i.e. pyridine more senior than benzene.

2-Sulfanylpropane-1,2,3-tricarboxylic acid, i.e. carboxylic acid more senior than thiol; however, several identical functional groups allowed.

Older, still commonly used names like tropolone (2-hydroxycyclohepta-2,4,6-trien-1-one), here with the two principal functional group suffixes 'ol' and 'one', disobey these rules and are thus no longer regarded as systematic.

4.3 Disused systematic names

Systematic names can over time be discarded because of homonymy with common trivial names or because of rule changes.

4.3.1 Systematic names never adopted by IUPAC

Azine (pyridine). The name pyridine was so well-established that a change to the systematic equivalent appeared unrealistic. Also, the class name azine had already been used for *N,N'*-dialkylidenehydrazines such as acetone azine [$(CH_3)_2C{=}N{-}N{=}C(CH_3)_2$].

Azole (pyrrole). It was considered impractical to replace the well-established name pyrrole.

Carbane (methane). It was considered impractical to replace the well-established name methane.

Cyclohexa-1,3,5-triene (benzene). This name, which does not do justice to benzene's aromatic properties, would be no more appropriate than the retained trivial name benzene.

1,2-Diazine (pyridazine), cf. azine.

1,3-Diazine (pyrimidine), cf. azine.

1,4-Diazine (pyrazine), cf. azine.

1,2-Diazole (pyrazole), cf. azole.

1,3-Diazole (imidazole), cf. azole.

Dicarbane (ethane), cf. carbane.

Dioxidanediyl (peroxy).

Dioxidanyl (hydroperoxy).

Hexalene (naphthalene), the 'missing link' between pentalene and heptalene, i.e. two fused benzene rings. The name naphthalene prevailed for historical reasons.

Oxane (oxidane), not adopted in order to avoid confusion with oxane (tetrahydropyran).

Oxidanediyl (oxy).

Oxine (pyran), because of homonymy with the semitrivial name oxine (quinolin-8-ol). However, the name oxane is used for the corresponding saturated ring system.

Oxole (furan), cf. pyrrole.

Pentacarbane etc. (pentane etc.), cf. carbane.

Tetracarbane (butane), cf. carbane.

Thiine (thiopyran), cf. pyrrole.

Thiole (thiophene), because of homophony with thiol.

Tricarbane (propane), cf. methane.

4.3.2 Systematic and retained trivial names made obsolete by IUPAC rule changes

Acetylene (now ethyne).

Amylene (now 2-methylbut-2-ene and/or its isomers).

Azobenzene (now diphenyldiazene).

Azomethane (now dimethyldiazene).

Azoxybenzene (now diphenyldiazene oxide).

Azoxymethane (now dimethyldiazene oxide).

Anethole {now 1-methoxy-4-[(1*E*)-prop-1-en-1-yl]benzene}.

Arsine (now arsane).

Arsonium (now arsanium), by analogy with ammonium.

Benzenesulfenic acid (now benzene-*S*,*O*-thioperoxol).

Benzyne (now 1,2-didehydrobenzene).

Biphenyl {now [1,1′-biphenyl]}, still allowed in general nomenclature.

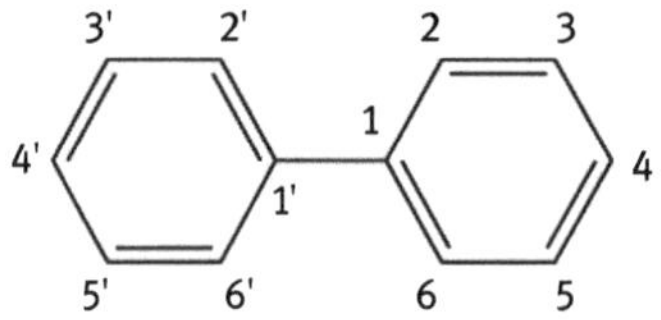

1,1′-biphenyl

Butylene (now butene).

Carbodiimide (now methanediimine).

Diphenyl (now 1,1′-biphenyl).

Dimethyl ether (now methoxymethane). Note the difference in the current naming of dimethyl ether and diphenyl ether.

Diphenyl ether (now 1,1′-oxydibenzene). Note the difference in the current naming of dimethyl ether and diphenyl ether.

Ethylene (now ethene).

Eugenol [now 2-methoxy-4-(prop-2-en-1-yl)phenol], after the genus *Eugenia* (tropical trees, shrubs), named for the Austrian Prince Eugen of Savoy (1663–1736).

Furazan (now 1,2,5-oxadiazole), named as nitrogen-containing furan analog.

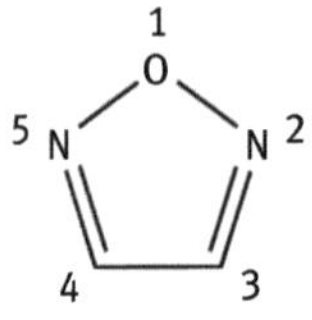

1,2,5-oxadiazole

Furoxan (now 1,2,5-oxadiazole 2-oxide), from furazan and oxide.

Guaiacol (now 2-methoxyphenol), after the genus *Guaiacum* (guaiac tree), from Taino, *guayacan*, guaiac.

Mercaptan (now thiol), from New Latin *mercurium captans*, seizing mercury, referring to the characteristic dark precipitates formed with mercury(II) ions.

Methanesulfenic acid (now methane-*S*,*O*-thioperoxol).

Naphthacene (now tetracene), from naphth(o)- and -acene.

Pentahelicene (now dibenzo[*c*,*g*]phenanthrene). The helical hexahelicene is the first true helicene. Its lower homolog pentahelicene lacks helicity and is therefore more appropriately named as a fused aromatic pentacycle.

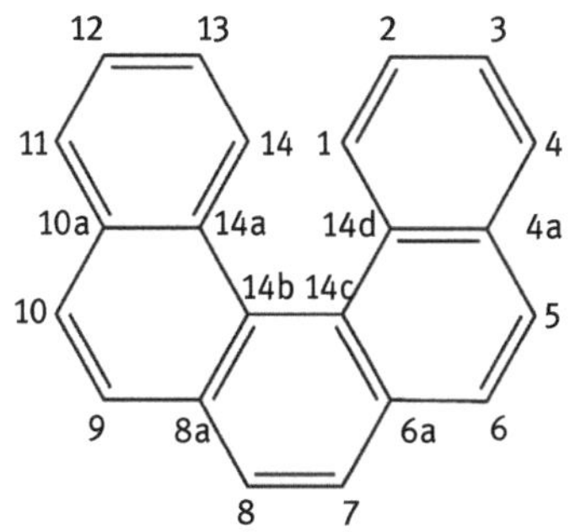

dibenzo[*c*,*g*]phenanthrene

Persulfurane (now λ^6-sulfane).

Phenetole (now ethoxybenzene).

Phosphine (now phosphane).

Phosphonium (now phosphanium), by analogy with ammonium.

Phosphorane (now λ^5-phosphane).

Propargyl- (now prop-2-yn-1-yl).

Propylene (now propene).

Sulfonium (now sulfanium), by analogy with ammonium.

Sulfurane (now λ^4-sulfane).

Veratrol (now 1,2-dimethoxybenzene), after the genus *Veratrum* (false hellebore, corn lily), from Latin *veratrum*, hellebore.

4.4 Class names

Class names are less subject to rule changes than compound names. Thus, while individual azo compounds after some IUPAC rule changes now are treated as disubstituted diazenes, azo compound and azo dye are still valid class names.

Acetal, with contraction, from acetaldehyde and alcohol.

Acylal, from acetal.

Acyloin, from acyl- and benzoin.

Aglycon, from a(n)- and glycose.

Alcohol, cf. Section 3.5.

Aldehyde, with contraction, from New Latin *alcohol dehydrogenatum*.

Aldimine, from aldehyde and imine.

Aldoxime, from aldehyde and oxime.

Alkaloid, 'alkali-like', referring to these compounds' basic properties.

Alkane, from alkyl and 'ane'.

Alkene, from alkyl and 'ene'.

Alkyne, from alkyl and 'yne'.

Amide, ultimately from ammonia.

Amidine, from amide and 'idine'.

Amine, ultimately from ammonia.

Anhydride, from Greek *a(n)-*, not, and *hydor*, water, referring to compounds formed by loss of water from a precursor.

Anil (obsolete), from aniline, referring to aromatic Schiff bases.

Anilide, from aniline, patterned after amide.

Ansa compound, from Latin *ansa*, handle.

Arene, from aromatic and 'ene'.

Azide, from 'az(o)'.

Azine, from hydrazine or pyrazine.

Azlactone, from 'az(o)' and lactone.

Betaine, from betaine [2-(trimethylazaniumyl)acetate)].

Bromohydrin, named as product of the addition of hypobromous acid to an alkene.

Carbanion, from carbon and anion.

Carbene (methylidene, also used as class name), from carbon and 'ene'.

Carbenium ion, from carbene and 'ium'.

Carbide, from carbon and 'ide'.

Carbocation, from carbon and cation. This more accurate name replaced the earlier term carbonium ion, q.v.

Carbohydrate, from carbon and hydrate.

Carbonium ion (obsolete), from carbon, patterned after ammonium.

Carborane, from carbon and boron.

Carboxylic acid, from carbon and oxygen.

Carbylamine (obsolete, isocyanide), from carbon and amine.

Carbyne (methylidyne, also used as class name), from carbon and 'yne'.

Carceplex, from Latin *carcer*, prison, and complex.

Carcerand, from Latin *carcer*, prison.

Cardiac glycoside, from Greek *kardiakos*, pertaining to the heart, from *kardia*, heart.

Carotenoid, from carotene.

Chlorohydrin, named as product of the addition of hypochlorous acid to an alkene.

Clathrand; clathrate, from Latin *clathratus*, furnished with lattice, ultimately from Greek *kleiein*, to shut.

Complex, from Late Latin *complexus*, totality, from Latin *complecti*, to embrace, from *plectere*, to weave, to braid, to twine, to entwine.

Corand, contraction of coronand.

Coraplex, by contraction of corand and complex.

Coronand, from corona-.

Crown ether. A.k.a. **lariat ether**.

Cumulene, from Latin *cumulus*, heap, mass.

Cyanide, from Greek *kyanos*, dark blue, from German *Blausäure*, hydrogen cyanide, literally blue acid, named after its preparation by acid treatment of Berlin blue.

Cyclitol, from cycl(o)- and 'itol'.

Cyclophane, from cycl(o)-, phenylene, and alkane. Cf. phane.

Depside, from Greek *depsein*, to knead.

Endorphin, a contraction of endogeous morphine.

Enol, from 'ene' and 'ol'.

Ester, cf. Chapter 3.

Ether, cf. Chapter 3.

Flavin, from Latin *flavus*, light yellow.

Furanose, from furan and 'ose'.

Glucosinolate, from glucose, Latin *sinapis*, mustard, and 'ol'.

Glycan, a contraction of glycosan, named, with contraction, as glycose anhydride.

Glycaric acid, from glycose and saccharic acid.

Glyconic acid, an arbitrary name derived from glycose.

Glycol, from Greel *glykys*, *glyker*-, sweet, and 'ol'.

Glycose, by alteration of glucose.

Glycoside, from glycose.

Helicene, from Greek *helix*, spiral.

Hemiacetal, from hemi- and acetal.

Hemiketal, from hemi- and ketal.

Hetarene, from heter(o)- and arene.

Hydrate, from Greek *hydor, hydro*-, water.

Hydrazide, from hydrazine, patterned after amide.

Hydrazidine, from hydrazine, patterned after amidine.

Hydrazone, from hydrazine and 'one'.

Hydroxamic acid, with contraction, from hydroxylamine.

Imide, from imine and 'ide'.

Imine, coined by alteration of amine.

Ion, from Greek *ion*, going, referring to the mobility of ions in solution.

Isocyanide, from 'iso' and cyanide.

Isoprenoid (terpene), from isoprene.

Ketal, from ketone and alcohol, patterned after acetal.

Ketazine, from 'ket(o)' and azine.

Ketide, a contraction of polyketide, from poly-, ket(o)-, and 'ide'.

Ketimine, from 'ket(o)' and imine.

Ketitol, from ketose and 'itol'.

Ketol, from 'ket(o)' and alcohol.

Ketone, from .

Ketoxime, from ketone and oxime.

Ketose, from 'ket(o)' and glycose.

Kinin, from Greek *kinein*, to move, to stimulate, to set in motion.

Lactam, from lactone and amide.

Lactim, from lactone and imide.

Lactone, from lactic acid.

Lariat ether, from Spanish *la reata*, the lasso. A.k.a. **crown ether**.

Leukotriene, from leukocyte, tri-, and 'ene'. A.k.a. **LT**.

Lipid, from Greek *lipos*, fat.

Luciferin, from Latin *lux, luc-*, light, and *ferre*, to carry.

Macrolide, named as large-ring lactone.

Meriquinone, from meri- and quinone, referring to the only partially quinoid character of these compounds.

Metallocene, from metal and ferrocene.

Moloxide, from molecule and oxide, referring to such compounds' formation by 1:1 reaction of a substrate with oxygen.

Molozonide, from molecule and ozonide, referring to such compounds' formation by 1:1 reaction of a substrate with ozone.

Monomer, from Greek *monos*, alone, and -mer.

Münchnone, after German *München*, Munich, Germany

Nitrile, from nitrogen and 'il', 'ile'.

Nitrolic acid, from 'nitr(o)' and 'ol'.

Nitrone, from nitrogen and 'one'.

Nitronic acid, from nitrone.

Osazone, from 'ose' and hydrazone.

Osone, from 'ose' and 'one'. A.k.a. **ketoaldose.**

Oxide, from French *oxyde*, oxide, coined by contraction of *oxygène*, oxygen, and *acide*, acid.

Oxime, from obsolete oxyimide.

Peptide, from Greek *pepsis*, *pept-*, digestion, from *peptein*, to digest.

Phane, short for cyclophane, but also used for acyclic acompounds.

Polymer, from Greek *polys*, much, many, and -mer.

Protein, from Greek *proteion*, the first place, the chief rank, referring to the proteins' key role in living organisms.

Pseudoacid, from pseudo- and acid, referring to the required tautomeric reorganization before these compounds' acidity can manifest itself.

Pyranose, from pyran and 'ose'.

Pyrethroid, from pyrethrin and 'oid'.

Quinodimethane (dimethylidenecyclohexadiene), named as carbon analog of quinone.

Quinol, from quinine and 'ol'.

Quinomethane (methylidenecyclohexadienone), named as carbon analog of quinone.

Quinone (arenedione), a class name, from quinine and 'one'. For unfathomable reasons IUPAC has excepted benzene from the general rule of naming quinones as arenediones, i.e. the preferred names for *p*-benzoquinone and *o*-benzoquinone are cyclohexa-2,5-diene-1,4-dione and cyclohexa-3,5-diene-1,2-dione, respectively.

Reductone, from reduction and 'one'.

Rotane, from Latin *rota*, wheel, and 'ane'.

Rotaxane, from Latin *rota*, wheel, *axis*, axis, and 'ane'.

Rotenoid, from rotenone and 'oid'.

Sapogenin, from saponin and -gen.

Saponin, from sapo(n)-, referring to these compound's emulsifying properties.

Schiff base (imine), after the German-Italian chemist Hugo Schiff (1834–1915).

Semicarbazone, from semicarbazide and 'one'.

Semidine, from semi- and benzidine.

Semiquinone, from semi- and quinone.

Speleand; speleate, from Greek *speleion*, cave.

Spherand, from sphere.

Spheraplex, a contraction of spherand and complex.

Sphingolipid, from Sphinx and lipid, referring to the enigmatic nature of these lipids. Sphinx is from Greek *sphinx*, strangler, from *sphingein*, to bind, to squeeze.

Sterane (perhydrocyclopenta[*a*]phenanthrene), from steroid and 'ane', found in sediments and named as reduction and defunctionalization products of steroids.

Steroid, from obsolete cholesterin, cholesterol, from Greek *chole*, bile, and *stereos*, solid.

Sugar, ultimately from Sanskrit *sarkara*, sugar, grit, gravel.

Sulfene, named as sulfur anlog of ketene.

Sulfenic acid, named as analog of sulfonic acid with divalent sulfur.

Sulfide, from sulfur.

Sulfine, named as λ^4 analog of sulfene.

Sulfinic acid, named as analog of sulfonic acid with tetravalent sulfur.

Sulfone, from sulfur and ketone, referring to an obsolete notion of analogy between sulfones and ketones.

Sulfonamide, from sulfonic acid and amide.

Sulfonic acid, ultimately from sulfuric acid.

Sulfoxide, from sulfide and oxide.

Sultam, from sulfonic acid and lactam.

Sultim, from sulfonic acid and lactim.

Sultone, from sulfonic acid and lactone.

Sydnone, after the city of Sydney, Australia.

Terpene, terpenoid (isoprenoid), from German *Terpentin*, turpentine.

Thiol, from thi(o)- and 'ol'.

Toxoid, from toxin and -oid, indicating that the toxophoric group of the original toxin has been removed.

4.5 Major changes over time of IUPAC recommendations

Whenever IUPAC recommendations are changed the old names will, of course, persist in the printed literature and often even in electronic media which may be updated unevenly. Thus, an awareness of such changes is useful when one is confronted with different looking, but in reality synonymous names.

4.5.1 The place of locants

According to an earlier rule locants were placed in front of a name, for instance in the case of 2,5-heptanediol. The current rule places the locants immediately before the functionality they belong to, i.e. heptane-2,5-diol. This change has made it easier to decode, especially longer, names, but at the expense of euphony. In the spoken name the phonetic value of the vowel before a locant can become uncertain.

4.5.2 The longest chain

A recent change has been made regarding the choice of the longest carbon chain as the parent structure. Earlier any unsaturation (i.e. double and/or triple bonds) should be included in the parent structure, even if that meant the choice of a shorter chain than otherwise possible. The current rule prescribes the longest carbon chain to be used as parent regardless of unsaturation. For example, limononaldehyde with the previous systematic name 4-methyl-3-(3-oxobutyl)pent-4-enal is now called 6-oxo-3-(prop-1-en-2-yl)heptanal.

4.5.3 Locants for side chains

According to earlier rules the carbon atoms of a side chain had to be numbered from the point of attachment, i.e. the systematic equivalent of isopropylcyclohexane was (1-methylethyl)cyclohexane. The current treatment of side chains numbers them regardless of the point of attachment, i.e. the current systematic name becomes (propan-2-yl)cyclohexane.

4.5.4 Methylene vs. methylidene

Earlier names did not discern between $-CH_2-$ and $CH_2=$, both being called methylene. According to the current IUPAC recommendations methylene is used for $-CH_2-$ and methylidene for $CH_2=$.

4.5.5 Mercapto and thio vs. sulfanyl

Organic derivatives of hydrogen sulfide, RSH, were earlier called mercaptans, from New Latin *mercurium captans*, mercury-seizing, referring to their reaction with mercury(II) ions to yield dark-colored, insoluble salts. The sulfur analogs of alkoxy (aryloxy) substituents were by analogy called alkylthio (arylthio). This led to now obsolete names like methyl mercaptan, 2-mercaptopropionic acid, and 2-(methylthio)propionic acid. The current IUPAC recommendations prescribe ‘thiol’ for the functional group SH and ‘sulfanyl’ for sulfur-centered substituents. Thus, the three abovementioned names changed to methanethiol, 2-sulfanylpropanoic acid, and 2-(methylsulfanyl)propanoic acid, respectively. By the same token the group –S– is now named ‘sulfanediyl’ which leads from the traditional, still acceptable, name diphenyl sulfide to the preferred name 1,1′-sulfanediyldibenzene. The name 1,1′-thiodibenzene is explicitly discouraged.

This policy change also affected the previous name ‘thioxo’ for the S= group which was accordingly renamed ‘sulfanylidene’. Thus, for instance, 2-thioxopropanal changed to 2-sulfanylidenepropanal. Such a change removes the onus of implying oxygen in the naming of groups which do not contain oxygen.

4.5.6 Ethers, sulfides, sulfoxides, etc.

Compounds earlier named as ethers, sulfides, disulfides, sulfoxides, etc. are now treated as substituted hydrocarbons. This has led to names like methoxymethane (earlier dimethyl ether), methoxyethane (earlier ethyl methyl ether), 1,

1′-oxydibenzene (earlier diphenyl ether), 1,1′-disulfanediyldibenzene (earlier diphenyl disulfide), (methanesulfinyl)methane (earlier dimethyl sulfoxide), (methylperoxy)ethane (earlier ethyl methyl peroxide), etc. However, ether, sulfide, etc. are still valid class names.

4.5.7 Names for ligands

Many coordination compounds contain a cationic central atom and anionic ligands. Previously the ligand name was derived from the name of the anion by deleting 'ide' and replacing it with 'o', i.e. tetrafluoroborate(1–) for BF_4^-. According to the current IUPAC rules this elision is no longer applicable which leads to the revised name tetrafluoridoborate(1–). Likewise the ligand names hydroxo and cyano have been changed to hydroxido and cyanido, respectively. Ligated hydrogen is always considered as anionic, i.e. hydrido, regardless of the actual polarization of a coordination compound.

4.5.8 Roman numerals

Earlier Roman numerals for oxidation numbers were normally written with lower-case small caps ɪ, ᴠ, and x. In their newer publication [4] IUPAC uses instead the upper-case roman letters I, V, and X, however without giving preference to the new spelling. The two spelling modes can be illustrated with iron(ɪɪ) diiron(ɪɪɪ) tetraoxide (Fe_3O_4) vs. iron(II) diiron(III) tetraoxide.

4.6 Computer-friendly names

When large amounts of chemical data are to be processed by computer systematic names can be translated into computer-friendly language such as Simplified Molecular-Input Line-Entry System (canonical and isomeric SMILES) [8], the International Chemical Identifier (InChI and InChI Key) [9], as well as Chemical Markup Language (CML) [10]. Among other possibilities such names allow the computer generation of structural formulas from systematic names and vice versa.

Aflatoxin B$_1$ [1162-65-8] can serve as an example:

IUPAC name: (6a*R*,9a*S*)-4-methoxy-2,3,6a,9a-tetrahydrocyclopenta[*c*]furo[2,3-*h*][1]benzopyran-1,11-dione

Canonical SMILES name: COC1=C2C3=C(C(=O)CC3)C(=O)OC2=C4C5C=COC5OC4=C1

Isomeric SMILES name: COC1=C2C3=C(C(=O)CC3C(=O)OC2=C4[C@@H]5C=CO[C@@H]5OC4=C1

InChI key name: OQIQSTLJSLGHID-WNWIJWBNSA-N.

InChI name: InChI=1S/C17H12O6/c1-20-10-6-11-14(8-4-5-21-17(8)22-11)15-13(10)7-2-3-9(18)12(7)16(19)23-15/h4-6,8,17H,2-3H2,1H3/t8-,17+/m0/s1.

4.7 Preferred IUPAC names (PIN)

Ideally, rationally interlinked nomenclature rules should always lead to one, and only one, valid name for a compound. IUPAC's recommendations for the naming of organic compounds historically have often allowed different name constructions for a given compound. This problem has been addressed in [4] by the introduction of 'preferred IUPAC names' (PIN), although other names are often given as 'still acceptable'. Similar problems in IUPAC's recommendations for the naming of inorganic compounds [3] still await to be addressed.

An example of the PIN approach is $(CH_3)_3N^+O^-$. Here the IUPAC name *N*,*N*-dimethylmethanamine oxide is PIN rather than the equally valid IUPAC name (trimethylazaniumyl)oxidanide.

4.8 The naming of tautomers and aromatic systems

Special problems arise when tautomers or aromatic substances are to be named.

4.8.1 The naming of tautomers

In the case of tautomeric equilibria each tautomer has, of course, its own systematic name and sometimes even its own trivial name, such as the tautomer pairs

Cyanuric acid (1,3,5-triazine-2,4,6-triol), named as urea derivative formed by oligomerization of hydrogen cyanide.

Isocyanuric acid [1,3,5-triazine-2,4,6(1*H*,3*H*,5*H*)-trione].

Urea (carbonic diamide).

Isourea (carbamimidic acid).

This is unproblematic as long as one deals with a single tautomer, but in many cases it is desirable to have one single name for a tautomeric equilibrium mixture where the focus is on stoichiometry only or on overall reactivity. In such situations it is common to let the name of one of the tautomers represent the total equilibrium mixture. Often this choice falls on the tautomer with the simplest name, regardless of its prevalence or otherwise in the equilibrium.

Thus, in practice acetone is simply named as propanone, without regard for the minute amount of the corresponding enol (prop-1-en-2-ol) in eqilibrium with the ketone, which, however, can play a crucial role in certain reactions of acetone. Generally, in compilations of compound names the keto isomer is often preferred to represent a keto-enol equilibrium because of its simpler name. At the same time the enol tautomer is omitted in the compilation to keep its size manageable.

One extreme case are the amino acids such as glycine (aminoethanoic acid) which is practically unidentifiable in its tautomeric equilibrium with the zwitterion azaniumylacetate, nor does it play any role in glycine chemistry. However, the overwhelming convenience of being able to use names without charged groups as well as the ease of drawing uncharged structural formulas has led to the exclusive use of uncharged amino acid names and structural formulas.

Another, no less important, case are the common monosaccharides. Thus, in the case of D-glucose in, say, aqueous solution three tautomers are to be considered, the acyclic aldehyde and the two hemiacetals α- and β-D-glucopyranose. Also here the simplest name and structure belong to the tautomer which there is least of in the equilibrium and again convenience suggests its name to be used for the equilibrium mixture, its exact nature just being understood. Moreover, since neither L-glucose nor DL-glucose are ever encountered in practical work, further simplification leads to the use of the name glucose as a synonym for D-glucose.

4.8.2 The naming of aromatic systems

Just as compounds need to be named there is also a need to name resonance formulas which, of course, is accomplished according to the relevant nomenclature rules. However, if a name is meant to represent a compound for which several resonance formulas can be drawn, it is impractical to assign to this compound as many names as there are resonance formulas. Practical considerations lead to two procedures: either the name of one of the resonance formulas is taken as the name of the compound or a trivial name is chosen to represent the full set of resonance formulas.

These two methods can be exemplified with the following two aromatic ions:

Cyclopentadienide (cyclopenta-2,4-dien-1-ide). This 'vernacular' locantless version of the fully systematic name has the advantage over the latter that it does not suggest localized single and double C–C bonds nor a localized negative charge. The fully systematic name corresponds to one of the five resonance formulas of this anion and is rather a placeholder for an in principle unavailable fully aromatic equivalent.

Tropylium (cyclohepta-2,4,6-trien-1-ylium), ultimately derived from atropine. This commonly used trivial name of this cation, like its systematic counterpart, is also a placeholder for an unavailable systematic name which could imply the composition and regular heptagonal shape of this aromatic cation. Retained trivial names like this are used, both in IUPAC and CAS systematic nomenclature, for most common aromatic compounds.

4.9 The naming of oxidized aromatic compounds

IUPAC's suffix 'one' and 'prefix 'oxo' normally imply that a methylene group of a parent compound has been converted to a carbonyl group or, in other words, that two geminal hydrogen atoms have been replaced by a double-bonded oxygen atom. As long as one is dealing with aliphatic systems this is a straightforward operation.

On the other hand, the introduction of an oxo group into an aromatic system is not possible as a one-step procedure for the obvious lack of geminal hydrogen atoms. The simplest way to obtain the desired oxo derivative would be to add hydrogen to the arene and subsequently oxidize a methylene groups thus created to a carbonyl group. In the case of, say, thiophene we could achieve oxidation at C-2 by first converting thiophene to 2,3-dihydrothiophene and finally oxidizing this to 2,3-dihydrothiophen-2-one.

However, this is not the way IUPAC has chosen. In their naming procedure the intermediate hydrogenation step is omitted as far as nomenclature is concerned and the above oxidized product called thiophene-2(3*H*)-one.

By the same token quinones are named as arenediones:

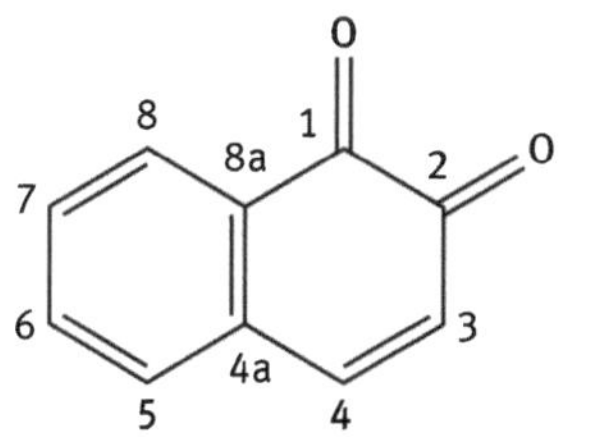

1,2-Naphthoquinone (naphthalene-1,2-dione).

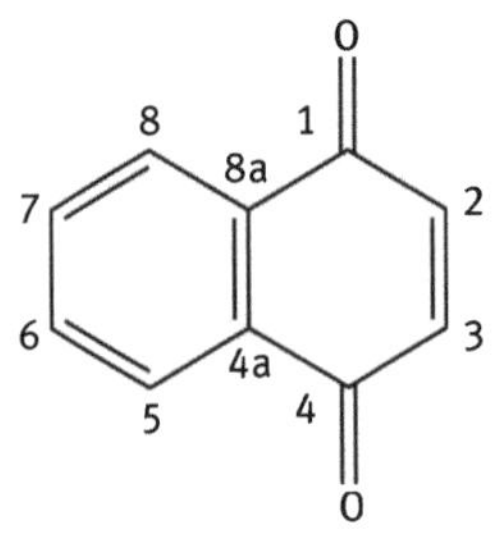

1,4-Naphthoquinone (naphthalene-1,4-dione).

1,2-Anthraquinone (anthracene-1,2-dione).

However, in the most recent edition of the Blue Book [3] the preferred names of the two benzoquinones are given as cyclohexa-3,5-diene-1,2-dione and cyclohexa-2,5-diene-1,4-dione, respectively, an intentional, but unexplained exemption of benzene from the general rule.

4.10 A caveat concerning Greek letters in systematic names

Systematic names may contain Greek letters which can easily be confused, or garbled by sloppy typography. Thus μ (lower-case Greek mu) can be confused with η (lower-case Greek eta), γ (lower-case Greek gamma) with the Roman letter y, and ν (lower-case Greek nu) with the Roman letter v. Also the letter pairs ξ (lower-case Greek xi) and ζ (lower-case Greek zeta) as well as φ (lower-case Greek phi) and ψ (lower-case Greek psi) can cause difficulties for the casual reader. The alas not uncommon misspelling of μg as mg moves the decimal point a whopping three places and could easily lead to a toxicological catastrophy [11].

Confusion may also arise when the German letter ß (German *Eszett*, pronounced ss) is taken for β, the lower-case Greek letter *beta*.

4.11 Incompatibilities

Once in a while names can be encountered which are illogical and appear as hybrids of mutually exclusive nomenclature systems.

Isopropanol may look innocuous at first glance, but it is actually an unfortunate hybrid of the traditional name isopropyl alcohol and the systematic name propan-2-ol. It should be avoided at all cost.

In commercial lingo **methyl hydrate** can be seen to raise its ugly head as a euphemism for methyl alcohol (methanol) [12]. Also here the obsolete name methylene hydrate and the traditional name methyl alcohol have been hybridized to generate a nomenclatural abomination.

References

[1] https://iupac.org (retrieved August 2, 2018); http://www.qmul.ac.uk/sbcs/iupac/ (retrieved March 14, 2019).

[2] https://iupac.org/what-we-do/nomenclature/ (retrieved August 2, 2018).

[3] 'The Red Book', N. G. Connelly, T. Damhus, R. M. Hartshorn, and A. T. Hutton, *Nomenclature of Inorganic Chemistry – IUPAC Recommendations 2005*, Royal Society of Chemistry, Cambridge, UK, 2005, ISBN 0-85404-438-8; free online version, https://www.iupac.org/cms/wp-content/uploads/2016/07/Red_Book_2005.pdf (retrieved August 2, 2018). Lists with corrections of misprints and errors are available as http://www.qmul.ac.uk/sbcs/iupac/bibliog/RBcorrect.html (retrieved March 14, 2019) and http://www.qmul.ac.uk/sbcs/iupac/bibliog/RBcorrect2.html (retrieved March 14, 2019). Cf. G. P. Moss (Ed.), *Brief Guide to Nomenclature of Inorganic Chemistry*, https://www.qmul.ac.uk/sbcs/iupac/BriefGuide/inorganic.html.

[4] 'The Blue Book', H. A. Favre and W. H. Powell, *Nomenclature of Organic Chemistry – IUPAC Recommendations and Preferred Names 2013*, Royal Society of Chemistry, Cambridge, UK, 2013, ISBN 978-0-85404-182-4. Extensive lists with corrections of misprints and errors are available as www.qmul.ac.uk/sbcs/iupac/bibliog/Bberrors.html (retrieved March 14, 2019) and www.qmul.ac.uk/sbcs/iupac/bibliog/BBminor.html (retrieved March 14, 2019).

[5] https://iupac.org/projects/project-details/?project_nr=2015-052-1-800 (retrieved August 2, 2018).

[6] G. P. Moss, *IUPAC Recommendations on Organic & Biochemical Nomenclature, Symbols & Terminology etc.*, https://www.qmul.ac.uk/sbcs/iupac/ (retrieved March 14, 2019).

[7] S. Allard and G. Ourisson, *Tetrahedron* **1**, 277–281 (1957); DOI: 10.1016/0040-4030(57)88002-8.

[8] https://en.wikipedia.org/wiki/Simplifiedmolecular-inputline-entrysystem (retrieved August 2, 2018).

[9] https://en.wikipedia.org/wiki/InternationalChemicalIdentifier (retrieved August 2, 2018).

[10] http://www.xml-cml.org (retrieved August 26, 2018).

[11] https://en.wikipedia.org/wiki/HypervitaminosisD (retrieved August 26, 2018).

[12] https://en.wikipedia.org/wiki/Methanol (retrieved April 5, 2019).

5 The Chemical Abstracts systematic nomenclature

The Chemical Abstracts Service (CAS) [1], founded 1907, maintains a comprehensive collection of chemical data such as abstracts of scientific publications and patents, as well as substance information. In order to serve these activities CAS has developed, and maintains, its own conventions for naming and indexing of chemical substances.

5.1 CAS names

While CAS nomenclature [2] by and large coincides with IUPAC nomenclature there are several significant differences. Most important is the fact that the CAS nomenclature was devised for in-house use, not as a tool for the chemical community at large. Most conspicuous is CAS's exclusive use of 'inverted names', i.e. the name of the parent compound is followed by a comma and then the rest of the name such as stereochemical descriptors, prefixes, infixes, etc. This feature facilitates the indexing of names in an order given by the name of the parent compound rather than alphabetically where, for instance, a near unmanageable string of unrelated names beginning with 'methyl' or 'chloro' would impede the recognition and selection of relevant entries.

While IUPAC in one of its rule changes has moved locants from the beginning of the name (such as in 2-butanol) to the place immediately before the relevant function (i.e. butan-2-ol) CAS has not made this change. Similarly, the stereochemical descriptors α and β which CAS places in parentheses (with the appropriate locants and together with other stereochemical descriptors) in the beginning of the full name, are in IUPAC names placed inside the name after the respective locants.

Simple inorganic compounds are dealt with unspecifically. Thus, in CAS names like 'hydrogen sulfide (H_2S)' and 'hydrogen sulfide (H_4S)' the constitution of the species is not defined by the name, but by an added molecular formula.

5.2 CAS registry numbers

A substance to be indexed by CAS receives not only a systematic name, but also a registry number (RN). The last digit of an RN is a check digit to prevent the use of mistyped RNs. A registry number implies no structural information, it can be compared to a telephone number. The first RN to be issued was [50-00-0] for formaldehyde. By now RNs have reached seven-digit size, such as [1196102-19-8] for the arsinite ion ($HAsO_2^{2-}$). An RN belongs to a substance, not to a specific name. When ununoctium (Uuo) changed its name to oganesson (Og) the RN [54144-19-3] remained the same.

https://doi.org/10.1515/9783110612714-005

Racemates and the corresponding enantiomers receive their registry numbers individually and independently. So does a substance with undisclosed stereochemistry. An example:

DL-**Alanine** [302-72-7], (*RS*)-2-aminopropanoic acid.

D-**Alanine** [338-69-2], (2*R*)-2-aminopropanoic acid.

L-**Alanine** [56-41-7], (2*S*)-2-aminopropanoic acid.

Once issued, RNs can be deleted if they have become redundant, for instance if the same substance has been registered twice under different names, or when structural information has been corrected after a registration.

5.3 CAS vs. IUPAC names

The natural compound taxol A, used as a drug with the INN paclitaxel, is an excellent example of different legitimate ways to construct systematic names. It would take a hardcore nomenclaturist to recognize the following three systematic names as synonyms. While the inverted names used by CAS are essential when compiling printed name indexes it seems far from obvious to anyone outside CAS that benzenepropanoic acid should be the parent heading for the diterpene taxol A.

Trivial name: taxol A, after the species *Taxus brevifolia* (Pacific yew tree). The genus name is from Latin *taxus*, yew tree.

CA index name: benzenepropanoic acid, β-(benzoylamino)-α-hydroxy-, (2a*R*,4*S*,4a*S*,6*R*,9*S*,11*S*,12*S*,12a*R*,12b*S*)-6,12b-bis(acetyloxy)-12-(benzoyloxy)-2a,3,4,4a,5,6,9,10,11,12,12a,12b-dodecahydro-4,11-dihydroxy-4a,8,13,13-tetramethyl-5-oxo-7,11-methano-1*H*-cyclodeca[3,4]benz[12-*b*]oxet-9-yl ester, (α*R*,β*S*)-.

IUPAC name (based on taxane as parent): 4,10β-diacetoxy-13α-{[(2*R*,3*S*)-3-(benzoylamino)-2α-hydroxy-3-phenylpropanoyl]oxy}-1,7β-dihydroxy-9-oxo-5β,20-epoxytax-11-en-2-yl benzoate.

IUPAC name (fully systematic): (1*S*,2*S*,3*R*,4*S*,7*R*,9*S*,10*S*,12*R*,15*S*)-4,12-bis(acetyloxy)-1,9-dihydroxy-15-({(2*R*,3S)-2-hydroxy-3-phenyl-3-[(phenylcarbonyl)amino]propanoyl}oxy)-10,14,17,17-tetramethyl-11-oxo-6-oxatetracyclo[11.3.1.$0^{3,10}$.$0^{4,7}$]heptadec-13-en-2-yl benzoate.

Among the differences between CAS and IUPAC nomenclature one could mention the following additional examples:

1) Stereoparents for fused ring compounds are chosen differently. A number of IUPAC's stereoparents are named systematically by CAS and vice versa.

2) A number of stereoparents acknowledged in both systems are numbered differently by IUPAC and CAS.
3) In IUPAC nomenclature anionic functional groups have seniority over cationic groups. The opposite is the case in CAS nomenclature.
4) CAS prefers conjunctive names, such as benzeneethanol (with Greek-letter locants for the chain, here α and β), while IUPAC prefers substitutive names, i.e. 2-phenylethan-1-ol.
5) IUPAC prefers the retained name aniline, CAS the name benzenamine.
6) CAS' kaurane structure corresponds to IUPAC's of the enantiomer *ent*-kaurane and vice versa. The CAS assignment is opposite to that agreed everywhere else, i.e. that the isomer with the 10β-methyl group (i.e. 10*R*-methyl) should be called kaurane and the enantiomer with a 10α-methyl group (i.e. 10*S*-methyl) *ent*-kaurane.

References

[1] https://www.cas.org (retrieved August 2, 2018).
[2] *Naming and Indexing of Chemical Substances for Chemical Abstracts, 2007 Edition*, American Chemical Society, Washington, DC, USA, 2008.

6 The naming of the elements

A somewhat unevenly researched list of element name etymologies is available in Wikipedia [1].

While a fully systematic naming of the elements based on their atomic numbers might in principle seem attractive, strong tradition and custom dictate otherwise. Thus, all 118 known elements have been assigned trivial names. However, IUPAC has adopted a systematic nomenclature for undiscovered elements as a placeholder function. Once a new element has been discovered the involved scientists have to prove the validity and priority of their claim to a joint IUPAC-IUPAP committee. Thereafter they submit their choice of a name which is then adopted or modified by IUPAC after a public hearing concerning possible objections [2].

6.1 The undiscovered elements

According to IUPAC's recommendations [1] undiscovered elements are named by atomic number, spelled out with a special set of numerals, followed by the ending 'ium'.

These numerals (of mixed Latin and Greek origin) are: 0 = nil, 1 = un, 2 = bi, 3 = tri, 4 = quad, 5 = pent, 6 = hex, 7 = sept, 8 = oct, 9 = enn.

Thus, element 119 would be called ununennium (with the element symbol Uue), element 125 unbipentium (Ubp), etc.

Earlier placeholder names used the prefixes 'eka' und 'dvi' (from the Sanskrit numerals 1 and 2, respectively) to show empty places in incomplete periodic tables. Thus, eka-manganese was the placeholder name for technetium, the next lower neighbor of manganese in group 7, and eka-rhenium for bohrium, rhenium's next lower neighbor. By the same token dvi-rubidium was a placeholder name for its second-next lower neighbor in group 1, francium.

6.2 The 'old' elements

Some elements have been known since antiquity and have, accordingly been named before the dawn of chemistry. They are:

Antimony (Sb, element 51), from Latin *antimonium*, possibly from Late Greek *stimmi*, powdered antimony, possibly from dialectal Arabic *uthmud*, ultimately from Ancient Egyptian *stm*, powdered antimony. Earlier the name antimony was used for antimony(III) sulfide, the metal being called regulus of antimony. The element

https://doi.org/10.1515/9783110612714-006

symbol Sb is derived from New Latin *stibium*, antimony, ultimately loaned from Ancient Egyptian. Both antimony and the New Latin name *stibium* are used for the construction of substance names.

Bismuth (Bi, element 83), a name of disputed origin. Probably ultimately derived from Greek *psimythion*, most often white lead [$2PbCO_3{\cdot}Pb(OH)_2$], however, also used as a name for bismuth subnitrate [$BiO(NO_3){\cdot}H_2O$], also known by the Latin name *magisterium bismuthi*. Greek *psimythion* is probably loaned from Ancient Egyptian. Etymological threads leading to the interpretation of the German common name *Wismut* as meaning 'white mass' or 'prospecting in the meadows' bear the hallmark of folk etymology and deserve hardly serious consideration.

Copper (Cu, element 29), from Late Latin *cuprum*, from Latin *aes cyprium*, metal of Cyprus, later *cuprum*. The name Cyprus is from Greek *Kypros*, a name with three suggested etymologies. From Greek *kyparissos*, cypress tree (*Cupressus sempervirens*), from Greek *kypros*, henna plant (*Lawsonia alba*) or from Sumerian *zubar*, copper or *kubar*, bronze. Thus, it is still unclear if Cyprus was named after copper or copper after Cyprus. Both copper and New Latin *cuprum* are used for the construction of substance names.

Gold (Au, element 79), from Proto-Indo-European *ghultham*, gold, from the root *ghel-*, to shine. The element symbol Au is derived from Latin *aurum*, gold. Both gold and New Latin *aurum* are used for the construction of substance names.

Iron (Fe, element 26), originally an adjective, probably borrowed from Celtic *isarnon*, from Proto-Indo-European *is-(e)ro-*, powerful, holy, i.e. holy metal, strong metal as opposed to softer bronze. The element symbol Fe is derived from Latin *ferrum*, iron, probably of Semitic origin. Both iron and New Latin *ferrum* are used for the construction of substance names.

Lead (Pb, element 82), from Proto-Germanic **lauda-*, lead, of uncertain origin. Another Proto-Germanic word for lead, **bliwa-* [2], is the source of the German word for lead, *Blei*. The element symbol Pb is from Latin *plumbum* (lead), most likely borrowed from a Mediterranian language. Both lead and New Latin *plumbum* are used for the construction of substance names.

Mercury (Hg, element 80), from Medieval Latin *mercurius*, mercury, named for the Roman god *Mercurius*, the god of speed and messenger to the gods, whose name is possibly related to Latin m*ercari*, to trade, from *merx, merc-*, merchandise. The element symbol Hg is derived from Latin *hydrargyrum*, mercury, from Greek *hydrargyros*, mercury, from *hydor*, water, and *argyros*, silver, literally liquid silver.

Silver (Ag, element 79), from Proto-Germanic **silubra-*, ultimately derived from Akkadian *sarpu*, silver, literally purified by melting, from Akkadian *sarapu*, to melt.

The element symbol Ag is derived from Latin *argentum*, silver, from a Proto-Indo-European root meaning white. Both silver and New Latin *argentum* are used for the construction of substance names.

Sulfur (S, element 16), from Early Latin *sulpur, sulphur*, and *sulfur*, sulfur, from Proto-Indo-European **swelp*, to burn [3]. The British spelling sulphur inappropriately suggests a Greek origin of the word. Both sulfur and the element's Greek name, *theion*, literally the divine, are used for the construction of substance names.

Tin (Sn, element 50), from Proto-Germanic **tinom*; a word only found in the Germanic languages. The element symbol Sn is derived from Latin *stannum*, tin, of Celtic origin. Both tin and New Latin *stannum* are used for the construction of substance names.

Zinc (element 30), from German *Zink*, from German *Zinke*, prong, referring to the spiky crystals which form when molten zinc solidifies. The New Latin equivalent is *zincum*.

6.3 The 'newer' elements

Their names are of known etymology and reflect the thinking of their discoverers.

6.3.1 Named after sources

Aluminum (Al, element 13), from Latin *alumen, alumin-*, alum [$KAl(SO_4)_2{\cdot}12H_2O$], meaning bitter salt. A.k.a. **aluminium**.

Arsenic (As, element 33), from Latin *arsenicum*, arsenic, ultimately from Persian *zarnikh*, yellow orpiment (As_2S_3) via Greek folk etymological *arsenikon*, male, virile. Orpiment is from Latin *auripigmentum*, from Latin *aurum*, gold, and *pigmentum*, pigment from *pingere, pinct-*, to paint. Also realgar (As_4S_4) and arsenic trioxide (As_2O_3) have been called arsenic. Here, as in the case of manganese, a compound name was taken as an element name. In German, like in many other languages, *Arsenik* is the name of arsenic trioxide, while arsenic is called *Arsen*.

Beryllium (Be, element 4), from Greek *beryllos*, beryl [$Be_3Al_2(SiO_3)_6$]. The Greek word is from a Sanskrit root meaning becoming pale. Earlier called **glucinium**, after the sweet taste of its compounds.

Boron (B, element 5), from the mineral borax ($Na_2B_4O_7{\cdot}10H_2O$), via Greek, and Latin *borax, borac-*, ultimately derived from Persian *burah*, borax. The suffix 'on' was taken from carbon. Crude borax is called tincal, ultimately from Sanskrit *tankana*.

Cadmium (Cd, element 48), after Latin *cadmia*, calamine, a mixture of zinc minerals in which cadmium is found as an impurity. The Latin name of the mineral is related (via Greek) to Cadmus, the founder and first king of Thebes, Greece around 2000 BC.

Calcium (Ca, element 20), from Latin *calx*, lime (CaO), possibly from Greek *chalix*, pebble.

Carbon (C, element 6), from Latin *carbo, carbon-*, coal.

Cobalt (Co, element 27), from German *Kobold*, goblin. A superstitious term used by miners trying and failing to extract valuable copper or nickel from cobalt ores. Upon roasting of arsenic-containing cobalt ores highly toxic arsenic oxide (As_2O_3) added to the miners' unpleasant experience, blamed on a goblin. Also Greek *kobalos*, mine, has been suggested as the origin of both cobalt and German *Kobold*.

Fluorine (F, element 9), named after the mineral fluorspar, fluorite (CaF_2), so called because of its use as flux in iron melting.

Krypton (Kr, element 36), from Greek *krypton*, neuter of *kryptos*, the hidden one, referring to the difficulty of isolating krypton from atmospheric air.

Lithium (Li, element 3), from Greek *lithos*, stone; to emphasize its occurrence in solid stone as opposed to potassium's occurrence in plant ashes, etc.

Magnesium (Mg, element 12), after the Latin mineral name *magnesia alba* ($MgCO_3$), white magnesia, of uncertain origin. Also magnesium oxide (MgO) has been called *magnesia alba*.

Manganese (Mn, element 25), after the Latin mineral name *magnesia negra*, of uncertain origin, black magnesia, pyrolusite (MnO_2). The element name is from Italian *manganese*, manganese dioxide. Here, as in the case of arsenic, a compound name was taken as an element name.

Molybdenum (Mo, element 42), named after the mineral molybdena (MoS_2), now called molybdenite. In antiquity this mineral was easily confused with galena (PbS) and therefore believed to contain lead, hence its name from Greek *molybdos*, lead, ultimately from the Proto-Indo-European root **mork*, dark. This etymology raises the intriguing question of how the naming of lead and molybdenum is handled in Modern Greek. Lead is called *molybdos*, just as in Ancient Greek, and molybdenum is *molybdainio*.

Nickel (Ni, element 28), after Swedish *kopparnickel* and German *Kupfernickel*, copper bogey, the nickname of the mineral nickeline (NiAs) which failed to yield the copper desired by miners. The New Latin name of the element is *niccolum*.

Potassium (K, element 19), from potash (K_2CO_3), also known as its New Latin equivalent potassa. The element symbol K is derived from the New Latin name *kalium*, from alkali, from Arabic *al-qali*, saltwort ashes.

Samarium (Sm, element 62), from samarskite, i.e. samarskite-(Y) [Y,Fe^{3+},U)(Nb,Ta)O $_4$] and samarskite-(Yb) [(Yb,Fe^{3+})$_2$(Nb,Ta)$_2$O$_8$], two minerals named after the Russian mine official colonel Vasiliy Samarsky-Bykhovets (1803–1870) which contain small amounts of samarium.

Silicon (Si, element 14), from silica (SiO_2), from Latin *silex. silic-*, pebble, flint.

Sodium (Na, element 11), from soda ($Na_2CO_3{\cdot}10H_2O$), from Medieval Latin *soda*, barilla, ultimately probably from Arabic *suwwad*, barilla. Soda was first found in saltwort ash. The element symbol Na is from the New Latin name *natrium*, from Greek *nitron*, soda, ultimately from Ancient Egyptian *netjeri*, soda, via Arabic *natrun*, soda.

Tungsten (W, element 74), after the Swedish name *tungsten*, literally heavy stone, of the mineral scheelite ($CaWO_4$). The element symbol is from the German name *Wolfram*, literally wolf's soot. Wolf refers to the fact that the presence of tungsten ores made the extraction of tin from tin ores difficult, it 'eats up the tin like a wolf'. The Middle High German word *rām*, soot, refers to the friability of the mineral wolframite [(Fe,Mn)WO_4] which, when powdered, resembles soot.

Xenon (Xe, element 54), from Greek *xenon*, neuter of *xenos*, foreigner, stranger, guest, referring to its minute quantity in the atmosphere.

Zirconium (Zr, element 40), named after the mineral zircon, a zirconium silicate, ($ZrSiO_4$) which in turn has its name from Persian *sargun*, gold-hued.

6.3.2 Named after chemical properties

Argon (Ar, element 18), from Greek *argon*, neuter of *argos*, idle, inactive, referring to argon's lack of chemical reactivity.

Bromine (Br, element 35), from Greek *bromos*, stench, referring to the obnoxious smell of bromine vapor. The suffix 'ine' was taken from chlorine.

Dysprosium (Dy, element 66), from Greek *dysprositos*, hard to get, referring to the difficulty of its separation from other closely related elements.

Hydrogen (H, element 1), from Greek *hydor*, water, and *-genes*, producer, referring to the fact that water is formed when hydrogen burns.

Lanthanum (La, element 57), from lanthana (La_2O_3), from Greek *lanthanein*, to lie hidden, referring to the difficulty of separating lanthana from ceria (CeO_2).

Nitrogen (N, element 7), from niter (KNO_3) and Greek *-genes*, producer, i.e. niter maker, referring to the fact that nitrogen is the essential part of niter. Both nitrogen and the element's French name *azote*, from Greek *a-*, without, and *zoein*, to live, referring to its inability to support respiration, are used in the construction of substance names.

Osmium (Os, element 76); from Greek *osme*, smell, referring to the foul smell of the volatile osmium tetraoxide (OsO_4).

Oxygen (O, element 6), from Greek *oxys*, sharp, acidic, and *-genes*, producer, thus named as acid maker. The existence of oxygen was postulated around 1600 by the Polish alchemist Michael Sendivogius (Michał Sędziwój) (1566–1636), roughly 170 years before its discovery 1773/1774 by the Swedish chemist Carl Wilhelm Scheele (1742–1786), the British scientist Joseph Priestley (1733–1804), and the French chemist Antoine-Laurent de Lavoisier (1743–1794). Contrary to what the name oxygen might suggest not all oxides are acids and not all acids contain oxygen.

Phosphorus (P, element 15), from New Latin *phosphorus mirabilis*, miraculous bearer of light, referring to the faint glow emitted by white phosphorus in contact with air. The word phosphorus is derived from Greek *phos, photo-*, light, and *phoros*, carrying forward, from *pherein*, to bear, to carry.

6.3.3 Named after physical properties

Actinium (Ac, element 89), from Greek *aktis, aktin-*, ray, referring to the radioactivity of this element. Cf. radium.

Astatine (At, element 85), from Greek *astatos*, unstable, referring to this element's radioactivity and lack of stable isotopes.

Barium (Ba, element 56), from the mineral name baryte ($BaSO_4$), from Greek *barys, baryt-*, heavy, referring to the high mass density of its compounds.

Cesium (Cs, element 55), also spelled caesium, from Latin *caesius*, bluish gray, akin to *caelum*, sky, referring to the color of its characteristic spectral lines.

Chlorine (Cl, element 17), from Greek *chloros*, yellow green, referring to the green color of elemental chlorine.

Chromium (Cr, element 24), from Greek *chroma, chromat-*, color, referring to chromium's many colored compounds.

Indium (In, element 49), from indigo, referring to an indigo-colored spectral line. While the element indium did not receive its name as an homage to India the name of the dyestuff indigo, ultimately from Greek *indikon pharmakon*, i.e. blue dye from India, does have a relationship with India.

Iodine (I, element 53), from Greek *iodes*, violet, referring to the color of iodine vapor.

Iridium (Ir, element 77), from Latin *iris, irid-*, rainbow, referring to the many strongly colored salts of this metal.

Platinum (Pt; element 78), from Spanish *platina del Pinto*, little silver from the river Pinto, from Spanish *plata*, silver. Platinum was earlier considered inferior to silver.

Praseodymium (Pr, element 59), a hard-to-separate mixture of two unknown elements was called didymium, from Greek *didymos*, twin. One of them, praseodymium, was called after the green color of its salts, from Greek *prasios*, green as a leek, from *prason*, leek, the other was neodymium.

Protactinium (Pa, element 91), from protoactinium, i.e. named as the parent of actinium in the decay chain of ^{235}U.

Radium (Ra, element 88), from Latin *radius*, ray, referring to this element's radioactivity. Cf. actinium.

Radon (Rn, element 86), a contraction of radium emanation, referring to radon's appearance in radium's decay chain. Other suggested names, i.e. acton (An) and thoron (Tn), were rejected by IUPAC.

Rhodium (Rh, element 45), from Greek *rhodon*, rose, probably from Old Iranian **urda-*, rose, referring to this element's rose-red compounds.

Rubidium (Rb, element 37), from Latin *rubidus*, deep red, referring to the color of a spectral line of rubidium.

Technetium (Tc, element 43), from Greek *technetos*, artificial, referring to the fact that technetium was the first artificially produced element.

Thallium (Tl, element 81), from Greek *thallos*, green shot or twig, referring to a characteristic green spectral line of thallium.

6.3.4 Eponyms

Curium (Cm, element 96), after the French-Polish radiochemist Marie Skłodowska-Curie (1867–1834) and her husband, the French physicist Pierre Curie (1859–1906).

Einsteinium (Es, element 99), after the German-American theoretical physicist Albert Einstein (1879– 1955).

Fermium (Fm, element 100), after the Italian-American physicist Enrico Fermi (1901–1954).

Gadolinium (Gd, element 64), after the Finnish chemist, physicist, and mineralogist Johan Gadolin (1760–1852).

Lawrencium (Lr, element 103), after the American physicist Ernest Orlando Lawrence (1901–1958).

Mendelevium (Md, element 101), after the Russian chemist Dmitri Ivanovich Mendeleyev (1834–1907). An etymologically correct eponym would have been mendeleyevium.

Nobelium (No, element 102), after the Swedish chemist, industrialist, and philanthropist Alfred Nobel (1833–1896). The name joliotium was also proposed, but without success.

6.3.5 Toponyms

Americium (Am, element 95), after the Americas, and referring to the fact that this element's position in the actinoid series corresponds to that of europium in the lanthanoidseries. The toponym America is after the given name of the Italian navigator Amerigo Vespucci (1454–1512).

Berkelium (Bk, element 97), after the Lawrence Berkeley Laboratory, Berkeley, CA, USA. The toponym Berkeley refers to the Anglo-Irish philosopher George Berkeley (1685–1753).

Californium (Cf, element 98), after the University of California. The toponym California is assumed to originate from California, an imaginary realm with this name described in Garci Ordóñez de Montalvo's 'Las sergas de Esplandián' (the exploits of Esplandián), published in 1510. Spanish explorers, reaching Baja California and assuming that it was such an island, thought they had reached this magic country.

Erbium (Er, element 68), after its oxide erbia (Er_2O_3) which in turn was named by cannibalization of the name ytterbia (Yb_2O_3). Cf. ytterbium, yttrium, and terbium.

Europium (Eu, element 63), after Europe. The toponym is from Greek *Europe* of uncertain origin, possibly from Phoenician *ereb*, evening, i.e. west.

Francium (Fr, element 87), after New Latin *Francia*, France, cf. gallium.

Gallium (Ga, element 31), after Latin *Gallia*, Gaul, named in honor of France, cf. francium. According to an urban legend the discoverer of gallium, Paul Emile Lecoq de Boisbaudran (1838–1912), named the element after himself, since French *le coq*, the rooster, is *gallus* in Latin.

Germanium (Ge, element 32), after New Latin *Germania*, Germany.

Hafnium (Hf, element 72), after New Latin *Hafnia*, Copenhagen, Denmark. The Danish name of the city, *København*, means ultimately merchants' harbor. The Danish physicist Niels Henrik David Bohr (1885–1962), the Dutch physicist Dirk Coster (1889–1950), and the Hungarian radiochemist George Charles de Hevesy (Hevesy-Bischitz György) (1885–1966), as the discoverers of this element, soon regretted their suggestion of the name hafnium and submitted a correction proposing the name danium, after New Latin *Dania*, Denmark. This correction, however, came too late. The first name suggestion had already been made public.

Holmium (Ho, element 67), after New Latin *Holmia*, Stockholm, Sweden.

Lutetium (Lu, element 71), earlier also spelled lutecium, named after its oxide lutecia (Lu_2O_3), which in turn was named after French *Lutèce*, ancient Paris. The name lutetium is after Latin *Lutetia*, ancient Paris, France. The Latin toponym might be derived from the Celtic root *luto-* or *luteuo-*, swamp, marsh.

Polonium (Po, element 84), after New Latin *Polonia*, Poland. Ironically, in 1898, when polonium was discovered and named, Poland, having been partitioned between Austria-Hungary, Germany, and Russia, did not exist as a sovereign state.

Rhenium (Re, element 75), from Latin *Rhenus*, the Rhine. The name of the river is from a Proto-Indo-European root **rei-*, to move, to flow, to run.

Ruthenium (Ru, element 44), from New Latin *Ruthenia*, Rus', a historical area that included present-day western Russia, Ukraine, Belarus and parts of Slovakia and Poland.

Scandium (Sc, element 21), after Scandinavia, ultimately, via Latin, from Proto-Germanic **skaÞiōnawjō* (from Proto-Germanic **aujo*, island), Scadia Island, ie. Scania. In antiquity Scandinavia was regarded as an island.

Strontium (Sr, element 38), after the Scottish village of Strontian (Gaelic *Srón an t-Sithein*, the point of the fairy hill). The mineral strontianite ($SrCO_3$), named as early as 1791 and with the same etymology as the element name, does not seem to have played a role in the 1808 naming of the element.

Terbium (Tb, element 65), named after its oxide terbia (Tb_2O_3) which in turn was named by cannibalization of the name ytterbia (Yb_2O_3). Cf. ytterbium, yttrium, and erbium.

Thulium (Tm, element 69), from thulia (Tm_2O_3), named after Latin *Ultima Thule*, Scandinavia, in antiquity regarded as a mythical island country. Contrary to popular belief the town of Thule (Inuit *Qaanaaq*) in Greenland, home of a well-known US air base, has nothing to do with the naming of this element. Both the element and the town owe their names to the same Latin source.

Ytterbium (Yb, element 70), named after its oxide ytterbia (Yb_2O_3) which in turn was named after the Swedish village of Ytterby, literally outer village, where the rare-earth metal-containing mineral ytterbite, a.k.a. gadolinite, was found. Cf. yttrium, terbium, and erbium. Other suggested names, never used, were neoytterbium, and aldebaranium, after the orange giant star Aldebaran, from Arabic *al-dabaran*, the follower.

Yttrium (Y, element 39), named after its oxide yttria (Y_2O_3) which in turn was named by cannibalization of the name ytterbia (Yb_2O_3). Cf. ytterbium, terbium, and erbium.

6.3.6 Named after mythological figures

Cerium (Ce, element 58), after the mineral cerite [$(Ce,La,Ca)_9(Mg,Fe^{3+})(SiO_4)_6(SiO_3OH)(OH)_3$], itself named after the dwarf planet Ceres which in turn was named after Ceres, the Roman goddess of fertility, from a Proto-Indo-Germanic root **ker-*, to grow.

Niobium (Nb, element 41), after Niobe, daughter of Tantalus. The etymology of the name Niobe remains unexplained. Quite intentionally the closely related element tantalum was named after Niobe's father Tantalus.

Palladium (Pd, element 46), after the asteroid Pallas which in turn was named after the Greek goddess of wisdom, handicraft, and warfare, Pallas Athena.

Promethium (Pm, element 61), after Prometheus who stole the fire of heaven and gave it to mankind. The name Prometheus is derived from Greek *prometheia*, foresight.

Selenium (Se, element 34), after the Greek moon goddess Selene, from Greek *selene*, moon, possibly from Greek *selas*, light. So named to show its close relationship with tellurium.

Tantalum (Ta, element 73), after Tantalus, referring to metallic tantalum's lack of reactivity. The name Tantalus is possibly derived from Greek *talantatos*, who has to bear much. Quite intentionally the closely related element niobium was named after Tantalus' daughter Niobe.

Tellurium (Te, element 52), after Latin *tellus, tellur-*, earth, and the Roman goddess Tellus Mater.

Thorium (Th, element 90), from thoria (ThO_2), named after the Norse god of thunder Thor.

Titanium (Ti, element 22), after the Titans of Greek mythology.

Vanadium (V, element 23), after the Norse goddess Vanadis, also known as Freyja, Gefn, Hörn, Mardöll, Skjálf, Sýr, Thröng, and Valfreyja. The name was chosen with regard to the beautifully multicolored compounds of this element.

6.3.7 Named after celestial bodies

Helium (He, element 2), from Greek Helios, the sun god, and Greek *helios*, sun. Helium was first identified by its characteristic emission lines in the Sun's spectrum. In hindsight being a noble gas helium ought to have been named helion.

Neptunium (Np, element 93), after the planet Neptune which in turn was named after the Roman god of the oceans Neptunus, a name of uncertain etymology.

Plutonium (Pu, element 94), after the dwarf planet Pluto which in turn was named after Pluto, the Greek god of the underworld.

Uranium (U, element 92), after the planet Uranus which in turn was named after the Greek god of the sky Uranus.

6.3.8 Named as new elements

Neodymium (Nd, element 60), from Greek *neos*, new and *didymos*, twin. The second element, after praseodymium, to be isolated from a mixture called didymium.

Neon (Ne, element 10), from Greek *neon*, neuter of *neos*, new.

6.4 The 'newest' elements

The transactinoids, all of them artificially produced, have been named after important scientists or after the place where they were discovered. The current IUPAC recommendations for the naming of new elements in English [2] propose that elements be named after a mythological concept, a mineral, a place or country, a property, or a scientist. The new names should end in 'ium', with the exception of group 17 elements whose names should end in 'ine', and group 18 elements whose names should end in 'on'.

6.4.1 Eponyms

Bohrium (Bh, element 107), after the Danish physicist Niels Henrik David Bohr (1885–1962). The originally suggested name nielsbohrium, which would have avoided confusion with the element boron, e.g. the homophonous derived names bohride and boride, was rejected by IUPAC as there was no precedent of a first name as part of an eponymous element name. During these discussions another proposal, by the Danish National Committee for Chemistry, to call this element danium (with the element symbol Da), from New Latin *Dania*, Denmark, came too late in the naming process to be considered. The name danium was a 'leftover' from the naming process of hafnium, q.v.

Copernicium (Cn, element 112), after the Polish astronomer Nicolaus Copernicus (Mikołaj Kopernik) (1473–1543).

Flerovium (Fl, element 114), named in honor of the Russian Flerov Laboratory of Nuclear Reactions, this in turn after the Russian physicist Georgi Flyorov (1913–1990). One should note that the etymologically correct eponym would have been flyorovium. A better known case of improperly transliterated Russian names is that of Prince Grigori Aleksandrovich Potemkin (1739–1791), of herostratic fame, whose proper pronunciation is [puh-**tyawm**-kyn].

Meitnerium (Mt, element 109), after the Austrian-Swedish physicist Lise Meitner (1878–1968).

Oganesson (Og, element 118), after the Armeinian-Russian-American physicist Yuri Tsolakovich Oganessian (born 1933), one of only two scientists ever to have an element named after him in his lifetime. The ending 'on' labels oganesson as a group 18 element.

Roentgenium (Rg, element 111), after the German physicist Wilhelm Conrad Röntgen (1845–1923). One should note that an etymologically correct eponym would have been röntgenium.

Rutherfordium (Rf, element 104), after the New Zealand-British physicist Ernest Rutherford (1871–1937). Two proposed names, i.e. dubnium, after the Soviet Joint Institute for Nuclear Research in Dubna, Moscow Oblast, Russia, and kurchatovium, in honor of the Russian physicist Igor Vasilyevich Kurchatov (1903–1960), were rejected by IUPAC.

Seaborgium (Sg, element 106), after the American nuclear chemist Glenn Theodore Seaborg (1912–1999), one of only two scientists ever to have an element named after him in his lifetime. A proposal to name this element rutherfordium, after the New Zealand-British physicist Ernest Rutherford (1871–1937), was unsuccessful.

6.4.2 Toponyms

While formally derived from place names the names of the following elements were chosen to recognize the institutions where they were discovered.

Darmstadtium (Ds, element 110), named in recognition of its discovery at the GSI Helmholtz Centre for Heavy Ion Research in Darmstadt, Hesse, Germany. The toponym Darmstadt is of uncertain and disputed origin.

Dubnium (Db, element 105), named in recognition of its discovery at the Soviet Joint Institute for Nuclear Research in Dubna, Moscow Oblast, Russia, cf. moscovium. The toponym Dubna is after the River Dubna. Other suggested names, i.e. hahnium, in honor of the German chemist Otto Hahn (1879–1968), joliotium, in honor of the French radiochemists Frédéric Joliot-Curie (1897–1956) and Irene Joliot-Curie (1900–1958), and nielsbohrium, in honor of the Danish physicist Niels Henrik David Bohr (1885–1962), were all rejected by IUPAC.

Hassium (Hs, element 108), named in recognition of its discovery at the GSI Helmholtz Centre for Heavy Ion Research in Darmstadt, Hesse, Germany; from New Latin *Hassia*, Hesse. The toponym Hesse is from German *Hessen*, derived from the Latin name *Chatti* of a Germanic tribe living there in antiquity. Another proposed name, hahnium, in honor of the German chemist Otto Hahn (1879–1968), was rejected by IUPAC

Livermorium (Lv, element 116), named in recognition of its discovery at the Lawrence Livermore Laboratory, at Livermore, CA, USA. The town of Livermore was named after the British-American rancher Robert Livermore (1799–1858).

Moscovium (Mc, element 115), from New Latin *Moscovia*, Muscovy, in recognition of its discovery at the Joint Institute for Nuclear Research, Dubna, Moscow Oblast, Russia, cf. dubnium; the city of Moscow and hence the Moscow Oblast were supposedly named after the Moskva River, Russia. Another suggested name, langevinium, in honor of the French physicist Paul Langevin (1872–1946), was rejected by IUPAC.

Nihonium (Nh, element 113), named in recognition of its discovery by the Japanese research organization RIKEN (Institute of Physical and Chemical Research, patterned after the German Kaiser-Wilhelm-Gesellschaft, now Max-Planck-Gesellschaft), after Japanese *Nihon, Nippon*, Japan, literally the sun's origin.

Tennessine (Ts, element 117), named in recognition of its discovery at the Oak Ridge National Laboratory, Oak Ridge, TN, USA. The name Tennessee is probably a Cherokee modification of a Yuchi word with uncertain meaning. The ending 'ine' labels tennessine as a group 17 element.

6.5 Previously used, but now discarded element and nuclide names

Most of these names are self-explanatory.

Actinium A (^{215}Po).

Actinium B (^{211}Pb).

Actinium B' (^{215}At).

Actinium C (^{211}Bi).

Actinium C' (^{211}Po).

Actinium C'' (^{207}Tl).

Actinium D (^{207}Pb), a.k.a. **actinium lead.**

Actinium K (^{223}Fr).

Actinium X (^{223}Ra).

Actinon (An; ^{219}Rn), a.k.a. **actinium emanation**, named as noble gas derived from actinium.

Actino-uranium (^{235}U).

Cassiopeium (Cp; now lutetium, Lu) after the constellation Cassiopeia, named after Cassiopeia, the mythological queen of Aethiopia. The name of the mythological figure is of uncertain etymology. The name cassiopeium was used in the German-language chemical literature until the 1950's.

Columbium (Cb; now niobium, Nb), after the mineral columbite [$(Fe,Mn)Nb_2O_6$], named after New Latin *Columbia*, the USA. The name columbium was used in the pre-WW II American chemical and especially geological literature.

Glucinium (Gl, G; now beryllium, Be), from Greek *glykys*, sweet, referring to the sweet taste of some of its salts. A.k.a. **glucinum.**

Ionium (Io; ^{230}Th), a name used until it was realized that this nuclide was a thorium isotope.

Meso-thorium I (^{228}Ra).

Meso-thorium II (^{228}Ac).

Niton (Nt; ^{222}Rn), from Latin *nitens*, shining, referring to radon's glowing in the dark.

Radio-actinium (^{227}Th).

Radio-thorium (^{228}Th).

Radium A (^{218}Po).

Radium B (^{214}Pb).

Radium B' (^{218}At).

Radium C (^{214}Bi).

Radium C' (^{214}Po).

Radium C'' (^{210}Tl).

Radium D (^{210}Pb).

Radium E (^{210}Bi).

Radium E'' (^{206}Tl).

Radium F (^{210}Po).

Radium G (^{206}Pb), a.k.a. **uranium lead.**

Radium emanation (^{222}Rn), a.k.a. **emanation** and **emanon.**

Thorium A (^{216}Po).

Thorium B (^{212}Pb).

Thorium B' (^{216}At).

Thorium C (^{212}Bi).

Thorium C' (^{212}Po).

Thorium C'' (^{208}Tl)

Thorium D (^{208}Pb), a.k.a. **thorium lead.**

Thorium X (^{224}Ra).

Thoron (Tn; ^{220}Rn), a.k.a. **thorium emanation**; named as noble gas derived from thorium.

Uranium I (^{238}U).

Uranium II (^{234}U).

Uranium X_1 (^{234}Th).

Uranium X_2 (^{234}Pa).

Uranium Y (^{231}Th).

6.6 Endings of element names

IUPAC recommends to let all new element names, with the exception of group 17 and 18 elements, end in 'ium'.

The English names of the halogens end in 'ine', a convention which has also been observed in the case of the latest newcomer, tennessine.

This convention applies only to the English names, thus in German the halogens are named *Fluor, Chlor, Brom, Iod, Astat*, and *Tenness*, the hallmark of German halogen names being the lack of an ending.

Apart from the first of the noble gases, helium, a common ending 'on' has been attached to all noble gas names, the newest being oganesson. The origin of the ending 'on' is the neuter ending of Greek adjectives, first apparent in the element name neon.

A homonymous ending 'on' can be seen in the name carbon, from the stem *carbon-* of Latin *carbo*, coal. The names of the neighboring elements boron and silicon were provided with the same ending in order to show the similar appearance and behavior of elemental carbon, boron, and silicon.

A number of element names carry the ending 'um' instead of 'ium', without any apparent reason. These elements are aluminum (also aluminium), lanthanum, molybdenum, platinum and tantalum.

6.7 Quasi-elements

As a matter of brevity and convenience certain nuclides and particles other than atoms have been named as quasi-elements.

6.7.1 Isotope names

Because of their great practical importance the three isotopes of hydrogen and one of radon have received individual names and element-like symbols.

Protium (^{1}H), from Greek *protos*, the first.

Deuterium (^{2}H, D), from Greek *deuteros*, the second.

Tritium (^{3}H, T), from Greek *tritos*, the third.

Thoron (Tn, ^{220}Rn), from thorium. While thoron (Tn) is obsolete as a synonym of radon (Rn), it is still in use for this specific isotope of radon.

6.7.2 Muonium

The exotic muonium particle, with a lifetime of 2.2 μs, is composed of an antimuon and an electron [6]. A more precise name would have been antimuonium. Chemically it behaves like an isotope of hydrogen about an order of magnitude lighter than ^{1}H. It has been assigned the symbol Mu and shown to form compounds such as muonium chloride (MuCl) and sodium muonide (NaMu). Muonium has been named after muon, after the English name mu of the Greek lower-case letter μ.

An earlier name of muon was μ meson. Greek letters were labels attached to the names of subatomic particles roughly in the order of their discovery, cf. Λ baryon and π meson (pion).

It should be noted that the name of the Greek letter μ is transcribed as my in German (and other languages) and thus muonium is called *Myonium* in German.

6.7.3 Positronium

The exotic positronium particle, with a lifetime of 142 ns, is composed of a positron and an electron [7]. It has been assigned the symbol Ps and can form both a hydride PsH and a cyanide PsCN. Also dipositronium, Ps_2, has been observed.

6.8 The Periodic Table

Over time many versions of the Periodic Table, with its grid of periods and groups, have been proposed. They typically operate with either eight main groups and eight side groups, cf. Table 6.1.A, with 18 groups, consecutively numbered, cf Table 6.1.B, or, in its large-cell version with 32 cells, 18 with the numbered groups and 14 unnumbered, all belonging to group 3 [5], cf. Table 6.1.C. The 32 group system, while strong on logic, is wasteful of empty space and therefore rather impractical both in book and in display format.

Even when 18 numbered groups are employed the terms main group and side group are still used as expedient class names.

Table 6.1.D comprises both the 8-group and the 18-group system and allows to explore their relative merits.

The name Periodic Table could be misunderstood, since the table is not exposed to nor does it cause periodic events. The table contains periods, which is much better expressed in its German name *Periodensystem*, literally system of periods.

6.8.1 Current versions

The Periodic Table recommended by IUPAC [4] is of the 18-group type, Table 6.1.B. All discussions in this book are based on this version.

6.8.2 Controversies

Even though all historical controversies concerning missing elements and the position of some elements in the Periodic Table have been resolved, controversies do

Table 6.1.A:

Combined Periodic Table of the Elements

		Group I				
First Period		1 H Hydrogen				
	Group 0	Group I	Group II		Group III	Group IV
Second Period	2 He helium	3 Li lithium	4 Be beryllium		5 B boron	6 C carbon
Third Period	10 Ne neon	11 Na sodium	12 Mg magnesium		13 Al aluminium	14 Si silicon
Fourth Period	18 Ar argon	19 K potassium	20 Ca calcium	Transition elements	31 Ga gallium	32 Ge germanium
Fifth Period	36 Kr krypton	37 Rb rubidium	38 Sr strontium		49 In indium	50 Sn tin
Sixth Period	54 Xe xenon	55 Cs caesium	56 Ba barium		81 Tl thallium	82 Pb lead
Seventh Period	86 Rn radon	87 Fr francium	88 Ra radium		113 Nh nihonium	114 Fl flerovium
	Group 0	Group I	Group II		Group III	Group IV

Transition elements

						21 Sc scandium	22 Ti titanium
26 Fe iron	27 Co cobalt	28 Ni nickel	29 Cu copper	30 Zn zinc	Main group elements	39 Y yttrium	40 Zr zirconium
44 Ru ruthenium	45 Rh rhodium	46 Pd palladium	47 Ag silver	48 Cd cadmium		57–71 La+ Lanthanoids lanthanum	72 Hf hafnium
76 Os osmium	77 Ir iridium	78 Pt platinum	79 Au gold	80 Hg mercury		89–103 Ac+ Actinoids actinium	104 Rf rutherfordium
108 Hs hassium	109 Mt meitnerium	110 Ds darmstadtium	111 Rg roentgenium	112 Cn copernicium			

Lanthanoids

57 La^{3+} lanthanum	58 Ce^{3+} cerium	59 Pr^{3+} praseodymium	60 Nd^{3+} neodymium	61 Pm^{3+} promethium	62 Sm^{3+} samarium	63 Eu^{3+} europium	64 Gd^{3+} gadolinium
64 Gd^{3+} gadolinium	65 Tb^{3+} terbium	66 Dy^{3+} dysprosium	67 Ho^{3+} holmium	68 Er^{3+} erbium	69 Tm^{3+} thulium	70 Yb^{3+} ytterbium	71 Lu^{3+} lutetium

			Group II	
			2 He helium	First Period
Group V	Group VI	Group VII	Group VIII	
7 N nitrogen	8 O oxygen	9 F fluorine	10 Ne neon	Second Period
15 P phosphorus	16 S sulfur	17 Cl chlorine	18 Ar argon	Third Period
33 As arsenic	34 Se selenium	35 Br bromine	36 Kr krypton	Fourth Period
51 Sb antimony	52 Te tellurium	53 I iodine	54 Xe xenon	Fifth Period
83 Bi bismuth	84 Po polonium	85 At astatine	86 Rn radon	Sixth Period
115 Mc moscovium	116 Lv livermorium	117 Ts tennessine	118 Og oganesson	Seventh Period
Group V	Group VI	Group VII	Group VIII	

23 V vanadium	24 Cr chromium	25 Mn manganese	26 Fe iron	27 Co cobalt	28 Ni nickel
41 Nb niobium	42 Mo molybdenum	43 Tc technetium	44 Ru ruthenium	45 Rh rhodium	46 Pd palladium
73 Ta tantalum	74 W tungsten	75 Re rhenium	76 Os osmium	77 Ir iridium	78 Pt platinum
105 Db dubnium	106 Sg seaborgium	107 Bh bohrium	108 Hs hassium	109 Mt meitnerium	110 Ds darmstadtium

Actinoids

89 Ac^{3+} actinium	90 Th^{3+} thorium	91 Pa^{3+} protactinium	92 U^{3+} uranium	93 Np^{3+} neptunium	94 Pu^{3+} plutonium	95 Am^{3+} americium	96 Cm^{3+} curium
96 Cm^{3+} curium	97 Bk^{3+} berkelium	98 Cf^{3+} californium	99 Es^{3+} einsteinium	100 Fm^{3+} fermium	101 Md^{3+} mendelevium	102 No^{3+} nobelium	103 Lr^{3+} lawrencium

Table 6.1.B:

IUPAC Periodic Table of the Elements

1	2	3	4	5	6	7	8	9
1 H hydrogen								
3 Li lithium	4 Be beryllium							
11 Na sodium	12 Mg magnesium							
19 K potassium	20 Ca calcium	21 Sc scandium	22 Ti titanium	23 V vanadium	24 Cr chromium	25 Mn manganese	26 Fe iron	27 Co cobalt
37 Rb rubidium	38 Sr strontium	39 Y yttrium	40 Zr zirconium	41 Nb niobium	42 Mo molybdenum	43 Tc technetium	44 Ru ruthenium	45 Rh rhodium
55 Cs caesium	56 Ba barium	57–71 lanthanoids	72 Hf hafnium	73 Ta tantalum	74 W tungsten	75 Re rhenium	76 Os osmium	77 Ir iridium
87 Fr francium	88 Ra radium	89–103 actinoids	104 Rf rutherfordium	105 Db dubnium	106 Sg seaborgium	107 Bh bohrium	108 Hs hassium	109 Mt meitnerium

57 La lanthanum	58 Ce cerium	59 Pr praseodymium	60 Nd neodymium	61 Pm promethium	62 Sm samarium
89 Ac actinium	90 Th thorium	91 Pa protactinium	92 U uranium	93 Np neptunium	94 Pu plutonium

10	11	12	13	14	15	16	17	18
								2 He helium
			5 B boron	6 C carbon	7 N nitrogen	8 O oxygen	9 F fluorine	10 Ne neon
			13 Al aluminium	14 Si silicon	15 P phosphorus	16 S sulfur	17 Cl chlorine	18 Ar argon
28 Ni nickel	29 Cu copper	30 Zn zinc	31 Ga gallium	32 Ge germanium	33 As arsenic	34 Se selenium	35 Br bromine	36 Kr krypton
46 Pd palladium	47 Ag silver	48 Cd cadmium	49 In indium	50 Sn tin	51 Sb antimony	52 Te tellurium	53 I iodine	54 Xe xenon
78 Pt platinum	79 Au gold	80 Hg mercury	81 Tl thallium	82 Pb lead	83 Bi bismuth	84 Po polonium	85 At astatine	86 Rn radon
110 Ds darmstadtium	111 Rg roentgenium	112 Cn copernicium	113 Nh nihonium	114 Fl flerovium	115 Mc moscovium	116 Lv livermorium	117 Ts tennessine	118 Og oganesson

63 Eu europium	64 Gd gadolinium	65 Tb terbium	66 Dy dysprosium	67 Ho holmium	68 Er erbium	69 Tm thulium	70 Yb ytterbium	71 Lu lutetium
95 Am americium	96 Cm curium	97 Bk berkelium	98 Cf californium	99 Es einsteinium	100 Fm fermium	101 Md mendelevium	102 No nobelium	103 Lr lawrencium

Table 6.1.C:

32-Group Periodic Table of the Elements

	IA	IIA														
	1	2														
First Period	1 H hydrogen															
Second Period	3 Li lithium	4 Be beryllium														
Third Period	11 Na sodium	12 Mg magnesium														
Fourth Period	19 K potassium	20 Ca calcium														
Fifth Period	37 Rb rubidium	38 Sr strontium														
Sixth Period	55 Cs caesium	56 Ba barium	57 La lanthanum	58 Ce cerium	59 Pr praseodymium	60 Nd neodymium	61 Pm promethium	62 Sm samarium	63 Eu europium	64 Gd gadolinium	65 Tb terbium	66 Dy dysprosium	67 Ho holmium	68 Er erbium	69 Tm thulium	70 Yb ytterbium
Seventh Period	87 Fr francium	88 Ra radium	89 Ac actinium	90 Th thorium	91 Pa protactinium	92 U uranium	93 Np neptunium	94 Pu plutonium	95 Am americium	96 Cm curium	97 Bk berkelium	98 Cf californium	99 Es einsteinium	100 Fm fermium	101 Md mendelevium	102 No nobelium

IIIB	IVB	VB	VIB	VIIB	VIIIB			IB	IIB	IIIA	IVA	VA	VIA	VIIA	VIIIA
3	4	5	6	7	8	9	10	11	12	13	14	15	16	17	18
															2 He helium
										5 B boron	6 C carbon	7 N nitrogen	8 O oxygen	9 F fluorine	10 Ne neon
										13 Al aluminium	14 Si silicon	15 P phosphorus	16 S sulfur	17 Cl chlorine	18 Ar argon
21 Sc scandium	22 Ti titanium	23 V vanadium	24 Cr chromium	25 Mn manganese	26 Fe iron	27 Co cobalt	28 Ni nickel	29 Cu copper	30 Zn zinc	31 Ga gallium	32 Ge germanium	33 As arsenic	34 Se selenium	35 Br bromine	36 Kr krypton
39 Y yttrium	40 Zr zirconium	41 Nb niobium	42 Mo molybdenum	43 Tc technetium	44 Ru ruthenium	45 Rh rhodium	46 Pd palladium	47 Ag silver	48 Cd cadmium	49 In indium	50 Sn tin	51 Sb antimony	52 Te tellurium	53 I iodine	54 Xe xenon
71 Lu lutetium	72 Hf hafnium	73 Ta tantalum	74 W tungsten	75 Re rhenium	76 Os osmium	77 Ir iridium	78 Pt platinum	79 Au gold	80 Hg mercury	81 Tl thallium	82 Pb lead	83 Bi bismuth	84 Po polonium	85 At astatine	86 Rn radon
103 Lr lawrencium	104 Rf rutherfordium	105 Db dubnium	106 Sg seaborgium	107 Bh bohrium	108 Hs hassium	109 Mt meitnerium	110 Ds darmstadtium	111 Rg roentgenium	112 Cn copernicium	113 Nh nihonium	114 Fl flerovium	115 Mc moscovium	116 Lv livermorium	117 Ts tennessine	118 Og oganesson

Table 6.1.D:

18- vs. 8-Group Periodic Table of the Elements

	0	I								
First Period		1 H Hydrogen								
b)	0a	Ia	IIa	IIIb	IVb	Vb	VIb	VIIb	VIIIb	or
a)	0	1	2	3	4	5	6	7	8	9
Second Period	2 He helium	3 Li lithium	4 Be beryllium							
Third Period	10 Ne neon	11 Na sodium	12 Mg magnesium							
Fourth Period	18 Ar argon	19 K potassium	20 Ca calcium	21 Sc scandium	22 Ti titanium	23 V vanadium	24 Cr chromium	25 Mn manganese	26 Fe iron	27 Co cobalt
Fifth Period	36 Kr krypton	37 Rb rubidium	38 Sr strontium	39 Y yttrium	40 Zr zirconium	41 Nb niobium	42 Mo molybdenum	43 Tc technetium	44 Ru ruthenium	45 Rh rhodium
Sixth Period	54 Xe xenon	55 Cs caesium	56 Ba barium	57 La lanthanum	72 Hf hafnium	73 Ta tantalum	74 W tungsten	75 Re rhenium	76 Os osmium	77 Ir iridium
Seventh Period	86 Rn radon	87 Fr francium	88 Ra radium	89 Ac actinium	104 Rf rutherfordium	105 Db dubnium	106 Sg seaborgium	107 Bh bohrium	108 Hs hassium	109 Mt meitnerium
a)	0	1	2	3	4	5	6	7	8	9
c)	0A	IA	IIA	IIIA	IVA	VA	VIA	VIIA	VIIIA	or

a) New IUPAC System

b) Chemical Abstracts System

c) Old IUPAC System

Lanthanoids	58 Ce cerium	59 Pr praseodymium	60 Nd neodymium	61 Pm promethium	62 Sm samarium	63 Eu europium
Actinoids	90 Th thorium	91 Pa protactinium	92 U uranium	93 Np neptunium	94 Pu plutonium	95 Am americium

								II	
								2 He helium	First Period
0b	Ib	IIb	IIIa	IVa	Va	VIa	VIIa	VIIIa	b)
10	11	12	13	14	15	16	17	18	a)
			5 B boron	6 C carbon	7 N nitrogen	8 O oxygen	9 F fluorine	10 Ne neon	Second Period
			13 Al aluminium	14 Si silicon	15 P phosphorus	16 S sulfur	17 Cl chlorine	18 Ar argon	Third Period
28 Ni nickel	29 Cu copper	30 Zn zinc	31 Ga gallium	32 Ge germanium	33 As arsenic	34 Se selenium	35 Br bromine	36 Kr krypton	Fourth Period
46 Pd palladium	47 Ag silver	48 Cd cadmium	49 In indium	50 Sn tin	51 Sb antimony	52 Te tellurium	53 I iodine	54 Xe xenon	Fifth Period
78 Pt platinum	79 Au gold	80 Hg mercury	81 Tl thallium	82 Pb lead	83 Bi bismuth	84 Po polonium	85 At astatine	86 Rn radon	Sixth Period
110 Ds darmstadtium	111 Rg roentgenium	112 Cn copernicium	113 Nh nihonium	114 Fl flerovium	115 Mc moscovium	116 Lv livermorium	117 Ts tennessine	118 Og oganesson	Seventh Period
10	11	12	13	14	15	16	17	18	a)
0B	IB	IIB	IIIB	IVB	VB	VIB	VIIB	VIIIB	c)

64 Gd gadolinium	65 Tb terbium	66 Dy dysprosium	67 Ho holmium	68 Er erbium	69 Tm thulium	70 Yb ytterbium	71 Lu lutetium
96 Cm curium	97 Bk berkelium	98 Cf californium	99 Es einsteinium	100 Fm fermium	101 Md mendelevium	102 No nobelium	103 Lr lawrencium

remain. While opinions about the relative stability of so far undiscovered superheavy elements in an 'island of stability' [8] are highly speculative the following two other controversies deserve serious consideration.

6.8.2.1 The position of the period 1 elements in the Periodic Table

Helium placed in group 2, rather than the customary group 18, as sometimes suggested, is rarely seen in current versions of the Periodic Table. However, the debate about hydrogen's proper place in the Periodic Table is still an issue [9]. In many earlier versions of the Periodic Table hydrogen was kept outside the main groups and placed in an unspecified position in the first period. In the Periodic Table currently recommended by IUPAC hydrogen is firmly placed in group 1. Other suggestions have placed hydrogen in group 4 or group 17, as well as in both group 1 and group 17, in keeping with its chemical resemblance of both the alkali metals and, even more so, the halogens. It seems that most of this confusion is due to an inappropriate projection of the eight main groups known from the second period, with their eight elements, to the first period, with its only two elements. It would seem that IUPAC's treatment of this problem, i.e. placing hydrogen simply in group 1, has been based more on the electron configuration of the hydrogen atom than on its chemical behavior.

6.8.2.2 The group 3 dilemma

The composition of group 3 of the Periodic Table is under debate. IUPAC has an ongoing project to resolve this controversy [10]. This project is expected to lead to a recommendation as to whether group 3 of the Periodic Table consists of the elements scandium, yttrium, lutetium, and lawrencium or the elements scandium, yttrium. lanthanum, and actinium. Preliminary reports have been given by Ball [11] and Öhrström [12].

6.8.3 Group names

Actinoid, actinium and the following 14 elements. The earlier name actinide has been discarded, the suffix 'ide' being reserved for anions, even if the current name actinoid, taken literally, suggests that actinium is actinium-like.

Alkali metal, the elements of group 1 of the Periodic Table. From alkali, from Arabic *al-qali*, saltwort ashes, thus referring to the occurrence of sodium and potassium in such ashes.

Alkaline earth metal, the elements of group 2 of the Periodic Table. A name based on the obsolete name earth for metal oxides and the pre-1800 conceptions of calcium oxide (CaO) and barium oxide (BaO) as elements.

Chalcogen, the elements of group 16 of the Periodic Table. From Greek *chalkos*, copper, ore, and *-genes*, producer, referring to these elements' as ore formers, i.e. their occurrence in (copper) ores.

Halogen, the elements of group 17 of the Periodic Table. From Greek *hals, halo-*, salt and *-genes*, producer. Thus named as salt makers.

Lanthanoid (Ln), lanthanum and the following 14 elements. The earlier name lanthanide has been discarded, the suffix 'ide' being reserved for anions, even if the current name lanthanoid, taken literally, suggests that lanthanum is lanthanum-like.

Main group, groups 1–2 and 13–18 of the Periodic Table. Although main groups and side groups are now numbered separately, the distinction between main groups and side groups is still useful and commonly applied.

Noble gas, the elements of group 18 of the Periodic Table, named on the basis of the earlier assumption that these elements were unable to form compounds. Although a number of compounds of the noble gases have now been prepared, first of the heavier ones, but recently also of the lighter ones, which form compounds like Na_2He and HArF, the traditional name still persists and is, for most practical purposes, not a misnomer. A.k.a. **rare gas.**

Pnictogen, the elements of group 15 of the Periodic Table. From the Greek root *pnig-*, *pnikt-*, choke, strangle, and *-genes*, producer, literally suffocation maker. The alternative name pnicogen is less common. The earlier name pnictide has been discarded, the suffix 'ide' being reserved for anions. A.k.a. **pentel**, from penta- and element.

Rare-earth metal (REM, unspecified element symbol Ln), an alternative name for the lanthanoids. Here earth is used in its obsolete meaning metal oxide. In spite of its label as rare-earth metal lanthanum is no rarer than the common elements cobalt, copper, and lead.

Side group, groups 3–12 of the Periodic Table. Although main groups and side groups are now numbered separately, the distinction between main groups and side groups is still useful and commonly applied.

Tetrel, the elements of group 14. From tetra- and element.

Triel, the elements of group 13. From tri- and element.

6.8.4 Special group names

Base metal, i.e. non-noble metal, metallic elements which corrode and oxidize readily, and which generate hydrogen by reaction with dilute strong acid. Common examples are copper (although this metal does not generate hydrogen when treated with dilute strong acid), iron, lead, nickel, and zinc.

Iron group, iron, cobalt, and nickel.

Metal, elements which are typically hard in the solid state, opaque, shiny, with good electrical and thermal conductivity, malleable, fusible, and ductile, with low ionization energy and electronegativity; from Greek *metallon*, mine, quarry, metal.

Metalloid, elements with both metallic and non-metallic properties such as elemental arsenic, silicon, and boron.

Noble metal, precious metal, the elements gold, iridium, palladium, platinum, rhodium, ruthenium, and silver, referring to these elements' resistance to corrosion and oxidation in moist air. While the elements niobium, tantalum, and titanium share this property, they are not included among the noble metals.

Non-metal, elements which are volatile in the elemental state, with low elasticity, good insulators of heat and electricity, with a high ionization energy, and high electronegativity.

Platinum metal, ruthenium, rhodium, palladium, osmium, iridium, and platinum.

Transactinoid, the 15 elements following the actinoids.

Transition metal, elements with a partially filled *d* sub-shell, or which can give rise to cations with an incomplete d sub-shell, i.e. the elements of groups 4–11 as well as scandium and yttrium.

6.9 Modified element symbols

All four 'corners' of an element symbol have been designated to carry specific information. The unmodified element symbol represents the natural isotopic composition of the element such as in H. The upper left corner is used to show the mass of the isotope in question such as in ^{1}H, ^{2}H, and ^{3}H, or ^{12}C, ^{13}C, and ^{14}C. The lower right corner is the place of the atomic number which, of course, is already implied by the element symbol itself. It is therefore mostly left unused, but may be practical as a mnemonic device such as in $_{118}Og$. The upper right corner is used to indicate electric charge (and/or unpaired electrons) such as in Ca^{2+}. Finally, the lower right corner is the place for the number of atoms to be indicated such as in O_2.

6.10 Fake news

Parallel with the development of the 'legitimate' Periodic Table the erroneous or frivolous announcement of the discovery of assumedly new elements was a thriving cottage industry until the advent of the complete 118-element Periodic System. The number of such illegitimate claims goes into the hundreds, thus surpassing the number of known elements. While many of these announcements were made by people of little renown and credibility even Nobel prize winners can be found among the proponents of unfounded claims. Thus the British chemist William Ramsay (1852–1916) announced the discovery of

Anglium, from New Latin *Anglia*, England,

Scotium, from New Latin *Scotia*, Scotland (the Latin name for Scotland is *Caledonia*),

Hibernium, from Latin *Hibernia*, Ireland in 1897 and of

Metargon (metaargon) in 1898. None of these claims could be substantiated.

As late as 1934 the Italian-American physicist Enrico Fermi (1901–1954) published his findings concerning the new, and soon discredited, elements

Ausonium, from Greek *Ausonia*, Italy, and

Hesperium, from Greek *Hesperia*, Italy, literally the land of the West.

A comprehensive history of discredited element claims and of naming controversies has been published [13].

References and comment

[1] https://en.wikipedia.org/wiki/List_of_chemical_element_name_etymologies (retrieved August 2, 2018).

[2] W. H. Koppenol, J. Corish, J. Garcia-Martinez, J. Meija, and J. Reedijk, *How to Name New Chemical Elements, (IUPAC Recommendations 2016), Pure Appl. Chem.* **88**, 401–405 (2016); http://dx.doi.org/10.1515/pac-2015-0802.

[3] A preceding asterisk indicates that the word in question has been reconstructed, not recorded as such.

[4] 'The Red Book', N. G. Connelly, T. Damhus, R. M. Hartshorn, and A. T. Hutton, *Nomenclature of Inorganic Chemistry – IUPAC Recommendations 2005*, Royal Society of Chemistry, Cambridge, UK, 2005, ISBN 0-85404-438-8; free online version, https://www.iupac.org/cms/wp-content/uploads/2016/07/Red_Book_2005.pdf (retrieved August 2, 2018).

[5] M. R. Leach, https://www.meta-synthesis.com/webbook/35_pt/pt_database.php (retrieved August 2, 2018).

[6] https://en.wikipedia.org/wiki/Muonium (retrieved August 2, 2018).

[7] https://en.wikipedia.org/wiki/Positronium (retrieved August 2, 2018).
[8] https://en.wikipedia.org/wiki/Island_of_stability (retrieved August 2, 2018).
[9] https://en.wikipedia.org/wiki/Period_1_element (retrieved August 2, 2018).
[10] E. Scerri, *Chem. Int.* **38**, 22–23 (2016); http://dx.doi.org/10.1515/ci-2016-0213. Cf. https://iupac.org/projects/project-details/?project_nr=2015-039-2-200 (retrieved March22, 2019).
[11] P. Ball, *Chem. World*, April 21, 2017; https://www.chemistryworld.com/opinion/the-group-3-dilemma/3007080.article (retrieved August 2, 2018).
[12] L. Öhrström, *Nature Chem.* **10**, 372 (2018); https://doi.org/10.1038/nchem.2938.article.
[13] M. Fontani, M. Costa, M. V. Orna, *The Lost Elements, The Periodic Table's Shadow Side*, Oxford University Press, Oxford, UK, etc., 2015, ISBN 978-0-19-938334-4

7 The naming of lipids and lipid constituents

Most lipids and naturally occurring acids have been named after their source, a few are eponyms, but also near-systematic or mixed trivial-systematic names are in use. A few non-naturally occurring isomers have been included to illustrate naming principles.

7.1 Aliphatic acids

Only unsubstituted acids are considered here.

Acetic acid (ethanoic acid), from Latin *acetum*, vinegar.

Acrylic acid (prop-2-enoic acid), from acrolein.

Adipic acid (hexanedioic acid), from Latin *adeps, adip-*, fat.

Adrenic acid [AdA, (7*Z*,10*Z*,13*Z*,16*Z*)-docosa-7,10,13,16-tetraenoic acid], from New Latin *glandula adrenalis*, adrenal gland, from Latin *ad*, at, to, *renalis*, pertaining to the kidney, from *ren*, kidney.

Agonandoic acid [(11*E*)-octadec-11-en-9-ynoic acid], after the genus *Agonandra* (tropical trees). The genus name is from Greek *agon*, assembly, gathering, collection, and *aner, andro-*, stamen.

Ajenoic acid [(3Z,5Z,7Z,9*Z*)-dodeca-3,5,7,9,11-tetraenoic acid], from Spanish *ajo*, garlic (*Allium sativum*) and 'ene'.

Alvaradoic acid (octadec-17-en-6-ynoic acid), after the species *Alvaradoa amorphoides* (Mexican alvaradoa). The genus name is after the Spanish conquistador Pedro de Alvarado (1485–1541).

Alvaradonic acid (octadec-17-en-8-ynoic acid), named as an isomer of alvaradoic acid.

Arachidic acid (icosanoic acid), after the genus *Arachis* (pea), from Greek *arachis*, diminutive of *arachos*, chickling vetch.

Arachidonic acid [AA, ARA (5*Z*,8*Z*,11*Z*,14*Z*)-icosa-5,8,11,14-tetraenoic acid], named as derivative of arachidic acid.

Asclepic acid [(11*Z*)-octadec-11-enoic acid], after the genus *Asclepias* (milkweeds). The genus name is after the Greek god of healing and medicine Asclepius.

Azelaic acid (nonanedioic acid), from az(o)- and oleic acid; the name refers to the fact that this acid can be prepared by oxidation of oleic acid with nitric acid.

https://doi.org/10.1515/9783110612714-007

Behenic acid (docosanoic acid), from ben or, from Arabic *ban*, seeds of any species of the genus *Moringa* (drumstick trees). The genus name is from Tamil *murungai*, drumstick.

Behenolic acid (docos-13-ynoic acid), named as derivative of behenic acid with a carbon-carbon triple bond.

Bishomocolumbinic acid [(7*Z*,11*Z*,14*E*)-icosa-7,11,14-trienoic acid], named as derivative of columbinic acid with two carbon atoms more. Current IUPAC rules specify the prefix 'dihomo' rather than 'bishomo'.

Bishomopinolenic acid [(7*Z*,11*Z*,14*Z*)-icosa-7,11,14-trienoic acid], named as derivative of pinolenic acid with two carbon atoms more. Current IUPAC rules specify the prefix 'dihomo' rather than 'bishomo'.

Bolekic acid [(13*Z*)-octadec-13-ene-9,11-diynoic acid], after its source, boleko oil, also called isano oil, obtained from the boleko tree (*Ongokea gore*).

Bosseopentaenoic acid [BPA, (5*Z*,8*Z*,10*E*,12*E*,14*Z*)-icosa-5,8,10,12,14-pentaenoic acid], after the genus *Bossiella* (red coralline algae). The genus name is after the Dutch botanist Anna Antoinette Weber-van Bosse (1852–1942).

Brassidic acid [(13*E*)-docos-13-enoic acid], after the genus *Brassica* (cabbage). The genus name is from Latin *brassica*, cabbage, of obscure origin.

Brassylic acid (tridecanedioic acid), as brassidic acid.

Butyric acid (butanoic acid), from Greek *butyron*, butter, from *bous*, cattle, and *tyros*, cheese.

Caleic acid [(3*E*,9*Z*,12*Z*)-octadeca-3,9,12-trienoic acid], after the genus *Calea* (flowering plants). The genus name is from Greek *kalos*, beautiful, referring to these plants showy flowers.

α-Calendic acid [(8*E*,10*E*,12*Z*)-octadeca-8,10,12-trienoic acid], after the genus *Calendula* (marigold). The genus name is from Latin *calendula*, diminutive of *calendae, kalendae*, the first day of each month, referring to these plants' long flowering period. A.k.a. **calendulic acid.**

β-Calendic acid [(8*E*,10*E*,12*E*)-octadeca-8,10,12-trienoic acid], named as isomer of α-calendic acid.

Capric acid (decanoic acid), from Latin *caper, capr-*, (goat), referring to its unpleasant smell.

Caproic acid (hexanoic acid), as capric acid. A.k.a. **capronic acid.**

Caproleic acid (dec-9-enoic acid), as capric acid and from oleic acid, labeling it as unsaturated compound.

Caprylic acid (octanoic acid), as capric acid.

Carboceric acid (heptacosanoic acid), from carbon and Greek *keros*, wax, referring to its occurrence in the mineral ozokerite a.k.a. ozocerite, q.v. A.k.a. **heptacosylic acid.**

Catalpic acid [(9*E*,11*E*,13*Z*)-octadeca-9,11,13-trienoic acid], after the species *Catalpa ovata* (yellow catalpa, Chinese catalpa). The genus name is from Creek Indian *katalpa*, from *iká, ka-*, head, and *talpa*, wing, referring to the shape of this tree's flowers.

Ceroplastic acid (pentatriacontanoic acid), from Greek *keroplastikos*, formed in wax, referring to this acid's waxy appearance.

Cerotic acid (hexacosanoic acid), from Greek *keroton*, pomade, from *keros*, wax. A.k.a. **ceric acid, cerotinic acid**, and **cerylic acid.**

Cervonic acid [DHA, (4*Z*,7*Z*,10*Z*,13*Z*,16*Z*,19*Z*)-docosa-4,7,10,13,16,19-hexaenoic acid], from Latin *cerebrum*, brain. A.k.a. **docosahexaenoic acid**, a simplified systematic name.

Cetelaidic acid [(11*E*)-docos-11-enoic acid], named as *trans*-isomer of cetoleic acid.

Cetoleic acid [(11*Z*)-docos-11-enoic acid], after spermaceti and from oleic acid, labeling it as unsaturated compound. Spermaceti, literally whale sperm, is actually a substance in the whale's head providing buoyancy, from Latin *sperma ceti*, whale's sperm, from *cetus*, whale, fish.

Cilienic acid [(6*Z*,11*Z*)-octadeca-6,11-dienoic acid], from cilium and 'ene'. Named for its occurrence in the ciliated protozoan *Tetrahymena pyriformis*. Cilium is from Latin *cilium*, eyelash.

Civetic acid (heptadec-8-enedioic acid), after civet, family *Viverridae*. Civet is ultimately from Arabic *zabad*, civet.

Clupadonic acid [DPA, (4*E*,8*E*,12*E*,15*E*,19*E*)-icosa-4,8,12,15,17-pentaenoic acid], after the fish genus *Clupanodon*, from Latin *clupea*, small river fish, and Greek *anodon*, toothless, from a(n)- and *odon, odont-*, tooth. A.k.a. **docosapentaenoic acid**, a simplified systematic name.

Clupanodonic acid, [DPA, (7*Z*,10*Z*,13*Z*,16*Z*,19*Z*)-docosa-7,10,13,16,19-pentaenoic acid], a variation of clupadonic acid. A.k.a. **docosapentaenoic acid**, a simplified systematic name.

Columbinic acid [(5*E*,9*Z*,12*Z*)-octadeca-5,9,12-trienoic acid], after the columbine plant, also called granny's bonnet (*Aquilegia*). A.k.a. **aquilegic acid**, after the genus *Aquilegia*, from Latin *aquilegus*, water-collecting, **isolinolenic acid**, from linolenic acid, and **ranunculeic acid**, after the genus *Ranunculus* (buttercup, spearworts, crowfoots). The genus name is from Latin *ranuncula*, diminutive of *rana*, frog, probably referring to these plants' habitat near water, like frogs.

Coniferonic acid [(5Z,9Z,12Z,15Z)-octadeca-5,9,12,15-tetraenoic acid], after conifer.

Corticrocin [(2*E*,4*E*,6*E*,8*E*,10*E*,12*E*)-tetradeca-2,4,6,8,10,12-hexaenoic acid], not explicitly named as an acid. After the fungal species *Corticium croceum*. The genus name is from Latin *cortex, cortic-*, bark, the specific epithet from Latin *croceus*, saffron yellow, blond.

Crepenynic acid [(9*Z*)-octadec-9-en-12-ynoic acid], after the genus *Crepis* (hawksbeard), from Greek *krepis*, slipper, sandal, referring to the shape of these plants' fruits, and with 'ene' and 'yne'.

Crotonic acid [(2*E*)-but-2-enoic acid], after the genus *Croton* (croton, rushfoil), from Greek *kroton* (dog tick), referring to the shape of these plants' seeds.

Dehydrocrepenynic acid [(9*Z*,14*Z*)-octadeca-9,14-dien-12-ynoic acid], from crepenynic acid.

Denticetic acid [(5*Z*)-dodec-5-enoic acid], after the genus *Denticetopsis* (catfish). The genus name is from Latin *dens, dent-*, tooth, Greek *ketos*, marine monster, whale, and *opsis*, appearance. A.k.a. **lauroleinic acid**, from Latin *laurus*, laurel, and oleic acid.

Dicramin [(9Z,12Z,15Z)-octadeca-9,12,15-trien-6-ynoic acid], not explicitly named as an acid. After the species *Herbertus dicranus* (liverwort). The specific epithet is from New Latin *dicranus*, two-pronged.

Dihomolinoleic acid [(11*Z*,14*Z*)-icosa-11,14-dienoic acid], from linoleic acid.

Dihomolinolenic acid [(11*Z*,14*Z*,17*Z*)-icosa-11,14,17-trienoic acid], from linolenic acid.

Dihomo-γ-linolenic acid [DGLA, (8*Z*,11*Z*,14Z)-icosa-8,11,14-trienoic acid], from linolenic acid.

Docosadienoic acid [(13*Z*,16*Z*)-docosa-13,16-dienoic acid], a simplified systematic name.

Eicosadienoic acid [(11*Z*,14*Z*)-icosa-11,14-dienoic acid], a simplified systematic name.

Eicosatetraenoic acid [ETA, (8*Z*,11*Z*,14*Z*,17*Z*)-icosa-8,11,14,17-tetraenoic acid], a simplified systematic name.

Eicosatrienoic acid [ETE, (11*Z*,14*Z*,17*Z*)-icosa-11,14,17-trienoic acid], a simplified systematic name.

Eicosenoic acid [(11*Z*)-icos-11-enoic acid], a simplified systematic name.

Elaidic acid [(9*E*)-octadec-9-enoic acid], from Greek *elaion*, oil. Named as *trans*-isomer of oleic acid.

α-Eleostearic acid [(9*Z*,11*E*,13*E*)-octadeca-9,11,13-trienoic acid], from Greek *elaion*, oil, and stearic acid, i.e. referring to an 18-carbon chain. A.k.a. **eleostearinic acid** and **margarolic acid**.

β-Eleostearic acid [(9*E*,11*E*,13*E*)-octadeca-9,11,13-trienoic acid], as α-eleostearic acid.

Enanthic acid (heptanoic acid), from Latin *oenanthe*, wild grape, from Greek *oine*, vine, and *anthe*, blossom. A.k.a. **oenanthic acid**.

Ephedrenic acid [(5*Z*,11*Z*)-octadeca-5,11-dienoic acid], after the genus *Ephedra* (joint-pine, jointfir, Mormon-tea, Brigham-tea). The genus name, given by Pliny to the horsetail (*Equisetum*), without explanation, is from Greek *ephedros*, sitting upon, from ep(i)- and *hedra*, chair. A.k.a. **ephedric acid**.

Equisetolic acid (triacontanedioic acid), after the genus *Equisetum* (horsetail, snake grass, puzzlegrass). The genus name is from Latin *equus*, horse, and *saeta*, bristle.

Eranthic acid [(5*Z*.13*Z*,16*Z*)-docosane-5,13,16-trienoic acid], after the genus *Eranthis* (winter aconite), from Greek *er*, spring, and *anthos*, flower.

Erucic acid [(13*Z*)-docos-13-enoic acid], after the genus *Eruca* (cruciferous plants), from Latin *eruca*, garden rocket (*Eruca sativa*).

Exocarpic acid [(13*E*)-octadec-13-ene-9,11-diynoic acid], after the genus *Exocarpos* (flowering shrubs and small trees). The genus name is from Greek *exokarpos*, outside fruit.

Formic acid (methanoic acid), from New Latin *acidum formicicum*, from Latin *formica*, ant, ultimately from Greek *myrmex*, ant.

Fumaric acid [(2*E*)-but-2-enedioic acid], after the genus *Fumaria* (fumitory), from Latin *fumus terrae*, smoke of the earth.

Gadelaidic acid [(9*E*)-icos-9-enoic acid], named as *trans*-isomer of gadoleic acid.

Gadoleic acid [(9*Z*)-icos-9-enoic acid], cf. hypogaeic acid.

Gaidic acid [(2*E*)-hexadec-2-enoic acid], cf. hypogaeic acid.

Geddic acid (tetratriacontanoic acid), after ghedda wax, oriental bees wax, from Telugu *gedda*, lump. A.k.a. **gheddic acid**.

Glutaric acid (pentanedioic acid), from Latin *gluten*, glue.

Gondoic acid [(11*Z*)-icos-11-enoic acid], of obscure etymology.

Gondoleic acid [(9*Z*)-icos-9-enoic acid], named as isomer of gondoic acid.

Goshuyic acid [(5*Z*,8*Z*)-octadeca-5,8-dienoic acid], after the Japanese tree *goshuyu* (medicinal evodia, *Euodia ruticarpa*).

Heneicosapentaenoic acid [HPA, (6*Z*,9*Z*,12*Z*,15*Z*,18*Z*)-henicosa-6,9,12,15,18-pentaenoic acid], a simplified systematic name.

Heneicosylic acid (henicosanoic acid), a simplified systematic name.

Hentriacosylic acid (hentriacosanoic acid), a simplified systematic name.

Hexadecatrienoic acid [HTA, (7*Z*,10*Z*,13*Z*)-hexadeca-7,10,13-trienoic acid], a simplified systematic name.

Hexatriacontylic acid (hexatriacontanoic acid), a simplified systematic name.

Hiragonic acid [(6*E*,10*E*,14*E*)-hexadeca-6,10,14-trienoic acid], named after the Japanese sardine, hirago (*Clupanodon melanostica*).

Hydrosorbic acid [(3*E*)-hex-3-enoic acid], from sorbic acid.

Hyenic acid (pentacosanoic acid), named after its first source, hyena feces. A.k.a. **pentacosylic acid.**

Hypogaeic acid [(7*Z*)-hexadec-7-enoic acid], after the species *Arachis hypogaea* (peanut). The specific epithet is from Greek *hypogaea*, under the earth.

Isanic acid (octadec-17-ene-9,11-diynoic acid), after isano oil, also called boleko oil, obtained from the boleko tree (*Ongokea gore*). A.k.a. **bolecic acid** and **erythrogenic acid.**

Isocrotonic acid [(2*Z*)-but-2-enoic acid], named as isomer of crotonic acid.

Isooleic acid [(10*Z*)-octadec-10-enoic acid], named as isomer of oleic acid.

Ixoric acid [(8*Z*,10*Z*,12*Z*,14E)-octadeca-8,10,12,12-tetraenoic acid], after the species *Ixora chinensis* (ixora). The genus name is from Sanskrit *isvara*, ixora, named after the deity Ishvara.

Jacaric acid [(8*Z*,10*E*,12*Z*)-octadeca-8,10,12-trienoic acid], after the species *Jacaranda mimosifolia* (jacaranda, blue jacaranda, black poui, fern tree). The genus name is from Tupi *jakaraná*, *jakarandá*, fragrant. A.k.a. **jacaranda acid.**

Japanic acid (henicosadioic acid), after Japan wax, plant fat of the genus *Rhus* (sumac).

Juniperonic acid [(5*Z*,11*Z*,14*Z*,17*Z*)-icosa-5,11,14,17-tetraenoic acid], after juniper (*Juniperus*). Juniper is from Latin *iuniperus*, juniper, of uncertain origin.

Keteleeronic acid [(5*Z*,11*Z*)-docosa-5,11-dienoic acid], after the genus *Keteleeria* (coniferous trees), named after the French nurseryman Jean Baptiste Keteleer (1813–1903).

Laballenic acid (octadeca-5,6-dienoic acid), after the plant family *Labiatae* (mint, deadnettle), a.k.a. *Lamiaceae*, named after the shape of the flowers, and after the presence of two cumulated carbon-carbon double bonds. The plant family name is from Latin *labium*, lip.

Lacceroic acid (dotriacontanoic acid), from a latinization of lacquer, named after its source, stick lac wax. A.k.a. **lacceric acid**.

Lamenallenic acid (octadeca-5,6-dienoic acid), after Lamium pupureum oil, obtained from the genus *Lamium* (dead-nettle, henbit), and after the presence of two cumulated carbon-carbon double bonds. The genus name is from Greek *lamion*, diminutive of *lamia*, monster, a sort of flatfish, referring to the resemblance of the flowers to the throat of this fish.

Lauric acid (dodecanoic acid), after the genus *Laurus* (laurel), from Latin *laurus*, laurel.

Lauroleic acid [(9*Z*)-dodec-9-enoic acid], from lauric acid and oleic acid, i.e. named as lauric acid derivative with one carbon-carbon double bond.

Lauroleinic acid [(5*Z*)-dodec-5-enoic acid], named as isomer of lauroleic acid.

Lignoceric acid (tetracosanoic acid), from Latin *lignum*, wood, and Greek *keros*, wax, referring to this acid's occurrence in wood tar. A.k.a. **carnaubic acid**, after carnauba palm (*Copernicia prunifera*). Carnauba is from Tupi *karnaiba*, from *karna*, a palm (*Mauritia carana* or *Mauritia flexuosa*), and *iba*, stem, plant, tree.

Linderic acid [(4*Z*)-dodec-4-enoic acid], after the species *Lindera obtusiloba* (blunt-lobe spice bush). The genus name is after the Swedish botanist Johan Linder (1676–1724).

Linelaidic acid [(9*E*,12*E*)-octadeca-9,12-dienoic acid], named as *all-trans* isomer of linoleic acid. A.k.a. **linolelaidic acid**.

Linoleic acid [LA, (9*Z*,12*Z*)-octadeca-9,12-dienoic acid], after the genus *Linum* (flax), from Greek *linon*, flax. A.k.a. **linolic acid** and **telfairic acid**, after the fungal species *Xylaria telfairii*. The specific epithet is after the Irish surgeon and botanist Charles Telfair (1778–1838).

α-Linoleic acid [ALA, (9*Z*,12*Z*,15*Z*)-octadeca-9,12,15-trienoic acid], named as a derivative of linoleic acid.

Linolenelaidic acid [(9*E*,12*E*,15*E*)-octadeca-9,12,15-trienoic acid], from linolenic acid and elaidic acid, i.e. named as *all-trans*-isomer of α-linolenic acid.

α-Linolenic acid [ALA, (9Z,12Z,15Z)-octadeca-9,12,15-trienoic acid], from linoleic acid and 'ene'. Named as a derivative of linoleic acid with an additional carbon-carbon double bond.

γ-Linolenic acid [GLA, (6*Z*,9*Z*,12*Z*)-octadeca-6,9,12-trienoic acid], as α-linolenic acid. A.k.a. **gamolenic acid**, a contraction of γ-linolenic acid.

Lumequeic acid [(21*Z*)-triacont-21-enoic acid], after lumeque seed, the seed of West African *Ximenia* species. Lumeque is an indigenous West African name. A.k.a. **lumequic acid.**

Lycopodic acid [(11*E*)-hexadec-11-enoic acid], after the genus *Lycopodium* (ground pines, creeping cedar). The genus name is from Greek *lykos*, wolf, and *podion*, diminutive of *pous*, foot, possibly referring to a resemblance of the roots to a wolf's paw or claws.

Maleic acid [(2*Z*)-but-2-enedioic acid], from malic acid, referring to the preparation of maleic acid by distillation of malic acid.

Malonic acid (propanedioic acid), from malic acid, referring to the preparation of malonic acid by oxidation of malic acid.

Mangold's acid [(9*E*,11*E*)-octadeca-9,11-dienoic acid], after the German chemist Helmut K. Mangold (1924–2012).

[1]**Margaric acid** (heptadecanoic acid), from Greek *margaron*, pearl, referring to this acid's appearance. A.k.a. **daturic acid**, after the species *Datura stramonium* (jimsonweed, devil's snare). The genus name is from Sanskrit *dhattura*, white thornapple (*Datura metel*).

[2]**Margaric acid** (a mixture of hexadecanoic acid and octadecanoic acid), erroneously believed to be heptadecanoic acid. Cf. [1]margaric acid.

Matricaric acid [(2*E*,8*E*)-deca-2.8-diene-4,6-diynoic acid], after the genus *Matricaria* (mayweed, chamomile), from Latin *matricaria*, feverfew (*Tanacetum parthenium*), from *matrix*, womb, referring to the ancient use of *Matricaria chamomilla* for treating uterus diseases.

Mead acid [(5*Z*,8*Z*,11*Z*)-icosa-5,8,11-trienoic acid], after the American biochemist James Franklyn Mead (1916–1987).

Megatomic acid [(3*E*,5Z)-tetradeca-3,5-dienoic acid], named after the species *Attagenus megatoma* (black carpet beetle). The specific epithet is from Greek *megatoma*, giant cutter.

Melissic acid (triacontanoic acid), from Greek *melissa*, bee.

von Mikusch's acid [(10*E*,12*E*)-octadeca-10,12-dienoic acid], after the German-American chemist Johannes Donatus von Mikusch-Buchberg (1908–2008).

Montanic acid (octacosanoic acid), after montan wax, a.k.a. lignite wax and OP wax.

Muconic acid (hexa-2,4-dienedioic acid), from mucic acid.

Mycomycin [(3*E*,5*Z*)-trideca-3,5,7,8-tetraene-10,12-diynoic acid], named as an antibiotic, not as a fatty acid. From Greek *mykes*, fungus, and the bacterial order *Actinomyces*.

Myristelaidic acid [(9*E*)-tetradec-9-enoic acid], from myristoleic acid and elaidic acid, i.e. named as *trans*-isomer of myristoleic acid.

Myristic acid (tetradecanoic acid), after the genus *Myristica* (nutmeg tree), ultimately from Greek *myron*, sweet oil, unguent, perfume.

Myristoleic acid [(9*Z*)-tetradec-9-enoic acid], from myristic acid and oleic acid, i.e. named as an unsaturated derivative of myristic acid.

Nervonic acid [(15*Z*)-tetracos-15-enoic acid], after the cerebroside nervone, from Latin *nervus*, nerve, sinew. A.k.a. **selacholeic acid**, from Greek *selachos*, cartilaginous fish.

Nisinic acid, [(6*Z*,9*Z*,12*Z*,15*Z*,18*Z*,21*Z*)-tetracosa-6,9,12,15,18,21-hexaenoic acid], from nisin, q.v.

Nonacosylic acid (nonacosanoic acid), a simplified systematic name.

Nonadecylic acid (nonadecanoic acid), a simplified systematic name.

Norlinoleic acid [(8*Z*,11*Z*)-heptadeca-8,11-dienoic acid], named as analog of linoleic acid with one carbon atom less.

Norlinolenic acid [(8Z,11Z,14Z)-heptadeca-8,11,14-trienoic acid], named as analog of linolenic acid with one carbon atom less.

Obtusilic acid [(4*Z*)-tetradeca-4-enoic acid], after the species *Lindera obtusiloba* (blunt-lobe spice bush). The specific epithet is from Latin *obtusus*, dulled, from *tundere*, to beat, and *lobus*, lobe.

Oleic acid [(9*Z*)-octadec-9-enoic acid], from Latin *oleum*, oil. A.k.a. **rapinic acid**, from rapeseed (*Brassica napus*).

Osbond acid [DPA, (4*Z*,7*Z*,10*Z*,13*Z*,16*Z*)-docosa-4,7,10,13,16-pentaenoic acid], after the British chemist John Mervyn Osbond (1924–2011). A.k.a. **osbond acid, Osbond's acid, ozubondo acid**, the Japanese version of the name osbond acid, and **docosapentaenoic acid**, a simplified systematic name.

Oxalic acid (ethanedioic acid), after the genus *Oxalis* (wood sorrels), from Latin *oxalis*, garden sorrel, ultimately from Greek *oxys*, sharp, acidic.

Palmitelaidic acid [(9*E*)-hexadec-9-enoic acid], from palmitic acid and elaidic acid, i.e. named as *trans*-isomer of palmitoleic acid.

Palmitic acid (hexadecanoic acid), named after its triglyceride palmitin. Palmitin is from French *palmite*, pith of the palm tree, ultimately from Latin *palma*, palm tree.

Palmitoleic acid [(9*Z*)-hexadec-9-enoic acid], from palmitic acid and oleic acid, i.e. named as unsaturated derivative of palmitic acid. A.k.a. **physetoleic acid**, after the genus *Physeter* (sperm whales), from Greek *physeter*, bellows, blowhole of a whale, from *physa*, bellows, and **zoomaric acid**.

Palmitolic acid (hexadec-7-ynoic acid), named as derivative of palmitic acid with a carbon-carbon triple bond.

Palmitvaccenic acid [(11*Z*)-hexadec-11-enoic acid], from palmitic acid and vaccenic acid, i.e. named as the C_{16} analog of the C_{18} acid vaccenic acid. A.k.a. **tanacetumoleic acid**, after the genus *Tanacetum* (tansies). The genus name is from Latin *tanazita*, tansy, akin to Greek *athanasia*, immortality, referring to the custom that tansy was placed among the winding sheets of the dead to repel vermin.

α-Parinaric acid [(9*Z*,11*E*,13*E*,15*Z*)-octadeca-9,11,13,15-tetraenoic acid], after the species *Parinarium laurinum*. The genus name is from Carib *parinari* (*Parinarium*).

β-Parinaric acid [(9*E*,11*E*,13*E*,15*E*)-octadeca-9,11,13,15-tetraenoic acid], named as isomer of α-parinaric acid.

Paullinic acid [(13*Z*)-icos-13-enoic acid], after the species *Paullinia cupana* (guarana). The genus name is after the Danish physician and botanist Simon Paulli (1603–1680).

Pelargonic acid (nonanoic acid), after the genus *Pelargonium* (geraniums, pelargoniums, storksbills), from Greek *pelargos*, stork.

Pentacosylic acid (pentacosanoic acid), a simplified systematic name.

Pentadecylic acid (pentadecanoic acid), a simplified systematic name.

Petroselaidic acid [(6*E*)-octadec-6-enoic acid], from petroselinic acid and elaidic acid, i.e. named as *trans*-isomer of petroselinic acid.

Petroselinic acid [(6*Z*)-octadec-6-enoic acid], after parsley (*Petroselinum*). The genus name is ultimately from Greek *petros*, stone, and *selinon*, celery.

Phellogenic acid (docosanedioic acid), from Greek *phellos*, cork, and -gen.

Phlomic acid (icosa-7,8-dienoic acid), after phlomis seed oil. The genus name *Phlomis* (Jerusalem sage, lampwick plant) is from Latin *phlomis*, mullein (*Verbascum*), ultimately from Greek *phlomis*, flame, referring to the ancient use of these plants' leaves as lamp wicks.

Physeteric acid [(5*Z*)-tetradeca-5-enoic acid], after the genus *Physeter* (sperm whales), from Greek *physeter*, bellows, blowhole of a whale, whale, from *physa*, bellows.

Pimelic acid (heptanedioic acid), from Greek *pimele*, lard.

Pinolenic acid [(5*Z*,9*Z*,12*Z*)-octadeca-5,9,12-trienoic acid], after pine nut oil.

Podocarpic acid [(5*Z*,11*Z*,14*Z*)-icosa-5,11,14-trienoic acid], after the genus *Podocarpus* (yellowwood), from Greek *pous, pod-*, foot, and *karpos*, fruit. A.k.a. **sciadonic acid**, after the genus *Sciadopitys* (umbrella pine), from Greek *skiado-*, shadow, and *pithys*, pine, **sciadopinolenic acid**, and **calthic acid**, after the species *Caltha palustris* (marsh marigold, kingcup). The genus name is from Latin *caltha*, pot marigold (*Calendula officinalis*).

Propiolic acid (propynoic acid), named as derivative of propionic acid with a carbon-carbon triple bond. A.k.a. **propargylic acid**, from Latin *argentum*, silver, referring to the propargyl group's weak acidity and hence ability to form silver salts.

Propionic acid (propanoic acid), with contraction, from Greek *proteon pion*, the first fat, referring to the fact that propionic acid is to be regarded as the first member of the homologous series of fatty acids judged by the soapy feel of its salts.

Pseudoeleostearic acid [(10*E*,12*E*,14*E*)-octadeca-10,12,14-trienoic acid], named as isomer of eleostearic acid.

Psyllic acid (tritriacontanoic acid), after psylla wax, produced by the genus *Psylla* (fleas). The genus name is from Greek *psylla*, flea. A.k.a. **ceromelissic acid.**

Punicic acid [(9*Z*,11*E*,13*Z*)-octadeca-9,11,13-trienoic acid], after the species *Punica granatum* (pomegranate), from Latin *malum punicum*, literally Carthaginian apple, pomegranate. A.k.a. **punicinic acid** and **trichosanic acid.**

Pyrulic acid [(10*E*)-heptadec-10-en-8-ynoic acid], after the species *Pyrularia pubera* (buffalo nut). The genus name is from Latin *pirulum*, diminutive of *pirum*, pear. A.k.a. **leptomeric acid,** after the genus *Leptomeria* (currant bush). The genus name is from Greek *leptos*, peeled, slender, small, from *lepein*, to peel, and -mer, referring to these plants' slender parts.

Ricinenic acid [(9*Z*,11*Z*)-octadeca-9,11-dienoic acid], named as formed from ricinoleic acid by loss of water.

Roughanic acid [(7Z,10Z,13Z)-hexadeca-7,10,13-trienoic acid], probably after roughage.

Rumelenic acid [(9*Z*,11*E*,15*Z*)-octadeca-9,11,15-trienoic acid], named as rumenic acid analog with one more carbon-carbon double bond.

Rumenic acid [(9*Z*,11*E*)-octadeca-9,11-dienoic acid], after ruminant fats.

Sapienic acid [(6*Z*)-hexadec-6-enoic acid], named after its occurrence in human sebum, after the species *Homo sapiens* (modern human). The specific epithet is from Latin *sapiens, sapient-*, wise, from *sapere*, to be wise, to understand.

Scoliodonic acid (tetracosapentaenoic acid), after the species *Scoliodon laticaudus* (spadenose shark). The genus name is from Greek *skolex*, worm, and *odon, odont-*, tooth.

Sebacic acid (decanedioic acid), after sebum, from Latin *sebum*, suet, grease. A.k.a. **iponic acid**.

Sebaleic acid [(5*Z*,8*Z*)-octadeca-5,8-dienoic acid], after its occurrence in human sebum, from Latin *sebum*, suet, grease, patterned after oleic acid.

Shibic acid [(11*Z*,14*Z*,17*Z*,20*Z*,23*Z*)-hexacosa-11,14,17,20,23-pentaenoic acid], after its occurrence in tunny oil. From Japanese *shibi*, bluefin tuna (*Thunnus thynnus*).

Sorbic acid [(2*E*,4*E*)-hexa-2,4-dienoic acid], after the genus *Sorbus* (whitebeam, rowan, service tree, mountain ash), from Latin *sorbus* (service tree).

Stearic acid (octadecanoic acid), named after its triglyceride, stearin, from Greek *stear*, fat, tallow, from *stereos*, hard. A.k.a. **talgic acid**, fro German *Talg*, tallow.

Stearidonic acid [SDA, (6*Z*,9*Z*,12*Z*,15*Z*)-octadeca-6,9,12,15-tetraenoic acid], named as unsaturated derivative of stearic acid. A.k.a. **moroctic acid**.

Stearolic acid (octadec-9-ynoic acid), named as stearic acid derivative with a carbon-carbon triple bond.

Stillingic acid [deca-2,4-dienoic acid], after the genus *Stillingia* (toothleaf). The genus name is after the British botanist Benjamin Stillingfleet (1702–1771).

Suberic acid (octanedioic acid), from Latin *suber*, cork.

Succinic acid (butanedioic acid), from Latin *succinum*, amber.

Tariric acid (octadec-6-ynoic acid), after the tariri plant, genus *Picramnia* (bitter-bushes). The plant name is ultimately of American Indian origin.

Taxoleic acid [(5*Z*,9*Z*)-octadec-5,9-dienoic acid], after the genus *Taxus* (yew) and from oleic acid. The genus name is from Latin *taxus*, yew.

^{1}**Tetracosapentaenoic acid** [(2*Z*,4*Z*,6*Z*,8*Z*,10*Z*)-tetracosa-2,4,6,8,10-pentaenoic acid], a simplified systematic name, also used for isomers.

^{2}**Tetracosapentaenoic acid** [(6*Z*,9*Z*,12*Z*,15*Z*,18*Z*)-tetracosa-6,9,12,15,18-pentaenoic acid], a simplified systematic name, also used for isomers.

^{3}**Tetracosapentaenoic acid** [(9*Z*,12*Z*,15*Z*,18*Z*,21*Z*)-tetracosa-9,12,15,18,21-pentaenoic acid], a simplified systematic name, also used for isomers.

Tetracosatetraenoic acid [(9*Z*,12*Z*,15*Z*,18*Z*)-tetracosa-9,12,15,18-tetraenoic acid], a simplified systematic name.

Thalictric acid [(5*E*)-octadec-5-enoic acid], after the genus *Thalictrum* (meadow rue). The genus name is from Greek *thaliktron*, meadow rue.

Thapsic acid (hexadecanedioic acid), after the deadly carrot (*Thapsia garganica*). The genus name is ultimately from Greek *Thapsos*, an ancient town and peninsula in Sicily.

Thynnic acid [(8*Z*,11*Z*,14*Z*,17*Z*,20*Z*,23*Z*)-hexacosa-8,11,14,17,20,23-hexaenoic acid], after tunny oil, obtained from *Thunnus thynnus* (northern bluefin tuna).

Timnodonic acid [EPA, (5*Z*,8*Z*,11*Z*,14*Z*,17*Z*)-icosa-5,8,11,14,17-pentaenoic acid], after the unaccepted genus *Temnodon* (fish), later renamed *Pomatomus*. The former genus name is from Greek *temnein*, to cut, and *odon, odont-*, tooth. A.k.a. **eicosapentaenoic acid**, a simplified systematic name.

Traumatic acid [(2*E*)-dodec-2-enedioic acid), from Greek *trauma*, wound, named as a plant hormone stimulating the healing of wounds of the plant. Traumatic is from Greek *trauma*, wound, from *titroskein*, to wound.

Tricosylic acid (tricosanoic acid), a simplified systematic name.

Tridecylic acid (tridecanoic acid), a simplified systematic name.

Tsuduic acid [(4Z)-tetradec-4-enoic acid]. This acid was obtained from the species *Lindera umbellata* (spicebush). The acid names are probably after local vernacular names of this plant. A.k.a. **tsuzuic acid** and **tuduic acid.**

Undecylenic acid [(10*E*)-undec-10-enoic acid], a simplified systematic name.

Undecylic acid (undecanoic acid), a simplified systematic name.

Vaccelenic acid [(11*E*,15*Z*)-octadeca-11,15-dienoic acid], named as vaccenic acid with an additional carbon-carbon double bond.

Vaccenic acid [(11*Z*)-octadec-11-enoic acid], from Latin *vacca*, cow, referring to this acid's occurrence in ruminant fats.

Valeric acid (pentanoic acid), after the genus *Valeriana* (valerian). The genus name is from Latin *Pannonia Valeria*, a Roman province, now part of Hungary, ultimately from the Roman name Valerius, from Latin *valere*, to be strong. A.k.a. **valerianic acid.**

Ximenic acid [(17*Z*)-hexacosa-17-enoic acid], after the species *Ximenia americana* (tallow wood, yellowplum, sea lemon, pi'ut). The genus name is after the Spanish priest Francisco Ximénez (1666–1729).

Ximenynic acid [(11*E*)-octadec-11-en-9-ynoic acid], named as ximenic acid with an additional carbon-carbon triple bond. A.k.a. **santalbic acid**, with contraction, after the species *Santalum album* (Indian sandalwood, white sandalwood). The genus name is from Persian *shandul*, sandal-wood tree, the specific epithet is from Latin *albus*, white.

Xionenynic acid [(10*Z*)-octadec-10-en-8-ynoic acid], named as isomer of ximenynic acid.

7.2 Other acids

Aleuritic acid (9,10,16-trihydroxyhexadecanoic acid), after the genus *Aleurites* (Indian walnut, candlenut tree, country walnut). The genus name is from Greek *aleuron*, wheaten flour, ground meal, referring to the appearance of the lower surface of the leaves.

Aleutiric acid (9,10,18-trihydroxyhexadecanoic acid), anagram of aleuritic acid.

Ambrettolic acid [(7*E*)-16-hydroxyhexadec-7-enoic acid], after amber seed oil, obtained from the species *Abelmoschus moschatus* (ambrette). Ambrette is from French *ambre*, amber.

Cerebronic acid (2-hydroxytetracosanoic acid), from Latin *cerebrum*, brain.

Diabolic acid (15,16-dimethyltriacontanedioic acid), occurring in rumen bacteria and named after the difficulty of its isolation and identification.

Juniperic acid (16-hydroxyhexadecanoic acid), from juniper (*Juniperus*). A.k.a. **juniperinic acid.**

Levulinic acid (4-oxopentanoic acid), named after levulose (D-fructose), referring to its formation from fructose.

Mevalonic acid [(3*R*)-3,5-dihydroxy-3-methylpentanoic acid], a contraction of dihydroxymethylvalerolactone, the corresponding lactone.

Mucic acid (D-galactaric acid, (2*S*,3*R*,4*S*,5*R*)-2,3,4,5-tetrahydroxyhexanedioic acid), from Latin *mucus*, nasal mucus, referring to its occurrence in plant slimes.

Phorbic acid (1,3-dihydroxypentane-1,3,5-tricarboxylic acid), after the genus *Euphorbia* (spurge), named after the Greek physician Euphorbos, the Greek physician of King Juba II of Numidia (52 BC–23 AD), who described these plants' laxative properties.

Phytanic acid [(3*R*,7*R*,11*R*)-3,7,11,15-tetramethylhexadecanoic acid], named after phytol, cf. Section 8.2.

Quisqualic acid [3-(3,5-dioxo-1,2,4-oxadiazolidin-2-yl)-L-alanine], after the genus *Quisqualis* (wood vines), from Latin *quis*, who, and *qualis*, of what kind, referring to these plants' changing appearance which makes their identification difficult.

Ricinoleic acid [(9*Z*,12*R*)-12-hydroxyoctadec-8-enoic acid], after the species *Ricinus communis* (castor oil plant). The genus name is from Latin *ricinus*, tick, referring to markings on the plant which resemble ticks.

Rosilic acid (10-hydroxyoctadecanoic acid), after rosemary (*Rosmarinus officinalis*).

Sabinic acid (12-hydroxydodecanoic acid), after the species *Juniperus sabena* (savin juniper), from Latin *herba sabina*, Sabine plant.

7.3 Lipids

Lecithin, a class name for glycerophospholipids, from Greek *lekithos*, egg yolk.

Olein {triolein, glycerol trioleate, glyceryl trioleate, trioctadecenoin, propane-1,2,3-triyl tri-[(9*Z*)-octadec-9-enoate]}, from Latin *oleum*, oil.

Palmitin (tripalmitin, glycerol tripalmitate, glyceryl tripalmitate, trihexadecanoin, propane-1,2,3-triyl trihexadecanoate), from French *palmite*, pith of the palm tree, ultimately from Latin *palma*, palm tree.

Ricinolein, after the species *Ricinus communis* (castor oil plant), cf. ricinoleic acid.

Stearin (tristearin, glycerol tristearate, glyceryl tristearate, trioctadecanoin, propane-1,2,3-triyl trioctadecanoate), from Greek *stear*, fat, tallow, from *stereos*, hard.

8 The naming of terpenes

The class name terpene is derived from German *Terpentin*, turpentine, ultimately from Greek *terebinthos*, terebinth tree (*Pistacia terebinthus*), suggested to be of Creto-Minoic origin.

Isoprenoids with 1, 2, 3, 4, etc. isoprene units are called hemiterpenes, monoterpenes, sesquiterpenes, diterpenes, etc. Strictly speaking the name terpene only applies to hydrocarbons while functionalized terpenes are called terpenoids. However, it is common practice to ignore this difference and also call terpenoids terpenes.

8.1 Hemiterpenes (C_5 compounds)

Hemiterpene is from Greek *hemisys*, half, and terpene.

Isoprene (2-methylbuta-1,3-diene), an arbitrary name. Not found in nature.

Isovaleric acid (3-methylbutanoic acid), named as isomer of valeric acid.

Prenol (3-methylbut-2-en-1-ol), back formation from isoprene. A.k.a. **prenyl alcohol**.

8.2 Monoterpenes (C_{10} compounds)

α-Amorphene [4,7-dimethyl-1-(propan-2-yl)-1,2,4a,5,6,8a-hexahydronaphthalene], after the genus *Amorpha* (herbs, and shrubs). The genus name is from Greek *amorphos*, shapeless, from *morphe*, shape. A.k.a. **zizanene**, after the species *Chrysopogon zizanioides* (vetiver). The specific epithet means resembling Canadian wild rice (*Zizania aquatica*). The latter genus name is from Latin *zizania*, cockle, tares.

Ascaridole {1-methyl-4-(propan-2-yl)-2,3-dioxabicyclo[2.2.2]oct-5-ene}, after the genus *Ascaris* (nematode worms), referring to this compound's anthelmintic properties. The genus name is ultimately from Greek *askaris, askarid-*, intestinal worm.

***endo*-Borneol** {(1*R*,2*S*,4*R*)-1,7,7-trimethylbicyclo[2.2.1]heptan-2-ol}, after the island of Borneo, Indonesia. A.k.a. **borneo camphor**.

α-Bulgarene [(1R,4aR,8aR)-7-methyl-4-methylidene-1-(propan-2-yl)-2,3,4a,5,6,8a-hexahydronaphthalene], after the fungal species *Bulgaria inquinans* (black bulgar).

Camphene {2,2-dimethyl-3-methylidenebicyclo[2.2.1]heptane}, from camphor.

Camphor (bornan-2-one), ultimately from Malay *kapur*, camphor tree (*Cinnamomum camphora*), from Old Malay *kapur Barus*, chalk of Barus, an ancient port in Sumatra, Indonesia.

https://doi.org/10.1515/9783110612714-008

β-Carene [car-3(10)-ene], after the genus *Carum* (caraway). The genus name is ultimately from Greek *karon*, caraway.

Carvacrol [2-methyl-5-(propan-2-yl)phenol)], after the species *Carum carvi* (caraway) and Latin *acer, acr-*, sharp, referring to its smell. The specific epithet is from Medieval Latin *carvi*, caraway. A.k.a. **cymophenol**, from *p*-cymene and phenol, and ***p*-cymen-2-ol.**

(–)-*cis*-Carveol [(1*R*,5*R*)-2-methyl-5-(propan-2-yl)cyclohex-2-en-1-ol], from Medieval Latin *carvi*, caraway.

Cineol (1,8-epoxy-*p*-menthane), after the species *Artemisia cina* (santonica, Levant wormwood). The specific epithet is from New Latin *cinus*, Chinese. A.k.a. **cajeputol**, after the cajeput tree (*Melaleuca cajuputi*), from Malay *kayu putih*, cajeput tree, literally white tree, and **eucalyptol**, after the species *Eucalyptus globulus* (Tasmanian bluegum, southern blue-gum, blue gum). The genus name *Eucalyptus* is from Greek *eu*, good, well, and *kalyptos*, covered.

Citral (3,7-dimethylocta-2,6-dien-1-al), after the genus *Citrus* (citrus), from Latin *citrus*, citrus tree, and with 'al'. A.k.a. **lemonal**, from lemon and with 'al'.

Citronellal (3,7-dimethyloct-6-en-1-al), after citronella (*Cymbopogon nardus*) and with 'al'. A.k.a. **rhodinal.**

Cosmene [(3*E*,5*E*)-2,6-dimethylocta-1,3,5,7-tetraene], after the genus *Cosmos* (sunflower), from Greek *kosmos*, order, ornament, universe.

***p*-Cymene** [1-methyl-4-(propan-2-yl)benzene], from Greek *kyminon*, cumin, ultimately from Hebrew *kammon*, cumin.

Fenchyl alcohol {fenchan-2-ol, (1*R*,2*R*,4*S*)-1,3,3-trimethylbicyclo[2.2.1]heptan-2-ol}, from German *Fenchel*, fennel, ultimately from Latin *foeniculum*, fennel, diminutive of *foenum*, hay.

Geraniol [(2*E*)-3,7-dimethylocta-2,6-dien-1-ol], after the genus *Geranium* (geranium). The genus name is from Greek *geranion* (geranium), diminutive of *geranos* (crane). The Greek name refers to the fact that the plant's fruit resembles a crane's bill.

Iridane [1,2-dimethyl-3-(propan-2-yl)cyclopentane], after the genus *Iridomyrmex* (rainbow ant), from Greek *iris, irid-*, rainbow, and *myrmex*, ant. A.k.a. **osmane**, from Greek *osme*, smell, and **nepetane**, after the genus *Nepeta* (catmint, catnip). The genus name is after Nepete, an ancient Etruscan city, today Nepi, Italy.

Iridomyrmecin {(4*S*,4a*S*,7*S*,7a*R*)-4,7-dimethylhexahydrocyclopenta[c]pyran-3(1*H*)-one}), as iridane.

Isomenthol [*rel*-(1*R*,2*S*,5*S*)-5-methyl-2-(propan-2-yl)cyclohexan-1-ol], named as isomer of menthol.

Lavandulol [5-methyl-2-(prop-2-en-2-yl)hex-4-en-1-ol], after the genus *Lavandula* (lavender). The genus name is from Medieval Latin *lavandula*, marjoram, lavender, from Latin *lavare*, to wash, referring to its use as a bath perfume.

Limonene [1-methyl-4-(prop-1-en-2-yl)cyclohex-1-ene], from French *limon*, lemon, referring to its source, citrus oil.

Linalool (3,7-dimethylocta-1,6-dien-3-ol), named after its source, linaloe oil from Latin *lignum aloes*, wood of the aloe.

Linden ether (3,6-dimethyl-2,4,5,7a-tetrahydro-1-benzofuran), after linden honey, obtained from the lime tree (*Tilia cordata*).

Lineatin {(1*R*,2*S*,5*R*,7*R*)-1,3,3-trimethyl-4,6-dioxatricyclo[3.3.1.0^{2,7}]nonane}, after the species *Trypodendron lineatum* (ambrosia beetle). The specific epithet is from Latin *lineatus*, lined.

Litsoeine {1,2,10-trimethoxy-5,6,6a,7-tetrahydro-4*H*-dibenzo[*de*,*g*]quinolin-9-ol}, after the genus *Litsea* (litsea). The genus name is from Chinese *lei tsai*, cherry, literally small plum. A.k.a. **litsoene**, and **laurotetanine**.

Menthol [*rel*-(1*R*,2*S*,5*R*)-5-methyl-2-(propan-2-yl)cyclohexan-1-ol], from Latin *mentha*, mint (genus *Mentha*).

Menthone (*p*-menthan-3-one), from menthol.

β-Myrcene (7-methyl-3-methylideneocta-1,6-diene), after the genus *Myrcia* (tropical flowering plants). The genus name is after Greek Myrcia, corresponding to Roman Venus.

(–)-Myrtenal [(1*R*)-pin-2-en-10-al], after the genus *Myrtus* (myrtle) and with 'al'.

Neoisomenthol (*rel*-(1*R*,2*R*,5*R*)-5-methyl-2-(propan-2-yl)cyclohexan-1-ol), named as isomer of isomenthol.

Neomenthol [*rel*-(1*R*,2*R*,5*S*)-5-methyl-2-(propan-2-yl)cyclohexan-1-ol], named as isomer of menthol.

Nerol [(2*Z*)-3,7-dimethylocta-2,6-dien-1-ol], after its source, neroli oil, named after its creator, Marie Anne de La Trémoille, duchess of Bracciano and princess of Nerola (1642–1722). Neroli oil is made from the blossoms of the bitter orange tree (*Citrus aurantium*).

α-Ocimene [(3*E*)-3,7-dimethylocta-1,3,7-triene], after the genus *Ocimum* (basil), from Greek *okimon*, basil.

Perillene [3-(4-methylpent-3-en-1-yl)furan], after the species *Perilla frutescens* (perilla). The genus name is possibly from a Hindu vernacular name.

α-Phellandrene (*p*-mentha-1,5-diene), after the species *Eucalyptus phellandra* (narrow-leaved peppermint, Forth River peppermint), later renamed *Eucalyptus radiata*. The former specific epithet is from Greek *phellandrion*, an ivy-leaved plant.

β-Phellandrene (*p*-mentha-1,7(2)-diene), named as isomer of α-phellandrene.

α-Pinene [pin-2(3)-ene], named after pine oil, ultimately from Latin *pinus*, fir.

β-Pinene [pin-2(10)-ene], named as isomer of α-pinene.

Pinocarveol (6,6-dimethyl-2-methylidenebicyclo[3.1.1]heptan-3-ol), after the genus *Pinus* (pine) and carveol. The genus name is ultimately from an Indo-European base **pit-*, resin.

Pulegone [(1*R*)-*p*-menth-4(8)-en-3-one], from *Mentha pulegium* (pennyroyal), from Latin *puleium, pulegium*, pennyroyal.

Sabinene [thuj-4(10)-ene], after savin oil, obtained from *Juniperus sabina* (savin). The specific epithet is from Latin *herba sabina*, literally Sabine plant.

Safranal (2,6,6-trimethylcyclohexa-1,3-diene-1-carbaldehyde), from German *Safran*, saffron (*Crocus sativus*), ultimately from Arabic *zafaran*, saffron, literally the yellow.

Tagetone (2,6-dimethylocta-5,7-dien-4-one), after the genus *Tagetes* (marigold). The genus name is probably after the Etruscan deity Tages, referring to these plants' proliferous growth, both from seeds and from stems.

α-Terpineol [2-(4-methylcyclohex-3-en-1-yl)propan-2-ol], from terpene and 'ol'.

α-Thujene (thuj-3-ene), after the genus *Thuja* (thuja), ultimately from Greek *thyia*, a kind of cedar.

β-Thujene (thuj-2-ene), named as isomer of α-thujene.

Thymol [5-methyl-2-(propan-2-yl)phenol], after the species *Thymus vulgaris* (common thyme, German thyme, garden thyme), ultimately from Greek *thymon*, thyme, probably from *thyein*, to make a burnt offering. A.k.a. ***p*-cymen-3-ol.**

Thymotic acid [2-hydroxy-6-methyl-3-(propan-2-yl)benzoic acid], as thymol. A.k.a. **thymotinic acid.**

Umbellulone (thuj-3-en-2-one), after the species *Umbellularia californica* (Oregon myrtle, California bay laurel, pepperwood. spicebush, cinnamon bush, spice tree, headache tree, maountain laurel, balm of heaven). The genus name is from Latin *umbella*, parasol, diminutive of *umbra*, shade.

8.3 Sesquiterpenes (C_{15} compounds)

Sesquiterpene is from Latin *sesqui*, one and a half, a contracted form of *semisque*, literally one half more, one and a half.

Aristolone (aristol-9-en-8-one), after the species *Aristolochia debilis* (birthwort, pipe-vine, Dutchman's pipe) and with 'one'. The genus name is from Greek *aristos*, best, and *locheia*, childbirth, referring to these plants' uterus shape and placenta-expelling properties.

Aromadendrene {(1a*R*,4a*R*,7*R*,7a*R*,7b*S*)-1,1,7-trimethyl-4-methylidenedecahydro-1*H*-cyclopenta[*c*]azulene}, after the species *Aromadendron elegans* (magnolia), later renamed *Magnolia elegans*. *Aromadendron* is from aroma and Greek *dendron*, tree.

Artemisinin {(3*R*,5a*S*,6*R*,8a*S*,9*R*,12*S*,12a*R*)-3,6,9-trimethyloctahydro-3,12-epoxy-12*H*-pyrano[4,3-*j*]-1,2-benzodioxepin-10(3*H*)-one}, after the species *Artemisia annua* (sweet wormwood). The genus name is after Queen Artemisia II of Caria (deceased 351 BCE), a botanist and medical researcher.

Bazzanene {(4*R*)-4-[(1*R*)-1,2-dimethylcyclopent-2-en-1-yl]-1,4-dimethylcyclohex-1-ene}, after the species *Bazzania trilobata* (greater whipwort). The genus name is after the Italian physician and naturalist Matteo Bazzani (1674–1749).

α-Bergamotene {2,5-dimethyl-6-(4-methylpent-3-en-1-yl)bicyclo[3.1.1]hept-2-ene}, after the species *Citrus bergamia* (Bergamot orange). Bergamot is from Italian *bergamotto*, influenced by the name of the Italian city of Bergamo, ultimately from Turkish *beg armudu*, prince's pear or prince of pears.

α-Bisabolol [6-methyl-2-(4-methylcyclohex-3-en-1-yl)hept-5-en-2-ol], after the gum resin bisabol, obtained from the African tree genus *Commiphora* (myrrhs). The name of the gum resin is from Wolof *bisap u ala*, bisabol.

β-Bourbonene {(1*S*,3a*S*,3b*R*,6a*S*,6b*R*)-3a-methyl-6-methylidene-1-(propan-2-yl)decahydrocyclopenta[3,4]cyclobuta[1,2]cyclopentene}, after bourbon whiskey from Bourbon County, KY, USA, in turn named after the French House of Bourbon in gratitude for Louis XVI of France's assistance during the American Revolutionary War.

Bulnesol [guai-1(10)-en-11-ol], after *Bulnesia sarmientoi* (Argentine lignum vitae, Paraguay lignum vitae, verawood). The genus name is after the Chilean military officer and politician Manuel Bulnes Prieto (1799–1866).

α-Cadinene [(1*S*,4a*R*,8a*R*)-4,7-dimethyl-1-(propan-2-yl)-1,2,4a,5,6,8a-hexahydronaphthalene], after the cade juniper (*Juniperus oxycedrus*). Cade is from Latin *catanus*, cade juniper.

Caryophyllene {(1*R*,4*E*,9*S*)-4,11,11-trimethyl-8-methylidenebicyclo[7.2.0]undec-4-ene}, named after the genus name *Caryophyllus* (clove), later renamed *Syzygium*. The former genus name is from Greek *karyophyllon*, clove tree, from *karion*, grain, seed, kernel, and *phyllon*, leaf.

Cedrenol [cedr-8(15)-en-8-ol], after the genus *Cedrus* (cedar) and with 'ol'.

α-Copaene {(1*R*,2*S*,6*S*,7*S*,8*S*)-1,3-dimethyl-8-(propan-2-yl)tricyclo[4.4.0.0^{2,7}]dec-3-ene}, named after its source, copaiba balsam oil, obtained from the genus *Copaifera* (leguminous trees), from Tupi *copaiba*, copaiba. A.k.a. **ylangene**, cf. α-ylangene.

Cubebol {(3*S*,3a*R*,3b*R*,4*S*,7*R*,7a*R*)-3,7-dimethyl-4-(propan-2-yl)octahydro-1*H*-cyclopenta[1,3]cyclopropa[1,2]benzen-3-ol}, after the species *Piper cubeba* (cubeb pepper, tailed pepper) and with 'ol'. The specific epithet is ultimately from Arabic *kababa* of unknown origin.

Cuparene {1-methyl-4-[(1*R*)-1,2,2-trimethylcyclopentyl]benzene}, after the order *Cupressales* (cypress).

α-Curcumene {1-methyl-4-[(2*R*)-6-methylhept-5-en-2-yl]benzene}, after the genus *Curcuma* (turmeric), from Arabic *kurkum*, turmeric.

δ-Elemene [(1*S*,2*R*)-1-ethenyl-2-(prop-1-en-2-yl)-*p*-menth-3-ene], after elemi, an oleoresin from the species *Canarium luzonicum* (elemi), ultimately from Arabic *al lāmi*, elemi.

Eremophilene [eremophila-1(10),11-diene], after the genus *Eremophila* (emu bush, poverty bush), from Greek *eremos*, lone, solitary, and *philos*, loving.

α-Farnesene [(3*E*,6*E*)-3,7,11-trimethyldodeca-1,3,6,10-tetraene], cf. farnesol.

Farnesol [(2*E*,6*E*)-2,7,11-trimethyldodeca-2,6,10-trien-1-ol], after the species *Acacia farnesiana* (sweet acacia) and with 'ol'. The specific epithet is after the Italian Cardinal Odoardo Farnese (1573–1626).

Germacrene D [1-methyl-5-methylidene-8-(propan-2-yl)cyclodeca-1,6-diene], with contraction, after the species *Geranium macrorhizum* (bigroot perennial). The genus name is from Greek *geranion*, geranium, cranesbill, diminutive of *geranos*, crane. The specific epithet is from macro- and Greek *rhiza*, root.

α-Guaiene [(1*S*,4*S*,7*R*)-1,4-dimethyl-7-(prop-1-en-2-yl)-1,2,3,4,5,6,7,8-octahydroazulene], from guaiane and 'ene'.

Guaiol {2-[(3*S*,5*R*,8*S*)-3,8-dimethyl-1,2,3,4,5,6,7,8-octahydroazulen-5-yl]propan-2-ol}, from guaiane. A.k.a. **champacol**, from champac (*Michelia champaca*) and 'ol' Champac is from Sanskrit *campaka*, champac, of Dravidian origin.

α-Gurjunene {(1a*R*,4*R*,4a*R*,7b*S*)-1,1,4,7-tetramethyl-1a,2,3,4,4a,5,6,7b-octahydro-1*H*-cyclopropa[*e*]azulene}, after gurjun balsam, obtained from the genus *Dipterocarpus* (flowering plants). Gurjun is from Bengali *garjan*, gurjun.

β-Gymnomitrene, after the family *Gymnomitriaceae* (liverwort). The family name is from Greek *gymnos*, naked, and *mitra*, headband, turban, referring to the absence of a calyx. A.k.a. **β-barbatene** and **β-pompene**.

α-Humulene [(1*E*,4*E*,8*E*)-2,6,6,9-tetramethylcycloundeca-1,4,8-triene], after the species *Humulus lupulus* (hop). The genus name is from New Latin *humulus*, hop, a name of Germanic origin, cf. Old Norse *humli*, hop. A.k.a. **α-caryophyllene**, cf. caryophyllene.

Longifolene {(1*R*,2*S*,7*S*,9*S*)-2,2,7-trimethyl-8-methylidenetricyclo[5.4.0.$0^{2,9}$]undecane}, after the species *Pinus longifolia*, later renamed *Pinus roxburghii* (chir pine, longleaf Indian pine). The former specific epithet is from Latin *longus*, long, and *folium*, leaf.

Machilol (eudesm-4-en-11-ol), after the species *Machilus kusanoi* (evergreen trees and shrubs). The genus name is Moluccan or after the insect genus *Machilis*. A.k.a. **γ-eudesmol**, after the genus name *Eudesmia* (eucalyptus), from eu- and desm(o)-. A.k.a. **selinenol**, after the genus *Selinum* (flowering plants); the genus name if from Greek *selinon*, parsley, celery. A.k.a. **uncineol**, after the species *Melaleuca uncinata* (tea-tree). The specific epithet is from Latin *uncinatus*, crooked.

Matricin {(3*S*,3a*R*,4*S*,9*R*,9a*S*,9b*S*)-9-hydroxy-3,6,9-trimethyl-2-oxo-2,3,3a,4,5,9,9a,9b-octahydroazuleno[4,5-*b*]furan-4-yl acetate}, after the genus *Matricaria* (mayweed, chamomile), The genus name is from Latin *matricarius*, of the womb, from *matrix*, womb, referring to the ancient use of *Matricaria chamomilla* for treating uterus diseases. A.k.a. **proazulene**, from pro- and azulene.

Nerolidol [(6*E*)-3,7,11-trimethyldodeca-1,6,10-trien-3-ol], from nerol.

Nootkatone [(4*R*,4a*S*,6*R*)-4,4a-dimethyl-4,4a,5,6,7,8-hexahydronaphthalen-2(3*H*)-one], named after the species *Chamaecyparis nootkatensis* (yellow cedar) and with ‘one’. The specific epithet is after Nootka, an American Indian people.

Patchouli alcohol {(1*R*,3*R*,6*S*,7*S*,8*S*)-2,2,6,8-tetramethyltricyclo[5.3.1.$0^{3,8}$]undecan-3-ol}, after patchouli oil, from *Pogostemon cablin* (dreadnettle, patchouli). Patchouli is from Tamil *pacculi*, woolly patchouli.

α-Selinene (eudesma-3,11-diene), after Greek *selinon*, celery (*Apium graveolens*).

Sesquiphellandrene [3-(6-methylhept-2-en-1-yl)-6-methylidenecyclohex-1-ene], from ‘sesqui’ and phellandrene.

α-Triticene [(2*E*)-4-methylidenetetradec-2-enal], not explicitly named as aldehyde, after the species *Triticum aestivum* (wheat). The genus name is from Latin *triticum*, wheat.

Valencene [(3*R*,4a*S*,5*R*)-4a,5-dimethyl-3-(prop-1-en-2-yl)-1,2,3,4,4a,5,6,7-octahydro naphthalene], after Valencia oranges (*Citrus*×*sinensis*). The toponym Valencia is from Latin *Valentia*, Valencia, meaning strength or valor, in recognition of the valor of former Roman soldiers settled there.

Viridiflorol {(1a*R*,4*S*,4a*S*,7*R*,7a*S*,7b*S*)-1,1,4,7-tetramethyldecahydro-1*H*-cycloprop[*e*]azulen-4-ol}, from Latin *viridis*, fresh, green, and *flos, flor-*, flower. A.k.a **himbaccol**, with contraction, after the species *Himantandra baccata* (northern pigeonberry ash), later renamed *Galbulimima baccata*. The specific epithet is from Latin *baccatus*, having berries, from *bacca*, berry.

α-Ylangene {(1*S*,2*R*,6*R*,7*R*,8*S*)-1,3-dimethyl-8-(propan-2-yl)tricyclo[4.4.0.$0^{2,7}$]dec-3-ene}, after ylang-ylang, perfume extracted from the cananga tree (*Canangium odoratum*). Ylang-ylang is from Tagalog *ilang-ilang*, a reduplication of *ilang*, wilderness.

Zingerone {4-(4-hydroxy-3-methoxyphenyl)butan-2-one). Cf. zingiberene. A.k.a. **vanilllylacetone.**

Zingiberene [2-methyl-5-(6-methylhept-5-en-2-yl)cyclohexa-1,3-diene), after the species *Zingiber officinale* (ginger). The genus name is ultimately of Indo-Aryan origin.

8.4 Diterpenes (C_{20} compounds)

Aphidicolin {(3*R*,4*R*,4a*R*,6a*S*,8*R*,9*R*,11a*S*,11b*S*)-4,9-bis(hydroxymethyl)-4,11b-dimethyltetradecahydro-8,11a-methanocyclohepta[*a*]naphthalene-3,9-diol}, after the fungal species *Cephalosporium aphidicola*. The New Latin specific epithet means aphid dweller.

Cafestol {(1*S*,4*S*,12*S*,13*R*,16*R*,17*R*)-17-(hydroxymethyl)-12-methyl-8-oxapentacyclo[14.2.1.$0^{1,13}$.$0^{4,12}$.$0^{5,9}$]nonadeca-5(9),6-dien-17-ol}, from French *café*, coffee, and sterol.

Caffeine (1,3,7-trimethyl-3,7-dihydro-1*H*-purine-2,6-dione), from French *café*, coffee. A.k.a. **theine**, from New Latin *thea*, tea plant. Interestingly, although the German word for coffee is *Kaffee*, caffeine has been named *Coffein* in German.

Cembrene A [(1*E*,5*E*,9*E*,12*R*)-1,5,9-trimethyl-12-(prop-1-en-2-yl)cyclotetradeca-1,5,9-triene], after the species *Pinus cembra* (Swiss pine, Swiss stone pine, Arolla pine, Austrian stone pine, stone pine). The species name is after Italian *pino cembro*, Swiss pine, named after the Cembra Valley, Italy.

Forskolin {(3*R*,4a*R*,5*S*,6*S*,6a*S*,10*S*,10a*R*,10b*S*)-5-(acetyloxy)-3-ethenyl-6,10,10b-trihydroxy-3,4a,7,7,10a-pentamethyl-1-oxododecahydro-1*H*-naphtho[2,1-*b*]pyran-1-one}, after the species *Coleus forskohlii* (Indian coleus), synonymous with *Plectranthus barbatus*. The genus name is from Greek *koleos*, sheath, the specific epithet is after the Swedish-Danish botanist Peter Forskål (1736–1763). A.k.a. **coleonol** and **colforsin.**

Kahweol {(1*S*,4*S*,12*S*,13*R*,16*R*,17*R*)-17-(hydroxymethyl)-12-methyl-8-oxapentacyclo [14.2.1.$0^{1,13}$.$0^{4,12}$.$0^{5,9}$]nonadeca-5(9),6,10-trien-17-ol}, from Arabic *qahwah*, coffee.

Neophytadiene (7,11,15-trimethyl-3-methylidenehexadec-1-ene), from phytane, q.v.

Patulin {4-hydroxy-4H-furo[3,2-*c*]pyran-2(6*H*)-one}, after the fungal species *Penicillium patulum*. The specific epithet is from Latin *patulus*, widespread. A.k.a. **clavacine**, after the fungal species *Aspergillus clavatus*. The specific epithet is from Latin *clavatus*, furnished with clubs, from *clava*, club.

Pecilocin {1-[(2*E*,4*E*,6*E*,8*R*)-8-hydroxy-6-methyldodeca-2,4,6-trienoyl]pyrrolidin-2-one}, after the fungal genus *Paecilomyces*. The genus name is from Greek *poikilos*, colorful.

Pederin {(2*S*)-*N*-[(2*S*,4*R*,6*R*)-6-[(2*S*)-2,3-dimethoxypropyl]-4-hydroxy-5,5-dimethyltetrahydro-2*H*-pyran-2-yl](methoxy)methyl]-2-hydroxy-2-[(2*R*,5*R*,6*R*)-2-methoxy-5,6-dimethyl-4-methylidenetetrahydro-2*H*-pyran-2-yl]acetamide}, after the genus *Paederus* (beetles). The genus name is from Greek *paideros*, probably meaning short-winged.

Phytane (2,6,10,14-tetramethylhexadecane), from Greek *phyton*, plant, from *phyein*, to bring forth, to grow, to generate.

Phytol [(2*E*,7*R*,11*R*)-3,7,11,15-tetramethylhexadec-2-en-1-ol], as phytane.

Pimaric acid (pimara-8(14),15-dien-18-oic acid), with contraction, after the species *Pinus maritima* (cluster pine). The genus name is from Latin *pinus*, pine tree, and *maritimus*, maritime.

Plaunotol {(2*Z*,6*E*)-2-[(3*E*)-4,8-dimethylnona-3,7-dien-1-yl]-6-methylocta-2,6-diene-1,8-diol}, after the Thai plant *plau-noi*, croton (*Croton sublyratus*).

Retinal, earlier called retinene, after retina, from Greek *rete*, net and 'al'. A.k.a. **vitamin A aldehyde**.

Retinol, after retina, from Greek *rete*, net, and 'ol'. A.k.a. **vitamin A, axerol**, and **axerophthol**, from Greek *xeros*, dry, and *ophthalmos*, eye, i.e. as preventing xerophthalmia.

Salvinorin A, after the species *Salvia divinorum* (sage of the diviners, seer's sage, salvia).

Taxadiene (taxa-4,11-diene), after the genus *Taxus* (yews), from Latin *taxus*, yew.

8.5 Sesterterpenes (C_{25} compounds)

Sesterterpene is from Latin *sestertius*, two and a half, literally two and a half times as great.

Geranylfarnesol [(2*E*,6*E*,10*E*,14*E*)-3,7,11,15,19-pentamethylicosa-2,6,10,14,18-pentaen-1-ol], from geraniol and farnesol.

8.6 Triterpenes (C_{30} compounds)

α-Amyrin (urs-12-en-3β-ol), after the genus *Amyris* (torchwoods). The genus name is from Greek *amyron*, scented oil.

Betulinic acid [3β-hydroxylup-20(29)-en-28-oic acid], after the species *Betula pubescens* (white birch), from Latin *betula*, birch. A.k.a. **betulic acid.**

Campesterol [(24*R*)-ergost-5-en-3β-ol], after the species *Brassica campestris* (rapeseed). The specific epithet is from Latin *campestris*, of the field.

Cholecalciferol, [(5*Z*,7*E*)-9,10-secocholesta-5,7,10(19)-trien-3β-ol], from Greek *chole*, bile, calcium, and Latin *ferre*, to bear, to carry. A.k.a. **vitamin D.**

Cholesterol [cholest-5-en-3β-ol], earlier called cholesterin, from Greek *chole*, bile, and sterol.

Cycloartenol (9β,19-cyclolanost-24-en-3β-ol), a contraction of its systematic name and the name of the species *Artocarpus integrifolia* (Jack fruit). The genus name is from Greek *artos*, bread, and *karpos*, fruit.

Friedelin [friedelan-3-one, (4*R*,4a*S*,6aS,6b*S*,8a*R*,12a*R*,12b*S*,14a*S*,14b*S*)-4,4a,6b,8a,11,11,12b,14a-octamethylicosahydropicen-3(2*H*)-one], after the French chemist and mineralogist Charles Friedel (1832–1899).

Lanosterol (lanosta-8,24-dien-3-ol), from Latin *lana*, wool, and sterol.

Oleanene (olean-1-ene), after oleander (*Nerium*), obscurely akin to Medieval Latin *laurandrum*, oleander, perhaps a conflation of Latin *laurus*, laurel, and *rhododendron.*

Moronic acid (3-oxoolean-18-en-28-oic acid), named as oxidation product of morolic acid, a secondary metabolite of the species *Mora excelsa* (black mora, red mora, sand mora). The genus name is after a native South American name.

β-Sitosterol (stigmast-5-en-3β-ol), from Greek *sitos*, grain, and sterol.

Squalene [(6*E*,10*E*,14*E*,18*E*)-hexamethyltetracosa-2,6,10,14,18,22-hexaene], after the genus *Squalus* (spurdog), from Latin *squalus*, sea fish.

Stigmasterol [(20*R*,22*E*,24*S*)-stigmasta-5,22-dien-3β-ol]. after the genus *Physostigma* (including the calabar bean), from Greek *physa*, bladder, and *stigma*, mark.

Ursolic acid (3β-hydroxyurs-12-ene-28-oic acid), from ursane, q.v. A.k.a. **malol, prunol**, and **urson.**

8.7 Sesquarterpenes (C_{35} compounds)

Sesquarterpene is from Latin *sesquartus*, three and a half times, and terpene.

β-Heptaprene [(6*E*,10*E*,14*E*,18*E*,22*E*)-7,11,15,19,23,27-hexamethyl-3-methylidene octacosa-1,6,10,14,18,22,26-heptaene], named as consisting of seven isoprene units.

8.8 Tetraterpenes (C_{40} compounds)

Astaxanthin (3,3′-dihydroxy-β,β-carotene-4,4′-dione), after the genus *Astacus* (crayfish), from Greek *astakos*, lobster, crayfish.

Canthaxanthin (β,β-carotene-4,4′-dione), after the species *Cantharellus cinnabarinus* (chantarelle). The genus name is from Latin *cantharellus*, chantarelle, diminutive of *cantharus*, large drinking cup with handles.

Capsanthin [(3*R*,3′*S*,5′*R*)-3,3′-dihydroxy-β,κ-caroten-6′-one], after the genus *Capsicum* (nightshades) and anth(o)-. The genus name is from Greek *kapsikon*, red pepper, of obscure etymology.

Capsorubin [(3*S*,3′*S*,5*R*,5′*R*)-3,3′-dihydroxy-κ,κ-carotene-6,6′-dione], cf. capsanthin, and from Latin *ruber*, red.

α-Carotene (β,ε-carotene), after Latin *carota*, carrot.

β-Carotene (β,β-carotene), as α-carotene.

γ-Carotene (β,ψ-carotene), as α-carotene.

δ-Carotene (ε,ψ-carotene), as α-carotene.

ε-Carotene (ε,ε-carotene), as α-carotene.

Cryptoxanthin (3*R*)-β,β-caroten-3-ol], from Greek *kryptos*, hidden, from *kryptein*, to hide.

Lutein (β,ε-caroten-3,3′-diol), from Latin *luteus*, yellow. A.k.a. **xanthophyll**.

Lycopene (ψ,ψ-carotene), after the genus *Lycopersicon* (tomato), from Greek *lykopersikon*, an Egyptian plant.

Neurosporene (7′,8′-dihydro-ψ,ψ-carotene), after the fungal genus *Neurospora*. The genus name refers to the characteristic striations on the spores of these fungi which resemble axons.

Phytoene (octahydro-ψ,ψ-carotene), from Greek *phyton*, plant, and with ‘ene’.

Phytofluene (15-*cis*-7,7′,8,8′,11,12-hexahydro-ψ,ψ-carotene), from Greek *phyton*, plant, and fluorescence, and with 'ene'.

Rubixanthin [(3*R*)-β,ψ-caroten-3-ol], from rubi- and xanth(o)-.

8.9 Higher terpenes

Balata {poly(*trans*-1,4-isoprene, poly[(3*E*)-2-methylbuta-1,3-diene]}, ultimately a Carib name. This latex is obtained from the species *Manilkara bidentata* (balata, bulletwood, cow tree).

Gutta-percha {poly(*trans*-1,4-isoprene, poly[(3*E*)-2-methylbuta-1,3-diene]}, from Malay gĕtah-pĕrcha, from gĕtah, sap, latex, and pĕrcha, scrap, rag.

Hevea rubber {poly(cis-1,4-isoprene, poly[(3*Z*)-2-methylbuta-1,3-diene]}, after the species *Hevea brasiliensis* (rubber tree). The genus name is from Spanish *jebe*, rubber tree, of American Indian origin.

9 The naming of carbohydrates

IUPAC's glycose nomenclature comes in three versions, each suitable for a different purpose, always based upon the acyclic tautomer. A shorthand notation is not based, as usual elsewhere, on parent hydrides, but on fully functionalized parent compounds with retained trivial names such as fructose and glucose. Both the length of the carbon chain, the substitution, and most of the stereochemistry is understood. The configuration at the penultimate carbon atom is given as D or L. A second system is based on an explicit chain length such as in pentose or hexose, but still with the full functionalization understood. Apart from the configuration at the penultimate carbon atom, also here given as D or L, the stereochemistry is shown with specific descriptors, i.e. '*glycero*' (from glycerol), '*erythro*' (from erythrose), '*threo*' (from threose), '*ribo*' (from ribose), '*arabino*' (from arabinose), '*xylo*' (from xylose), '*lyxo*' (from lyxose), '*allo*' (from allose), '*altro*' (from altrose), '*gluco*' (from glucose), '*manno*' (from mannose), '*gulo*' (from gulose), '*ido*' (from idose), '*galacto*' (from galactose), and '*talo*' (from talose). Altogether three methods are used here to denote stereochemistry. The third way to name carbohydrates, based on parent hydrides, is explicit with regard to functionality and stereochemistry, but generates lengthy and unwieldy names, and is therefore seldom, if ever, used.

The dextrorotatory D-glucose was also called dextrose and the levorotatory D-fructose went by the name of levulose. These names are the source of the suffixes 'ose' and 'ulose' used in systematic names of aldoses and ketoses, respectively. It should be noted that the observed optical rotation of D-glucose in aqueous solution is the weighted mean of the optical rotations of the anomers α-D-glucopyranose and β-D-glucopyranose which are in equilibrium with each other. D-Fructose behaves correspondingly.

9.1 Monosaccharides

The class name for monosaccharides is glycose (an alteration of glucose). Other class names are aldose, ketose, triose, tetrose, pentose, hexose, etc. Systematic aldose names end in 'ose' (from glucose), ketose names in 'ulose' (from levulose).

Pentoses and hexoses, as well as other monosaccharides, form cyclic hemiacetals almost to the exclusion of the acyclic tautomer. For convenience, when the hemiacetal is not at issue, names are based on the acyclic tautomer. Class names for monosaccharide hemiacetals are furanose for five-membered rings and pyranose for six-membered rings. The configuration of the stereogenic center created by the ring closure is given as α or β.

https://doi.org/10.1515/9783110612714-009

For instance D-glucose [(2*R*,3*S*,4*R*,5*R*)-2,3,4,5,6-pentahydroxyhexanal] forms the two diastereomeric hemiacetals α-D-glucopyranose [(2*S*,3*R*,4*S*,5*S*,6*R*)-6-(hydroxymethyl)oxane-2,3,4,5-tetrol] and β-D-glucopyranose [(2*R*,3*R*,4*S*,5*S*,6*R*)-6-(hydroxymethyl)oxane-2,3,4,5-tetrol].

When the name of a compound with one stereogenic center is used without a stereodescriptor this usually means that it refers to the racemate or to an enantiomer of unknown configuration. In the case of the common monosaccharides with their multiple stereogenic centers the situation is fundamentally different. Thus, the simple name glucose means the naturally occurring D-glucose, never the exotic enantiomer L-glucose nor the unheard-of racemate DL-glucose. Unless otherwise specified it can also be used to mean the equilibrium mixture of α-D-glucopyranose and β-D-glucopyranose.

9.1.1 Parent monosaccharides

Allose (All, *allo*-hexose), from Greek *allos*, other, different, referring to this monosaccharide's stereoisomerism with glucose.

Altrose (Alt, *altro*-hexose), from Latin *alter, altr-*, other, referring to this monosaccharide's stereoisomerism with glucose.

Arabinose (Ara, *arabino*-pentose), after gum arabic.

Erythrose (*erythro*-tetrose), from erythritol, ultimately from erythrin, named after the red color observed upon erythrin's oxidation, from Greek *erythros*, red. Ironically, erythrose is neither red nor does it take part in reactions which generate a red color.

Erythrulose (*glycero*-tetrulose), from erythrose.

Fructose (Fru, *arabino*-hex-2-ulose), from Latin *fructus*, fruit. A.k.a. **levulose** and **laevulose**.

Galactose (Gal, *galacto*-hexose), from Greek *gala, galakt-*, milk. A.k.a. **cerebrose**, from Latin *cerebrum*, brain.

Glucose (Glc, *gluco*-hexose), from Greek *gleukos*, must, sweet wine, which thus provided both the word stem and the suffix 'ose' which has become a standard in carbohydrate nomenclature. The alternative derivation from Greek *glykys*, sweet, as claimed occasionally, is mistaken. Derived names are:

Glucoside (glucopyranoside), from glucose i.e. a substitution product of a glucose hemiacetal.

Glycoside, an alteration of glucoside, a substitution product of a glycose hemiacetal.

Glyceraldehyde (*glycero*-triose, 2,3-dihydroxypropanal), from glycerol.

Glycerone (1,3-dihydroxypropan-2-one), from glycerol.

Glycose, the class name for monosaccharides, by alteration of glucose.

Gulose (Gul, *gulo*-hexose), anagrammatically derived from glucose.

Idose (Ido, *ido*-hexose), a name formed by contraction of Latin *idem*, the same, and gulose, referring to the stereoisomerism between idose and gulose.

Lyxose (Lyx, *lyxo*-pentose), anagrammatically derived from xylose.

Mannose (Man, *manno*-hexose), from manna as in manna tree (*Fraxinus ornus*), ultimately from Hebrew *man*, gift.

Psicose (Psi, *ribo*-hex-2-ulose), a contraction and alteration of pseudofructose, i.e. named as an isomer of fructose.

Ribose (Rib, *ribo*-pentose), from ribonic acid, a name derived from arabinonic acid by transpositional contraction.

Ribulose (Rul, *erythro*-pent-2-ulose), from ribose.

Sorbose (Sor, *xylo*-hex-2-ulose), after the service tree (*Sorbus*). Both the vernacular and the genus name are ultimately from Latin *sorbus*, service tree.

Tagatose (Tag, *lyxo*-hex-2-ulose), anagrammatically derived from galactose.

Talose (Tal, *talo*-hexose), from talonic acid, anagrammatically derived from galactonic acid. A derivation of talose from the Greek mythical figure Talos, a giant automaton made of bronze, has been considered, but without rationale.

Threose (*threo*-tetrose), anagrammatically derived from erythrose.

Xylose (Xyl, *xylo*-pentose), from Greek *xylon*, wood.

Xylulose (Xul, *threo*-pent-2-ulose), from xylose.

9.1.2 Modified monosaccharides

Substitution of a monosaccharide at, say, position 3 can take place either at the oxygen atom or at the carbon atom. For instance, a methyl substituent would accordingly appear either as 3-*O*-methyl or 3-*C*-methyl.

9.1.2.1 Substituted, anhydro-, and deoxymonosaccharides

Abequose (Abe, 3,6-dideoxy-D-*xylo*-hexose), with contraction, after the bacterium *Salmonella abortus equi*. The Latin specific epithet means horse's miscarriage.

Amicetose (2,3,6-trideoxy-D-*erythro*-hexose), after the antibiotic amicetin.

Antiarose (6-deoxy-D-gulose), named after α-antiarin, a glycoside from the species *Antiaris toxicaria* (bark cloth tree, antiaris tree, false iroko, false mvule, upas tree). The genus name is from Javanese *ancar*. Another Javanese name is *upas*.

Apiose (Api, 3-*C*-(hydroxymethyl)-*glycero*-tetrose), from Latin *apium*, parsley, celery, ultimately from *apis*, bee.

Arcanose (2,6-dideoxy-3-*C*-methyl-3-*O*-methyl-*xylo*-hexose), supposedly named after Latin *arcanum*, deep secret, referring to the fact that this sugar was difficult to isolate and to identify.

Ascarylose (3,6-dideoxy-L-*arabino*-hexose), after the nematode *Parascaris equorum*. The genus name is from Greek *para*, beside, past, beyond, and *askaris*, intestinal worm.

Axenose (2,6-dideoxy-3-*C*-methyl-*xylo*-hexose), after axenomycin.

Boivinose (2,6-dideoxy-D-gulose), after the species *Strophanthus boivinii* (wood shaving flower). The specific epithet is after the French botanist Louis Hyacinthe Boivin (1808–1852).

Chalcose (4,6-dideoxy-3-*O*-methyl-D-xylo-hexose), after the antibiotic chalcomycin.

Chitose (2,5-anhydro-D-mannose), from chitin, q. v.

Cladinose (2,6-dideoxy-3-*C*-methyl-3-O-methyl-L-*ribo*-hexose), after the antibiotic cladomycin.

Colitose (Col, 3,6-dideoxy-L-*xylo*-hexose), after the bacterial species *Escherichia coli*. The specific epithet is from Greek *kolon*, large intestine

Cymarose (2,6-dideoxy-3-*O*-methyl-*ribo*-hexose), after the cardiac glycoside cymarin, which has its name after a trademark. The trademark is derived from the genus *Apocynaceae* (dogbanes), from Latin *apocynum*, dogbane, from Greek *kyon*, dog.

Diginose (2,6-dideoxy-3-*O*-methyl-*lyxo*-hexose), ultimately named after the genus *Digitalis purpurea* (foxglove), from Latin *digitalis*, of the finger, from *digitus*, finger, toe. The genus name refers to the ease with which a flower of foxglove can be fitted over a fingertip.

Digitalose (6-deoxy-3-*O*-methyl-D-galactose), as diginose.

Digitoxose (Dig, 2,6-dideoxy-D-*ribo*-hexose), named after the cardiac glycoside digitoxin, cf. diginose.

D-Evalose (6-deoxy-3-*C*-methyl-D-mannose), after the antibiotic evaminomycin.

Fucose (Fuc, 6-deoxygalactose), named as a constituent of the polysaccharide fucoidan, ultimately from Latin *fucus*, orchil, red dye, rouge, deceit, from Greek *phykos*, seaweed, rouge, of Semitic origin.

Hamamelose [2-*C*-(hydroxymethyl)-D-ribose], after the genus *Hamamelis* (witch hazel, winterbloom). The genus name is from Greek *hamamelis* (medlar), from *hama*, together with, and *melon*, apple, fruit.

Isosorbide (1,4:3,5-dianhydro-D-glucitol), from is(o)- and sorbide.

Levoglucosan (1,6-anhydro-β-D-glucopyranose), so named as levorotatory intramolecular anhydride of glucose.

Mycarose (2,6-dideoxy-3-*C*-methyl-L-ribo-hexose), probably a contracted anagram of carbomycin, an arbitrarily named antibiotic, ultimately from Greek *mykes*, fungus.

Mycinose (6-deoxy-2,3-di-*O*-methyl-D-allose), after the antibiotic chalcomycin, ultimately from Greek *mykes*, fungus.

Nigerose [α-D-glucopyranosyl-(1→3)-D-glucose], named as a constituent of the polysaccharide nigeran found in *Aspergillus niger* (black mold), from Latin *niger*, black.

Noviose (6-deoxy-5-*C*-methyl-4-*O*-methyl-L-*lyxo*-hexose, after the antibiotic novobiocin, from Latin *novus*, new.

Oleandrose (2,6-dideoxy-3-*O*-methyl-*arabino*-hexose), after the species *Nerium oleander* (nerium, oleander). The specific epithet is from Medieval Latin *oleander*, oleander, from Late Latin *laurandrum*, oleander, of uncertain etymology, possibly a conflation of Latin *laurus*, laurel, and Greek/Latin *rhododendron*, rhododendron.

Oliose (2,6-dideoxy-D-*lyxo*-hexose), from olivomycin, after the bacterial species *Actinomyces olivoreticuli*.

Olivose (Oli, 2,6-dideoxy-D-*arabino*-hexose), as oliose.

Quinovose (Qui, 6-deoxy-D-glucose), after the glycoside quinovin, obtained from cinchona bark. The glycoside's name is from New Latin *quina nova*, false cinchona (*Cinchona oblongifolia*), later renamed *Ladenbergia oblongifolia*.

Rhamnose (Rha, 6-deoxymannose), after the genus *Rhamnus* (buckthorn), from Greek *rhamnos*, buckthorn.

Rhodinose (2,3,6-trideoxy-*threo*-hexose), ultimately from Greek *rhodon*, rose, probably from Old Iranian **urda-*, rose.

Sarmentose (2,6-dideoxy-3-*O*-methyl-*xylo*-hexose), after the steroid cardiac glycoside sarmentocymarin, ultimately after the species *Strophanthus sarmentosus* (spider tresses, poison arrow vine). The specific epithet is from Latin *sarmentosus*, full of twigs, from *sarpere*, to trim, to cut off.

Sedoheptulose (D-*altro*-hept-2-ulose), after stonecrop (*Sedum acre*). The genus name is from Latin *sedum*, houseleek.

Sedoheptulosan (2,7-anhydro-β-D-*altro*-hept-2-ulopyranose), so named as intramolecular anhydride of sedoheptulose.

Sorbide (a class name for anhydrosorbitols), from sorbitol, ultimately from sorbose, and 'ide'.

Streptose (5-deoxy-3-*C*-formyl-L-lyxose), after the antibiotic streptomycin, after the bacterial genus *Streptomyces*, from Greek *streptos*, twisted, easy to bend, pliant, from *strephein*, to twist, to turn.

L-Thevetose (3-*O*-methyl-6-deoxy-L-glucose), after the species *Thevetia nereifolia* (yellow oleander), later renamed *Cascabela thevetia*. The former genus name and the latter specific epithet are after the French Franciscan priest, traveler, and author André Thévet (1516–1590).

Thyminose (2-deoxy-D-ribose, 2-deoxy-D-*erythro*-pentose), from thymidine, q.v.

Tyvelose (Tyv, 3,6-dideoxy-D-*arabino*-hexose), after the bacterial species *Salmonella typhi*. The specific epithet is from Greek *typhos*, stupor caused by fever, literally smoke.

9.1.2.2 Amino sugars and derivatives

The class name is glycosamine.

Bacillosamine (Bac, 2,4-diamino-2,4,6-trideoxy-D-glucose), after the bacterial species *Bacillus subtilis*, The genus name is from Medieval Latin *bacillus*, diminutive of *baculus*, from Latin *baculum*, stick, staff, walking stick.

Daunosamine (3-amino-2,4,6-trideoxy-L-*lyxo*-hexose), after the antibiotic daunorubicin and with -amine. The antibiotic was simultaneously discovered by a French and an Italian research team, who together coined its name from Dauni, a.k.a. Apulians, a pre-Roman tribe who occupied the area of Italy where the compound was isolated, and French *rubis*, ruby.

Desosamine [3,4,6-trideoxy-3-(dimethylamino)-D-*xylo*-hexose], a contracted anagram of the systematic name.

Fucosamine (FucN, 2-amino-2,6-dideoxygalactose), from fucose.

Galactosamine (GalN, 2-amino-2-deoxygalactose), from galactose. A.k.a. **chondrosamine**, from chondroitin, q.v.

Glucosamine (GlcN, 2-amino-2-deoxyglucose), from glucose. A.k.a. **chitosamine**, from chitin, q.v.

Gulosamine (GulN, 2-amino-2-deoxygulose), from gulose.

Kanosamine (3-amino-3-deoxy-D-glucose), from kanamycin and -amine.

Kansosamine (4-amino-4,6-dideoxy-3-*C*-methyl-2-*O*-methyl-D-mannose), from kanamycin and -amine.

Mannosamine (ManN, 2-amino-2-deoxymannose), from mannose.

Muramic acid {2-amino-3-*O*-[(1*R*)-1-carboxyethyl]-2-deoxy-D-glucose}, from Latin *murus*, wall, and glucosamine, referring to its occurrence in bacterial cell walls.

Mycaminose [3,6-dideoxy-3-(dimethylamino)-D-glucose], from myc(o)-, -amine, and 'ose'.

Perosamine (4-amino-2,6-dideoxy-D-mannose), of obscure etymology.

D-Pneumosamine (2-amino-2,6-dideoxy-D-talose), after the bacterial species *Streptococcus pneumoniae*, ultimately from Greek *pneumon*, lung.

Quinovosamine (2-amino-2,6-dideoxy-D-glucose), from quinovose.

L-Rhamnosamine (2-amino-2,6-dideoxy-L-mannose), from rhamnose.

Streptamine [(1*R*,2*r*,3*S*,4*R*,5*s*,6*S*)-4,6-diaminocyclohexane-1,2,3,5-tetrol], from streptose.

9.1.2.3 Sugar alcohols and derivatives

Their systematic names are constructed with the suffix 'itol', a modification of earlier 'ite', from the Latin suffix *-itus*, denoting origin or relationship. The class name is glycitol.

D-Allitol [(2*R*,3*R*,4*S*,5*S*)-hexanehexol], from allose.

D-Altritol [(2*R*,3*R*,4*S*,5*R*)-hexanehexol], from altrose. A.k.a. **D-talitol**.

D-Arabitol [(2*R*,4*R*)-pentane-1,2,3,4,5-pentol], from arabinose.

D-Fucitol (D-Fuc-ol, 6-deoxy-D-galactitol), from fucose.

L-Fucitol (L-Fuc-ol, 1-deoxy-D-galactitol), from fucose.

D-Galactitol [(2*R*,3*S*,4*R*,5*S*)-hexane-1,2,3,4,5,6-hexol], from galactose. A.k.a. **dulcitol**, from Latin *dulcis*, sweet.

D-Glucamine (1-amino-1-deoxy-D-glucitol), from glucitol and amine.

Glycerol (propane-1,2,3-triol), earlier called glycerine, from Greek *glykys, glyker-*, sweet. A.k.a. **glyceritol**.

D-Iditol [(2R,3S,4S,5R)-hexane-1,2,3,4,5,6-hexol], from idose.

Inositol (cyclohexane-1,2,3,4,5,6-hexol), not a sugar alcohol *stricto sensu*, but usually placed in this context, from Greek *is, in-*, fiber, muscle, sinew, tendon. For the naming of the isomeric inositols cf. Section 2.6.

D-Mannitol [(2*R*,3*R*,4*R*,5*R*)-hexane-1,2,3,4,5,6-hexol], from mannose.

D-Ribitol [D-*erythro*-pentitol, (2*R*,3*s*,4*S*)-pentane-1,2,3,4,5-pentol], from ribose. A.k.a. **adonitol**, after the plant genus *Adonis* (adonis), after the Greek myth that the goddess Aphrodite created the anemone by sprinkling nectar on Adonis' blood, and **D-xylitol**, from xylose.

D-Sorbitol [(2*R*,3*R*,4*R*,5*S*)-hexane-1,2,3,4,5,6-hexol], from sorbose. A.k.a. **D-glucitol**, from glucose, and **L-gulitol**, from gulose.

Volemitol [D-*glycero*-D-*manno*-heptitol, (2R,3R,5R,6R)-heptane-1,2,3,4,5,6,7-heptol], after the species *Lactarius volemus* (a mushroom). The specific epithet is from Latin *volemus*, filling a hand, from *vola*, the hollow of the hand, referring to this mushroom's copious content of latex.

D-Xylitol [(2*R*,3*r*,4*S*)-pentane-1,2,3,4,5-pentol], from xylose.

9.1.2.4 Oxidized sugars

Oxidation at the terminal carbon atoms of a sugar chain leads to a variety of possible products all of which possess retained names.

From D-erythrose one can derive

D-Erythronic acid [(2*R*,3*R*)-2,3,4-trihydroxybutanoic acid).

D-Threaric acid [(2*S*,3*S*)-tartaric acid, (2*S*,3*S*)-2,3-dihydroxybutanedioic acid].

Starting with D-glucose we are looking at four compounds:

D-Glucaraldehyde [(2*S*,3*S*,4*S*,5*R*)-2,3,4,5-tetrahydroxyhexanedial]. A.k.a. **D-*gluco*-hexodialdose.**

D-Gluconic acid [(2*S*,3*S*,4*S*,5*R*)-2,3,4,5,6-pentahydroxyhexanoic acid]. A.k.a. **dextronic acid** and **maltonic acid.**

D-Glucuronic acid [(2*S*,3*S*,4*S*,5*R*)-2,3,4,5-tetrahydroxy-6-oxohexanoic acid].

D-Glucaric acid [(2*S*,3*S*,4*S*,5*R*)-2,3,4,5-tetrahydroxy hexanedioic acid]. A.k.a. **L-gularic acid** and **D-saccharic acid.**

From D-fructose the following oxidation products can be derived:

D-Fructosone [D-glucosone, D-*arabino*-hexos-2-ulose, (3*S*,4*R*,5*R*)-3,4,5,6-tetrahydroxy-2-oxohexanal]. A.k.a. **D-mannosone.**

D-Fructosonic acid [D-glucosonic acid, D-*arabino*-hex-2-ulosonic acid, (3*S*,4*R*,5*R*)-3,4,5,6-tetrahydroxy-2-oxohexanoic acid]

D-Fructuronic acid [D-*lyxo*-hex-5-ulosonic acid, (2*S*,3*S*,4*S*)-2,3,4,6-tetrahydroxy-5-oxohexanoic acid].

D-Galactose can be oxidized to the following acids:

D-Galactonic acid [(2*R*,3*S*,4*S*,5*R*)-2,3,4,5,6-pentahydroxyhexanoic acid].

D-Galactaric acid [(2*S*,3*R*,4*S*,5*R*)-2,,3,4,5-tetrahydroxyhexanedioic acid]. A.k.a. **mucic acid.**

And D-talose to:

D-Talonic acid [(2*S*,3*S*,4*S*,5*R*)-2,3,4,5,6-pentahydroxyhexanoic acid].

D-Altraric acid [(2*S*,3*S*,4*S*,5*R*)-2,3,4,5,6-tetrahydroxyhexanedioic acid]. A.k.a. **D-talaric acid.**

From the D-pentoses the following four D-pentonic acids can be derived:

D-Arabinonic acid [(2*S*,3*R*,4*R*)-2,3,4,5-tetrahydroxypentanoic acid], from arabinose. A.k.a. **D-arabonic acid.**

D-Lyxonic acid [(2*S*,3*S*,4*R*)-2,3,4,5-tetrahydroxypentanoic acid], from lyxose.

D-Ribonic acid [(2*R*,3*R*,4*R*)-2,3,4,5-tetrahydroxypentanoic acid], a name derived from arabinonic acid by transpositional contraction.

D-Xylonic acid [(2*R*,3*S*,4*R*)-2,3,4,5-tetrahydroxypentanoic acid], from xylose.

9.2 Disaccharides and derivatives

The term **invert sugar** refers to the levorotatory equimolar mixture of fructose and glucose which is obtained by the hydrolysis of dextrorotatory saccharose.

Cellobiose [β-D-glucopyranosyl-(1→4)-D-glucose], named as a disaccharide obtained by partial hydrolysis of cellulose.

Chitobiose [2-amino-4-*O*-(2-amino-2-deoxy-β-D-glucopyranosyl)-2-deoxy-D-glucose)], from chitin, q.v.

Gentiobiose [β-D-glucopyranosyl-(1→6)-D-glucose], after gentian (*Gentiana*). The genus name is probably after its discoverer, Gentius, 2nd century BC King of Illyria. A.k.a. **amygdalose**, from amygdalin.

Isomaltose [α-D-glucopyranosyl-(1→6)-D-glucose], from maltose.

Kojibiose [α-D-glucopyranosyl-(1→2)-D-glucose], after Japanese *koji*, a rice enzyme preparation.

Lactitol (4-*O*-β-D-galactopyranosyl-D-glucitol), from lactose.

Lactose [β-D-galactopyranosyl-(1→4)-D-glucose], from Latin *lac, lact-*, milk.

Maltitol (4-*O*-α-D-glucopyranosyl-D-glucitol), from maltose.

Maltose [α-D-glucopyranosyl-(1→4)-D-glucose], from malt.

Mannobiose [β-D-mannopyranosyl-(1→4)-D-mannose], from mannose.

Melibiose [α-D-galactopyranosyl-(1→6)-D-glucose], from Greek *meli*, honey.

Primeverose [β-D-xylopyranosyl-(1→6)-D-glucose], after primeverin and primulaverin, both from French *primevère*, cowslip (Primula veris), literally the firstling of spring).

Rutinose [α-L-rhamnopyranosyl-(1→6)-D-glucose], after the flavonoid glycoside rutin. Rutin is after the species *Ruta graveolens* (rue). The genus name is from Latin *ruta*, rue.

Saccharose (β-D-fructofuranosyl α-D-glucopyranoside), ultimately from Greek *sakcharon*, sugar, gravel, ultimately from Sanskrit *sarkara*, sugar, gravel, grit. A.k.a. **sucrose**, from French *sucre*, sugar.

Scillabiose [6-deoxy-4-*O*-β-D-glucopyranosyl-L-mannose], after the species name *Scilla maritima* (squill, sea squill, sea onion, maritime squill), later renamed *Drimia maritima*. The former genus name is from Greek *skilla*, squill.

Solabiose (β-D-glucopyranosyl-(1→3)-D-galactose), from solanine, from which it can be obtained by hydrolysis.

Sophorose [β-D-glucopyranosyl-(1→2)-D-glucose], after *Sophora japonica* (Japanese pagoda tree), later renamed *Styphnolobium japonicum*. The former genus name *Sophora* is from Arabic *sophera*, pea-flowered tree.

Swietenose (α-D-galactopyranosyl-(1→6)-D-galactose), after the genus *Swietenia* (tropical trees). The genus name is after the Dutch-Austrian physician and botanist Gerard van Swieten (1700–1772).

α,α-Trehalose (α-D-glucopyranosyl α-D-glucopyranoside), after trehala, a sugary substance secreted by beetles of the genus *Larinus*, ultimately from Persian *tighal*, a kind of manna.

Turanose [α-D-glucopyranosyl-(1→3)-D-fructose], after manna fromTurkestan.

9.3 Oligosaccharides

Class names for oligosaccharides and polysaccharides are biose, triose, and, without elision, tetraose, pentaose, etc., indicating the number of monosaccharide units. The corresponding elided names tetrose, pentose, etc. are used for monosaccharides where they indicate the length of the carbon chain. Thus, the class name triose is ambiguous, but on the other hand, since the three three-carbon monosaccharides are rarely encountered, its meaning is seldom an issue. Nonetheless once in a while trisaccharides are called triaoses for clarity, without etymological justification.

Gentianose [β-D-fructofuranosyl β-D-glucopyranosyl-(1→6)-α-D-glucopyranoside], after gentian (*Gentiana*). The genus name is probably after its discoverer, Gentius, 2nd century BC King of Illyria.

Kestose [*O*-β-D-fructofuranosyl-(2→6)-β-D-fructofuranosyl α-D-glucopyranoside], after the location of the Tate and Lyle Research Laboratories, Ravensbourne, Keston, Kent, UK where this trisaccharide was discovered.

Melezitose [α-D-glucopyranosyl-(1→3)-β-D-fructofuranosyl α-D-glucopyranoside], from French *mélèze*, larch (*Larix*). A.k.a. **melecitose.**

Planteose [α-D-galactopyranosyl-(1→6)-β-D-fructofuranosyl α-D-glucopyranoside], after blond plantain, also called desert indianwheat, blond psyllium, and isphagul (*Plantago ovata*). The genus name is from Latin *plantago*, sprout.

Raffinose [β-D-fructofuranosyl α-D-galactopyranosyl-(1→6)-α-D-glucopyranoside], from French *raffiner*, to refine. A.k.a. **gossypose**, from the genus name *Gossypium* (cotton plant), from Greek *gossypion*, cotton plant, ultimately from Sanskrit *karpasa*, cotton, **melitose**, and **melitriose**, both from Greek *meli*, honey.

Stachyose [β-D-fructofuranosyl α-D-galactopyranosyl-(1→6)-α-D-galactopyranosyl-(1→6)-α-D-glucopyranoside], after the plant genus *Stachys*, from Greek *stachys*, downy woundwort (*Stachys germanica*).

Umbelliferose [β-D-fructofuranosyl α-D-galactopyranosyl-(1→2)-α-D-glucopyranoside], after the plant family *Umbelliferae* (umbellifers), a.k.a. *Apiaceae*. The former genus name is from Latin *umbella*, parasol, diminutive of *umbra*, shadow, and *ferre*, to bear, to carry.

Verbascose [β-D-*arabino*-hex-2-ulofuranosyl α-D-*galacto*-hexopyranosyl-(1→6)-α-D-*galacto*-hexopyranosyl-(1→6)-α-D-*galacto*-hexopyranosyl-(1→6)-α-D-*gluco*-hexopyranoside], named after mullein, a.k.a. velvet plant (*Verbascum thapsus*), from Latin *verbascum*, mullein, akin to *verbena*, foliage.

9.4 Polysaccharides

The class name for polysaccharides is glycan (an alteration of glucan).

Agar (a mixture of agarose and agaropectin), from Malay *agar-agar*, the red algae of the genus *Gracillaria* which produce agar.

Agarose (3,6-anhydro-α-L-galacto-β-D-galactan), from agar.

Amylopectin [α-(1→4)/α-(1→6)-D-glucan], from Greek *amylon*, starch, literally not ground; starch was obtained from unground grain, and *pektos*, fixed, congealed, from *pegnynai*, to fasten together, to coagulate

Amylose [α-(1→4)-D-glucopyranan], as amylopectin.

Callose [β-(1→3)-D-glucopyranan], from Latin *callus*, weal, referring to the formation of this polysaccharide in the healing process of injured plants.

Carrageenan (sulfated polysaccharide), after the red alga carrageen [Irish moss, carrageen moss (*Chondrus crispus*)]. Carrageen is from Irish *carraigin*, little rock, referring to its occurrence on rocky Atlantic shores. A.k.a. **carrageenin.**

Cellulose [β-(1→4)-D-glucopyranan], from New Latin *cellula*, living cell, diminutive of Latin *cella*, small room, store room, hut.

Chitin [β-(1→4)-2-acetamido-2-deoxy-D-glucopyranan], from Greek *chiton*, tunic, of Semitic origin, referring to this polymer's occurrence in the hard outer integument of insects, arachnids, and crustaceans.

Chitosan [β-(1→4)-2-amino-2-deoxy-D-glucopyranan], from chitin and 'ose'.

Chondroitin sulfate (sulfated glycosaminoglycan), from chondr(o)-.

Chrysolaminarin [β-(1→3)/β-(1→6)-D-glucopyranan], after laminarin and its source, the golden-brown algae *Chrysophyceae*, ultimately from Greek *chrysos*, gold and *phykos*, seaweed. A.k.a. **chrysolaminaran** and **leucosin.**

Curdlan [β-(1→3)-D-glucopyranan], from curdle, referring to this polysaccharide's ability to curdle when heated. A.k.a. **paramylon.**

Dextran [α-(1→6)-D-glucopyranan with α-(1→3)-branches], a contraction of dextrose anhydride.

Dextrin (a class name for polysaccharides obtained by partial degradation of starch), from dextrose.

Fructan (a class name), from fructose. A.k.a. **fructosan** and **levulin.**

Fucan (a class name), from fucose. A.k.a. **fucosan.**

Furcellaran, after the genus *Furcellaria* (algae). The irregular genus name is from Latin *furcilla*, diminutive of *furca*, fork. A.k.a. **Baltic agar** and **Danish agar**.

Glycogen [α-(1→4)/α-(1→6)-D-glucan], so named as sugar-forming substance.

Guaran (galactomannan), after guar (*Cyamopsis tetragonaloba*). Guar is from Hindi *guar*. A.k.a. **guar gum**.

Hemicellulose (a matrix polysaccharide). From hemi- and cellulose, referring to it being less complex than cellulose. A.k.a. **polyose**.

Heparin (a mucopolysaccharide sulfuric acid ester), from Latin *hepar, hepat-*, liver.

Hyaluronic acid (a glycosaminoglycan), from Greek *hyalos*, glass, referring to its occurrence in vitreous humor. A.k.a. **HA** and **hyaluronan**.

Inulin [β-(2→1)-D-fructofuranan], after the genus *Inula* (elecampane). The genus name is from Greek *helenion*, elecampane, ultimately maybe from *helene*, wicker basket, from *eilein*, to wind, to roll. A.k.a. **dahlin**, after the genus *Dahlia* (dahlia), after the Swedish botanist Anders Dahl (1751–1789).

Laminarin (β-(1→3)/β-(1→6)-D-glucan), after kelp, brown algae of the genus *Laminaria*, from Latin *lamina*, sheet, leaf. A.k.a. **laminaran**,

Lentinan (β-(1→6)/β-(1→4)-D-glucan), after the species *Lentinula edodes* (shiitake mushroom). The genus name is from Latin *lentulus*, rather tough, diminutive of *lentus*, tough.

Lichenin [β-(1→3)/β-(1→4)-D-glucan], after its source. Lichen is ultimately from Greek *leichen*, what eats around itself, from *leichein*, to lick.

Pectin (a heteropolysaccharide), from Greek *pektos*, congealed, from *pegnynai*, to fasten together, to coagulate. A.k.a. **homogalacturonan, rhamnogalacturonan I**, and **rhamnogalacturonan II**.

Pleuran [β-(1→3)/β-(1→6)-D-glucan], after the oyster mushroom (*Pleurotus ostreatus*). The genus name is from Greek *pleure*, side, and *ous, outo-*, ear, and the specific epithet from Latin *ostreatus*, oyster shaped, from *ostrea*, oyster.

Pullulan (α-(1→4)/α-(1→6)-D-glucan), after the fungal species *Aurobasidium pullulans* which generates it from starch. The specific epithet *pullulans, pullulant-*, sprouting, is derived from Latin *pullulus*, diminutive of *pullus*, young animal, from *putus*, boy.

Schardinger dextrin [cyclic oligo-(1→4)-D-glucopyranosides], a class name, after the German biochemist Franz Schardinger (1853–1917).

Starch (α-D-glucan), ultimately from the proto-Indo-European root **ster-*, stiff.

Tunicin (a cellulose-like polysaccharide), after the subphylum *Tunicata* (tunicates), referring to its occurrence in the test of tunicates. Tunicate is from Latin *tunica*, tunic, of Semitic origin.

Zymosan (β-(1→3)-D-glucan), from Greek *zyme*, yeast, referring to the occurrence of zymosan in the yeast cell wall.

9.5 Shorthand vs. other IUPAC carbohydrate names

It goes without saying that explicit systematic names become highly impractical when dealing with oligo- and polysaccharides. This can be seen in the following example:

(2*S*,3*R*,4*S*,5*S*,6*R*)-2-[[(2*R*,3*S*,4*R*,5*R*,6*R*)-4,5-dihydroxy-2-(hydroxymethyl)-6-methoxyoxan-3-yl]oxy]-6-(hydroxymethyl)oxane-3,4,5-triol (systematic IUPAC name).

Methyl-β-D-glucopyranosyl-(1→4)-β-D-glucopyranose (IUPAC carbohydrate name).

β-D-Glc*p*-(1→4)-β-D-Glc*p*-OMe (IUPAC shorthand carbohydrate name).

9.6 Limits of IUPAC carbohydrate nomenclature

With its many parent names and permitted modifications IUPAC carbohydrate nomenclature, if used indiscriminately, could in theory even be used for compounds one would not consider as carbohydrates or even as remotely related derivatives. Thus, the proteinogenic amino acid L-serine could, without breaking any formal rules, be called 2-deoxy-2-amino-D-glyceric acid.

10 The naming of amino acids, oligo- and polypeptides

The term amino acid is a contraction of 2-aminoalkanoic acid. It is always used in this sense to the exclusion of other amino acids such as anthranilic acid (2-aminobenzoic acid).

10.1 Amino acids

Both in the solid state and in aqueous solution amino acids exist as zwitterions, i.e. inner salts. For the sake of convenience this fact is either ignored or implied and the uncharged, easier-to-name, tautomer is used in presentations. If the zwitterionic nature of an amino acid is focused upon, the usual name of, for instance, alanine (2-aminopropanoic acid) becomes 2-ammoniopropanoate (2-ammoniumylpropanoate, 2-azaniumylpropanoate).

10.1.1 Proteinogenic amino acids

For the twenty standard and two non-standard proteinogenic amino acids, retained trivial names are used, together with three-letter and one-letter codes. The stereochemistry is indicated with the descriptors D- and L-. A plain name such as alanine and its three-letter code such as Ala usually stand for the L-enantiomer, i.e. L-alanine. In the rare cases of D-amino acids this stereochemistry must be explicitly stated, both in the name such as D-alanine and in the three-letter code such as D-Ala. The amino acids' descriptor L- appears as (*S*)- in all corresponding systematic names, except in the case of cysteine whose systematic name contains an (*R*)-descriptor, due to the relatively high priority of the sulfur atom in the CIP system. An amino acid's three-letter code (as well as its one-letter code) does not stand for the amino acid, but for the corresponding amino acid residue in a peptide chain. If an amino acid is at the N-terminal of a peptide, this is specifically shown as in H_2N-Ala, the same amino acid at the C-terminal of a peptide is shown as Ala-COOH.

10.1.1.1 Standard proteinogenic amino acids

L-Alanine [Ala, A, (2*S*)-2-aminopropanoic acid)], from acetaldehyde, referring to the racemic amino acid's synthesis from acetaldehyde.

L-Arginine [Arg, R, (2S)-2-amino-5-(carbamimidamido)pentanoic acid], from New Latin *argentum*, silver, referring to the silvery white appearance of arginine nitrate.

https://doi.org/10.1515/9783110612714-010

L-Asparagine [Asn, N, (2*S*)-2,4-diamino-4-oxobutanoic acid], from asparagus (*Asparagus officinalis*), ultimately from Greek *spargan*, to swell, referring to the isolation of this amino acid from asparagus juice.

L-Aspartic acid (Asp, D, (2*S*)-2-aminobutanedioic acid), from asparagine, referring to its formation upon hydrolysis of asparagine.

L-Cysteine [Cys, C, (2*S*)-2-amino-3-sulfanylpropanoic acid], by modification of cystine, the corresponding disulfide, from Greek *kystis*, bladder, referring to the isolation of L,L-cystine from urinary bladder stones.

L-Glutamic acid [Glu, E, (2*S*)-2-aminopentanedioic acid], a contraction of aminoglutaric acid; glutaric acid from Latin *gluten*, glue.

L-Glutamine [Gln, Q, (2*S*)-2-amino-4-carbamoylbutanoic acid], from glutamic acid.

Glycine (Gly, G, aminoacetic acid), the only achiral compound in this group, from Greek *glykys, glyker-*, sweet, referring to this amino acid's sweet taste. A.k.a. **glycocoll**, from Greek *glykys, glyker-*, sweet, and *kolla*, glue.

L-Histidine [His, H, (2*S*)-2-amino-3-(1*H*-imidazol-4-yl)propanoic acid], from Greek *histos*, web, from *histanai*, to stand, referring to this amino acid's perceived importance for tissue function, especially in infants.

L-Isoleucine (Ile, I, (2*S*,3*S*)-2-amino-3-methylpentanoic acid], from leucine.

L-Leucine [Leu, L, (2*S*)-2-amino-4-methylpentanoic acid], from Greek *leukos*, white, referring to the appearance of the crystalline substance.

L-Lysine [Lys, K, (2*S*)-2,6-diaminohexanoic acid], earlier called lysatine, from Greek *lysis*, dissolution, from *lyein*, to loosen, to dissolve, referring to its isolation from protein hydrolysates.

L-Methionine [Met, M, (2*S*)-2-amino-4-(methylsulfanyl)butanoic acid], a contraction of γ-methylthio-α-aminobutyric acid.

L-Phenylalanine (Phe, F, (2*S*)-2-amino-3-phenylpropanoic acid], from alanine.

L-Proline [Pro, P, (2*S*)-pyrrolidine-2-carboxylic acid], the only amino acid without a primary amino group, named by contraction of the systematic name.

L-Serine [Ser, S, (2*S*)-2-amino-3-hydroxypropanoic acid], after sericin, silk gum protein, from Medieval Latin *sericum*, silk, from Latin *sericus*, Chinese, silken, from *Seres*, the Chinese.

L-Threonine (Thr, T, (2*S*)-2-amino-3-hydroxybutanoic acid), named after its structural similarity with D-erythrose.

L-Tryptophan [Trp, W, (2*S*)-2-amino-3-(1*H*-indol-3-yl)propanoic acid], from Greek *tribein*, to rub, to powder, to grind, and *phainein*, to be visible, to be obvious, to show, literally to be broken and to reappear, referring to the isolation of this amino acid from tryptic digests of proteins.

L-Tyrosine [Tyr, Y, (2*S*)-2-amino-3-(4-hydroxyphenyl)propanoic acid], from Greek *tyros*, cheese, referring to this amino acid's isolation from cheese casein hydrolysate.

L-Valine [Val, V, (2*S*)-2-amino-3-methylbutanoic acid], from valeric acid, referring to its structural relationship with valeric acid.

10.1.1.2 Non-standard proteinogenic amino acids

The following two amino acids also occur, but only rarely, in proteins.

L-Pyrrolysine {Pyl, O, N^6-{[(2*R*,3*R*)-3-methyl-3,4-dihydro-2*H*-pyrrol-2-yl]carbonyl}-L-lysine}, from pyrrole and lysine.

L-Selenocysteine [Sec, earlier Se-Cys, U, (2*R*)-3-selanyl-2-aminopropanoic acid], from selenium and cysteine. An unusual name construction since replacement names are normally based on the replacement of carbon or oxygen by other atoms, not on that of other atoms such as sulfur.

10.1.2 Other amino acids

L-Abrine (*N*-methyl-L-tryptophan), after the polypeptide abrin, q.v.

Agaritine {L-glutamic acid [2-[4-(hydroxymethyl)phenyl]hydrazide]}, after the species *Agaricus bisporus* (portobello mushroom). The genus name is ultimately from Greek *agarikon*, a toponym referring to Agaria, a district in Sarmatia.

β-Alanine (βAla, 3-aminopropanoic acid), named as an isomer of alanine.

L-Alanosine {(2*S*)-2-amino-3-[hydroxy(nitroso)amino]propanoic acid}, from alanine and a contraction of the systematic name.

Albizziin [3-(carbamoylamino)-L-alanine], after the genus *Albizia* (silk plants). The genus name is after the Italian naturalist Filippo degli Albizzi (1724–1789).

L-Alloisoleucine [alle, (2*S*,3*R*)-amino-3-methylpentanoic acid)], named as isomer of isoleucine.

L-Allothreonine [aThr, (2*S*,3*S*)-2-amino-3-hydroxybutanoic acid), named as isomer of threonine.

L-Allysine (6-oxo-L-norleucine), from 'allo' and lysine, i.e. named as isomer of lysine.

Anserine (*N*-β-alanyl-3-methyl-L-histidine), from Latin *anser*, goose.

L-Canaline {(2*S*)-2-amino-4-[(amino)oxy]butanoic acid}, back formation from canavanine, q.v.

L-Canavanine {(2*S*)-2-amino-4-[[(diaminomethylidene)amino]oxy}butanoic acid}, after the species *Canavalia ensiformis* (jack beans). The genus name is from Malabar *kanavali*, jack bean, from *kanam*, forest, and *valli*, climber.

L-Carnosine (*N*-β-alanyl-L-histidine), from carnos-.

Chondrine [(3*R*)-thiomorpholine-3-carboxylic acid 1-oxide], after the genus *Chondria* (red algae). The genus name is from Greek *chondros*, grain, cartilage.

L-Citrulline [Cit, (2*S*)-2-amino-5-(carbamoylamino)pentanoic acid, from Latin *citrullus*, water melon.

Connatin [N^5-(dimethylcarbamoyl)-N^5-hydroxy-L-ornithine], after the fungal species *Lyophyllum connatum*. The specific epithet is from Latin *connatus*, double, twin.

Coprine [N^5-(1-hydroxycyclopropan-1-yl)-L-glutamine], after the fungal species *Coprinus atramentarius* (inky cap). The genus name is from Greek *koprinos*, of dung, from *kopros*, dung, referring to some of these fungi's habitat.

Creatine (*N*-carbamimidoyl-*N*-methylglycine), from Greek *kreas, kreat-*, meat, referring to its isolation from basified water extract of skeletal muscle.

L-Cystathionine [Ala-Hey, *S*-(2*R*)-2-amino-2-carboxyethyl-L-homocysteine], from cysteine.

L-Cysteic acid (Cya, 3-sulfo-L-alanine), from cysteine.

L,L-Cystine (Cys-Cys, 3,3′-disulfanediyldi-L-alanine), from Greek *kystos*, bladder, referring to its occurrence in bladder stones.

Dopa (3-hydroxytyrosine), an initialism taken from the obsolete name dioxyphenylalanine. 'Oxy' was earlier used instead of the current 'hydroxy'. The L-isomer is also known as **levodopa**.

L-Homocysteine [Hey, (2*S*)-2-amino-4-sulfanylbutanoic acid], from cysteine.

L-Homoserine [Hse, (2*S*)-2-amino-4-hydroxybutanoic acid], from serine.

Isoserine (3-amino-2-hydroxypropanoic acid), from serine, not found in nature.

Isovaline (α-ethylalanine, 2-amino-2-methylbutanoic acid), from valine. This amino acid, together with glycine, alanine, glutamic acid, and pseudoleucine, the three chiral ones with slight excess of the L-enantiomer, was found on the Murchison meteorite, found in Australia in 1969. It has been speculated that this suggests an extraterrestrial origin of the homochirality of life on Earth.

Kynurenine [(2*S*)-2-amino-4-(2-aminophenyl)-4-oxobutanoic acid], from Greek *kynos ouron*, dog's urine, and Latin *ren*, kidney.

L-Lanthionine (Ala-Cys, 3,3′-sulfanediyldi-L-alanine), from Latin *lana*, wool and with 'thi(o)'. Not to be confused with lenthionine.

Lathyrine [(2*S*)-2-amino-3-(2-aminopyrimidin-4-yl)propanoic acid], after the genus *Lathyrus* (peavines, vetchlings). The genus name is from Greek *lathyros*, chickling (*Lathyrus sativus*).

Mimosine [3-(-3-hydroxy-4-oxopyridin-1(4*H*)-yl)-L-alanine], after the genus *Mimosa* (mimosa). The genus name is from Greek *mimos*, actor, and the feminine suffix *-osa*, referring to the fact that these plants' sensitive leaves seem to mimic conscious life.

Nopaline {*N*-[(1*R*)-1-carboxy-4-[(carbamimidoyl)amino]butyl]-L-glutamic acid}, from French *nopal*, Barbary fig, common prickly pear (*Opuntia vulgaris*).

L-Norleucine [Nle, (2*S*)-2-aminohexanoic acid], named as the straight-chain isomer of leucine. Here the prefix 'nor' is used in the now obsolete sense of an unbranched, 'normal', carbon chain. Considering the present-day use of the prefix 'nor' to label a compound with one carbon atom less than the parent compound, names in tune with contemporary practice would be neoleucine, i.e. an isomer of leucine and isoleucine, or trihomoalanine.

L-Norvaline [Nva, (2*S*)-2-aminopentanoic acid], named as the straight-chain isomer of valine. Here the prefix 'nor' is used in the now obsolete sense of an unbranched, 'normal', carbon chain. Considering the present-day use of the prefix 'nor' to label a compound with one carbon atom less than the parent compound, names in tune with contemporary practice would be neovaline, i.e. an isomer of valine and isovaline, or dihomoalanine.

L-Ornithine [Orn, (2*S*)-2,5-diaminopentanoic acid], named after its formation by hydrolysis of ornithuric acid (N^2,N^5-dibenzoyl-L-ornithine), from Greek *ornis, ornith-*, bird, and urine.

5-Oxo-L-proline [Glp, (2*S*)-5-oxopyrrolidine-2-carboxylic acid], from proline. The three-letter code is from the name pyroglutamic acid. A.k.a. **L-pyroglutamic acid.** i.e. named as the product formed by intramolecular loss of water from L-glutamic acid.

Phosphinothricin {(*RS*)-2-amino-4-[hydroxy(methyl)phosphinoyl]butanoic acid}, named as phosphinic acid. No rationale has been offered for the remainder of the name. A.k.a. **glufosinic acid** and **PT**.

Pseudoleucine (2-amino-3,3-dimethylbutanoic acid), named as isomer of leucine.

Sarcosine [Sar, 2-(methylamino)acetic acid], from Greek *sarx, sark-* flesh, referring to its isolation from meat extract.

Statine [(3*S*,4*S*)-4-amino-3-hydroxy-6-methylheptanoic acid], named as component of the oligopeptide pepstatin, q.v.

Stizolobic acid {4-[(2*S*)-2-amino-2-carboxyethyl]-2-oxo-2*H*-pyran-6-carboxylic acid}, after the species *Stizolobium hassjoo* (velvet bean, cowage, cowitch, Lyon bean), later renamed *Mucuna pruriens*. The former genus name is from Greek *stizein*, to tattoo, and *lobos*, lobe.

Stizolobinic acid {3-[(2*S*)-2-amino-2-carboxyethyl]-2-oxo-2*H*-pyran-6-carboxylic acid}, named as isomer of stizolobic acid.

Theanine (N^5-ethyl-L-glutamine), from New Latin *thea*, tea plant (*Camellia sinensis*).

L-Thyronine [*O*-(4-hydroxyphenyl-L-tyrosine], as L-thyroxine.

L-Thyroxine [Thx, *O*-(4-hydroxy-3,5-diiodophenyl)-3,5-diiodo-L-tyrosine], from thyroid gland. The name of the gland is ultimately from Greek *thyreoeides*, shield-shaped, from *thyreos*, shield shaped like a door, from *thyra*, door. A.k.a. $\mathbf{T_4}$.

Tricholomic acid [(2S)-2-amino-2-[(5S)-3-oxo-1,2-oxazolidin-5-yl]acetic acid], after the genus *Tricholoma* (mushrooms). The genus name is from trich(o)- and Greek *loma*, hem, fringe.

Willardiine {3-[2,4-dioxo-3,4-dihydropyrimidin-1(2*H*)-yl]-L-alanine}, after the species *Mariosousa willardiana* (Palo Blanco tree, palo liso, guinola, Willard acacia). The specific epithet is after Alexander Willard, US consul in Guaymas, Mexico, who helped the British-American physician, botanist, and archeologist Edward Palmer (1829/31–1911) to collect his specimens.

10.2 Enzymes

The term enzyme, earlier ferment, from Latin *fermentum*, yeast, ultimately from *fervere*, to boil, is from Greek *enzymos*, leavened, from *zyme*, yeast. The suffix 'ase' for systematic enzyme names is from diastase.

10.2.1 Traditional enzyme names

Adipsin (E.C. 3.4.21.46), from adiposity, ultimately from Latin *adeps, adip-*, fat. A.k.a. **factor D**.

Ancrod (ophidian L-amino acid oxidase, EC 3.4.21.28), coined by contraction of the species name *Agkistrodon rhodostoma* (Malayan pitviper), irregularly from Greek *angkistron*, fishhook, *odon, odont-*, tooth, *rhodon*, rose, probably from Old Iranian **urda-*, rose, and *stoma, stomat-*, mouth.

Apopain (EC 3.4.22.56), from apoptosis and -ain(e).

Apyrase (EC 3.6.1.5), a contraction of adenylpyrophosphatase.

Aromatase (EC 1.14.14.1), from aromatic, referring to the fact that this enzyme catalyzes the aromatization of androgens to estrogens.

Bromelain [EC 3.4.22.32 (stem bromelain), EC 3.4.22.33 (fruit bromelain)], after the genus name *Bromelia* (pineapple), after the Swedish physician and botanist Olaf Bromelius (1639–1705).

Calpain-1 (EC 3.4.22.52), a contraction of calmodulin, q.v., and papain.

Catalase (EC 1.11.1.6), from catalysis. A tautology since an enzyme, with the systematic label 'ase', is by definition a catalyst.

Catecholase (EC 1.10.2.1), a contraction of catechol and oxidase.

Cathepsin (proteases, EC 3.4.), from Greek *kathepsein*, to digest, from cat(a)- and *hepsein*, to boil. A.k.a. **catheptic enzyme**.

Cellulase (EC 3.2.1.4), from cellulose.

Chitinase (EC 3.2.1.14), from chitin.

Chymosin (EC 3.4.23.4), from Greek *chymos*, juice. A.k.a. **rennin**, from rennet, from Old English *gerennan*, to coagulate, and **FPC**.

Chymotrypsin (EC 3.4.21.1), from Greek *chymos*, juice, and trypsin.

Cyclophilin (EC 5.2.1.8), from cyclosporin A, referring to the fact that this enzyme binds to this immunosuppressant.

Diastase (amylase, EC 3.2.1.), from Greek *diastasis*, separation, interval, from *diistanai*, to separate. This name refers to the action of the enzyme on beer mash. The starch is quickly hydrolyzed to soluble sugars and thus the husk separates readily from the rest of the barley seed.

Dynamin (an enzyme family, EC 3.6), from Greek *dynamis*, power.

Dioxygenase (EC 1.14), from dioxygen.

Elastase (EC 3.4.21.37), from elastin, an elastic protein fiber.

Endonuclease (EC 3.1), from nucleic acid, now a class name.

Erepsin (a mixture of enzymes), from Latin *eripere*, to take away, from *rapere*, to seize, to rob; patterned after pepsin. An obsolete term for exopeptidases.

Esterase (EC 3.1), from ester, now a class name.

Exonuclease (EC 3.1), from nucleic acid, now a class name.

Ficin (EC 3.4.22.3), after the genus *Ficus* (fig tree). The genus name is from Latin *ficus*, fig tree, fig. A.k.a. **ficain** and **debricin**, from debridement.

β-Glucosidase (EC 3.2.1.21). A.k.a. **emulsin**, from emulsion.

Hementin (an anticoagulant protease, EC 3.4), after the genus *Haementeria* (leeches). The genus name is from haem(o)- and enter(o)-.

Invertase (EC 3.2.1.26), from inversion, referring to the fact that this enzyme converts a dextrorotatory solution of sucrose into a levorotatory mixture of fructose and glucose. A.k.a. **saccharase**, from saccharose.

Kallikrein (a group of serine proteases, EC 3.4.21), from Greek *kallikreas*, sweetbread, pancreas, from *kallos*, beauty, from *kalos*, beautiful, and *kreas, kreat-*, flesh.

Laccase (EC 1.10.3.2), from lacquer, referring to its occurrence in the sap of the Japanese lacquer tree (*Toxicodendron vernicifluum*) where it helps to form lacquer by catalyzing phenol oxidation.

Lactase (EC 3.2.1.108), from lactose.

Lipase (EC 3.1), from lipid, now a class name.

Luciferase (EC 1.13.12.7), from luciferin, referring to the bioluminescence upon luciferase catalyzed oxidation of luciferins.

Lysozyme (EC 3.2.1.17), a contraction of lysis and enzyme. A.k.a. **muramidase** and **muraminidase**, from Latin *murus*, wall, referring to this enzyme's destruction of bacterial cell walls.

Maltase (EC 3.2.1.20), from maltose.

Monooxygenase (EC 1.13), from mon(o)- and oxygen.

Notatin (glucose oxidase, GOx, EC 1.1.3.4), after the fungal species *Penicillium notatum*. The specific epithet is from Latin *notatus*, marked with spots or lines, from *nota*, mark.

Nuclease (EC 3.1), from nucleic acid, now a class name.

Papain (EC 3.4.22.2), after the species *Carica papaya* (green papaya, melon fruit). The specific epithet is from Carib *ababai*, papaya.

Parvulin (EC 5.2.1.8), from Latin *parvulus*, diminutive of *parvus*, small, referring to the fact that this enzyme is much smaller than other prolyl isomerases.

Pepsin (endopeptidase, EC 3.4.23.1), from Greek *pepsis, pept-*, digestion.

Phaseolin (EC 3.4.16.5), after the species *Phaseolus vulgaris* (French bean). The genus name is from Latin *phaseolus*, diminutive of *phaselus*, ultimately from Greek *phaselos*, cowpea (*Vigna unguiculata*).

Phosphatase (EC 3.1.3), from phosphate, now a class name.

Pinguinain (EC 3.4.99.18), after the species *Bromelia pinguin* (pinguin). The specific epithet is a native West Indian word. Patterned after papain.

Protease (EC 3.4), from protein, now a class name.

Ptyalin (α-amylase, EC 3.2.1.1), from Greek *ptyalon*, saliva.

Renin (EC 3.4.23.15), from Latin *ren*, kidney.

Rennet (EC 3.4.23.4), from Old English *gerennan*, to coagulate.

Rubredoxin (a class name, EC 1.14, 1.15, and 1.18), from rubi- and redox, referring to these enzymes' reddish color in the oxidized state.

Schardinger enzyme (xanthine oxidase, EC 1.17.3.2), after the German biochemist Franz Schardinger (1853–1917).

Steapsin (EC 3.1.1.3), from Greek *stear*, fat, patterned after pepsin.

Streptodornase (EC 3.1.21.1), a contraction of streptococcal deoxyribonuclease.

Thrombin (EC 3.4.21.5), from thrombus, referring to its role in blood clotting. Thrombus is from Greek *thrombos, thrombot-*, clot, lump.

Trypsin (serine protease, EC 3.4.21.4), from Greek *tripsis*, friction, from *tribein*, to rub; alternatively from Greek *tryein*, to wear out. Trypsin was first obtained by rubbing fresh pancreas with glass powder and alcohol.

Urease (EC 3.5.1.5), from urea.

Urokinase (EC 3.4.21.73), from urine and kinase.

Warburg's respiratory enzyme (cytochrome C oxidase, EC 1.9.3.1), after the German biochemist Otto Heinrich Warburg (1883–1970).

Zymase (an enzyme complex), from Greek *zyme*, yeast.

10.2.2 Systematic enzyme names

With the advance of biochemistry the number of fully characterized enzymes went into the thousands and thus the traditional trivial enzyme names had to be replaced with systematic names, based on the type of reaction an enzyme catalyzes. Under the auspices of the Enzyme Commission [1] these reactions are assigned EC numbers while enzymes are provided with systematic names based on the reaction catalyzed by the enzyme in question. If two or more enzymes catalyze the same reaction they are regarded as isozymes, i.e. isomeric enzymes, and linked to the same EC number.

The Enzyme Commission operates with seven classes of enzymes:

Hydrolases (EC 3), catalyze the formation of two products from a substrate by hydrolysis, i.e. AB + H_2O → AOH + BH.

Isomerases (EC 5), catalyze intramolecular rearrangements, i.e. ABC → BCA.

Ligases (EC 6), catalyze the formation of a new molecule from two substrates by formation of new C-O, C-S, C-N, or C-C bonds, i.e. X + Y + ATP →XY + ADP + P_1.

Lyases (EC 4), catalyze the non-hydrolytic addition or removal of groups from substrates, i.e. RCOCOOH → RCHO + CO_2 or [X-A + B-Y] → [A=B + XY].

Oxidoreductases (EC 1), catalyzing oxidation/reduction reactions, transferring hydrogen or oxygen atoms, and electrons, i.e. AH + B → A+ BH or A + O → AO.

Transferases (EC 2), catalyze the transfer of a funrtional group from one substrate to another, i.e. AB + C → A + BC.

Translocases (EC 7), catalyze the movement of ions or molecules across membranes, or their separation within membranes.

Typical systematic enzyme names are:

Acetylcholinesterase (EC 3.1.1.7).

Choline acetyltransferase (EC 2.3.1.6).

Cytochrome c peroxidase (EC 1.11.1.5).

Dihydrobenzophenanthridine oxidase (EC 1.5.3.12).

Hypoxanthine-guanine phosphoribosyltransferase (EC 2.4.28).

Methylenetetrahydrofolate reductase (EC 1.5.1.20).

10.3 Other oligo- and polypeptides

Abrin, after the genus *Abrus* (tropical vines). The genus name is from Greek *habros*, graceful, delicate. Not to be confused with abrine.

Actin, from Latin *agere, act-*, to set in motion, to drive forward, to do, to perform, to keep in movement, referring to its role in muscle contraction.

Activin, from activation.

Aequorin, after the genus *Aequorea* (jellyfish), from Latin *aequoreus*, marine, from *aequor*, surface of the sea.

Albumin, from Latin *albumen, albumin-*, egg white, from *albus*, white.

β-Amyloid, from Greek *amyloides*, starch-like, referring to this protein's 'wrong' starch-like folding.

Apoferritin, apo- and ferritin.

Calmodulin, a contraction of calcium-modulated protein.

Canavalin, after the species *Canavalia ensiformis* (jack beans). The genus name is from Malabar *kanavali*, jack bean, from *kanam*, forest, and *valli*, climber. Cf. concanavalin.

Chalone, with contraction, from Greek *chalasis*, a slackening, from *chalan*, to slacken, and hormone.

Chaperonin, named as molecular chaperons which assist correct folding, and prevent misfolding, of other proteins. Chaperon is from French, *chaperon*, literally hood.

Clupeine, after the genus name *Clupea* (herring). The genus name is from Latin *clupea*, a small river fish.

Colicin, after the bacterial species *Escherichia coli*. The specific epithet is from Greek *kolon*, large intestine.

Collagen, from Greek k*olla*, glue, referring to its use as glue.

Concanavalin, from con- and canavalin. Cf. canavalin.

Conotoxin, after the genus *Conus* (cone snail). The genus name is from Latin *conus*, cone.

Coronin, from Latin *corona*, crown. This actin binding protein was named after its strong immunolocalization in the actin rich crown-like extension of the cell cortex in *Dictyostelium discoideum* (slime mold).

Dystrophin, after dystrophy, referring to its absence or deficiency in muscular dystrophy patients.

Edestin, from Greek *edestos*, edible.

Elastin, from elastic, referring to its role in body tissues' resumption of shape after contraction or stretching.

Endothelin, from endothelium, ultimately from Greek *thele*, nipple.

Enkephalin (pentapeptides), from Greek *enkephalos*, brain.

Ferritin, from Latin *ferrum*, iron, referring to its transport function for soluble iron.

Fibrin, from fiber, referring to its function in the clotting of blood.

Fibrinogen, from fibrin and -gen.

Fibroin, from fiber.

Flagellin, from flagellum, from Latin *flagellum*, diminutive of *flagrum*, whip.

Gelatin, from French *gélatine*, edible jelly, from Italian *gelatina*, jelly, from *gelato*, frozen.

Gliadin, from gli(o)-.

Gluten, from Latin *gluten*, glue.

Hemoglobin, also spelled haemoglobin, earlier called hematoglobulin, from Greek *haima, haimat-*, blood, and Latin *globulus*, diminutive of *globus*, globe.

Histone, from hist(o)-.

Inhibin, from inhibition.

Integrin, a contraction of integral protein.

Interferon (a group of signaling proteins), from interference.

Interleukin (a class name), from inter- and leukocyte, referring to these proteins' action on leukocytes.

Iridine, named after the rainbow trout (*Oncorhynchus mykiss*), from Greek *iris*, rainbow.

Keratin, from Greek *keras, kerat-*, horn, referring to keratin's occurrence in hair and nails.

King Kong peptide, so called because this oligopeptide causes small crabs to abandon their usual submissive behavior towards larger members of the species, to erect their head parts and to curl their tails up like a scorpion. A.k.a. **conotoxin KK-0**.

Lantibiotics (a class name), referring to these proteins' content of lanthionine.

Lectin, from Latin *legere, lect-*, to gather, to select, referring to these proteins' recognition of carbohydrates.

Leghemoglobin, a contraction of leguminous hemoglobin. A.k.a. **legoglobin**.

Livetin, an anagram of vitellin.

Magainin, from Hebrew *maghen*, shield, referring to these polypeptides antimicrobial properties.

Myoglobin, from Greek *mys*, myo-, muscle, and Latin *globulus*, diminutive of *globus*, globe, referring to this protein's role in the oxygen supply of muscle.

Nebulin, from Latin *nebula*, mist cloud, referring to this protein's histological localization in the N_2 line of the sarcomere, a nebulous striation within the I band.

Neurophysin, a class name, from neur(o)- and Greek *hypophysis*, outgrowth.

Nociceptin, from Latin *nocere*, to harm, and *receptare*, to receive, referring to its anti-analgesic properties. A.k.a. **orphanin**, named as agonist of an 'orphan' opiate receptor-like G protein-coupled receptor.

Orexin, from Greek *orexis*, appetite, referring to its action on appetite. A.k.a. **hypocretin**, from hypothalamus and secretin, referring to its formation in the hypothalamus and its resemblance to secretin.

Palladin (a class name), after the Italian architect Andrea Palladio (Andrea Di Pietro della Gondola) (1508–1580), referring to the localization of these proteins to architectural elements of the cell.

Pepstatin, named as inhibitor of aspartyl proteases.

Perforin, named as pore forming cytolytic protein.

Phaseolin, after the species *Phaseolus vulgaris* (French bean). The genus name is from Latin *phaseolus*, diminutive of *phaselus*, bean, ultimately from Greek *phaselos*, cowpea (*Vigna unguiculata*).

Plant defensin, earlier called gamma-thionin, from defense.

Prolamine (a class name), a contraction of proline and glutamine.

Promine, a name referring to the cancer growth-promoting activity of this protein.

Protamine, a contraction of protein and amine, referring to the high content of arginine residues in these strongly basic proteins.

Properdin, from pro- and Latin *perdere*, to destroy, referring to this serum protein's destructive action on bacteria, neutralization of viruses, and lysis of red blood cells in the presence of complement and magnesium ions.

Protamine, from protein and amine, referring to the basic properties of these proteins due to the preponderance of basic amino acids.

Redoxin (a class name), from redox.

Resilin, from resilience, referring to the rubber-like elasticity of this protein.

Reticulin, named as a constituent of reticular tissue.

Rhodopsin, from Greek *rhodon*, rose, probably from Old Iranian **urda-*, rose, and *opsis*, vision, sight, named as visual purple.

Ricin, after the species *Ricinus communis* (castor plant). The genus name is from Latin *ricinus*, tick, because some markings on the plant resemble ticks.

Salmine, after the family *Salmonidae* (salmon).

Saporin, after the genus *Saponaria* (soapworts). The genus name is from Latin *saponarius*, of soap, from *sapo, sapon-*, soap.

Sarafotoxin (a class of snake venoms), from Hebrew *saraf*, a biblical poisonous snake, possibly the mole viper (*Atractaspis engaddensis*), from which these venoms were obtained. A.k.a. **SRTX.**

Scotophobin, from Greek *skotos*, dark, and *phobos*, fear, referring to this peptide's occurrence in the brain of rats with an acquired fear of the dark.

Scyliorhinine, named after dogfish (*Scyliorhinus*). The genus name is from Greek *skylion*, dogfish, and *rhis, rhin-*, nose.

Secalin (a prolamin glycoprotein), after the genus *Secale* (rye). The genus name is from Latin *secale*, rye.

Stelline A, named after starry sturgeon (*Acipenser stellatus*). The specific epithet is from Latin *stellatus*, starry, shiny, from *stella*, star.

Subtilin, after the bacterial species *Bacillus subtilis*. The specific epithet is from Latin *subtilis*, finely woven, fine, thin, subtle, from sub- and *tela*, web, akin to *texere, text-*, to weave.

Survivin, from survival, referring to its negative regulation of apoptosis. A.k.a. **BIRC5.**

Tensin, from Latin *tensio, tension-*, tension, from *tendere*, to tighten.

Tetranectin, from tetra- and Latin *nectere*, to bind, to tie, named as consisting of four identical noncovalently bound polypeptide chains.

Thaumatin, after the species *Thaumatococcus daniellii* (miracle berry, miracle fruit, katamfe, Yoruba soft cane, African serendipity berry). The genus name is from Greek *thauma*, miracle, wonder, and *kokkos*, grain, kernel, kermes berry.

Thinnine, named after the genus *Thunnus* (tunafish). The genus name is ultimately from Greek *thynnos*, tunny-fish, from *thynein*, to rush, to dart.

Thionin, named as cysteine-rich protein.

Thiopeptin, named as cysteine-rich protein.

Thioredoxin, from thi(o)- and redox. A.k.a. **TRX.**

Titin (a class name), after the Titans, Greek mythical giants, referring to the 'titanic' size of these proteins.

Tubulin, from tubule, referring to its role in mitosis. Tubule is from Latin *tubulus*, diminutive of *tubus*, tube.

Ubiquitin, from Latin *ubique*, everywhere, referring to this protein's universal occurrence.

Vitellin, from Latin *vitellus*, egg yolk.

Zein, from ze(a)-.

Reference

[1] https://en.wikipedia.org/wiki/Enzyme_Commission_number (retrieved August 2, 2018).

11 The naming of other natural products

Obviously, only the surface of this monumental subject can be scratched here, given the limitations of a book as the present one.

11.1 Vitamins [1] and cofactors [2]

Vitaminology, by now a mature science, began with a misnomer. In 1912 the Polish-American biochemist Casimir Funk (Kazimierz Funk) (1884–1967) coined the name 'vitamine' by contraction of 'vital amine' and clearly defined the nature and function of the vitamins known at that time. This, scientifically less correct, name was soon favored above another suggestion, 'accessory food factor', proposed the same year by the English biochemist Frederick Gowland Hopkins (1861–1947). When it became clear that not all vitamins were in fact amines, as assumed by Funk, the final e in vitamine was deleted, giving the term vitamin its present meaning.

Adenosine triphosphate, from aden(o)- and 'ose'. A.k.a. **ATP**.

Aneurin {2-[3-[(4-amino-2-methylpyrimidin-5-yl)methyl]-4-methyl-1,3-thiazol-3-ium-5-yl]ethanol, a cation, vitamin B_1}, with contraction, from a(n)- and polyneuropathy, referring to this vitamin's ability to prevent beri-beri. A.k.a. **thiamine**, from thi(o)- and amine.

Ascorbic acid (L-*threo*-hex-2-enono-1,4-lactone, vitamin C), from a(n)- and New Latin *scorbutus*, scurvy, ultimately from Old Norse *skyr*, curdled milk, and *bjugr*, edema, referring to the ancient belief that scurvy was caused by the sailors' diet of old curdled milk.

Biotin {5-[(3a*S*,4*S*,6a*R*)-2-oxohexahydro-1*H*-thieno[3,4-*d*]imidazol-4-yl]pentanoic acid, vitamin B_7}, from Greek *biotos*, life, sustenance, from *bios, biot-*, life. A.k.a. **vitamin H** and **coenzyme R**.

Cholecalciferol [(3*S*,5*Z*,7*E*)-9,10-secocholesta-5,7,10-trien-3-ol, vitamin D_3] from cholesterol and calciferous.

Cobalamin (vitamin B_{12}), with contraction, from cobalt and vitamin.

Cobamamide (coenzyme B_{12}), with contraction, from cobalt, vitamin, and amide.

Coenzyme A. The A stands for 'activation of acetate' or 'acetylation'.

Cyanocobalamin (vitamin B_{12}), with contraction, from cyan(o)-, cobalt, and vitamin.

Ergocalciferol [(3*S*,5*Z*,7*E*,20*R*,22*E*,24*R*)-9,10-secoergosta-5,7,10,22-tetraen-3-ol, vitamin D_2] from ergosterol and calciferous.

https://doi.org/10.1515/9783110612714-011

Folic acid {(2*S*)-2-[[[4-[(2-amino-4-oxo-1*H*-pteridin-6-yl)methyl]amino]benzoyl]amino] pentanedioic acid, vitamin B_9} from Latin *folium*, leaf.

Folinic acid, from folic acid, referring to its close relationship with folic acid. A.k.a. **citrovorum factor**, after the bacterial species *Leuconostoc citrovorum*, later renamed *Leuconostoc cremoris*. The former specific epithet is from citr(o)- and Latin *-vorus*, devouring, from *vorare*, to devour.

α-Lipoic acid {5-[(3*R*)-1,2-dithiolan-3-yl]pentanoic acid}, from lipid. A.k.a. **thioctic acid**, a contraction of 6,8-dithiooctanoic acid.

Menaquinone {2-methyl-3-[(2*E*,6*E*,10*E*,14*E*)-3,7,11,15,19-pentamethylicosa-2,6,10,14,18-pentaen-1-yl]naphthalene-1,4-dione, vitamin K_2,}, with contraction, from 2-methyl-1,4-naphthoquinone.

Niacin (pyridine-3-carboxylic acid, vitamin B_3], with contraction, from nicotinic acid. A.k.a. **nicotinic acid**, from nicotin.

Pantothenic acid {*N*-[(2*R*)-2,4-dihydroxy-3,3-dimethylbutanoyl]-β-alanine, vitamin B_5}, from Greek *pantothen*, from everywhere, referring to this compound's ubiquitous occurrence.

Phylloquinone {2-methyl-3-[(2*E*,7*R*,11*R*)-3,7,11,15-tetramethylhexadec-2-en-1-yl] naphthalene-1,4-dione, vitamin K_1} from Greek *phyllon*, leaf, and quinone.

Prephenic acid [*cis*-1-(2-carboxy-2-oxoethyl)-4-hydroxycyclohexa-2,5-diene-1-carboxy lic acid], from phene (benzene), referring to this acid's intermediacy in the biosynthesis of aromatic compounds.

Pyridoxal 5′-phosphate [(4-formyl-5-hydroxy-6-methylpyridin-3-yl)methyl phosphate, vitamin B_6], pyridoxal from pyridoxine and 'al'. A.k.a. **codecarboxylase**, from coenzyme and decarboxylase and **PLP**.

Pyridoxamine, [4-(aminomethyl)-5-(hydroxymethyl)-2-methylpyridin-3-ol, vitamin B_6], from pyridoxine and amine. A.k.a. **PM**.

Pyridoxamine 5′-phosphate {4-[(aminomethyl)-5-hydroxy-6-methylpyridin-3-yl] methyl dihydrogen phosphate, vitamin B_6}. A.k.a. **PMP**.

Pyridoxine [4,5-bis(hydroxymethyl-2-methylpyridin-2-ol, vitamin B_6], from pyridine and oxygen. A.k.a. **adermine**, a contraction of antidermatitis vitamin. A.k.a. **PN**.

Pyridoxine 5′-phosphate {5-hydroxy-4-(hydroxymethyl)-6-methylpyridin-3-yl]methyl dihydrogen phosphate, vitamin B_6}. A.k.a. **P5P**.

Retinoic acid [(2*E*,4*E*,6*E*,8*E*)-3,7-dimethyl-9-(2,6,6-trimethylcyclohex-1-en-1-yl)nona-2,4,6,8-tetraenoic acid, vitamin A acid], from retinol.

Retinol [(2*E*,4*E*,6*E*,8*E*)-3,7-dimethyl-9-(2,6,6-trimethylcyclohex-1-en-1-yl)-nona-2,4,6,8-tetraen-1-ol, vitamin A], from retina, from Latin *rete*, net.

Riboflavin {1-dexoxy-1-(7,8-dimethyl-2,4-dioxo-3,4-dihydrobenzo[*g*]pteridin-10(2*H*)-yl)-D-ribitol, vitamin B_2}, from ribitol and Latin *flavus*, light yellow, referring to its color. A.k.a. **lactoflavin**, from lact(o)- and flavin.

Riboflavin adenosine diphosphate (vitamin B_2), a.k.a. **FAD**.

Thiamine pyrophosphate (thiamine diphosphate, vitamin B_1). A.k.a. **cocarboxylase**, from coenzyme and carboxylase.

α-Tocopherol {(2*R*)-3,4-dihydro-2,5,7,8-tetramethyl-2-[(4*R*,8*R*)-4,8,12-trimethyltridecan-1-yl]-2*H*-1-benzopyran-6-ol, vitamin E}, from Greek *tokos*, childbirth, and *pherein*, to bear, referring to its importance for the development of fertilized mouse eggs.

Tocotrienol {2-methyl-2-[(3*E*,7*E*)-4,8,12-trimethyltrideca-3,7,11-trien-1-yl)-3,4-dihydro-2*H*-1-benzopyran-6-ol, vitamin E}, from tocopherol.

11.2 Nucleic acid components

Nucleic acid components have also received trivial names which have been retained in IUPAC and CAS nomenclature. It should be noted that cytosine is a nucleobase and not a ribonucleoside, as the name may suggest. As a consequence, its corresponding ribonucleoside, cytidine, lacks the usual 'osine' suffix.

Adenine (A, 9*H*-purin-6-amine), from Greek *aden*, gland.

Adenosine (9-β-D-ribofuranosyladenine), named as ribosyladenine.

Cytidine (C, 1-β-D-ribofuranosylcytosine), named as ribosylcytosine. As a consequence of the irregular name cytosine, the name cytidine becomes equally irregular.

Cytosine [4-aminopyrimidin-2(1*H*)-one], from Greek *kytos*, cell, hollow vessel, referring to this compound's first isolation from calf thymus tissues. Despite its name not a ribonucleoside.

Deoxyadenosine [dA, 9-β-D-(2-deoxyribofuranosyl)adenine)], named as deoxyribosyladenine.

Deoxycytidine [dC, 1-β-D-(2-deoxyribofuranosyl)cytosine], named as deoxyribosylcytidine.

Deoxyguanosine [dG, 9-β-D-(2-deoxyribofuranosyl)guanine], named as deoxyribosylguanine.

Deoxyuridine [dU, 1-β-D-(2-deoxyribofuranosyl)uracil], named as deoxyribosyluracil.

Guanine [2-amino-9*H*-purin-6(1*H*)-one], after guano. Guano is ultimately from Quechua *wanu*, guano.

Guanosine (G, 9-β-D-ribofuranosylguanine), named as ribosylguanine.

Hypoxanthine (1*H*-purin-6(9*H*)-one), named as biosynthetic precursor of xanthine.

Inosine (9-β-D-ribofuranosylhypoxanthine), from Greek *is, inos-*, sinew.

Ribothymidine (m^5U, T, 1-β-D-ribofuranosylthymine), named as ribosylthymidine. A.k.a. **5-methyluridine.**

Thymidine (T, dT, 2′-deoxy-5-methyluridine, thymine 2′-deoxyriboside), named as deoxyribosylthymine.

Thymine [5-methylpyrimidine-2,4(1*H*,3*H*)-dione], after its isolation from calf thymus tissue. Thymus gland is from Greek *thymos*, thymus gland, warty excrescence.

Uracil [pyrimidine-2,4(1*H*,3*H*)-dione], from urea and acrylic acid, referring to its synthesis from these starting materials.

Uridine (U, 1-β-D-ribofuranosyluracil), named as ribosyluracil.

Xanthine (3,7-dihydro-1*H*-purine-2,6-dione), from xanthosine.

Xanthosine (9-β-D-ribofuranosylxanthine), from Greek *xanthos*, yellow.

11.3 Hormones

The class name hormone is from Greek *hormon*, that which sets in motion, from *horman*, to impel, to urge on, from *horme*, impulse, assault, akin to *ornynai*, to rouse.

Abscisic acid {(2*Z*,4*E*)-5-[(1*S*)-1-hydroxy-2,6,6-trimethyl-4-oxocyclohex-2-en-1-yl]-3-methylpenta-2,4-dienoic acid}, after Latin *abscissio- abscission-*, abscission, referring to this compound's perceived role in plant abscission. A.k.a. **ABA.**

Adrenaline {4-[(1*R*)-1-hydroxy-2-(methylamino)ethyl]benzene-1,2-diol}, from Latin *ad*, to, at, and *ren*, kidney, referring to the adrenal gland. A.k.a. **epinephrine**, from epi- and *nephron*, kidney.

Aldosterone (11β,21-dihydroxy-3,20-dioxopregn-4-en-18-al), from aldehyde, ketone, and steroid.

Amylin (a polypeptide), a contraction of islet amyloid polypeptide.

Androgen, from Greek *aner, andr-*, man, a class name.

Androsterone (3α-hydroxy-5α-androstan-17-one), from androgen and steroid.

Auxin [2-(1*H*-indol-3-yl)acetic acid], from Greek *auxein*, to increase, referring to this hormone's regulation of growth processes in plants.

Bradykinin (an oligopeptide), from Greek *bradys*, slow, heavy, and kinin. The name refers to its blood pressure lowering effect.

Calcitonin (a polypeptide), from calci- and -tonin, referring to this hormone's lowering of the calcium level in the blood plasma.

Cholecystokinin (CCK, polypeptide), from chol(e)-, cyst(o)-, and kinin.

Corticoliberin (CRF, CRH, corticotropin-releasing hormone), from corticotropin and Latin *liberare*, to set free.

Cortisol (11β,17α,21-trihydroxypregn-4-ene-3,20-dione), from Latin *cortex, cortic-*, bark, referring to this hormone's biosynthesis in the adrenal cortex.

Cortisone (17a,21-dihydroxypregn-4-ene-3,11,20-trione), from Latin *cortex, cortic-*, bark, referring to this hormone's biosynthesis in the adrenal cortex. A.k.a. **Kendall's compound F**, after the American chemist Edward Calvin Kendall (1886–1972), **Reichstein's substance F_α**, after the Polish-Swiss chemist Tadeusz Reichstein (1897–1996) and **Wintersteiner's compound F**, after the Austrian-American chemist Oskar Wintersteiner (1898–1971).

Cytokinin, from Greek *kytos*, cell, hollow vessel, and *kinein*, to move, to stimulate, to set in motion, a class name.

Ecdysone {(2*S*,3*R*,5*R*,9*R*,10*R*,13*R*,14*S*,17*R*)-17-[(2*S*,3*R*)-3,6-dihydroxy-6-methylheptan-2-yl]-2,3,14-trihydroxy-10,13-dimethyl-2,3,4,5,9,11,12,15,16,17-decahydro-1*H*-cyclopenta[*a*]phenanthren-6-one}, named as insect molting hormone, from ecdysis, from Greek *ekdysis*, act of getting out.

Elcatonin (a synthetic modification of calcitonin), an alteration of calcitonin.

Estradiol, from estrogen, with di- and 'ol'.

Estrogen, from Greek *oistros*, gadfly, frenzy, a class name.

Ethylene (ethene), not specifically named as a plant hormone.

Florigen (a hypothetical plant hormone), from Latin *flos, flor-*, flower, and -gen.

Follitropin, from follicle, from Latin *folliculus*, diminutive of *follis*, leather bag, and -tropin.

Gastrin (a polypeptide), from Greek *gaster, gastr-*, stomach, ultimately from *gran*, to gnaw, to eat.

Gibberellin (a group of plant hormones), after the fungal genus *Gibberella*, from Latin *giberella*, diminutive of *gibber*, hump on the back.

Ghrelin (a polypeptide), a contraction of growth hormone release inducing (hormone).

Glucagon (a polypeptide), from glucose, and Greek *agon*, leading, driving, from *agein*, to lead, to drive.

Hippulin [3-hydroxyestra-1,3,5(19),8-tetraen-17-one], from Greek *hippos*, horse, referring to its isolation from the urine of pregnant mares, and patterned after folliculin.

Insulin (a polypeptide), after New Latin *insulae Langerhans* (pancreatic islets of Langerhans), after the German pathologist, physiologist, and biologist Paul Langerhans (1847–1888).

Juvenile hormone I {methyl (2*E*,6*E*)-7-ethyl-9-[(2*R*,3*S*)-3-ethyl-3-methyloxiran-2-yl]-3-methylnona-2,6-dienoate}.

Kairomone (a class name), from Greek *kairos*, right time, right measure, right place, and hormone, referring to the fact that these hormones are beneficial to species other than those producing them.

Leptin (a polypeptide), from Greek *leptos*, slender, referring to this hormone's appetite suppressing action.

Lipotropin (a polypeptide), from lip(o)- and -tropin, referring to its perceived lipid-mobilizing action.

Luliberin (a polypeptide), a contraction of lutropin and liberin.

Lutropin (a glycoprotein), from New Latin *corpus luteum*, literally yellow body, and -tropin. A.k.a. **luteinizing hormone**.

Mammotropin (a polypeptide), from Latin *mamma*, female breast, and -tropin. A.k.a. **prolactin**, from pro- and lactation, **PRL**, and **luteotropin**.

Medullipin (of unknown structure), from medullary lipid.

Melanotropin (a group of peptides), from melanin and -tropin. A.k.a. **intermedin**, a contraction of New Latin *pars intermedia*, intermediate part.

Melatonin {*N*-[5-methoxy-1*H*-indol-3-yl)ethyl]acetamide}, from mela(no)- and -tonin, referring to its being secreted by the pineal gland in response to darkness.

Motilin (a polypeptide), from motility, referring to this hormone's stimulating effect on intestinal motility.

Oxytocin (a polypeptide), from Greek *oxys*, sharp, quick, and *toketos*, birth, i.e. quick birth.

Parotin (a polypeptide), after the parotid gland, from Greek *parotis*, near the ear, from para- and *ous, ot-*, ear.

Progesterone (pregn-4-en-3,20-dione), from pro-, Latin *gestatio, gestation-*, pregnancy, and with 'one'.

Prostaglandin (a class name), after the prostate gland, ultimately from Greek *proistanai*, to put in front, and from Latin *glans, gland-*, acorn.

Relaxin, from Latin *relaxare*, to relax, from *laxus*, loose, referring to this hormone's relaxing action on the symphysis and the cervix.

Somatoliberin, from somat(o)- and Latin *liberare*, to set free. A.k.a. **GRF**, gonadotropin-releasing hormone, and **somatocrinin**, from somat(o)- and Greek *krinein*, to separate, to distinguish, to secrete.

Somatotropin, from somat(o)- and -tropin. A.k.a. **GH**, growth hormone, and **phyone**.

Somatostatin, from somat(o)- and statin. A.k.a. **GH-RIF**.

Secretin, from secretion, referring to its osmoregulatory activity.

Testosterone (17β-hydroxyandrost-4-en-3-one), from Latin *testis*, testicle, steroid, and with 'one'.

Thyrotropin (a glycoprotein), a.k.a. **thyrotrophin, thyroid-stimulating hormone**, and **TSH**.

Thyrotropine-realeasing hormone (Glp-His-Pro-NH_2), a.k.a. **TRH** and **protitrelin**, from prolactin, thyrotropin, and release.

L-Thyroxine [Thx, *O*-(4-hydroxy-3,5-diiodophenyl)-3,5-diiodo-L-tyrosine], from thyroid gland. The name of the gland is ultimately from Greek *thyreoeides*, shield-shaped, from *thyreos*, shield shaped like a door, from *thyra*, door.

Vasopressin (a polypeptide), originally a trademark, from Latin *vas*, vessel, and pressure, referring to its effect of increased blood pressure. A.k.a. **adiuretin**, ^{2}**ADH, antidiuretic hormone,** referring to its effect of reduced urine flow, and **antidiuretin**.

References

[1] *https://en.wikipedia.org/wiki/Vitamin*
[2] http://www.ebi.ac.uk/thornton-srv/databases/CoFactor/ (retrieved August 26, 2018).

12 Named reactions and named concepts

The following examples illustrate the practice of naming reactions and concepts. Extensive listings of named and unnamed reactions can be found online [1–5].

Eponyms can be subject to Stigler's law of eponymy [6, 7] which states that no scientific discovery is named after its original discoverer.

Abeggs's rule, after the German chemist Richard Abegg (1869–1910).

Abel-Pensky method, after the British chemist Frederick Augustus Abel (1827–1902) and the German engineer Berthold Pensky (1850–1930).

Abramov reaction, after the Russian chemist Vasiliy Abramov (1904–1968).

Abramovitch-Shapiro tryptamine synthesis, after the American-Egyptian chemist Rudolph A. Abramovitch (1930–2013) and the American chemist D. Shapiro.

Acheson process, after the American inventor Edward Goodrich Acheson (1856–1931).

Achmatowicz reaction, after the Polish chemist Osman Achmatowicz Jr. (born 1931).

Adams decarboxylation, after the American chemist Roger Adams (1889–1971).

Adkins-Peterson reaction, after the American chemists Homer Burton Adkins (1892–1949) and Wesley R. Peterson.

Akabori amino acid reactions, after the Japanese chemist Shiro Akabori (1900–1992).

Alder ene reaction, after the German chemist Kurt Alder (1902–1958).

Alder rule, as Alder ene reaction.

Allan-Robinson reaction, after the British chemists James Allan and Robert Robinson (1885–1975).

Allmann-Waugh anion structure, after the German crystallographer Rudolf Allmann (born 1931) and the British chemist John Ludovick Thomson Waugh (1922–1996).

Amadori rearrangement, after the Italian chemist Mario Amadori (1886–1941).

Amontons' law, after the French physicist Guillaume Amontons (1663–1705). A.k.a. **Charles' law**, after the French inventor, scientist, mathematician, and balloonist Jacques Alexandre César Charles (1746–1823).

Anderson-Evans ion structure, after the British-Australian chemist John Stuart Anderson (1908–1990) and the American chemist Howard Tasker Evans, Jr. (1920–2000).

Andrussow process, after the Latvian-German chemical engineer Leonid Andrussow (1896–1988).

https://doi.org/10.1515/9783110612714-012

Angeli-Rimini reaction, after the Italian chemists Angelo Angeli (1864–1931) and Enrico Rimini (1874–1917).

Ångström (Å), a unit named for the Swedish physicist Anders Jonas Ångström (1814–1874).

Arbuzov-Michaelis reaction, after the Russian chemist Aleksandr Ermingeldovich Arbuzov (1877–1968) and the German chemist August Karl Arnold Michaelis (1847–1916). A.k.a. **Michaelis-Arbuzov reaction.**

Arens-van Dorp synthesis, after the Dutch chemists Josef Ferdinand Arens (1914–2001) and David Adriaan van Dorp (1915–1995).

van Arkel and de Boer process, after the Dutch chemists Anton Eduard van Arkel (1893–1976) and Jan Hendrik de Boer (1899–1971).

Arndt-Eistert synthesis, after the German chemists Fritz Arndt (1885–1969) and Bernd Eistert (1902–1978).

Arrhenius equation, after the Swedish chemist Svante August Arrhenius (1859–1927).

Auger electron appearance spectroscopy, cf. AEAPS. The Auger effect has been named after its rediscoverer, the French physicist Pierre Victor Auger (1899–1993). The original discovery was made by the Austrian-Swedish physicist Lise Meitner (1878–1968).

Auger electron spectroscopy, cf. AES, as Auger electron appearance spectroscopy and after the French physicist Pierre Victor Auger (1899–1993).

Auwers synthesis, after the German chemist Karl Friedrich von Auwers (1863–1939).

Auwers-Skita rule, as Auwers synthesis, and after the Austrian-German chemist Aladar Skita (1876–1953).

Avogadro's law, after the Italian chemist Lorenzo Amedeo Carlo Avogadro Conte de Quaregno e Cerreto (1776–1856).

Avogadro's number, as Avogadro's law. A.k.a. **Loschmidt's number**, after the Austrian high school chemistry teacher Johann Josef Loschmidt (1821–1895).

Babo's law, after the German chemist Lambert Heinrich Joseph Anton Konrad Freiherr von Babo (1818–1899).

Badger rules, after the American chemist Richard McLean Badger (1896–1974).

Baeyer-Drewson indigo synthesis, after the German chemist Johann Friedrich Wilhelm Adolf von Baeyer (1835–1917) and V. Drewson.

Baeyer strain, after the German chemist Johann Friedrich Wilhelm Adolf von Baeyer (1835–1917). Cf. Pitzer strain and Prelog strain.

Baeyer-Villiger oxidation, as Baeyer strain and after the Swiss chemist Victor Villiger (1868–1934).

Baizer synthesis, after the American chemist Manuel Mannheim Baizer (1914–1988).

Baker-Nathan effect, after the British chemists John William Baker (1898–1967) and Wilfred Samuel Nathan (1910–1961).

Baker-Venkatamaran rearrangement, after the British chemist Wilson Baker (1900–2002) and the Indian chemist Krishnaswami Venkatamaran (1901–1981).

Baljet reaction, after the Dutch-Swiss pharmacist Henri Baljet (1887–1961).

Balz-Schiemann reaction, after the German chemists Günther Balz (1902–1988) and Günther Robert Arthur Schiemann (1899–1969). A.k.a. **Schiemann reaction.**

Bamberger rearrangement, after the German chemist Eugen Bamberger (1857–1932).

Bamford-Stevens reaction, after the British chemists William Randall Bamford and Thomas Stevens Stevens (1900–2000).

Barbier reactions, after the French chemist Philippe Antoine Barbier (1848–1922).

Barbier-Wieland reaction, as Barbier reactions and after the German chemist Heinrich Otto Wieland (1877–1957).

Barfoed's test, after the Danish chemist Christen Thomsen Barfoed (1815–1889).

Bartlett-Condon-Schneider reaction, after the American chemists Paul Doughty Bartlett (1907–1997), Francis Edward Condon (born 1919), and Abraham Schneider (1919–1997).

Barton reaction, after the British-American chemist Derek Harold Richard Barton (1918–1998).

Barton-Kellogg reaction, after the British-American chemist Derek Harold Richard Barton (1918–1998) and the American-Dutch chemist Richard Morrison Kellogg (born 1939). A.k.a. **Staudinger-type diazo-thioketone coupling**, after the German chemist Hermann Staudinger (1881–1965).

Barton-McCombie reaction, as Barton reaction and after the British chemist Stuart Walter McCombie (born 1947).

Barton-Zard reaction, as Barton reaction and after the Nigerian-French chemist Samir Z. Zard (born 1955).

BASF process, after the chemical company Badische Anilin- & Soda-Fabrik AG, Ludwigshafen, Germany.

Bashkirov oxidation, after the Russian chemist Andrei Nikolayevich Bashkirov (1903–1982).

Basolo rule, after the American chemist Fred Basolo (1920–2007).

Baudisch reaction, after the Austrian-American chemist Oskar Baudisch (1881–1950).

Baumé scale, after the French pharmacist Antoine Baumé (1728–1804).

Bayer process, after the German chemist Karl Josef Bayer (1847–1904).

Baylis-Hillman reaction, after the British-American chemist Anthony Basil Baylis (born 1937) and the American chemist Melville Ernest Douglas Hillman (1926–1994). A.k.a. **Morita-Baylis-Hillman reaction.**

BCS theory, after the American physicists John Bardeen (1908–1991), Leon Neil Cooper (born 1930), and John Robert Schrieffer (born 1931).

Béchamp reduction, after the French chemist Pierre Jacques Antoine Béchamp (1816–1908).

Beckmann fragmentation, after the German chemist Ernst Otto Beckmann (1853–1923).

Beilstein test, after the German-Russian chemist Friedrich Konrad Beilstein (1838–1906).

Bell's rule, after the British chemist Ronald Percy Bell (1907–1996).

Belousov-Zhabotinsky reaction, after the Russian chemist Boris Pavlovich Belousov (1893–1970) and the Russian biophysicist Anatol Markovich Zhabotinsky (1938–2008). A.k.a. **Briggs-Rauscher reaction.**

Bénary reaction, after the German chemist Erich Otto Max Bénary (1881–1941).

Benkeser reduction, after the American chemist Robert Anthony Benkeser (1920–2017).

Benson's additivity rule. after the American chemist Sidney William Benson (1918–2011).

Bergius process, after the German chemist Friedrich Bergius (1884–1949). A.k.a. **Bergius-Pier process.**

Bergman cyclization, after the American chemist Robert George Bergman (born 1942).

Bergmann azlactone peptide synthesis, after the German-American chemist Max Bergmann (1886–1944).

Bergmann-Zervas carbobenzoxy method, as Bergmann azlactone peptide synthesis and after the Greek chemist Leonidas Zervas (1902–1980).

Bernthsen acridine synthesis, after the German chemist August Bernthsen (1855–1931).

Berry pseudorotation, after the American chemist Richard Steven Berry (born 1931).
Berthelot reaction, after the French chemist Marcellin Pierre Eugène Berthelot (1827–1907).

Berzelius test, after the Swedish chemist Jöns Jacob Berzelius (1779–1848).

Bessemer process, after the British inventor Henry Bessemer (1813–1898), earlier also developed by the American inventor William Kelly (1811–1888) and others.

BET method, after the American chemists Stephen Brunauer (1903–1986) and Paul Hugh Emmett (1900–1985), and the Hungarian-American physicist Edward Teller (Teller Ede) (1908–2003).

Biginelli reaction, after the Italian chemist Pietro Biginelli (1860–1937).

Billiter process, after the Austrian-French chemist Jean Billiter (1877–1965).

Bingham fluid, after the American chemist Eugene Cook Bingham (1878–1945).

Biot's law, after the French mathematician and physicist Jean-Baptiste Biot (1774–1862).

Birch reduction, after the Australian chemist Arthur John Birch (1915–1995).

Birkeland-Eyde process, after the Norwegian chemist Kristian Olaf Bernhard Birkeland (1867–1917) and the Norwegian engineer Samuel Eyde (1866–1940).

Bischler-Möhlau indole synthesis, after the Russian-German-Swiss chemist August Bischler (1865–1957) and the German chemist Richard Bernhard Julius Möhlau (1857–1940).

Bischler-Napieralski reaction, after the Russian-German-Swiss chemist August Bischler (1865–1957) and the Polish chemist Bernard Napieralski (born 1861).

Blaise ketone synthesis, after the French chemist Edmond Emile Blaise (1872–1939). A.k.a. **Blaise-Maire reaction**.

Blanc reaction, after the French chemist Gustave Louis Blanc (1872–1927).

Bodroux reaction, after the French chemist Fernand Bodroux (1873–1968).

Bodroux-Chichibabin aldehyde synthesis, as Bodroux reaction and after the Russian chemist Aleksei Yevgenievich Chichibabin (1871–1945).

Boger pyridine synthesis, after the American chemist Dale Lester Boger (born 1953).

Bogert-Cook synthesis, after the American chemists Marston Taylor Bogert (1868–1954) and J. W. Cook.

Bohn-Schmidt reaction, after the German chemists René Bohn (1862–1922) and Robert Emanuel Schmidt (1864–1938).

Boltzmann's constant, after the Austrian physicist Ludwig Eduard Boltzmann (1844–1906).

Boord olefin synthesis, after the American chemist Cecil Ernest Boord (1884–1969).

Borch reduction, after the American chemist Richard Frederic Borch (born 1941).

Born-Haber cycle, after the German physicist Max Born (1882–1970) and the German chemist Fritz Haber (1868–1934).

Borodin reaction, after the Georgian-Russian chemist and musical composer Aleksandr Porfieryevich Borodin (1833–1887). A.k.a. **Hunsdiecker reaction.**

Borsche-Drechsel cyclization, after the German chemists Walter Borsche (1877–1950) and Edmund Drechsel (1843–1897). Drechsel discovered this reaction in 1888 and Borsche rediscovered it in 1908.

Boudouard equilibrium, after the French chemist Octave Leopold Boudouard (1872–1923).

Boughton system for the naming of isotopically labeled molecules, after the American chemist W. A. Boughton (1885–1977).

Bouveault aldehyde synthesis, after the French chemist Louis Bouveault (1864–1909).

Bouveault-Blanc reduction, as Bouveault aldehyde synthesis and after the French chemist Gustave Louis Blanc (1872–1927).

Boyland-Sims oxidation, after the British chemists Eric Boyland (1905–2002) and Peter Sims (1919–1983).

Boyle's law, after the Anglo-Irish chemist Robert Boyle (1627–1691).

Bradsher cyclization, after the American chemist Charles Kilgo Bradsher (1912–2000).

von Braun reaction, after the German chemist Julius von Braun (1875–1939).

Bray-Liebhafsky reaction, after the American chemist William Crowell Bray (1879–1946) and the Hungarian-American chemist Herman A. Liebhafsky (1905–1982).

Bredt's rule, named after the German chemist Konrad Julius Bredt (1855–1937).

Briggs-Rauscher reaction, after the American chemists Thomas S. Briggs and Warren Carlton Rauscher (born 1939). A.k.a. **Belousov-Zhabotinsky reaction**.

Brin process, after the French-British chemists and industrialists, the twin brothers Leon Quentin Brin (1840–1890) and Emile Auguste Arthur Brin (1840–1894).

Brønsted-Bjerrum equation, after the Danish chemists Johannes Nicolaus Brønsted (1879–1947) and Niels Janniksen Bjerrum (1879–1958).

Brook rearrangement, after the Candian chemist Adrian Gibbs Brook (1924–2013).

Bucherer reaction, after the German chemist Hans Theodor Bucherer (1869–1949).

Buchner method of ring enlargement, after the German chemist Eduard Buchner (1860–1917).

Buchner-Curtius-Schlotterbeck reaction, after the German chemists Eduard Buchner (1860–1917), Theodor Curtius (1857–1928), and Fritz Schlotterbeck (1876–1940). Buchner and Curtius discovered this reaction in 1885, Schlotterbeck rediscovered it in 1907. The same reaction was also described in 1885 by the German chemist Hans von Pechmann (1850–1902).

Buchwald-Hartwig amination, after the American chemists Stephen L. Buchwald (born 1955) and John F. Hartwig (born 1964).

Bunnett-Olsen equation, after the American chemists Joseph Frederik Bunnett (1921–2015) and Frederic Phillip Olsen (born 1937).

Bunsen-Roscoe law, after the German chemist Robert Wilhelm Eberhard Bunsen (1811–1899) and the British chemist Henry Enfield Roscoe (1833–1915).

Cadiot-Chodkiewicz coupling, after the French chemist Paul Cadiot (born 1923) and the Polish-French chemist Wladimir Chodkiewicz (1921–2017).

Calutron process, a contraction of California University Cyclotron process.

Carius method, after the German chemist Georg Ludwig Carius (1829–1875).

Carnot's cycle, after the French physicist Nicolas Leonard Sadi Carnot (1796–1832).

Carr-Price reaction, after the British chemist Francis Howard Carr (1874–1969) and the British biochemist Ernest Arthur Price (born 1882).

Casale process, after the Italian chemist Luigi Casale (1882–1927).

Castner process, after the American chemist Hamilton Young Castner (1858–1899).

Castner-Kellner process, after the American chemist Hamilton Young Castner (1858–1899) and the Austrian chemist Carl Kellner (1851–1905).

Castro-Stephens coupling, often misspelled as Castro-Stevens coupling, after the American chemists Charles Edward Castro (1931–2007) and Robert D. Stephens.

Celsius temperature scale, after the Swedish astronomer and natural scientist Anders Celsius (1701–1744).

Chapman mechanism, after the British mathematician and physicist Sydney Chapman (1888–1970).

Chapman rearrangement, after the British chemist Arthur William Chapman (1898–1980).

Charles' law, after the French physicist Jacques Alexandre César Charles (1746–1823). A.k.a. **Amontons' law**, after the French inventor and physicist Guillaume Amontons (1663–1705).

Chevrel phase, after the French chemist R. Chevrel (born 1944).

Chichibabin pyridine synthesis, after the Russian-French chemist Aleksei Yevgenievich Chichibabin (1871–1945).

Chugaev reaction, after the Russian chemist Lev Aleksandrovich Chugaev (1873–1922).

CIP, Cahn-Ingold-Prelog priority rule, created by the British chemists Robert Sidney Cahn (1899–1981) and Christopher Kelk Ingold (1893–1970), and the Croatian-Swiss chemist Vladimir Prelog (1906–1998).

Claisen condensation, after the German chemist Ludwig Claisen (1851–1930).

Claisen-Ireland rearrangement, as Claisen condensation and after the American chemist Robert Ellsworth Ireland (1929–2012).

Claus process, after the German-British chemist Carl Friedrich Claus (1827–1900).

Clausius-Clapeyron equation, after the German physicist Rudolf Julius Emanuel Clausius (1822–1888) and the French engineer Bénoit Paul Emile Clapeyron (1799–1864).

Clausius cycle, after the German physicist Rudolf Julius Emanuel Clausius (1822–1888).

Cleland rules, after the American chemist William Wallace Cleland (1930–2013).

Clemmensen reduction, after the Danish-American chemist Erik Christian Clemmensen (1876–1941).

Collins oxidation, after the American chemist Joseph Charles Collins, Jr. (born 1931).

Conia reaction, after the French chemist Jean-Marie Conia (1921–1995).

Conrad-Limpach cyclization, after the German chemists M. Conrad and Leonhard Limpach (1853–1933).

Cooper pair, after the American physicist Leon Neil Cooper (born 1930).

Cope elimination, after the American chemist Arthur Clay Cope (1909–1966).

Corey-Kim oxidation, after the American chemist Elias James Corey (born 1928) and the Korean-Japanese-American chemist Choung Un Kim (born 1942).

Corey-Pauling-Koltun coloring of molecular models, cf. C-P-K, created by the American biochemist Robert Brainard Corey (1897–1971), the American chemist Linus Carl Pauling (1901–1994), and the American biochemist Walter Lang Koltun (1929–2010).

Corey-Winter olefin synthesis, after the American chemist Elias James Corey (born 1928) and the Estonian-American chemist Roland Arthur Edwin Winter (born 1935).

Cornforth rearrangement, after the Australian-British chemist John Warcup Cornforth (1917–2013).

Cotton effect, after the French physicist Aimé Auguste Cotton (1869–1951).

Cotton-Mouton effect, as Cotton effect and after the French physicist Henri Mouton (1869–1935).

Cottrell process, after the American chemist Frederick Gardner Cottrell (1877–1948).

Coulomb, a unit named after the French physicist Charles Augustin de Coulomb (1736–1806).

Coulometry, as coulomb.

Couper-Brown-Butlerov structure, after the British chemists Archibald Scott Couper (1831–1892) and Alexander Crum Brown (1838–1922), and the Russian chemist Aleksandr Mikhailovich Butlerov (1828–1886).

Craig method, after the American chemist Lyman Creighton Craig (1906–1974).

Cram's rule, after the American chemist Donald James Cram (1919–2001).

Criegee reaction, after the German chemist Rudolf Criegee (1902–1975).

Curie point, after its rediscoverer, the French physicist Pierre Curie (1859–1906). The original discovery was made by the French physicist Claude Pouillet (1790–1868) [1].

Curtius rearrangement, after the German chemist Theodor Curtius (1857–1928).

Czochralski process, after the Polish metallurgist Jan Czochralski (1885–1953).

Dakin oxidation, after the British chemist Henry Drysdale Dakin (1880–1952).

Dakin-West reaction, as Dakin oxidation and after the American chemist Randolf West (1890–1949).

Dalton's law, after the British meteorologist and chemist John Dalton (1766–1844).

Danishefsky reaction, after the American chemist Samuel Joseph Danishefsky (born 1936).

Darzens condensation, after the Russian-French chemist George Darzens (1867–1954).

Dawson anion structure, after the British-Australian chemist Barrie Dawson (1925–1974).

Debye-Clausius-Mossotti equation, after the Dutch-American physicist Peter Joseph Wilhelm Debye (1884–1966), the German physicist Rudolf Julius Emanuel Clausius (1822–1888), and the Italian physicist Ottaviano-Fabrizio Mossotti (1791–1863).

Debye-Hückel law, after the Dutch-American physicist Peter Joseph Wilhelm Debye (1884–1966) and the German chemist and physicist Erich Armand Arthur Joseph Hückel (1896–1980).

Debye-Scherrer method, after the Dutch-American physicist Peter Joseph Wilhelm Debye (1884–1966) and the Swiss physicist Paul Scherrer (1890–1969).

Degussa process, after the chemical company Degussa AG (short for Deutsche Gold- und Silber-Scheide-Anstalt AG), Frankfurt am Main, Germany.

Delépine reaction, after the French chemist Stéphane Marcel Delépine (1871–1965).

de Mayo reaction, after the British-Canadian chemist Paul José de Mayo (1924–1994).

Demyanov rearrangement, after the Russian chemist Nikolai Jakovlevich Demyanov (1861–1938).

Dess-Martin oxidation, after the American chemists Daniel Benjamin Dess (born 1955) and James Cullen Martin (1928–1999).

Dexter-Silverton anion structure, after the American chemists David Dyer Dexter (1940–2016) and J. V. Silverton.

Diamond process, after the chemical company Diamond Alkali Company, Pittsburgh, PA, USA.

Dieckmann condensation, after the German chemist Walter Dieckmann (1869–1925).

Diels-Alder reaction, after the German chemists Otto Paul Hermann Diels (1876–1954) and Kurt Alder (1902–1958).

Dimroth rearrangement, after the German chemist Otto Dimroth (1872–1940).

DLVO theory, after the Russian chemist Boris Vladimirovich Deryagin (1902–1994), the Russian physicist Lev Davidovich Landau (1908–1968), and the Dutch chemists Evert Johannes Willem Verwey (1905–1981) and Jan Theodoor Gerard Overbeek (1911–2007).

Dobson unit, DU, a unit named after the British physicist Gordon Miller Bourne Dobson (1889–1975).

Doebner-Miller reaction, after the German chemists Oscar Döbner (1850–1907) and Wilhelm von Miller (1848–1899).

Doering-LaFlamme allene synthesis, after the American chemists William von Eggers Doering (1917–2011) and Paul M. LaFlamme (1927–1990).

Donnan equilibrium, after the British chemist Frederick George Donnan (1870–1956).

Dowd-Beckwith ring expansion reaction, after the American chemist Paul Dowd (1936–1996) and the Australian chemist Athelstan Laurence Johnson Beckwith (1930–2010).

Duff reaction, after the British chemist James Cooper Duff (1888–1971).

Duhem-Margules equation, after the French physicist Pierre Maurice Marie Duhem (1861–1916) and the Austrian physicist and meteorologist Max Margules (1856–1920).

Dulong-Petit law, after the French chemists Pierre Louis Dulong (1785–1838) and Alexis Thérèse Petit (1791–1820).

Dumas method, after the French chemist Jean Baptiste André Dumas (1800–1884).

Dutt-Wormall reaction, after the Indian chemist Pavitra Kumar Dutt and and the British chemist Arthur Wormall (1900–1964).

Edeleanu process, after the Romanian chemist Lazăr Edeleanu (1861–1941).

Edman degradation, after the Swedish biochemist Pehr Edman (1916–1977).

Eglinton reaction, after the British chemist Geoffrey Eglinton (1927–2016).

Ehrlich-Sachs reaction, after the German physician and bacteriologist Paul Ehrlich (1854–1915) and the German chemist Franz Sachs (1877–1919).

Einhorn-Brunner reaction, after the German chemist Alfred Einhorn (1856–1917) and the Austrian chemist Karl Brunner (1855–1935).

Elbs anthracene synthesis, after the German chemist Karl Elbs (1858–1933).

Ellis carbonylate synthesis, after the American chemist John Emmett Ellis (born 1943).

El-Sayed's rule, after the American-Egyptian chemical physicist Mostafa Amr El-Sayed (born 1933).

Embden-Meyerhof-Parnas path, after the German chemists Gustav Embden (1874–1933) and Otto Fritz Meyerhof (1884–1951), and the Polish-Russian chemist Jakub Karol Parnas (1884–1949). A.k.a. **EMP**.

Emde reaction, after the German chemist Hermann Karl Christian Maximilian Emde (1880–1935).

Emmert reaction, after the German chemist Bruno Emmert (1880–1951).

Entner-Doudoroff pathway, after the American biochemists Nathan Entner (1920–1984) and Michael Doudoroff (1911–1975).

Epton titration, after the British chemist Sidney Robert Epton (1895–1955).

Erlenmeyer-Plöchl azlactone and amino acid synthesis, after the German chemists Richard August Karl Emil Erlenmeyer (1825–1909) and Josef Plöchl (1853–1923).

Ernst angle, after the Swiss chemist Richard Robert Ernst (born 1933).

Eschenmoser coupling reaction, after the Swiss chemist Albert Eschenmoser (born 1925).

Eschenmoser-Tanabe fragmentation, as Eschenmoser coupling reaction and after the American chemist Masato Tanabe (born 1925). A.k.a. **Eschenmoser-Ohloff fragmentation**, as Eschenmoser coupling reaction and after the German-Swiss chemist Günther Ohloff (1924–2005).

Eschweiler-Clarke reaction, after the German chemist Wilhelm Eschweiler (1860–1936) and the British chemist Hans Thatcher Clarke (1888–1972).

Etard reaction, after the French chemist Alexandre Léon Etard (1852–1910).

Evans aldol reaction, after the American chemist David A. Evans (born 1941).

Evans-Polanyi reaction, after the British chemist Meredith Gwynne Evans (1904–1952) and the Hungarian-British physicist Michael Polanyi (Polânyi Mihâly) (1891–1976).

Evans principle, after the British chemist Meredith Gwynne Evans (1904–1952).

Evans-Saksena reduction, after the American chemist David A. Evans (born 1941) and the Indian chemist Anil Kumar Saksena.

Evans-Tishchenko reaction, after the American chemist David A. Evans (born 1941) and the Russian chemist Vyacheslav Yevgenyevich Tishchenko (1861–1941).

Ewens-Bassett system, after the British chemists Ronald Victor George Ewens (1913–1948) and Henry Bassett (1881–1965).

Eyring equation, after the American-Mexican chemist Henry Eyring (1901–1981).

Farad, a unit named after the British chemist Michael Faraday (1791–1867).

Faraday, as farad.

Faraday constant, as farad.

Fauser process, after the Italian engineer Giacomo Fauser (1892–1871).

Favorsky-Babayan synthesis, after the Russian chemist Aleksei Yefgrafovich Favorsky (1860–1945) and the Armenian chemist Araxie Babayan (1906–1993).

Favorsky rearrangement, after the Russian chemist Aleksei Yefgrafovich Favorsky (1860–1945).

Feist-Bénary synthesis, after the German chemists Franz Feist (1864–1941) and Erich Otto Max Bénary (1881–1941).

Fenton reaction, after the British chemist Henry John Horstman Fenton (1854–1929).

Fieser-Woodward rules, after the American chemists Louis Frederick Fieser (1899–1977), Mary Peters Fieser (1909–1997), and Robert Burns Woodward (1917–1979).

Fischer esterification, after the German chemist Emil Hermann Fischer (1852–1919).

Fischer-Hafner synthesis, after the German chemists Ernst Otto Fischer (1918–2007) and Walter Hafner (1927–2004).

Fischer-Hepp rearrangement, after the German chemists Philipp Otto Fischer (1852–1932) and Eduard Hepp (1851–1917).

Fischer-Speier esterification, after the German chemists Emil Hermann Fischer (1852–1919) and Arthur Speier (born 1869).

Fischer-Tropsch synthesis, after the German chemists Franz Julius Emil Fischer (1877–1947) and Hans Tropsch (1889–1935).

Fittig reaction, after the German chemist Rudolph Fittig (1835–1910).

Flory equation, after the American chemist Paul John Flory (1910–1985).

Forster diazoketone synthesis, after the British chemist Martin Onslow Forster (1872–1945).

Fourier transform, after the French mathematician and physicist Jean Baptiste Joseph Baron de Fourier (1768–1830).

Franchimont reaction, after the Dutch chemist Antoine Paul Nicolas Franchimont (1844–1919).

Franck-Condon principle, after the German-American physicist James Franck (1882–1964) and the American physicist Edward Uhler Condon (1902–1974).

Franck-Rabinovitch cage effect, after the German-American physicist James Franck (1882–1964) and the Russian-American chemist Eugene Rabinovitch (1901–1973).

Frank-Caro process, after the German chemists Adolph Frank (1834–1916) and Nikodem Caro (1871–1935).

Frankland-Duppa reaction, after the British chemists Edward Frankland (1825–1899) and Baldwin Francis Duppa (1828–1873).

Frankland synthesis, after the British chemists Edward Frankland (1825–1899).

Frasch process, after the German-American chemist Herman Frasch (Hermann Frasch) (1851–1914).

Free-Wilson analysis, after the American chemists Spencer Michael Free, Jr. (born 1923) and J. W. Wilson (born 1919).

Frenkel defect, after the Russian physicist Yakov Ilyich Frenkel (1894–1952).

Freund reaction, after the Austrian chemist August Freund (1835–1892).

Friedel-Crafts reaction, after the French chemist Charles Friedel (1832–1899) and the American chemist James Mason Crafts (1839–1917).

Friedländer synthesis, after the German chemist Paul Friedländer (1857–1923).

Fries rearrangement, after the German chemist Karl Theophil Fries (1875–1962).

Fritsch-Buttenberg-Wiechell rearrangement, after the German chemists Paul Ernst Moritz Fritsch (1859–1913), Wilhelm Paul Buttenberg (1864–1946), and Heinrich G. Wiechell (born 1865).

Frumkin effect, after the Russian chemist Aleksandr Naumovich Frumkin (1895–1976).

Fujimoto-Belleau reaction, after the American chemist George Iwao Fujimoto (born 1920) and the Canadian chemist Bernard Belleau (1925–1989).

Gabriel synthesis, after the German chemist Siegmund Gabriel (1851–1924).

Galvanization, after the Italian physiologist Luigi Galvani (1737–1798).

Gattermann reaction, after the German chemist Ludwig Gattermann (1860–1920).

Gauss orbital, after the German mathematician and astronomer Carl Friedrich Gauss (1777–1855). A.k.a. **GO.**

Gay-Lussac's law, after the French chemist Joseph Louis Gay-Lussac (1778–1850).

Geneva nomenclature, named after the 1892 International Chemistry Committee congress in Geneva, Switzerland.

Gibbs-Duhem equation, after the American mathematician and physicist Josiah Willard Gibbs (1839–1903) and the French physicist Pierre Maurice Marie Duhem (1861–1916).

Gibbs energy, after the American mathematician and physicist Josiah Willard Gibbs (1839–1903).

Gibbs-Helmholtz equation, as Gibbs energy and after the German physicist and physiologist Hermann Ludwig Ferdinand von Helmholtz (1821–1894).

Gibbs reaction, after the American chemist Harry Drake Gibbs (1872–1934).

Gif oxidation, after Gif-sur-Yvette, France, where this reaction was discovered by the CNRS (Centre National de la Recherche Scientifique). The toponym Gif is of obscure etymology.

Gillespie-Nyholm model, after the British-Canadian chemist Ronald James Gillespie (born 1924) and and the British chemist Ronald Sydney Nyholm (1917–1971). A.k.a. **VSEPR**, q.v.

Girbotol prcoess, with contraction, after the Girdler Corporation, Louisville, KY, USA and the inventor, the American chemist Robert Roger Bottoms (1891–1962).

Glaser coupling, after the German chemist Carl Andreas Glaser (1841–1935).

Gmelin test, after the German mineralogist and chemist Leopold Gmelin (1788–1853).

Goldschmidt process, after the German chemist Hans Goldschmidt (1861–1923).

Goldschmidt rule, after the Norwegian chemist Victor Moritz Goldschmidt (1888–1947).

Gomberg-Bachmann reaction, after the Ukrainian-American chemist Moses Gomberg (1866–1947) and the American chemist Werner Emmanuel Bachmann (1901–1951).

Gouy-Chapman double layer, after the French physicist Louis Georges Gouy (1854–1926) and the British chemist David Leonard Chapman (1869–1958).

Graebe-Ullmann synthesis, after the German chemist Carl James Peter Graebe (1841–1927) and the German-Swiss chemist Fritz Ullmann (1875–1939).

Graham reaction, after the American chemist William Hardin Graham (1932–2004).

Griess-Ilosvay reaction, after the German-British chemist Johann Peter Griess (1829–1888) and the Hungarian chemist Lajos Ilosvay (Ilosvay Lajos) (1851–1936).

Griess reaction, after the German-British chemist Johann Peter Griess (1829–1888).

Grignard reaction, after the French chemist Victor François Auguste Grignard (1871–1935).

Grimm's hydride rule, after the German chemist Hans Georg Grimm (1887–1958).

Grimm-Sommerfeld rule, as Grimm's hydride rule and after the German physicist Arnold Sommerfeld (1868–1951).

Grob fragmentation, after the Swiss chemist Cyril A. Grob (1917–2003).

Grossheintz-Fischer-Reissert aldehyde synthesis, after the Swiss-Canadian chemist Jean Manfred Grossheintz (Grosheintz) (born 1910) and the German chemists Hermann Otto Laurenz Fischer (1888–1960), and Carl Arnold Reissert (1860–1945).

Grotthuss-Draper law, after the Germna-Russian naturalist Theodor Christian Johann Dietrich Freiherr von Grotthuß (1785–1822) and the British-American chemist and naturalist John William Draper (1811–1882).

Grunwald-Winstein equation, after the German-American chemist Ernest Max Grunwald (1923–2002) and the Canadian-American chemist Saul Winstein (1912–1969).

Guareschi-Thorpe condensation, after the Italian chemist Icilio Guareschi (1847–1918) and the British chemist Jocelyn Field Thorpe (1872–1940).

Gudden-Pohl effect, after the German physicists Bernhard Friedrich Adolf Gudden (1892–1945) and Robert Wiechert Pohl (1884–1976).

Guerbet reaction, after the French chemist Marcel Guerbet (1861–1938).

Guldberg-Waage law, after the Norwegian mathematician and chemist Cato Maximilian Guldberg (1836–1902) and the Norwegian chemist Peter Waage (1833–1900).

Gustavson reaction, after the Russian chemist Gavriil Gavriilovich Gustavson (1842–1908).

Gutzeit test, after the German chemist Heinrich Wilhelm Theodor Gutzeit (1845–1888).

Haber-Bosch process, after the German chemists Fritz Haber (1868–1934) and Carl Bosch (1874–1940).

Haber-Weiss reaction, after the German chemist Fritz Haber (1868–1934) and the Austrian-British chemist Joseph Joshua Weiss (1905–1972).

Hahn's rule, after the German physicist Otto Hahn (1879–1968).

Haines' test, after the American chemist Walter Stanley Haines (1850–1923).

Haller-Bauer reaction, after the French chemists Albin Haller (1849–1925) and Edouard Bauer (1879–1915).

Hammett equation, after the American chemist Louis Plack Hammett (1894–1987).

Hammick reaction, after the British chemist Dalziel Llewellyn Hammick (1887–1966).

Hanes plot, after the Canadian biochemist Charles Samuel Hanes (1903–1990).

Hansch analysis, after the American chemist Corwin Herman Hansch (1918–2011).

Hantzsch pyridine synthesis, after the German chemist Arthur Rudolf Hantzsch (1857–1935).

Hantzsch-Widman name, as Hantzsch pyridine synthesis and after the Swedish chemist Karl Oskar Widman (1852–1930).

Harkins rule, after the American chemist William Draper Harkins (1873–1951).

Harries ozonolysis reaction, after the German chemist Carl Dietrich Harries (1866–1923).

Hartree, a unit named after the British physicist Douglas Rayner Hartree (1897–1958).

Hartree-Fock method, as hartree and after the Russian physicist Vladimir Aleksandrovich Fock (1898–1974).

Hartree-Fock-Roothaan method, as Hartree-Fock method and after the Dutch-American physicist Clemens Carel Johannes Roothaan (1918–2019).

Hass cyclopropane process, after the American chemist Henry Bohn Hass (1902–1987).

Haworth methylation, after the British chemist Walter Norman Haworth (1883–1950). **Haworth phenanthrene synthesis,** after the British chemist Robert Downs Haworth (1898–1990).

Haworth projection, as Haworth methylation.

Hayashi rearrangement, after the Japanese chemist Mosuke Hayashi (1895–1990).

Heck reaction, after the American chemist Richard Frederick Heck (1931–2015).

Hedvall effect, after the Swedish chemist Johan Arvid Hedvall (1888–1974).

Heisenberg's principle, after the German physicist Werner Karl Heisenberg (1901–1976).

Heitler-London-Slater-Pauling method, after the German physicist Walter Heinrich Heitler (1904–1981), the German-American physicist Fritz Wolfgang London (1900–1954), the American physicist John Clarke Slater (1900–1978), and the American chemist Linus Carl Pauling (1901–1994).

Helferich method, after the German chemist Burckhard Helferich (1887–1982).

Hell-Volhard-Zelinsky reaction, after the German chemists Carl Magnus von Hell (1849–1926) and Jacob Volhard (1834–1910), and the Russian chemist Nikolai Dmitrievich Zelinsky (1861–1953).

Helmholtz equation, after the German physicist Hermann Ludwig Ferdinand von Helmholtz (1821–1894).

Henderson-Hasselbalch equation, after the American chemist Lawrence Joseph Henderson (1878–1942) and the Danish physician Karl Albert Hasselbalch (1874–1962).

Henkel process, after the chemical company Henkel & Cie., Düsseldorf, Germany. A.k.a. **Raecke process**.

Henry reaction, after the Belgian chemist Louis Henry (1834–1913).

Henry's law, after the British chemist William Henry (1775–1836).

Herz reaction, after the German chemist Richard Herz (1867–1936).

Herzig-Meyer method, after the Austrian chemists Josef Herzig (1853–1924) and Hans Leopold Meyer (1871–1942).

Hess' law, after the Swiss-Russian chemist Germain Henri Hess (1802–1850).

Hevesy-Paneth analysis, after the Hungarian radiochemist George Charles de Hevesy (Hevesy-Bischitz György) (1885–1966) and the Austrian-British chemist Friedrich Adolf Paneth (1866–1958).

Hilbert-Johnson reaction, after the American chemists Guido Edward Hilbert (1901–1982) and Treat Baldwin Johnson (1875–1947).

Hildebrand scale, after the American chemist Joel Henry Hildebrand (1881–1983).

Hill system (for the formatting of chemical formulas), after the American chemist Edwin Allston Hill (1850–1929).

Hinsberg thiophene synthesis, after the German chemist Oscar Heinrich Daniel Hinsberg (1857–1939).

Hinshelwood equation, after the British chemist Cyril Norman Hinshelwood (1897–1967).

Hittorf transfer number, after the German chemist and physicist Johann Wilhelm Hittorf (1824–1914).

Hock cleavage, after the German chemist Heinrich Hock (1887–1971).

Hofmann reaction, after the German chemist August Wilhelm von Hofmann (1818–1892).

Hofmann's rule, as Hofmann reaction.

Hofmeister series, after the German physiologist Franz Hofmeister (1850–1922).

Hooker oxidation, after the British-American chemist Samuel Cox Hooker (1854–1935).

Horecker pathway, after the American biochemist Bernard Leonard Horecker (1914–2010).

Horner reaction, after the German chemist Leopold Horner (1911–2005).

Horner-Wadsworth-Emmons reaction, as Horner reaction and after the American chemists William Steele Wadsworth, Jr. (born 1927) and William David Emmons (1924–2001).

Hosomi-Sakurai reaction, after the Japanese chemists Akira Hosomi (born 1943) and Hideki Sakurai (born 1931).

Houben-Hoesch synthesis, after the German chemists Josef Houben (1875–1940) and Kurt Hoesch (1882–1932).

Houdry cracking process, after the French-American chemist Eugene Jules Houdry (1892–1962).

Huang-Minlon reduction, after the Chinese chemist Huang Minlon (1898–1979).

Hückel-Möbius concept, after the German chemist Erich Armand Arthur Joseph Hückel (1896–1980) and the German mathematician August Ferdinand Möbius (1790–1868).

Hückel molecular orbital, cf. HMO, after the German physicist and physical chemist Erich Armand Arthur Joseph Hückel (1896–1980).

Hückel rule, as Hückel molecular orbital.

Hudson's rules, after the American chemist Claude Silbert Hudson (1881–1952).

Hughes-Ingold rule, after the British chemists Edward David Hughes (1906–1963) and Christopher Kelk Ingold (1893–1970).

Hume-Rothery phase, after the British metallurgist William Hume-Rothery (1899–1968).

Hund's rule, after the German physicist Friedrich Hermann Hund (1896–1997).

Hunsdiecker reaction, after the German chemists Heinz Hunsdiecker (1904–1981) and Cläre Hunsdiecker (1903–1995). A.k.a. **Borodin reaction.**

Imperial Smelting process, after the chemical company Imperial Smelting Corporation, Avonmouth, UK.

Ivanov reaction, after the Bulgarian chemist Dimitar Ivanov Popov (1894–1975). This scientist published his work under his middle name rather than under his family name.

Jabłoński diagram, after the Ukrainian-Polish physicist Aleksander Jabłoński (1898–1980).

Jacobsen rearrangement, after the German chemist Oscar Georg Friedrich Jacobsen (1840–1889).

Jaffé reaction, after the German physician Max Jaffé (1841–1911).

Jaffé-Schlesinger reaction, as Jaffé reaction and after the Austrian physician Wilhelm Schlesinger (1869–1947).

Jahn-Teller effect, after the British mathematician and chemist Hermann Arthur Jahn (1907–1979) and the Hungarian-American physicist Edward Teller (Teller Ede) (1908–2003).

Janovsky reaction, after the Czech chemist Jaroslav V. Janovský (1850–1907).

Japp-Klingemann reaction, after the British chemist Francis Robert Japp (1848–1925) and the German-British chemist Felix Klingemann (1863–1944).

Jones oxidation, after the British chemist Ewart Ray Herbert Jones (1911–2002).

Joule, a unit named after the British physicist James Prescott Joule (1818–1889).

Joule-Thomson effect, as joule and after the British physicist William Thomson (from 1892 William 1st Baron Kelvin of Largs) (1824–1907).

Jourdan-Ullmann-Goldberg synthesis, after the German chemist Friedrich Jourdan (born 1857), the German-Swiss chemist Fritz Ullmann (1875–1939), and the Russian-Swiss chemist Irma Ullmann (née Goldberg, born 1871). Actually, Jourdan was the first to describe this method in 1885, while Ullmann published his results in 1903 and Goldberg hers 1906/1907.

Julia synthesis, after the French chemist Marc Julia (1922–2010).

Kaptein-Closs rules, after the Durch chemist Robert Kaptein (born 1941) and the German-American chemist Gerhard Ludwig Closs (1928–1992).

Karplus equation, after the Austrian-American theoretical chemist Martin Karplus (born 1930).

Kasha's rule, after the American chemist Michael Kasha (1920–2013).

Kasha-Vavilov rule, as Kasha's rule and after the Russian chemist Sergei Ivanovich Vavilov (1891–1951).

Kedde test, after the Dutch chemist Derk Leonard Kedde (1910–1995).

Keggin anion structure, after the British chemist James Fargher Keggin (1905–1993).

Kekulé structures, after the German chemist Friedrich August Kekulé von Stradonitz (1829–1896).

Keller-Kiliani reaction, after the German chemists C. C. Keller and Heinrich Kiliani (1855–1945).

Kellogg process, after the chemical company M. W. Kellogg Co., New York, NY, USA, founded 1901 by Morris Woodruff Kellogg.

Kelvin temperature scale, after the British physicist William Thomson, 1st Baron Kelvin of Largs (1824–1897).

Kemler number, after the French engineer H. Kemler (born 1910).

Kemp reaction, after the American chemist Daniel Schaeffer Kemp (born 1936).

Kendall-Mattox reaction, after the American chemists Edward Calvin Kendall (1886–1972) and Vernon Ross Mattox (1914–2008).

Kharasch cyclization, after the Ukrainian-American chemist Morris Selig Kharasch (1895–1957).

Kiliani-Fischer synthesis, after the German chemists Heinrich Kiliani (1855–1945) and Emil Hermann Fischer (1852–1919).

Kirchhoff's laws, after the German physicist Gustav Robert Kirchhoff (1824–1887).

Kirkendall effect, after the American metallurgist Ernest Oliver Kirkendall (1914–2005).

Kirkwood-Onsager equation, after the American chemist John Gamble Kirkwood (1907–1959) and the Norwegian-American chemist Lars Onsager (1903–1976).

Kjeldahl determination, after the Danish chemist Johan Gustav Christoffer Thorsager Kjeldahl (1849–1900).

Knoevenagel condensation, after the German chemist Heinrich Emil Albert Knoevenagel (1865–1921).

Knoop's rule, after the German biochemist Franz Knoop (1875–1946).

Knoop synthesis, as Knoop's rule.

Knorr pyrrole synthesis, after the German chemist Ludwig Knorr (1859–1921).

Kober reaction, after the Dutch chemist Salomon Kober (1903–1945).

Koch-Haaf synthesis, after the German chemists Herbert Koch (1904–1967) and Wolfgang Haaf (born 1928).

Kochi reaction, after the American chemist Jay Kazuo Kochi (1927–2008).

Koch synthesis, after the German chemist Herbert Koch (1904–1967).

Koenigs-Knorr synthesis, after the German chemists Wilhelm Koenigs (1851–1906) and Ludwig Knorr (1859–1921).

Kohler reaction, after the American chemist Elmer Peter Kohler (1865–1938).

Kohlrausch's law, after the German physicist Friedrich Wilhelm Georg Kohlrausch (1840–1887).

Kölbel-Engelhardt process, after the German chemists Herbert Kölbel (1908–1995) and Friedrich Engelhardt (1913–1994).

Kolbe-Schmitt reaction, after the German chemist Adolf Wilhelm Hermann Kolbe (1818–1884) and Rudolf Schmitt (1830–1898).

Kolbe synthesis, after the German chemist Adolf Wilhelm Hermann Kolbe (1818–1884).

Kondakov reaction, after the Russian chemist Ivan Lavrentiyevich Kondakov (1857–1931).

Koopmans' theorem, after the Dutch-American physicist and economist Tjalling Charles Koopmans (1910–1985).

Kopp-Neumann rule, after the German chemists Hermann Franz Moritz Kopp (1817–1892) and Franz Ernst Neumann (1798–1895).

Kopp's rule, after the German chemist Hermann Franz Moritz Kopp (1817–1892).

Kornblum's rule, after the American chemist Nathan Kornblum (1914–1993).

Körner-Contardi reaction, after the Italian chemists G. Körner and Angelo Contardi (1877–1951).

Kostanecki acylation, after the German chemist Stanislaus von Kostanecki (1860–1923).

Krafft degradation, after the German chemist Friedrich Krafft (1852–1923).

Krafft point, as Krafft degradation.

Kramer effect, after the German physicist Johannes Kramer (1905–1975).

Krapcho decarbalkoxylation, after the American chemist Andrew Paul Krapcho (born 1932).

Kreis reaction, after the Swiss chemist Hans Kreis (1861–1931).

Kremer-Stein mechanism, after the Israeli chemists Mordechai Ladislaus Kremer (born 1930) and Gabriel Stein (1920–1976).

Kröger-Vink notation, after the Dutch chemists Ferdinand Anne Kröger (1915–2006) and Hendrik Jan Vink (1915–2009).

Kröhnke synthesis, after the German chemist Fritz Kröhnke (1903–1981).

Kroll process, after the Luxembourgish metallurgist William Justin Kroll (1889–1973).

Kucherov reaction, after the Russian-German chemist Mikhail Gregoryevich Kucherov (1850–1911).

Kuhn-Winterstein reaction, after the German chemist Richard Kuhn (1900–1967) and the Swiss chemist Alfred Winterstein (1889–1969).

Kulinkovich reaction, after the Belorussian chemist Oleg Grigoryevich Kulinkovich (born 1948).

Kumada cross-coupling reaction, after the Japanese chemist Makoto Kumada (1920–2007).

Ladenburg rearrangement, after the German chemist Albert Ladenburg (1842–1911).

Landolt reaction, after the Swiss chemist Hans Heinrich Landolt (1831–1910).

Langmuir-Blodgett film, after the American chemist Irving Langmuir (1881–1957) and the American physicist and chemist Katharine Burr Blodgett (1898–1979).

Langmuir's equation, after the American chemist Irving Langmuir (1881–1957).

Laporte's rule, after the German-American physicist Otto Laporte (1902–1971).

Lapworth reaction, after the British chemist Arthur Lapworth (1872–1941).

Lassaigne test, after the French chemist Jean Louis Lassaigne (1800–1859).

Laves phase, after the German crystallographer and mineralogist Fritz Henning Emil Paul Berndt Laves (1906–1978).

Lavoisier's law, after the French chemist Antoine-Laurent de Lavoisier (1743–1794).

LD process, an abbreviation of Linz-Donawitz process, after the location of the steel mill VOEST (Vereinigte Österreichische Eisen- und Stahlwerke) Alpine AG, Austria.

Lebedev process, after the Russian chemist Sergei Vasilyevich Lebedev (1874–1934).

Leblanc soda process, after the French physician and chemist Nicolas Leblanc (1742–1806).

Le Chatelier's principle, after the French chemist Henri Le Chatelier (1850–1936). A.k.a. **Le Chatelier-Braun principle.**

Lederer-Manasse reaction, after the German chemists Leonhard Lederer and Otto Manasse (1861–1942).

Legal test, after the German physician Emmo Legal (1859–1922).

Lehmstedt-Tanasescu reaction, after the German chemist Kurt Lehmstedt (1891–1969) and the Romanian chemist Ioan Tanăsescu (1892–1959).

Lennard-Jones potential, after the British chemist John Edward Lennard-Jones (1894–1954).

Letts nitrile synthesis, after the British chemist Edmund Albert Letts (1851–1918).

Leuckart reaction, after the German chemist Carl Louis Rudolf Alexander Leuckart (1854–1889).

Leuckart-Wallach reaction, as Leuckart reaction and after the German chemist Otto Wallach (1847–1931).

Lewis acid and base, after the American chemist Gilbert Newton Lewis (1875–1946).

Liebau classification of silicates, after the German crystallographer Friedrich Liebau (1926–2011).

Lieben iodoform reaction, after the Austrian chemist Adolf von Lieben (1836–1914).

Liebermann-Burchard reaction, after the German chemists Carl Theodor Liebermann (1842–1914) and H. R. F. Burchard (1865–1900).

Liebeskind-Srogl coupling, after the American chemist Lanny Steven Liebeskind (born 1950) and the American-Czech chemist Jiří Srogl (born 1966).

Liege nomenclature, named after the 1930 IUPAC congress in Liège (Luik), Belgium.

Liesegang rings, after the German chemist Raphael Eduard Liesegang (1869–1947).

Linde process, after the engineering company Linde AG, Wiesbaden, Germany, founded by the German engineer Carl von Linde (1842–1934).

Lindqvist anion structure, after the Swedish chemist Ingvar Lindqvist (1921–1991).

Lineweaver-Burk plot, after the American biochemists Hans Lineweaver (1907–2009) and Dean Burk (1904–1988).

Linz-Donawitz process, named after the location of the steel mill VOEST (Vereinigte Österreichische Eisen- und Stahlwerke) Alpine AG, Austria.

Lipinski's rule of five, after the American medicinal chemist Christopher A. Lipinski (born 1945).

Lobry de Bruyn-van Ekenstein transformation, after the Dutch chemists Cornelis Adriaan Lobry van Troostenburg de Bruyn (1857–1904) and Willem Alberda van Ekenstein (1858–1937).

Lohmann transphosphorylation, after the German biochemist Karl Heinrich Adolf Lohmann (1898–1978).

London force, after the German-American physicist Fritz Wolfgang London (1900–1954).

Longwell-Maniece method, after the Irish-British chemists John Longwell (born 1901) and William D. Maniece.

Loschmidt's number, after the Austrian high school chemistry teacher Josef Loschmidt (1821–1895). A.k.a. **Avogadro's number**.

Lossen rearrangement, after the German chemist Wilhelm Lossen (1838–1906).

Lowry concept, after the British chemist Thomas Martin Lowry (1874–1936).

Lowry method, after the American biochemist Oliver Howe Lowry (1910–1996).

Lucas test, after the American chemist Howard Johnson Lucas (1885–1963).

Madelung rule, rediscovered by the German physicist Erwin Madelung (1881–1972). This rule was originally discovered by the French engineer Charles Janet (1849–1932) and the Russian agricultural chemist Vsvevolod Mavrikievich Klechkovsky (1900–1972) [1].

Madelung synthesis, after the German chemist Walter Madelung (1879–1963).

Magnéli phase, after the Swedish crystallographer Arne Magnéli (1914–1996).

Maillard browning reaction, after the French chemist Louis Camille Maillard (1878–1936).

Malaprade reaction, after the French chemist Léon André Jean Eugène Malaprade (1903–1982).

Mannich base, a class name, after the German chemist Carl Ulrich Franz Mannich (1877–1947).

Mannich reaction, as Mannich base.

Marcus equation, after the Canadian-American chemist Rudolph Arthur Marcus (born 1923).

Mark-Houwink equation, after the Austrian-American chemist Herman Francis Mark (Hermann Franz Mark) (1895–1992) and the Dutch chemist Roelof Houwink (1897–1988).

Markovnikov's rule, after the Russian chemist Vladimir Vasilyevich Markovnikov (1838–1904).

Markush formula, after the Hungarian-American chemist Eugene Armand Markush (1888–1968).

Marschalk reaction, after the Swiss chemist Charles Henri Marschalk (1885–1968).

Marsh test, after the British chemist James Marsh (1794–1846).

Martinet dioxindole synthesis, after the French chemist J. Martinet.

Mattauch's rules, after the Austrian-German physicist Josef Mattauch (1895–1976).

McFadyen-Stevens reaction, after the Canadian-British chemist John S. McFadyen (1908–ca. 1990) and the British chemistThomas Stevens Stevens (1900–2000).

McLafferty rearrangement, after the American chemist Fred Warren McLafferty (born 1923).

McMurry reaction, after the American chemist John Edward McMurry (born 1942).

McReynolds constants, after the American chemist William Overton McReynolds (1917–1976). A.k.a **Rohrschneider constants**, after the German chemist Lutz Rohrschneider (born 1927) and **Rohrschneider-McReynolds constants.**

Meerwein-Ponndorf-Verley reduction, after the German chemists Hans Leberecht Meerwein (1879–1965) and Wolfgang Ponndorf (1894–1948), and the French chemist Albert Verley (1867–1959).

Meerwein rearrangement, after the German chemist Hans Leberecht Meerwein (1879–1965).

Meisenheimer complex, a class name, after the German chemist Jakob Meisenheimer (1876–1934).

Meisenheimer rearrangement, as Meisenheimer complex.

Menshutkin reaction, after the Russian chemist Nikolai Aleksandrovich Menshutkin (1842–1907).

Meyers aldehyde synthesis, after the American chemist Albert Irving Meyers (1933–2007).

Meyer-Schuster rearrangement, after the German chemists Karl Heinrich Meyer (1883–1952) and Kurt Eduard Schuster (1892–1990).

Meyer synthesis, after the German chemist Viktor Meyer (1848–1897).

Michael addition, after the American chemist Arthur Michael (1853–1942).

Michaelis-Arbuzov reaction, after the German chemist Karl Arnold Michaelis (1847–1916) and the Russian chemist Aleksandr Erminingeldovich Arbuzov (1877–1968). A.k.a. **Arbuzov-Michaelis reaction.**

Miescher degradation, after the Swiss chemist Karl Miescher (1892–1974).

Mignonac reaction, after the French chemist Georges Mignonac (1889–1993). A.k.a. **Moureau-Mignonac reaction.**

Milas hydroxylation of olefins, after the Greek-American chemist Nicholas A. Milas (1897–1971).

Millon reaction, after the French physician and chemist Auguste Nicolas Eugène Millon (1812–1867).

Mislow-Evans rearrangement, after the German-American chemist Kurt Mislow (1923–2017) and the American chemist David A. Evans (born 1941).

Mitscherlich's rule, after the German chemist Eilhard Mitscherlich (1794–1863).

Mitsunobu reaction, after the Japanese chemist Oyo Mitsunobu (1934–2003).

Möbius aromaticity, after the German mathematician August Ferdinand Möbius (1790–1868).

Möbius compound, a class name, as Möbius aromaticity.

Möbius electrolysis, after the German chemist Hans-Heinrich Möbius (1929–2011).

Mohs hardness, after the German chemist and mineralogist Friedrich Mohs (1773–1839).

Molisch reaction, after the German botanist Hans Molisch (1856–1937).

Mond process, after the German-British inventor and industrialist Ludwig Mond (1839–1909).

Monsanto process, after the chemical company Monsanto Company, St. Louis, MO, USA, named after its founder's, the American businessman John Francis Queeny's (1859–1933) wife's, Olga Mendez Monsanto, maiden name.

Mont-Cenis process, after the Mont Cenis coal mine, Herne, Germany, referring to the first customer to acquire an ammonia plant designed for this process.

Moore cyclization, after the American chemist Harold Wesley Moore (born 1936).

Morgan-Walls reaction, after the British chemists Gilbert Thomas Morgan (1870–1940) and Leslie Percy Walls (1905–1987).

Moseley's law, after the British physicist Henry George Jeffreys Moseley (1887–1915).

Mössbauer effect, after the German physicist Rudolf Ludwig Mössbauer (1929–2011).

Muetterties' rule, after the American chemist Karl Leonard Muetterties (1927–1984).

Mukaiyama aldol reaction, after the Japanese chemist Teruaki Mukaiyama (born 1927).

Murahashi reaction, after the Japanese chemist Shin-Ichi Murahashi (born 1937).

Myers cyclization, after the American chemist Andrew Gordon Myers (born 1959).

Nagata hydrocyanation, after the Japanese chemist Wataru Nagata (1922–1995).

Nametkin rearrangement, after the Russian chemist Sergei Semyonovich Nametkin (1876–1950).

Neber rearrangement, after the German chemist Peter Neber (1883–1960).

Néel temperature, after the French physicist Louis Eugène Félix Néel (1904–2000).

Nef reaction, after the Swiss-American chemist John Ulric Nef (1862–1915).

Negishi cross coupling, after the Japanese-Americam chemist Ei-ichi Negishi (born 1935).

Nencki reaction, after the German chemist Marcel von Nencki (1847–1901).

Nenitzescu indole synthesis, after the Romanian chemist Costin D. Nenițescu (1902–1970).

Nernst equation, after the German chemist Hermann Walther Nernst (1864–1941).

Nesmeyanov reaction, after the Russian chemist Aleksandr Nikolayevich Nesmeyanov (1899–1980).

Neumann-Kopp rule, after the German chemists Franz Ernst Neumann (1798–1895) and Hermann Franz Moritz Kopp (1817–1892).

Newtonian liquid, after the British physicist Isaac Newton (1642–1727).

Nicholas reaction, after the American chemist Kenneth Matthew Nicholas (born 1947).

Niementowski quinazoline synthesis, after the Polish chemist Stefan Niementowski (1866–1925).

Nierenstein reaction, after the British chemist Maximilian Nierenstein (1877–1946).

Nieuwland enyne synthesis, after the Belgian-American chemist Julius Arthur Nieuwland (1878–1936).

Niggli formula, after the Swiss crystallographer Paul Niggli (1888–1953).

NIH shift, after the National Institutes of Health, Bethesda, MD, USA, where this reaction was discovered.

Normann process, after the German chemist and industrialist Wilhelm Normann (1870–1939).

Northern blot, jocularly so named with reference to Southern blot. Cf. **Southern blot**.

Norrish reactions, after the British chemist Ronald George Wreyford Norrish (1897–1978).

Noyori hydrogenation, after the Japanese chemist Ryoji Noyori (born 1938).

Nozaki-Hiyama coupling reaction, after the Japanese chemists Hitoshi Nozaki (born 1922) and Timejiro Hiyama (born 1946). A.k.a. **Nozaki-Hiyama-Kishi reaction**.

Nuclear Overhauser effect, after the American physicist Albert Warner Overhauser (1925–2011).

Nylander's test, after the Swedish chemist Claus Wilhelm Gabriel Nylander (1835–1907). In his publication this scientist used the alias Emil Nylander.

Obermayer's test, after the Austrian physician Friedrich Obermayer (1861–1925).

Oppenauer oxidation, after the Austrian chemist Rupert Viktor Oppenauer (1910–1969).

Orton rearrangement, after the British chemist Kennedy Joseph Previté Orton (1872–1930).

Ostwald ripening, after the German biologist and physical chemist Carl Wilhelm Wolfgang Ostwald (1883–1943).

Ostwald rule, after the German chemist Friedrich Wilhelm Ostwald (1853–1932).

Ostwald-Volmer rule, as Ostwald rule and after the German chemist Max Volmer (1885–1965).

Outokumpu process, after the metallurgical company Outokumpu Oy, Espoo, Finland.

Overhauser effect, after the American physicist Albert Warner Overhauser (1925–2011).

Overman rearrangement, after the American chemist Larry Eugene Overman (born 1943).

Paal-Knorr pyrrole synthesis, after the Austrian chemist Carl Paal (1860–1935) and the German chemist Ludwig Knorr (1859–1921).

Parham cyclization, after the American chemist William Eugene Parham (1922–1976).

Pariser-Parr-Pople method, after the American chemists Rudolph Pariser (born 1923) and Robert Ghormley Parr (1921–2017), and the British-American chemist John Anthony Pople (1925–2004). A.k.a. **PPP**.

Parkes process, after the British metallurgist and inventor Alexander Parkes (1813–1890).

Passerini reaction, after the Italian chemist Mario Torquato Luigi Passerini (1891–1962).

Paternò-Büchi reaction, after the Italian chemist Emanuele Paternò, Marchese di Sessa (1847–1935) and the Swiss-American chemist Georg Hermann Büchi (1921–1998).

Pattinson process, after the British chemist Hugh Lee Pattinson (1796–1858).

Pauling complex theory, after the American chemist Linus Pauling (1901–1994).

Pauling process, after the German chemist Harry Pauling (1875–1956).

Pauli principle, after the Austrian-Swiss physicist Wolfgang Pauli (1900–1958).

Pauly reaction, after the German chemist Hermann Pauly (1870–1950).

Pauson-Khand reaction, after the German-British chemist Peter Ludwig Pauson (1925–2013) and the Pakistani-British chemist Ihsan Ullah Khand (1935–1980).

Payne rearrangement, after the American chemist George B. Payne.

Peacock-Weakley anion structure, after the British chemist Robert Davis Peacock (born 1945) and the British-American crystallographer Timothy John Ruffer Weakley (born 1933).

Pearson concept (HSAB concept), after the American chemist Ralph Gottfried Pearson (born 1919).

Pearson symbol, created by the British-Canadian crystallographer William Burton Pearson (born 1921).

Pechiney-Ugine-Kuhlmann process, after the industrial company Pechiney-Ugine-Kuhlmann, Paris, France.

Pechmann condensation, after the German chemist Hans Freiherr von Pechmann (1850–1902).

Pellizzari reaction, after the Italian chemist Guido Pellizzari (1858–1938).

Penning effect, after the Dutch physicist Frans Michel Penning (1894–1953).

Perkin reaction, after the British chemist William Henry Perkin (1838–1907).

Perkow reaction, after the German chemist Werner Perkow (1915–1994).

Pesez reaction, after the French chemist Maurice Pesez (1914–2009).

Petrenko-Kritschenko piperidone synthesis, after the Ukrainian chemist Pavel Ivanovich Petrenko-Kritschenko (1866–1944).

Pfau-Plattner azulene synthesis, after the Swiss chemists Alexandre Stanislav Pfau (1889–1938) and Placidus Andreas Plattner (1904–1975).

Pfeiffer effect, after the German chemist Paul Pfeiffer (1875–1951).

Pfitzinger reaction, after the German chemist Friedrich Wilhelm August Pfitzinger (1864–1939). A.k.a. **Pfitzinger-Borsche reaction**.

Pfitzner-Moffatt reaction, after the American chemist K. E. Pfitzner and the Canadian-American chemist John Gilbert Moffatt (born 1930).

Pickering emulsion, after the British chemist Percival Spencer Umfreville Pickering (1858–1920). While Pickering made his discovery in 1907 the phenomenon had already been recognized in 1903 by the British chemist Walter Ramsden (1868–1947).

Pictet-Spengler isoquinoline synthesis, after the Swiss chemists Amé Pictet (1857–1937) and Theodor Spengler.

Pictet-Trouton rule, after the Swiss physicist Raoul-Pierre Pictet (1846–1929) and the Irish-British physicist Frederick Thomas Trouton (1863–1922).

Piloty-Robinson synthesis, after the German chemist Oskar Piloty (1866–1915) and the British chemist Robert Robinson (1885–1975).

Pinner reaction, after the German chemist Adolf Pinner (1842–1909).

Piria reaction, after the Italian chemist Raffaele Piria (1815–1865).

Pitzer strain, after the American chemist Kenneth Sanborn Pitzer (1914–1997). Cf. Baeyer strain and Prelog strain.

Planck's constant, after the German physicist Max Karl Ernst Ludwig Planck (1858–1947).

Polonovski reaction, after the Russian-French chemist Max Polonovski (1861–1939) and his son, the French chemist Michel Polonovski (1889–1954).

Polzeniusz-Krauss process, after the Austro-Hungarian chemist Ferdinand Eduard Polzeniusz (1862–1918) and the German chemist Konstantin Krauß (1864–1928).

Pomeranz-Fritsch reaction, after the Austrian chemist Cäsar Pomeranz (1860–1926) and the German chemist Paul Fritsch (1859–1913).

Ponzio reaction, after the Italian chemist Giacomo Ponzio (1870–1945).

Porter-Silber reaction, after the American chemists Curt Culwell Porter (1914–2007) and Robert Howard Silber (1915–1980).

Pott-Broche process, after the German chemists Alfred Pott (1881–1951) and Hans Broche (1896–1963).

Prelog strain, after the Bosnian-Swiss chemist Vladimir Prelog (1906–1998). Cf. Baeyer strain and Pitzer strain.

Prévost reaction, after the French chemist Charles Prévost (1899–1983).

Prilezhaev reaction, after the Russian chemist Nikolai Aleksandrovich Prilezhayev (1872–1944).

Prins reaction, after the Dutch chemist Hendrik Jacobus Prins (1889–1958).

Pummerer rearrangement, after the Austrian-German chemist Rudolf Pummerer (1882–1973).

Purdie methylation, after the British chemist Thomas Purdie (1843–1916).

Quelet reaction, after the French chemist Raymond Quelet (1897–1967). A.k.a. **Blanc-Quelet reaction**.

Racah parameter, after the Italian-Israeli physicist Giulio (Yoel) Racah (1909–1965).

Raecke process, after the German chemist Bernhard Raecke. A.k.a. **Henkel process.**

Raman effect, after the Indian physicist Chandrasekhara Venkata Raman (1888–1970).

Ramberg-Bäcklund rearrangement, after the Swedish chemists Ludvig Ramberg (1874–1940) and Birger Bäcklund (1908–1997).

Rankine temperature scale, after the British physicist William John Macquorn Rankine (1820–1872).

Raoult's law, after the French chemist François Marie Raoult (1830–1901).

Raschig phenol process, after the German chemist Friedrich August (Fritz) Raschig (1863–1928).

Raymond reaction, after the British chemist W. D. Raymond (1901–1988).

Réaumur temperature scale, after the French naturalist René Antoine Ferchault de Rëaumur (1683–1757).

Reformatsky reaction, after the Russian chemist Sergei Nikolayevich Reformatskiy (1860–1934).

Regitz diazo transfer reaction, after the German chemist Manfred Regitz (born 1935).

Reichert-Meissl number, after the German chemist Emil Reichert (1838–1894) and the Austrian chemist Emerich Meißl (1855–1905).

Reimer-Tiemann reaction, after the German chemists Karl Ludwig Reimer (1845–1883) and Johann Karl Wilhelm Ferdinand Tiemann (1848–1899). Karl Ludwig Reimer (1845–1883) has been occasionally confused with his namesake, his cousin Karl Ludwig Reimer (1856–1921) who also worked on the same project, although in a more subordinate role [8].

Reinhardt-Zimmermann titration, after the German chemists C. Reinhardt (1859–1905) and Julius Clemens Zimmermann (1856–1896).

Reissert reaction, after the German chemist Arnold Carl Reissert (1860–1945).

Reppe chemistry, after the German chemist Julius Walter Reppe (1892–1969).

Reverdin reaction, after the Swiss chemist Frédéric Reverdin (1849–1931).

Rice-Herzfeld mechanism, after the British chemist Francis Owen Rice (1890–1989) and the Austrian-American physicist Karl Ferdinand Herzfeld (1892–1978).

Rice-Ramsberger-Kassel-Marcus theory, after the American chemists Oscar Knefler Rice (1903–1978), Herman Carl Ramsberger (1896–1923), and Louis Stevenson Kassel (1905–1973), and the Canadian chemist Rudolph Arthur Marcus (born 1923).

von Richter cinnoline synthesis, after the German chemist Victor von Richter (1841–1891).

Richter's law, after the German chemist Jeremias Benjamin Richter (1762–1807).

Riehm quinoline synthesis, after the German chemist P. Riehm.

Riemschneider thiocarbamate synthesis, after the German chemist Randolph Riemschneider (born 1920).

Riley oxidation, after the British chemist Harry Lister Riley (1899–1986).

Ritter reaction, after the American chemist John Joseph Ritter (1895–1975).

Robinson annulation, after the British chemist Robert Robinson (1885–1975).

Robinson-Schöpf reaction, as Robinson annulation and after the German chemist Clemens Schöpf (1899–1970).

Rochow-Müller process, after the American chemist Eugene George Rochow (1909–2002) and the German chemist Richard Gustav Müller (1903–1999).

Rockwell hardness, after the American metallurgists Hugh M. Rockwell (1890–1957) and Stanley P. Rockwell (1886–1940).

Roelen reaction, after the German chemist Otto Roelen (1897–1993).

Rohrschneider constants, after the German chemist Lutz Rohrschneider (born 1927).

Rosenmund reduction, after the German chemist Karl Wilhelm Rosenmund (1884–1965).

Rosenmund-von Braun synthesis, as Rosenmund reduction and after the German chemist Julius von Braun (1875–1939).

Rosiwal hardness, after the Austrian mineralogist August Karl Rosiwal (1860–1923).

Rothe-Frank-Caro process, after the German chemists Fritz Rothe (1867–1958), Adolf Frank (1834–1916), and Nikodem Caro (1871–1935).

Rothemund reaction, after the American chemist Paul Wilhelm Karl Rothemund (born 1972).

Rubottom oxidation, after the Amerian chemist George M. Rubottom (born 1940). A triple discovery, reported by the Canadian chemist Adrian Gibbs Brook

(1924–2013) on June 10, 1974, by the Israeli-American chemist Alfred Hassner (born 1930) on September 4, 1974, and by Rubottom on September 24, 1974.

Ruff-Fenton degradation, after the German chemist Otto Ruff (1871–1939) and the British chemist Henry John Horstman Fenton (1854–1929).

Ruggli dilution principle, after the Swiss chemist Paul Ruggli (1884–1945).

Rupe rearrangement, after the Swiss chemist Hans Rupe (1866–1951).

Russell-Saunders coupling, after the American astronomer Henry Norris Russell (1877–1957) and the Canadian physicist Frederick Albert Saunders (1875–1963).

Ruzicka large-ring synthesis, after the Croatian-Swiss chemist Leopold Ruzicka (1887–1976).

Rydberg's constant, after the Swedish physicist Johannes Robert Rydberg (1854–1919).

Sabatier-Senderens reduction, after the French chemists Paul Sabatier (1854–1941) and Jean-Baptiste Senderens (1856–1937).

Sachse-Mohr theory, after the German chemists Hermann Sachse (1862–1893) and Ernst Mohr (1873–1926).

Saegusa oxidation, after the Japanese chemist Takeo Saegusa (born 1927).

Salkowski reaction, after the German biochemist Ernst Leopold Salkowski (1844–1923).

Sandmeyer reaction, after the Swiss chemist Traugott Sandmeyer (1854–1922).

Schadt-Helfrich effect, after the Swiss physicist Martin Schadt (born 1938) and the German physicist Wolfgang Helfrich (born 1932).

Schardinger test, after the German biochemist Franz Schardinger (1853–1917).

Schenck mechanism, after the German chemist Günther Otto Schenck (1913–2003).

Schiemann reaction, after the German chemist Günther Schiemann (1899–1969). A.k.a. **Balz-Schiemann reaction.**

Schlegel diagram, after the German mathematician Stanislaus Ferdinand Victor Schlegel (1843–1905).

Schlenk equilibrium, after the German chemist Wilhelm Schlenk (1879–1943).

Schlesinger process, after the American chemist Hermann Irving Schlesinger (1882–1960).

Schlittler-Müller modification, of the Pomeranz-Fritsch reaction, after the Swiss chemists Emil Schlittler (1906–1979) and Johannes Müller.

Schmidt reaction, after the German chemist Karl Friedrich Schmidt (1887–1971).

Schmidt's double bond rule, after the German chemist Otto Schmidt (1874–1943).

Schoenflies symbol, after the German mathematician Arthur Moritz Schoenflies (1853–1928).

Schöllkopf bis-lactim amino acid synthesis, after the German chemist Ulrich Schöllkopf (1927–1998).

Scholl reaction, after the German chemist Roland Scholl (1865–1945).

Schomaker-Stevenson equation, after the American chemists Verner Schomaker (1914–1997) and David Paul Stevenson (1914–1999).

Schönberg reaction, after the German chemist Alexander Schönberg (1892–1985).

Schönherr process, after the German chemist Otto Schönherr (1861–1926).

Schotten-Baumann reaction, after the German chemists Carl Ludwig Johannes Schotten (1853–1910) and Eugen Baumann (1846–1896).

Schottky defect, after the German physicist Walter Schottky (1886–1976).

Schrödinger equation, after the Austrian physicist Erwin Schrödinger (1887–1961).

Schulze-Hardy rule, after the German chemist Hans Oscar Schulze (1853–1892) and the British chemist William Bate Hardy (1864–1934).

Seliwanoff's test (for aldoses and ketoses), after the Russian chemist Fyodor Fyodorovich Selivanov (1859–1938).

Semmler-Wolff reaction, after the German chemists Friedrich Semmler (1860–1931) and Ludwig Wolff (1857–1919).

Sendzimir process, after the Polish-American industrialist Tadeusz Sendzimir (1894–1989).

Serini reaction, after the German chemist Arthur Serini (1897–1945).

Serpek process, after the Czech-Austrian chemist Ottokar Serpek (born 1864).

Shannon-Prewitt ion radius, after the American chemists Robert Day Shannon (born 1935) and Charles Thompson Prewitt.

Shapiro reaction, after the American chemist Robert Howard Shapiro (1935–2004).

Sharpless epoxidation, after the American chemist Karl Barry Sharpless (born 1941).

Sharpless-Jacobsen dihydroxylation, as Sharpless epoxidation and after the American chemist Eric Niels Jacobsen (born 1960).

Shell-Deacon process, after the energy and petrochemical company Royal Dutch Shell plc, Den Haag, Netherlands, and the Deacon process.

Shell process, after the energy and petrochemical company Royal Dutch Shell plc, Den Haag, Netherlands.

Sherardization, after the British metallurgist Sherard Osborn Cowper-Coles (1866–1936).

Shoolery NMR rule, after the American chemist James Nelson Shoolery (1925–2015).

Shore hardness, after the American metallurgist Albert Ferdinand Shore (1876–1936).

Siemens-Martin process, after the German-British industrialists and inventors Friedrich Siemens (1826–1904) and Charles William Siemens (1823–1883), and the French industrialists François Marie Emile Martin (1794–1871) and Pierre-Emile Martin (1824–1915).

Simmons-Smith reaction, after the American chemists Howard Ensign Simmons, Jr. (1929–1997) and Ronald Dean Smith (born 1930).

Simonini reaction, after the Austrian-American chemist Angelo Simonini (1867–1913).

Simonis chromone cyclization, after the German chemist Hugo Simonis (1874–1949).

Skita's rule, after the Austrian-German chemist Aladar Skita (1876–1953).

Skraup reaction, after the Czech chemist Zdenko Hans Skraup (1850–1910).

Slater orbital, after the American physicist John Clarke Slater (1900–1976).

Smekal-Raman effect, after the Austrian theoretical physicist Adolf Gustav Stephan Smekal (1895–1959) and the Indian physicist Chandrasekhara Venkata Raman (1888–1970).

Smiles rearrangement, after the British chemist Samuel Smiles (1877–1953).

Sommelet-Hauser rearrangement, after the French chemist Marcel Sommelet (1877–1952) and the American chemist Charles Roy Hauser (1900–1970).

Sommelet reaction, after the French chemist Marcel Sommelet (1877–1952).

Sonn-Müller method, after the German chemists Adolf Sonn (1867–1942) and Ernst Müller (1881–1945).

Sonogashira coupling, after the Japanese chemist Kenkichi Sonogashira (born 1931).

Sonogashira-Hagihara cross-coupling, as Sonogashira coupling and after the Japanese chemist Nobue Hagihara (1916–2009).

Southern blot, after the British molecular biologist Edwin Mellor Southern (born 1938).

Staudinger reaction, after the German chemist Hermann Staudinger (1881–1965).

Stefan-Boltzmann constant, after the Austrian physicists Josef Stefan (1835–1893) and Ludwig Eduard Boltzmann (1844–1906).

Stelzner nomenclature, after the German chemist Robert Stelzner (1869–1943).

Stephen aldehyde synthesis, after the British chemist Henry Stephen (1889–1965).

Stern double layer, after the German-American physicist Otto Stern (1888–1969).

Stetter reaction, after the German chemist Hermann Stetter (1917–1993).

Stevens rearrangement, after the British chemist Thomas Stevens Stevens (1900–2000).

Stieglitz rearrangement, after the American chemist Julius Stieglitz (1867–1937).

Stille coupling reaction, after the American chemist John Kenneth Stille (1930–1989).

Still-Wittig rearrangement, after the American chemist William Clark Still, Jr. (born 1946) and the German chemist Georg Wittig (1897–1987).

Stobbe condensation, after the German chemist Johann Hans Hermann August Adolf Stobbe (1860–1938).

Stock system, after the German chemist Alfred Stock (1876–1946).

Stokes' law, after the British physicist George Gabriel Stokes (1819–1903).

Stollé synthesis, after the German chemist Robert Stollé (1869–1938).

Stoltz oxidative etherification, after the American chemist Brian Mark Stoltz (born 1970).

Stoney's electron hypothesis, after the Anglo-Irish physicist George Johnstone Stoney (1826–1911).

Stork enamine reaction, after the Belgian-American chemist Gilbert Stork (1921–2017).

Strandberg anion structure, after the Swedish chemist Rolf Strandberg (born 1933).

Strecker amino acid synthesis, after the German chemist Adolf Friedrich Ludwig Strecker (1822–1871).

Suarez reaction, after the Spanish chemist Ernesto Suárez.

Sugasawa reaction, after the Japanese chemist Tsutomu Sugasawa (born 1926).

Süs reaction, after the German chemist Oskar Süs (1903–1978).

Suzuki coupling, after the Japanese chemist Akira Suzuki (born 1930).

Swain-Lupton equation, after the American chemists Charles Gardner Swain (1917–1988) and Elmer Cornelius Lupton, Jr. (born 1945).

Swain-Scott equation, after the American chemists Charles Gardner Swain (1917–1988) and Carlton Browne Scott (born 1924).

Swarts reaction, after the Belgian chemist Frédéric Jean Edmond Swarts (1866–1940).

Swern oxidation, after the American chemist Daniel Swern (1916–1982).

Szilard-Chalmers effect, after the Hungarian-American physicist Leo Szilard (Leo Spitz, Szilárd Leó) (1898–1964) and the British physicist Thomas A. Chalmers.

Tafel rearrangement, after the Swiss chemist Julius Tafel (1862–1918).

Taft equation, after the American chemist Robert Wheaton Taft, Jr. (1922–1996).

Takata-Ara reaction, after the Japanese physicians Maki Takata (1892–1978) and Kiyoshi Ara (born 1894).

Tammann's rule, after the German chemist Gustav Tammann (1861–1938).

Teuber reaction, after the German chemist Hans-Joachim Teuber (born 1918).

Texaco process, after the energy and petrochemical company Texaco, White Plains, NY, USA. Texaco is a contraction of The Texas Company.

Thiele reaction, after the German chemist Johannes Thiele (1865–1918).

Thiele-Winter reaction, as Thiele reaction and after the Estonian chemist E. A. Winter (1867–1921).

Thomsen-Berthelot principle, after the Danish chemist Hans Peter Jørgen Julius Thomsen (1826–1909) and the French chemist Marcellin Pierre Eugène Berthelot (1827–1907).

Thorpe reaction, after the British chemist Jocelyn Field Thorpe (1872–1940).

Tiemann rearrangement, after the German chemist Carl Wilhelm Ferdinand Tiemann (1848–1899).

Tiffeneau-Demyanov rearrangement, after the French chemist Marc Emile Pierre Adolphe Tiffeneau (1873–1945) and the Russian chemist Nikolai Jakovlevich Demyanov (1861–1938).

Tishchenko reaction, after the Russian chemist Vyacheslav Yevgenyevich Tishchenko (1861–1941).

Toms effect, after the British physicist Bryan Atkinson Toms (1916–1987).

Traube purine synthesis, after the German chemist Wilhelm Traube (1866–1942).

Traube's rule, after the German chemist Isidor Traube (1860–1943).

Trommer's test, after the German chemist Karl August Trommer (1806–1879).

Trommsdorff-Norrish effect, after the German chemist Ernst Trommsdorff (1905–1996) and the British chemist Ronald George Wreyford Norrish (1897–1978).

Trost allylation, after the American chemist Barry Martin Trost (born 1940). A.k.a. **Tsuji-Trost reaction.**

Trouton constant, after the Irish physicist Frederick Thomas Trouton (1863–1922).

Truce-Smiles rearrangement, after the American chemist William Everett Truce (1940–2009) and the British chemist Samuel Smiles (1877–1953).

Tscherniac-Einhorn reaction, after the German chemists J. Tscherniac and Alfred Einhorn (1857–1917).

Twitchell process, after the American chemist Ernst Twitchell (1863–1929).

Uemura oxidation, after the Japanese chemist Sakae Uemura (born 1941).

Ugi reaction, after the Estonian-German chemist Ivar Karl Ugi (1930–2005).

Ullmann reaction, after the German-Swiss chemist Fritz Ullmann (1875–1939).

Ultée cyanohydrin method, after the Dutch chemist Arnoldus Johannes Ultée (1878–1964).

Upjohn reaction, after the chemical company The Upjohn Company, Kalamazoo, MI, USA, founded by the American physician and inventor William Erastus Upjohn (1853–1932).

Urech cyanohydrin method, after the German chemist Friedrich Urech (1844–1904).

van Arkel-de Boer process, after the Dutch chemists Anton Eduard van Arkel (1893–1976) and Jan Hendrik de Boer (1899–1971).

van der Waals equation of state, after the Dutch physicist Johannes Diderik van der Waals (1837–1923).

van Slyke method, after the American biochemist Donald Dexter van Slyke (1883–1971).

van't Hoff's law, after the Dutch chemist Jacobus Henricus van't Hoff (1852–1911).

van Urk reaction, after the Dutch pharmacist Hendrik Willem van Urk (1886–1945).

Varrentrapp reaction, after the German chemist Franz Varrentrapp (1815–1877).

Varrentrapp-Will method, as Varrentrapp reaction and after the German chemist Heinrich Will (1812–1890).

Vasella reaction, after the Swiss chemist Andrea Vasella (born 1943).

Verneuil process, after the French chemist Auguste Victor Louis Verneuil (1856–1913).

Vickers hardness scale, after the British industrialist Edward Vickers (1804–1897).

Vilsmeier-Haack reaction, after the German chemists Anton Vilsmeier (1894–1962) and Albrecht Haack (1898–1976).

Vilsmeier reaction, after the German chemist Anton Vilsmeier (1894–1962).

Vitali-Morin reaction, after the Italian pharmacist Dioscoride Vitali (1832–1917) and the Swiss pharmacist Antoine Morin (1800–1879).

Voight amination, after the German chemist K. Voight.

Volhard-Erdmann cyclization, after the German chemists Jacob Volhard (1834–1910) and Hugo Erdmann (1862–1910).

Volhard titration, after the German chemist Jacob Volhard (1834–1910).

Volhard-Wolff titration, as Volhard titration and after the German chemist N. Wolff.

Vorbrüggen glycosylation, after the German chemist Helmut Vorbrüggen (born 1930).

Wacker oxidation, after the chemical company Wacker-Chemie GmbH, Munich, Germany, founded by the German chemist Alexander Wacker (1846–1922).

Wacker-Tsuji oxidation, as Wacker oxidation and after the Japanese chemist Jiro Tsuji (born 1927).

Wade's rules, after the British chemist Kenneth Wade (1932–2014).

Wadsworth-Emmons reaction, after the American chemists William Steele Wadsworth, Jr. (born 1927) and William David Emmons (1924–2001).

Wagner-Jauregg reaction, after the Austrian chemist Theodor Wagner-Jauregg (1903–1992).

Wagner-Meerwein rearrangement, after the Russian chemist Georg Yegorovich Vagner (1849–1903) and the German chemist Hans Leberecht Meerwein (1879–1965).

Walden inversion, after the German chemist Paul Walden (1863–1957).

Wallach reaction, after the German chemist Ottto Wallach (1847–1931).

Walsh diagram, after the British chemist Arthur Donald Walsh (1916–1977).

Watson-Crick model, after the American biochemist James Dewey Watson (born 1928) and the British biochemist Francis Harry Compton Crick (1916–2004).

Wawzonek-Yeakey rearrangement, after the American chemists Stanley Wawzonek (1914–1998) and Ernest Leon Yeakey (born 1934).

Weakley anion structure, after the British-American crystallographer Timothy John Ruffer Weakley (born 1933).

Weakley-Yamase anion structure, as Weakley anion structure and after the Japanese chemist Toshihiro Yamase (born 1943).

Weerman degradation, after the Dutch chemist Rudolf Adrian Weerman (1880–1931).

Weidenhagen synthesis, after the German chemist Rudolf Weidenhagen (1900–1979).

Weiss reaction, after the Czech-American chemist Ulrich Weiss (1908–1989).

Weitz-Scheffer reaction, after the German chemists Ernst Weitz (1883–1954) and Alfred Scheffer.

Weldon process, after the British chemist Walter Weldon (1832–1885).

Wells-Dawson anion structure, after the British-American chemist Alexander Frank "Jumbo" Wells (1912–1994) and the British-Australian chemist Barrie Dawson (1925–1974).

Werner complex, after the Swiss chemist Alfred Werner (1866–1919).

Wessely oxidation, after the Austrian chemist Friedrich Wessely (1897–1967).

Wessely-Moser rearrangement, after the Austrian chemists Friedrich Wessely (1897–1967) and Georg H. Moser.

Western blot, jokingly so named with reference to Southern blot.

Westphalen-Lettré rearrangement, after the German chemists Theodor Westphalen (1889–1915) and Hans Heinrich Lettré (1908–1971).

Wharton reaction, after the British-American chemist Peter Stanley Wharton (born 1931).

Wheland intermediate, after the American chemist George Willard Wheland (1907–1972).

Whiting reaction, after the British chemist Mark Crosby Whiting (born 1925).

Wichterle reaction, after the Czech chemist Otto Wichterle (1913–1998).

Wickbold method, after the German chemist Reinhold Wickbold (1910–1991).

Widman-Stoermer synthesis, after the Swedish chemist Karl Oskar Widman (1852–1930) and the German chemist Richard Stoermer (1870–1940).

Widmark method, after the Swedish physiologist Erik Matteo Prochet Widmark (1889–1945).

Wiesner reaction, after the Austrian chemist Julius Ritter von Wiesner (1838–1916).

Wigner rule, after the Hungarian-American physicist Eugene Paul Wigner (Wigner Jenö Pál) (1902–1995).

Wilke process, after the German chemist Günther Wilke (1925–2016).

Willgerodt-Kindler reaction, after the German chemists Conrad Heinrich Christoph Willgerodt (1841–1930) and Karl Kindler (1891–1967).

Williamson ether synthesis, after the British chemist Alexander William Williamson (1824–1904).

Willstätter reaction, after the German chemist Richard Martin Willstätter (1872–1942).

Wilzbach method, after the American chemist Kenneth E. Wilzbach (1920–2004).

Winstein-Holness equation, after the Canadian-American chemist Saul Winstein (1912–1969) and the British chemist Norris John Holness (1927–1986).

Wisswesser line notation, after the American chemist William Joseph Wisswesser (1914–1989).

Witt diazotization method, after the German chemist Otto Nikolaus Witt (1853–1915).

Wittig reaction, after the German chemist Georg Wittig (1897–1987).

Wohl degradation, after the German chemist Alfred Wohl (1863–1939).

Wöhler's urea synthesis, after the German chemist Friedrich Wöhler (1800–1882).

Wohl-Ziegler reaction, after the German chemists Alfred Wohl (1863–1939) and Karl Ziegler (1878–1973).

Wolffenstein-Böters reaction, after the German chemists Richard Wolffenstein (1864–1929) and Oskar Böters.

Wolff-Kishner reduction, after the German chemist Ludwig Wolff (1857–1919) and the Russian chemist Nikolai Matveyevich Kishner (1867–1935).

Wolff rearrangement, after the German chemist Ludwig Wolff (1857–1919).

Wood-Bonhoeffer method, after the American physicist Robert William Wood (1868–1955) and the German chemist Karl Friedrich Bonhoeffer (1899–1957).

Woodward-Hoffmann rules, after the American chemist Robert Burns Woodward (1917–1979) and the Polish-American chemist Roald Hoffmann (born 1937).

Woodward *cis*-hydroxylation, after the American chemist Robert Burns Woodward (1917–1979).

Wood-Werkman reaction, after the American biochemists Harland Goff Wood (1907–1991) and Chester Hamlin Werkman (1893–1962).

Wulff-Dötz reaction, after the American chemist William D. Wulff (born 1949) and the German chemist Karl Heinz Dötz (born 1943).

Wurtz-Fittig reaction, after the French chemist Charles Adolphe Wurtz (1817–1884) and the German chemist Rudolph Fittig (1835–1910).

Wurtz reaction, after the French chemist Charles Adolphe Wurtz (1817–1884).

Wurzschmidt method, after the German chemist Bernhard Wurzschmidt (1895–1975).

Young method, after the British chemist Sydney Young (1857–1937).

Zachariasen's rule, after the Norwegian-American crystallographer Fredrik William Houlder Zachariasen (1906–1979).

Zaitsev-Rosenmund reaction, after the Russian chemist M. M. Zaitsev (1845–1904) and the German chemist Karl Wilhelm Rosenmund (1884–1965).

Zaitsev's rule, after the Russian chemist Aleksandr Mikhailovich Zaitsev (1841–1910). A.k.a. **Saytsev's rule** and **Saytzeff's rule**.

Zeisel method, after the American chemist Simon Zeisel (1854–1933).

Zemplén degradation of sugars, after the Hungarian chemist Géza Zemplén (1883–1956).

Zerewitinoff method, after the Russian chemist Fjodor Vasilyevich Zerevitinov (1874–1947). A.k.a. **Zerevitinov method.**

Zerner intermediate neglect of differential overlap, after the American chemist Michael Charles Zerner (1940–2000). A.k.a. **ZINDO.**

Ziegler method, after the German chemist Karl Ziegler (1898–1973).

Zimmermann reaction, after the German chemist Wilhelm Zimmermann (1910–1982).

Zincke-König reaction, after the German chemists Ernst Carl Theodor Zincke (1843–1928) and W. König (1878–1964).

Zincke nitration, after the German chemist Ernst Carl Theodor Zincke (1843–1928).

Zincke-Suhl reaction, as Zincke nitration and after the German chemist R. Suhl.

Zinin reaction, after the Russian chemist Nikolai Nikolayevich Zinin (1812–1880).

Zintl phases, after the German chemist Eduard Zintl (1898–1941).

Zintl rule, as Zintl phases.

Zwikker test, after the Dutch chemist and physicist Cornelis Zwikker (1900–1985).

References

[1] https://en.wikipedia.org/wiki/List_of_examples_of_Stigler%27s_law (retrieved August 18, 2018).
[2] https://en.wikipedia.org/wiki/List_of_organic_reactions (retrieved August 29, 2018).
[3] https://en.wikipedia.org/wiki/Category:Name_reactions (retrieved September 17, 2018).
[4] http://www.chem.yorku.ca/NAMED/ (retrieved September 18, 2018).
[5] Z. Wang, *Comprehensive Organic Name Reactions and Reagents*, John Wiley & Sons, Inc., Hoboken, NJ, USA, 2010, ISBN 978-0-47-170450-8.
[6] https://en.wikipedia.org/wiki/Stigler%27s_law_of_eponymy (retrieved August 18, 2018).
[7] https://en.wikipedia.org/wiki/List_of_examples_of_Stigler%27s_law (retrieved August 18, 2018).
[8] A. J. Rocke and A. J. Ihde, *J. Chem. Education* **63**, 309–310 (1986).

13 The naming of minerals

An exhaustive coverage of minerals is available online [1–4]. The International Mineralogical Association (IMA) [5] is responsible for the recognition of new minerals and new mineral names with a portfolio of approximately 6,500 names. However, 1,289 minerals discovered before 1959 were grandfathered, i.e. did not pass an official naming procedure. This chapter is devoted to a discussion of the principles used in the naming of minerals.

Many mineral names have been coined with little regard for euphony as perceived by English-speakers. The following portfolio also includes similar sounding names for unrelated minerals.

Petrography deals with the naming and systematization of rocks, such as basalt, gabbro, and granite [6]. This subject is outside the scope of this book.

13.1 The vocabulary used in mineral names

Crystal shape and physical properties are the dominating consideration in mineral nomenclature apart from epomyms and toponyms.

13.1.1 Word stems commonly used in mineral names

'clase', as in rhomboclase [$H_5Fe^{3+}O_2(SO_4)_2{\cdot}2H_2O$], from Greek *klasis*, cleavage, breaking, referring to this mineral's crystal form and perfect basal cleavage.

'clino', as in clinoptilolite [$(Na,K,Ca)_{2-3}Al_3(Al,Si)_2Si_{13}O_{36}{\cdot}12H_2O$], from Greek *kline*, couch, from *klinein*, to bend, to cause to slope. In this example from Greek *ptylon*, feather, referring to this mineral's appearance.

'hyper', as in hypercinnabar (hexagonal HgS), referring to this mineral's isomorphism with cinnabar (trigonal HgS) and metacinnabar (cubic HgS).

'meso', as in mesolite [$Na_2Ca_2(Al_2Si_3O_{10})_3{\cdot}8H_2O$] from Greek *mesos*, middle. The name mesolite indicates that this mineral's composition lies between that of natrolite ($Na_2Al_2Si_3O_{10}{\cdot}2H_2O$) and scolecite ($CaAl_2Si_3O_{10}{\cdot}3H_2O$).

'meta', as in metacinnabar (cubic HgS), referring to this mineral's isomorphism with cinnabar (trigonal HgS) and hypercinnabar (hexagonal HgS).

'ortho', as in orthoclase ($KAlSi_3O_8$), from Greek *ortho*, straight, right, true. The name orthoclase is from Greek *orthoklasis*, straight fracture.

'para', as in paracelsian, referring its isomorphism with celsian.

https://doi.org/10.1515/9783110612714-013

'pyro', as in pyromorphite [$Pb_5(VO_4)_3Cl$], from Greek *pyr, pyro-* fire. The name pyromorphite is from Greek *morphe*, form, shape, an allusion to the recrystallization reaction of the molten mineral.

13.1.2 Modification of existing mineral names

New mineral names can be derived from the name of a 'parent' mineral by a modification which indicates the relationship between the parent mineral and the new mineral. Thus, paracelsian (monoclinic-prismatic) and celsian (monoclinic) have the same composition, but are different dimorphs. The same is true of variscite (orthorhombic) and metavariscite (monoclinic).

The same prefix can be used in more than one meaning. For instance, metazeunerite is a less hydrated version of zeunerite.

The suffix 'oid' is another option to express similarity. Thus, chloritoid is to a large extent similar to chlorite.

13.2 Traditional names

While the etymology of modern mineral names is well documented the etymology of traditional mineral names can at times be ambiguous or even obscure.

Agate (SiO_2), according to ancient sources named after the Achates River (now Dirillo) in Southern Sicily, Italy. According to current opinion the river might well have been named after the mineral, which makes the etymology obscure.

Alabaster ($CaSO_4{\cdot}2H_2O$), of uncertain etymology. Suggested sources of the name have been the ancient Egyptian goddess of warfare Bast and the ancient Egyptian town Alabastron, with unknown location.

Alum [$KAl(SO_4)_2{\cdot}12H_2O$], from Latin *alumen, alumin-*, alun, literally bitter salt.

Anatase (TiO_2), from Greek *anatasis*, extension, referring to the fact that the vertical axis of the crystals is longer than in rutile.

Asbestos (silicate minerals), from Greek *asbestos*, unextinguishable, from *a(n)-*, non-, and *sbestos*, quenched, from *sbennynai*, to extinguish. The name of this mineral refers to an alchemistic belief that this material, once set afire, would be just as difficult to extinguish as it was to ignite.

Beryl [$Be_3Al_2(SiO_3)_6$], from Greek *beryllos*, beryl]. The Greek word is from a Sanskrit root meaning becoming pale.

Borax ($Na_2B_4O_7{\cdot}10H_2O$), ultimately from Persian *burah*, borax, literally white. A.k.a. **tincal**, from Malay *tingkal*, to tinkle.

[1]**Calamine** ($ZnCO_3$), so used in the UK; in the USA called smithsonite, after the British mineralogist James Smithson (1765–1829). Calamine is from Medieval Latin *calamina*, a corruption of Latin *cadmia*, from Greek *kadmeia ge*, Cadmean earth, cf. cadmium.

[2]**Calamine** ($Zn_2SiO_4{\cdot}H_2O$), so used in the USA. Calamine is from Medieval Latin *calamina*, a corruption of Latin *cadmia*, from Greek *kadmeia ge*, Cadmean earth, cf. cadmium.

Caliche (usually $NaNO_3$, but also $CaCO_3$), from Spanish *caliche*, flake of lime, pebble in a brick, ultimately from Latin *calx*, *calc-*, lime. A.k.a. **Chile saltpetre**.

Carborundum (SiC), from carbon and corundum, referring to this mineral's hardness which matches that of corundum. A.k.a. **moissanite**, after the French chemist Henri Moissan (1852–1907).

Cassiterite (SnO_2), from Greek *kassiteros*, tin, possibly from Elamite *kassi-ti-ra*, coming from the land of the Kassites.

Cerussite ($PbCO_3$), from Latin *cerussa*, white lead (lead carbonate), possibly of Greek origin.

Chrysocolla [$(Cu^{2+},Al)_2H_2Si_2O_5{\cdot}nH_2O$], from Greek *chrysos*, gold, and *kolla*, glue, referring to its flux in soldering gold.

Cinnabar (HgS), from Greek *kinnabari*, cinnabar, of obscure origin, possibly from Persian *shangarf*, cinnabar.

Corundum (Al_2O_3), ultimately derived from Sanskrit *kuruvinda*, ruby.

Electrum (Au,Ag), from Greek *elektron*, pale gold. Both this word and Greek *elektron*, amber, refer to their similar colors.

Galena (PbS), from Latin *galena*, lead ore, dross from smelting lead; of unknown origin.

Graphite (C), from Greek *graphein*, to write, referring to the use of this mineral in pencils etc.

Gypsum ($CaSO_4{\cdot}2H_2O$), from Greek *gypsos*, chalk, gypsum, of Semitic origin, akin to Arabic *jibs*, plaster, mortar.

Halite (NaCl), from Greek *hals*, *hal(o)-*, salt.

Jade, from Spanish *piedra de ijada*, loin stone, referring to its reputed efficacy in healing ailments of the loins and kidneys. The name jade applies to two minerals:

jadeite $[Na(Al,Fe^{3+})(SiO_3)_2]$, from jade, and **nephrite** $[Ca_2(Mg.Fe)_5Si_8O_{22}(OH)_{21}]$, from New Latin *lapis nephriticus*, literally kidney stone, used as a charm against diseases of the kidney.

Lime (CaO), from the Proto-Indo-European root **(s)lei-*, slime, slimy, sticky.

Litharge (PbO), from Greek *lithargyros*, spume of silver, literally stone silver, i.e. the mineral residue from silver refining.

Malachite $[Cu_2(CO_3)(OH)_2]$, from Greek *molochites*, malachite, from *moloche, malache*, mallow (*Malvia*), referring to this mineral's green color.

Massicot (PbO), from Italian *marzacotta*, potter's glaze, ultimately from Arabic *sabb qubti*, Egyptian alum, i.e. iron and/or aluminum sulfate. A.k.a. **massicotite.**

^{1}**Minium** (HgS), after the river Minius in Iberia (now Spanish Miño, Portuguese Minho) close to the main Roman cinnabar mines. The name of the river is ultimately from Iberian.

^{2}**Minium** (Pb_3O_4), used as adulterant of ^{1}minium and given the same name.

^{1}**Niter** (KNO_3), ultimately from Ancient Egyptian *netjeri.*

^{2}**Niter** (Na_2CO_3), ultimately from Ancient Egyptian *netjeri.*

Orpiment (As_2S_3), from Latin *auripigmentum*, orpiment, from *aurum*, gold, and *pigmentum*, coloring matter, pigment, paint, from *pingere, pict-*, to paint.

Pitchblende (UO_2), from German *Pechblende*, pitchblende, from *Pech*, pitch, and *Blende*, minerals looking deceptively like galena, from *blenden*, to blind, to deceive, from *blind*, blind.

Plasma (SiO_2), from Greek *plasma*, something molded, from *plassein*, to mold. A.k.a. **green chalcedony.**

Quartz (SiO_2), from German *Quarz*, possibly from Czech *tvrdý*, hard.

Realgar (As_4S_4), ultimately from Arabic *rahj al-ghar*, powder of the mine.

Rutile (TiO_2), from Latin *rutilus*, reddish, referring to a red color observed when the crystal is viewed by transmitted light.

Sard (SiO_2), probably named after the town of Sardis, the capital of the ancient kingdom of Lydia, Asia Minor.

Sardonyx (SiO_2), from sard, q.v., and onyx, q.v.

Schorl $[NaFe_3Al_6(BO_3)_3Si_6O_{18}(OH)_4]$, from German *Schörl*, waste.

Schorlomite $\{Ca_3(Ti,Fe)_2[(Si,Fe)O_4]_3\}$, from schorl, q.v., and homo-, referring this mineral's resemblance to schorl.

Talc [$Mg_3Si_4O_{10}(OH)_2$], from Medieval Latin *talcum*, from Arabic *talq*, from Persian *talk*, talc. Historically this name has also been used for related minerals such as mica and selenite.

Zircon ($ZrSiO_4$), ultimately from Persian *zar*, gold, and *gun*, colored.

13.2.1 Gemstones

The names of gemstones such as amethyst can be based on mythological concepts such as their use as charms.

Amethyst (SiO_2), from Greek *amethystos*, remedy against drunkenness, from *a(n)-*, not, and *methystos*, drunk, from *methyskein*, to make drunk, from *methy*, wine. This gem was named according to the ancient belief that it protected its owner from drunkenness.

Diamond (C), from Medieval Latin *adiamas, adiamant-*, from Greek *adamas, adamant-*, diamond, steel. The commonly assumed relationship of the Greek name with Greek *adamas*, proper, unalterable, unbreakable, untamed, could well be folk etymology. A loan from Akkadian has been suggested. As used over time the name diamond refers to the hardness of this mineral. Black diamonds are called **carbonado**, from Portuguese, literally carbonated. Non-gem grade or industrial diamonds are known as **ballas**, ultimately from balas, a rose red variety of spinel, from Persian *Badakshan*, a district near Samarkand where this spinel is found, and **bort**, ultimately from Old Norse *brot*, fragment.

Emerald [$Be_3Al_2(SiO_3)_6$], via Vulgar Latin *esmeraldus* from Greek *smaragdos*, emerald, ultimately from Semitic *baraq*, to shine.

Garnet (a group of nesosilicates), from Latin *granatus*, garnet, from *granum*, grain, seed. The Latin name of the pomegranate (*Punica granatum*) was *malum granatum* and *pomum granatum* in Medieval Latin, literally seedy apple, and the mineral called after the color of this fruit.

Onyx (SiO_2), from Greek *onyx*, claw, fingernail, referring to this mineral's flesh-colored bands.

Opal ($SiO_2 \cdot nH_2O$), via Latin *opalus*, opal, possibly ultimately from a source akin to Sanskrit *upala*, precious stone. A.k.a. **girasol**, from Italian *girare*, to turn, and *sole*, sun.

Peridot [$(Fe,Mg)_2SiO_4$], of obscure etymology. Latin *paederos, paederot-*, opal, and Arabic *faridat*, gem, have been suggested as source.

Ruby (Al_2O_3,Cr), from Latin *ruber*, dark red.

Sapphire (Al_2O_3), via Latin *sapphirus, sappirus* from Greek *sappheiros*, from Hebrew *sappir*, sapphire. The ultimate source might well be Sanskrit *shanipriya*, emerald, literally dear to Saturn, from *shani*, Saturn, and *priya*, dear.

Topaz [$Al_2SiO_4(F,OH)_2$], a name related to the Red Sea island Zabargad, Egypt, called *Topazios* by the Greeks, and known in antiquity as the source of this gem. However, the island could just as likely have been named after the gem. The name topaz could also ultimately be related to Sanskrit *tapas*, heat, fire.

13.2.1.1 Famous named diamonds

An extensive list of named diamonds is available in Wikipedia [7].

Cullinan diamond (3106.75 carat), the largest gem-quality rough diamond ever found. It was found on January 26, 1905 in the Premier no. 2 mine in Cullinan, South Africa. The diamond was named after the mine's chairman, Thomas Cullinan (1862–1936).

Hope diamond (≈ 112 carat), first recorded as purchased in India in 1666 by the French gem merchant Jean-Baptiste Tavernier (1605–1689) and called the Blue Tavernier. The diamond was sold to King Louis XIV of France (1638–1715) in 1668, and stolen in 1791. In 1839 the diamond reappeared in the collection of the London banking family Hope where it acquired its name.

Koh-I-Noor (≈ 793 carat), from *Persian koh-i-noor*, mountain of light.

Orlov diamond (189.62 carat). This diamond received its name when the Russian Count Grigori Grigoriyevich Orlov (1734–1783) bought it from a dealer in Amsterdam and gave it, unsuccessfully, as a gift to the Russian empress Catherine the Great (1729–1796) to regain her love after she had forsaken him for her new lover Grigori Aleksandrovich Potemkin (1739–1791).

13.3 Modern names

Most modern mineral names have the ending -lite (from Greek *lithos*, stone) or -ite, from the Latin suffix *-itus*, denoting origin or relationship. Interestingly, both endings have been used to create eponyms in honor of the same scientist, i.e. in the cases of the French mineralogist, petrologist, and meteoriticist Gabriel Auguste Daubrée (1814–1896] and the Swedish chemist Jöns Jacob Berzelius (1779–1848). The current naming rules prescribe that only one mineral may be named after a given person.

Daubréeite [BiO(OH,Cl)], chemically unrelated with daubréelite.

Daubréelite [$(Fe^{2+},Cr^{3+})S_4$], chemically unrelated with daubréeite.

Berzelianite (Cu_2Se), chemically unrelated with berzeliite.

Berzeliite [$(Ca,Na)_3(Mg,Mn)_2(AsO_4)_3$], chemically unrelated with berzelianite.

Three lunar minerals, armalcolite, pyroxferroite, and tranquillityite, were brought back by the crews of the Apollo missions. Two of them received names which specifically mark their lunar origin. However, all of these three minerals have subsequently been found in remote locations as native terrestrial material. This should play into the hands of conspiracy theorists who consider the Apollo program a monumental hoax.

13.3.1 Eponyms and demonyms

Abelsonite ($C_{31}H_{32}N_4Ni$), after the American geochemist Philip Hauge Abelson (1913–2004).

Adamite [$Zn_2AsO_4(OH)$], after the French mineralogist Gilbert Joseph Adam (1795–1881).

Adamsite [$NaY(CO_3)_2{\cdot}6H_2O$], after the Canadian geologist Frank Dawson Adams (1859–1942).

Aegirine ($Na\ Fe^{3+}Si_2O_6$), after Ægir, the Old Norse god of the sea. A.k.a. **acmite**, from Greek *achme*, spear point, referring to this mineral's habit. Cf. neptunite.

Alexandrite ($BeAl_2O_4$), named in honor of the Russian Czar Alexander I (1777–1825). This mineral displays the Russian imperial colors red and green.

Allanite [$(Ce,Ca,Y,La)_2(Al,Fe^{3+})_3(SiO_4)_3(OH)$], after the British mineralogist Thomas Allan (1777–1833). A.k.a. **orthite**, from orth(o)-, referring to this mineral's straight radii.

Armalcolite [$(Mg,Fe^{2+})Ti_2O_5$], first found on the Moon and named after the three American astronauts Neil Armstrong (1930–2012), Buzz Aldrin (born 1930), and Michael Collins (born 1930). Later this mineral was also found in Australia.

Armstrongite ($CaZrSi_6O_{15}{\cdot}3H_2O$), after the American astronaut Neil Armstrong (1930–2012), but not a lunar mineral.

Arupite [$Ni_3(PO_4)_2{\cdot}8H_2O$], after the Danish scientist Hans Henning Arup (1928–2012).

Avicennite (Tl_2O_3), after the Persian-Arab physician and philosopher Avicenna (Abu Ali Ibn Sina) (980–1037).

Avogadrite [$(K,Cs)BF_4$], after the Italian chemist Lorenzo Amedeo Carlo Avogadro Conte de Quaregno e Cerreto (1776–1856).

Azoproite [$(Mg,Fe^{2+})(Fe^{3+},Ti,Mg)BO_5$], after the abbreviation AZOPRO for the Russian equivalent of International Association for the Study of Deep Zones of the Earth's Crust in commemoration of this organization's 1969 meeting at Lake Baikal, Russia.

Baddeleyite (ZrO_2), after the British engineer and amateur geologist Joseph Baddeley.

Bayerite [$Al(OH)_3$], after the German chemist Karl Josef Bayer (1847–1904).

Becquerelite {$Ca[(UO_2)_3O_2(OH)_3]\cdot 8H_2O$}, after the French physicist Antoine Henri Becquerel (1852–1908).

Bindheimite [$Pb_2Sb_2O_{6.75}(OH)_{0.25}$], after the German chemist Johann Jacob Bindheim (1750–1825).

Biotite [$K(Mg,Fe)_3(AlSi_3O_{10})(F,OH)_2$], after the French physicist Jean-Baptiste Biot (1774–1862).

Böhmite [$AlO(OH)$], after the German geologist and paleontologist Johannes Böhm (1857–1938).

Braggite [$(Pt,Pd,Ni)S$], after the British physicist William Henry Bragg (1862–1942) and his son, the British physicist William Lawrence Bragg (1890–1971). This mineral was the first one to be discovered with assistance of their X-ray method.

Brownmillerite [$Ca_2(Al,Fe)_2O_5$], after the American chemist Lorrin Thomas Brownmiller (1902–1990). A.k.a. **celite.**

Bunsenite (NiO), after the German chemist Robert Wilhelm Eberhard Bunsen (1811–1899).

Bustamite [$(Mn,Ca)_3Si_3O_9$], after the Mexican general Anastasio Bustamente (1780–1853).

Cannizzarite ($Pb_4Bi_5S_{11}$), after the Italian chemist Stanislaus Cannizzaro (1826–1910).

Carlsbergite (CrN), after the Carlsberg Foundation, Copenhagen, Denmark.

Carnallite ($KMgCl_3\cdot 6H_2O$), after the Prussian mining engineer Rudolf von Carnall (1804–1874). A.k.a. **carnalite.**

Carnotite [$K_2(UO_2)_2V_2O_8\cdot 3H_2O$], contrary to popular belief named after the French mining engineer, chemist, and politician Marie-Adolphe Carnot (1839–1920), not after the far more famous French physicist Nicolas Léonard Sadi Carnot (1796–1832).

Celsian ($BaAl_2Si_2O_8$), after the Swedish astronomer, physicist, and mathematician Anders Celsius (1701–1744).

Chaoite (C_n), after the Chinese-American petrologist Edward C. T. Chao (1919–2008). A.k.a. **white carbon.**

Coesite (SiO_2), after the American chemist Loring Coes, Jr. (1915–1978).

Cordierite ($Mg_2Al_4Si_5O_{18}$), after the French mining engineer and geologist Pierre Louis Antoine Cordier (1777–1861). A.k.a. **dichroite**, from dichroism, and **iolite**, from Greek *ion*, the flower violet.

Covellite (CuS), after the Italian mineralogist Niccolò Covelli (1790–1829).

Curite $\{Pb_3[(UO_2)_8(OH)_6]\cdot 3H_2O\}$, after the French physicist Pierre Curie (1859–1906).

Dinite (1-ethyl-1,5,5,13-tetramethyltetradecahydrophenanthrene), after the Italian physicist Olinto Dini (1802–1866).

Dolomite $[CaMg(CO_3)_2]$, after the French naturalist and geologist Déodat Gratet de Dolomieu (1750–1801). This is the rare case where the type locality, the Dolomites, Italy has been named after a mineral found there, and not the other way around.

Eskimoite ($Ag_7Pb_{10}Bi_{15}S_{36}$), after the Eskimo people.

Friedelite $[(Mn^{2+})_8Si_6O_{15}(OH,Cl)_{10}]$, after the French chemist and mineralogist Charles Friedel (1832–1899).

Gadolinite $[(Ce,La,Nd,Y)_2FeBe_2Si_2O_{10}]$, after the Finnish chemist, physicist, and mineralogist Johan Gadolin (1760–1852). A.k.a. **ytterbite**, after the locality Ytterby, Sweden.

Gahnite ($ZnAl_2O_4$), after the Swedish chemist Johan Gottlieb Gahn (1745–1818).

Gaylussite $[Na_2Ca(CO_3)_2\cdot 5H_2O]$, after the French physicist and chemist Joseph Louis Gay-Lussac (1778–1850).

Geikielite ($MgTiO_3$), after the British geologist Archibald Geikie (1835–1924).

Glauberite $[Na_2Ca(SO_4)_2]$, after the related Glauber's salt ($Na_2SO_4\cdot 10H_2O$).

Joliotite $[(UO_2)(CO_3)\cdot 2H_2O$, named in honor of the French radiochemists Frédéric Joliot-Curie (1897–1956) and Irene Joliot-Curie (1900–1958).

Keatite (SiO_2), after the American chemist Paul Powell Keat (born 1929), who synthesized this mineral before it was found in nature.

Kieserite ($MgSO_4\cdot H_2O$), named in honor of the German professor of medicine Dietrich Georg von Kieser (1779–1862).

Kurchatovite $[Ca(Mg,Mn^{2+},Fe^{2+})B_2O_5]$, named in honor of the Russian physicist Igor Vasilyevich Kurchatov (1903–1960).

Laurite (RuS_2), after Laura R. Joy, the wife of the American chemist Charles Arad Joy (1823–1891).

Lechatelierite (SiO_2), after the French chemist Henry Louis Le Chatelier (1850–1936).

Lewisite [$(Ca,Fe^{3+},Na)_2(Sb,Ti)_2O_7$], after the British geologist William James Lewis (1847–1926).

Liebigite [$Ca_2(UO_2)(CO_3)_3 \cdot 11H_2O$], after the German chemist Justus von Liebig (1803–1873).

Linnaeite [$Co^{2+}(Co^{3+})_2S_4$], after the Swedish botanist Carl von Linné (Carolus Linnaeus) (1707–1778).

Livingstonite ($HgSb_4S_8$), after the British missionary and explorer David Livingstone (1813–1873).

Lonsdaleite (C_n), after the British crystallographer Kathleen Lonsdale (1903–1971).

Lorandite ($TlAsS_2$), after the Hungarian physicist Baron Loránd von Eötvös (Eötvös Loránd Ágoston) (1848–1919).

Meyerhofferite [$Ca_2B_6O_6(OH)_{10} \cdot 2H_2O$], after the German chemist Wilhelm Meyerhoffer (1864–1906) who synthesized this mineral before it was found in nature.

Mgriite (Cu_3AsSe_3), after the Russian initialism for the Moscow Geological Exploration Institute, MGRI.

Morganite ($Be_3Al_2Si_6O_{18}$), mamed in honor of the American industrialist John Pierpont Morgan (1837–1913).

Neptunite [$KNa_2Li(Fe,Mg,Mn)_2Ti_2Si_8O_{24}$], after Neptunus, the Roman god of the sea, of Etruscan origin, referring to the fact that this mineral was found associated with aegirine, q.v.

Nosean [$Na_8Al_6Si_6O_{24}(SO_4) \cdot H_2O$], after the German mineralogist Karl Wilhelm Nose (1753–1835).

Olympite [$LiNa_5(PO_4)_2$], named in honor of the 1980 Olympic Games in Moscow, Russia.

Parisite [$Ca(Ce,La)_2(CO_3)_2F_2$], after its discoverer José J. Paris (deceased 1849), the proprietor of an emerald mine in Muzo, Colombia.

Patronite (VS_4), after the Peruvian metallurgist Antenor Rizo-Patron (1866–1948).

Pavonite [$(Ag,Cu)(Bi,Pb)_3S_5$], from Latin *pavo, pavon-*, peacock. After the Canadian mineralogist Martin Alfred Peacock (1898–1950).

Pinnoite ($MgB_2O_4 \cdot 3H_2O$), after the German mining official Johann Friedrich Hermann Pinno (1831–1902).

Prehnite [$Ca_2Al(AlSi_3O_{10})(OH)_2$], after the Dutch colonel and governor of Cape of Good Hope (Cape Colony) Hendrik von Prehn (1733–1785).

Priceite [$Ca_2B_5O_7(OH)_5 \cdot H_2O$], after the American metallurgist, chemist, and assayer Thomas Price (1837–1912).

Proustite (Ag_3AsS_3), after the French chemist Joseph Louis Proust (1754–1826).

Ramsdellite (MnO_2), after the American mineralogist Lewis Stephen Ramsdell (1895–1975).

Roscoelite [$K(V,Al,Mg)_2AlSi_3O_{10}(OH)_2$], after the British chemist Henry Enfield Roscoe (1833–1915) who was the first to prepare pure vanadium.

Russellite (Bi_2WO_6), after the British mineralogist Arthur Edward Ian Montagu Russell (1878–1964).

Scheelite ($CaWO_4$), after the Swedish chemist and apothecary Carl Wilhelm Scheele (1742–1786).

Schoepite ($UO_3 \cdot 2H_2O$), after the Belgian mineralogist Alfred Schoep (1881–1966).

Scholzite [$CaZn_2(PO_4)_2 \cdot 2H_2O$], after the German chemist and mineral collector Adolf Scholz (1894–1950).

Semseyite ($Pb_9Sb_8S_{21}$), after the Hungarian mineralogist Andor von Semsey (Semsey Andor) (1833–1923).

Senarmontite (Sb_2O_3), after the French mineralogist Henri Hureau du Sénarmont (1808–1862).

Sillénite ($Bi_{12}SiO_{20}$), after the Swedish chemist Bo Lars Gunnar Sillén (1916–1970).

Simonellite [1,1-dimethyl-7-(propan-2-yl)-1,2,3,4-tetrahydrophenanthrene], after the Italian geologist Vittorio Simonelli (1860–1929).

Sklodowskite {$Mg[(UO_2)(SiO_3OH)]_2(H_2O)_6$}, after the Polish-French physicist and chemist Marie Skłodowska Curie (1867–1934).

Sørensenite [$Na_4Be_2Sn(Si_3O_9) \cdot 2H_2O$], after the Danish mineralogist and petrographer Henning Sørensen (1926–2013).

Sperrylite ($PtAs_2$), after the American chemist Francis Louis Sperry (1861–1906).

Stenhuggarite ($CaFeSbAs_2O_7$), after the American scientist Brian Harold Mason (1917–2009), jocularly from Swedish *stenhuggare*, mason.

Stenonite [$(Sr,Ba,Na)_2Al(CO_3)F_5$], after the Danish physician, anatomist, and naturalist Nicolaus Steno (Niels Steensen) (1638–1686).

Stishovite (SiO_2), after the Russian physicist Sergei Mikhailovich Stishov (born 1937) wwho synthesized this mineral before it was found in nature.

Stokesite ($CaSnSi_3O_9·2H_2O$), after the British physicist George Gabriel Stokes (1819–1903).

Stolzite ($PbWO_4$), either after the Czech scientist Johann Anton Stolz (1778–1855) and/or after the Czech mineralogist Joseph Alexis Stolz (1803–1896), probably father and son.

Stromeyerite (AgCuS), after the German mineralogist and chemist Friedrich Stromeyer (1776–1835).

Struvite ($NH_4MgPO_4·6H_2O$), after the German geographer and geologist Heinrich Christian Gottfried von Struve (1772–1851).

Stumpflite [$Pt(Sb,Bi)_2$], after the Austrian mineralogist Eugen Friedrich Stumpfl (1931–2004).

Sylvite (KCl), after the German-Dutch physician and chemist Franciscus Sylvius (Franz de le Boë) (1614–1672).

Szaibelyite ($MgHBO_3$), after the Hungarian mine surveyor Stephan Szaibely (1777–1855).

Tapiolite [$(Fe,Mn)(Ta,Nb)_2O_6$], after Tapio, the Finnish god of forests, animals, and hunting. The meaning of this name is unknown.

Tennantite [$Cu,Fe)_{12}As_4S_{13}$], after the British chemist Smithson Tennant (1761–1815).

Tenorite (CuO), after the Italian botanist Michele Tenore (1780–1861).

Thalenite [$Y_3Si_3O_{10}(OH)$], after the Swedish physicist Tobias Robert Thalén (1827–1905).

Thenardite (Na_2SO_4), after the French chemist Louis Jacques Thenard (1777–1857).

Thomsenolite ($NaCaAlF_6·H_2O$), after the Danish chemist Hans Peter Jørgen Julius Thomsen (1826–1909).

Thomsonite [$NaCa_2(Al_5Si_5)_{20}·6H_2O$], after the British chemist Thomas Thomson (1773–1852).

Thoreaulite ($SnTa_2O_5$), after the Belgian nineralogist Jacques Thoreau (1886–1973).

Thorianite (ThO_2), from thoria (thorium dioxide).

Thorite [$(Th,U)SiO_4$], named for Thor, the Old Norse god of thunder.

Thortveitite [$(Sc,Y)_2Si_2O_7$], after the Norwegian engineer Olaus Thortveit (1872–1917).

Tiettaite [$(Na,K)_{17}Fe^{3+}TiSi_{16}O_{29}(OH)_{30}·2H_2O$], from Lappish *tietta*, science, knowledge, honoring the USSR Academy of Science.

Torbernite [$Cu^{2+}(UO_2)_2(PO_4)_2 \cdot 8{-}12H_2O$], after the Swedish chemist and naturalist Torbern Olof Bergman (1735–1784).

Trevorite ($NiFe^{3+}O_4$), after the South African major and mining inspector Tudor Gruffydd Trevor (1865–1958).

Troilite (FeS), after the Italian astronomer Domenico Troili (1722–1792).

Twinnite [$Pb(Sb,As)_2S_4$], after the Canadian mineralogist Robert Mitchell Thompson (1918–1967). The name was arrived at by equaling Thompson with 'son of Thomas' and considering that the name Thomas is from Aramaic where it means twin.

Ulexite [$NaCaB_5O_6(OH)_6 \cdot 5H_2O$], after the German chemist Georg Ludwig Ulex (1811–1883) who discovered it.

Ullmannite (NiSbS), after the German chemist and mineralogist Johann Christoph Ullmann (1771–1821).

Uvarovite [$Ca_3Cr_2(SiO_4)_3$], after the Russian statesman and amateur mineral collector Count Sergei Semyonovich Uvarov (1765–1855).

Valentinite (Sb_2O_3), after the unattested 15th century German monk and alchemist Basilius Valentinus. Publications ascribed to Valentinus appeared no earlier than 1599. It is likely that the renowned name Valentinus has been used as an alias by later writers such as the German alchemist Johann Tölde (1565–1614). The name Valentinus is from Latin *Valens*, strong, vigorous, healthy.

Vanthoffite [$Na_6Mg(SO_4)_4$], after the Dutch chemist Jacobus Henricus van 't Hoff Jr. (1852–1911).

Vaterite (μ-$CaCO_3$), after the German soil and forestry scientist Heinrich Vater (1859–1930) who synthesized this mineral.

Villiaumite (NaF), after the French colonial artillery officer and explorer Maxime Charles Villiaume (born 1858) who discovered this mineral in 1908.

Vrbaite ($Hg_3Tl_4Sb_2As_8S_{20}$), after the Czech mineralogist Karel Vrba (1845–1922).

Wavellite [$Al_3(PO_4)_2(OH,F)_3 \cdot 5H_2O$], after the British physician, botanist, historian, and naturalist William Wavell (1750–1829) who discovered this mineral.

Whewellite ($CaC_2O_4 \cdot H_2O$), after the English polymath, naturalist, and scientist William Whewell (1794–1866).

Willemite (Zn_2SiO_4), named in honor of King Willem I of the Netherlands (1772–1843).

Witherite ($BaCO_3$), after the British physician and mineralogist William Withering (1741–1799).

Wöhlerite [$NaCa_2(Zr,Nb)Si_2O_7(O,OH,F)_2$], after the German chemist Friedrich Wöhler (1800–1882). A.k.a. **wohlerite.**

Wollastonite ($CaSiO_3$), after the British mineralogist and chemist William Hyde Wollaston (1766–1828).

Wulfenite ($PbMoO_4$), after the Austrian mineralogist Franz Xaver von Wulfen (1728–1805).

Wurtzite (ZnS), after the French chemist Charles Adolphe Wurtz (1817–1884).

Wüstite (FeO), after the German geologist, paleontologist, and botanist Ewald Wüst (1875–1934).

13.3.2 Toponyms

It is interesting to note that no less than three minerals have been named after Sri Lanka, drawing on its modern, its Arabic, and its Sanskrit name.

Admontite ($MgB_6O_{10}\cdot 7H_2O$), after the locality Admont, Styria, Austria.

Adularia [$(Na,K)AlSi_3O_8$], after the Adula Mountains, Switzerland.

Alabandite (MnS), after the ancient town of Alabanda, Caria, Asia Minor.

Alacranite (As_4S_4), after its locality Alacran, Pampa Larga, Chile.

Amosite {$(Al,Fe,Mg)_2[(Al,Si)_4O_{11}(OH)_2]$}, after its locality, the town of Amosa, South Africa, which in turn was named after AMOSA, Asbestos Mine of South Africa. A.k.a. **grünerite**, after the Swiss-French chemist and mining engineer Emmanuel Ludwig Grüner (1809–1883).

Andalusite (Al_2SiO_5), after its type locality Ronda Massif, Malaga, Andalusia, Spain.

Andesine [$(Na,K)(Si,Al)_4O_8$], after the Andes Mountains, South America.

Anilite (Cu_7S_4), after its locality Ani mine, Akita Prefecture, Japan.

Aragonite (λ-$CaCO_3$), after its type locality Molina de Aragón, Spain.

Bentonite ($Al_2O_3\cdot 4SiO_2\cdot H_2O$), after its locality Fort Benton, WY, USA.

Boléite [$Pb_{26}Ag_{10}(Cu^{2+})_{24}Cl_{62}(OH)_{48}\cdot 3H_2O$], after its discovery locality, Boléo, Mexico.

Carpathite (coronene), after the Carpathian Mountains.

Ceylonite [$Al_2(Fe,Mg)O_4$], after this mineral's locality, Ceylon. A.k.a. **pleonaste**, from Greek *pleonastos*, abundant, referring to the many faces of the crystals of this mineral.

Chalcedony (SiO_2), after the ancient Greek town of Chalcedon in Asia Minor.

Columbite [$(Fe,Mn,Mg)Nb_2O_6$], from New Latin *Columbia*, North America.

Cristobalite (SiO_2), after its discovery locality, Cerro San Cristobal, Mexico.

Cymrite [$BaAl_2Si_2(O,OH)_8{\cdot}H_2O$], from Welsh *Cymru*, Wales, thus named after Wales.

Diaboleite [$Cu^{2+}Pb_2Cl_2(OH)_4$]; nothing diabolical implied here! We have to do with a negative toponym, from Greek *dia*, distinct from, and boléite, referring to the difference between this mineral and boléite.

Epsomite ($MgSO_4{\cdot}7H_2O$), after the deposits associated with the mineral waters at Epsom, Surrey, UK. A.k.a. **Epsom salt**.

Evenkite (tetracosane), after the Evenkia district, Russia.

Fichtelite [(1*S*,4a*S*,4b*S*,7*S*,8a*S*,10a*S*)-1,4a-dimethyl-7-(propan-2-yl)tetradecahydrophenanthrene]. One of the few organic minerals. After the German Fichtel Mountains where it was first found.

Galaxite ($MnAlO_4$), after this mineral's locality, Galax, VA, USA.

Hatrurite (Ca_3SiO_5), after its discovery locality, Hatrurim Formation, Negev, Israel. A.k.a. **alite.**

Hercynite ($FeAl_2O_4$), from hercyn-, thus named after its locality Poběžovice, Czech Republic.

Illite {$(K,H_3O)(Al,Mg,Fe)_2(Si,Al)_4O_{10}[(OH)_2,(H_2O)]$}, named after the state of Illinois, referring to its first description as occurring in the Maquoketa shale, in Calhoun County, IL, USA.

Ilmenite ($FeTiO_3$), after its discovery locality, Lake Ilmen, Russia. The toponym might be from its Finnish name *Ilmajärvi*, literally air lake.

Kaolinite [$Al_2Si_2O_5(OH)_4$], after the Chinese village Gaoling, literally high ridge, where this mineral was first found. A.k.a. **lithomarge**, from Greek *lithos*, stone, and Latin *marga*, marl.

Karelianite (V_2O_5), after its discovery locality, the Karelian schist belt, Finland. The toponym is after the Karelian people, possible named as herdsmen, from Finnish *karja*, herd.

Kernite [$Na_2B_4O_6(OH)_2{\cdot}3H_2O$], after its discovery locality, Kern County, CA, USA. The toponym is after the Kern River, named after the American cartographer Edward Meyer Kern (1822/1823–1863).

Kladnoite (phthalimide), after its type locality, the Kladno coal basin, Czech Republic.

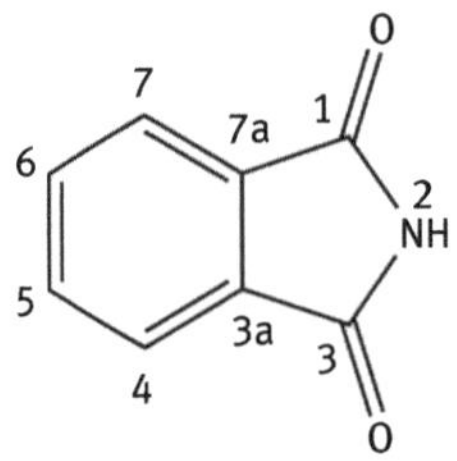

phthalimide

Larnite (Ca_2SiO_4), after its discovery location Scawt Hill, Larne, Northern Ireland. A.k.a. **belite**.

Lauta mass [$Fe(OH)_3$], after an aluminum plant in the town of Lauta, Saxony, Germany.

Lautarite [$Ca(IO_3)_2$], after this mineral's type locality, Oficina Lautaro, Chile.

Magnesite ($MgCO_3$), after Magnesia, the coastal district of ancient Thessaly, Greece.

Marcasite (FeS_2), ultimately from Assyrian *Markashitu*, an ancient province thought to be located in the northeastern part of Persia.

Mbobomkulite [$(Ni,Cu^{2+})Al_4(NO_3,SO_4)_2(OH)_{12}\cdot 3H_2O$], after its discovery locality, the Mbobo Mkulu cave, South Africa.

Montmorillonite [$(Al,Mg)_2Si_4O_{10}(OH)_2\cdot nH_2O$], after this mineral's locality, Montmorillon, Département Vienne, France. A.k.a. **smectite**, from Greek *smektos*, smeared, from *smechein*, to wipe off, to cleanse, referring to its appearance.

Sassolite (H_3BO_3), after Sasso Pisano, Italy, where its was found. Italian *sasso* means stone, rock, thus the mineral name is literally stonestone.

Serendibite [$Ca_2(Mg,Al)_6(Si,Al,B)_6O_{20}$], from Arab *Serendib*, Sri Lanka.

Sinhalite ($MgAlBO_4$), from Sanskrit *Sinhala*, Sri Lanka.

Skutterudite ($CoAs_3$), after this mineral's locality, Skutterud, Norway.

Spessartine [$Mn_3Al_2(SiO_4)_3$], after the German mountain range Spessart, the discovery locality. The name Spessart, earlier Spechtshardt, means literally woodpecker forest.

Srilankite [$(Ti,Zr)O_2$], after its discovery locality Sabaragamuva, Rakwana, Sri Lanka.

Sverigeite [$NaMn^{2+}MgSn^{4+}Be_2Si_3O_{12}(OH)$], after Swedish *Sverige*, Sweden, a tongue-twister for non-Swedish speakers.

Sylvanite [$(Au,Ag)_2Te_4$], after this mineral's locality Transylvania, Romania. Transylvania is ultimately from Latin *trans*, beyond, and *silva*, forest.

Tranquillityite [$(Fe^{2+})_8(Zr,Y)_2Ti_3Si_3O_{24}$], after the Moon's Sea of Tranquility, where this mineral was first found. Later it was also found in several locations on Earth.

Turquoise [$CuAl_6(PO_4)_4(OH)_8·4H_2O$], from French *turquois*, Turkish.

Variscite ($AlPO_4·2H_2O$), from Medieval Latin *Variscia*, Vogtland, Germany, the locality of this mineral.

Vuonnemite [$Na_{11}Nb_2TiSi_4O_{17}(PO_4)_2F_2$], after this mineral's locality, the Vuonnemi River, Kola Peninsula, Russia.

Weddellite (calcium oxalate dihydrate), after its discovery locality, the floor of the Weddell Sea, Antarctica. The Weddell Sea was named after the British sailor James Weddell (1787–1834).

Wittichenite ($Cu_6Bi_2S_6$), after this mineral's locality Wittichen, Baden-Württemberg, Germany.

Xonotlite [$Ca_6Si_6O_{17}(OH)_2$], after its locality Tetela de Xonotla, Puebla, Mexico. A.k. a. **eakleite**, after the American mineralogist Arthur Starr Eakle (1862–1931), and **jurupaite**, after the Jurupa Mountains, Riverside, CA, USA.

13.3.3 Names based on properties

Acanthite (Ag_2S), from Greek *akantha*, thorn, referring to the shape of the crystals. A.k.a. **argentite**, from Latin argentum, silver.

Actinolite [$Ca_2(Mg,Fe)_5Si_8O_{22}(OH)_2$], from [1]actino-, referring to this mineral's often fibrous, radiated, or columnar crystal shape.

Acumninite [$SrAlF_4(OH)·H_2O$], from Latin *acumen, acumin-*, spear head, referring to the characteristic crystal shape of this mineral.

Agalmatolite [$Al_2Si_4O_{10}(OH)_2$], from agalma, a primitive Greek statue of a god, from Greek *agallein*, to adorn, referring to this mineral's use in stone carving.

Albite ($NaAlSi_3O_8$), from Latin *albus*, white, referring to this mineral's commonly white color.

Amblygonite [$(Li,Na)Al(PO_4)(F,OH)$], from Greek *ambly*, blunt, and *gonos*, angle, referring to the obtuse cleavage angle.

Analcime ($NaAlSi_2O_6·H_2O$), from Greek *analkimos*, weak, referring to the weak electrostatic charge that develops when it is heated or rubbed. A.k.a. **analcite**.

Ancylite [$Sr_3Ce_4(CO_3)_7(OH)\cdot 3H_2O$], from Greek *ankylos*, crooked, referring to the rounded and distorted character of this mineral's crystals.

Anorthite ($CaAl_2Si_2O_8$), from Greek *anorthos*, not right, referring to its oblique crystals.

Anthophyllite, [$\square(Mg,Fe^{2+})_7Si_8O_{22}(OH)_2$], from Latin a*nthophyllum*, clove, from Greek *anthos*, flower, and *phyllon*, leaf, referring to this mineral's usually brownish color.

Apatite [$Ca_5(PO_4)_3(F,OH,Cl)$], from Greek *apate*, deceit, referring to the easy confusion of this mineral with other minerals of similar appearance.

Aphthitalite [$(K,Na)_3Na(SO_4)_2$], from Greek *aphthetos*, indestructible, from *phthiein*, to waste away, and *hals, hal(o)-*, salt, referring to this mineral's stability in air. A.k.a. **aphthalose, glaserite**, after the Swiss-French chemist Christophe Glaser (1615–1672), and **Vesuvian salt**, after Mount Vesuvius, Italy.

Aquamarine [$Be_3Al_2(SiO_3)_6$], from Latin *aqua marina*, sea water, referring to its bluish green color.

Arcanite (K_2SO_4), from New Latin *arcanum duplicatum*, double secret, the name used by alchemists for this salt.

Augite [$(Ca,Mg,Fe)(Si,Al)_2O_6$], from Greek *auge*, bright, referring to the appearance of its cleavage planes. A.k.a. **fassaite**, after the Fassa Valley, Trento-Alto Adige, Italy.

Azurite [$Cu_3(CO_3)_2(OH)_2$], from Medieval Latin *azura*, ultramarine, from Arabic *al-lazaward*, the town of Lajward, Turkestan, known for the blue rock, lapis lazuli, found there, referring to this mineral's deep blue color.

Barite ($BaSO_4$), from Greek *barys, baryt-*, heavy, referring to its high specific density.

Barysilite ($Pb_3Si_2O_7$), from Greek *barys, baryt-*, heavy, and silicon.

Calomel (Hg_2Cl_2), from Greek *kalos*, beautiful, and *melos*, black, referring to its turning black upon reaction with ammonia.

Carnelian (SiO_2), from Medieval French *cornelle*, cherry, probably influenced by Latin *caro, carn-*, flesh, referring to this mineral's red color. A.k.a. **cornelian.**

Celadonite {[$K(Mg,Fe^{2+})(Fe^{3+},Al)[Si_4O_{10}](OH)_2$]}, from French *céladon*, sea-green, referring to this mineral's color. The French adjective is an eponym referring to a character in L'Astrée, a tale by the French writer Honoré d'Urfé (1568–1625), the shepherd Céladon, who adorns himself with green ribbons.

Celestine ($SrSO_4$), from Latin *caelestis*, pertaining to the sky, referring to its sky-blue color.

Chabazite ($CaAl_2Si_4O_{12}$ $6H_2O$), erroneously derived from Greek *chabazios*, tune, melody, a misspelling of *chalazios*, literally of hail, a precious stone resembling a hail-stone, from *chalaza*, hail-stone, one of the twenty stones mentioned in the ancient Greek poem *Peri lithos*, About stones, ascribed to Orpheus. Thus, this mineral should have been named chalazite.

Chalcanthite ($CuSO_4 \cdot 5H_2O$), from chalc(o)- and anth(o)-, referring to this mineral's copper content and flowery appearance.

Chiolite ($Na_5Al_3F_{14}$), from Greek *chion*, snow, referring to its color and appearance.

Chlorapatite [$Ca_5Cl(PO_4)_3$], from chlorine and apatite.

Chlorite [$(Mg,Fe)_3(Si,Al)_4O_{10}(OH)_2 \cdot (Mg,Fe)_3(OH)_6$], from Greek *chloritis*, a light-green stone, from *chloros*, green, referring to this mineral's color.

Chrysotile [$Mg_3(Si_2O_5)(OH)_4$], from Greek *chrysos, chryso-* gold, and *tilos*, something plucked, fiber, referring to this mineral's soft and fibrous appearance.

Crocoite ($PbCrO_4$), from Greek *krokos*, saffron, referring to its color.

Cryolite (Na_3AlF_6), from Greek *kryos*, ice, referring to the ice-like appearance of this mineral.

Cryptomelane [$K(Mn^{4+},Mn^{2+})_8O_{16}$], from Greek *kryptos*, hidden, from *kryptein*, to hide, and *melas*, black, referring to the difficulty of telling many similar black manganese oxide minerals from each other.

Cylindrite ($Pb_3Sn_4FeSb_2S_{14}$), referring to this mineral's unique cyclindrical crystal shape.

Cymophane ($BeAl_2O_4$), from Greek *kyma*, wave, and -phan(e), referring to this mineral's opalescence.

Datolite ([$CaBSiO_4(OH)$], from Greek *dateisthai*, to divide, referring to the granular habit of some massive specimens.

Diallage ($CaMgSi_2O_6$), from Greek *diallage*, change, from *diallassein*, to interchange, to exchange, to change, from *allassein*, to change, referring to this mineral's bronzy sheen.

Diaspore [$AlO(OH)$], from Greek *diasporein*, to scatter, referring to this mineral's decrepitation upon heating.

Diopside ($MgCaSi_2O_6$), from Greek *dis, dia-*, twice, and *opse*, face, referring to this mineral's two ways of orienting the vertical prism.

Dioptase ($CuSiO_3 \cdot H_2O$), from dia- and Greek *optos*, visible, referring to this mineral's two cleavage directions that are visible inside unbroken crystals.

Dyscrasite (Ag_3Sb), from Greek *dyskrasis*, bad alloy, referring to the undesirable antimony content of this silver ore.

Emplectite ($CuBiS_2$), from Greek *emplektos*, interwoven, referring to this mineral's intimate association with quartz,

Enargite (Cu_3AsS_4), from Greek *enarges*, visible, referring to its excellent cleavage.

Enigmatite ($Na_2Fe_5TiSi_6O_{20}$), from enigma, from Greek *ainigma*, riddle, from *ainissesthai*, to speak in riddles, from *ainos*, tale, fable, referring to a perceived incomplete knowledge of this mineral.

Enstatite ($Mg_2Si_2O_6$), from Greek *enstates*, opponent, referring to this mineral's refractory nature.

Epidote [$(Ca_2,Al_2,Fe^{3+})(Si_2O_7)(SiO_4)O(OH)$], from Greek *epidote*, increase, referring to a crystal structure with one longer side at the base of the prism.

Eriochalcite ($CuCl_2{\cdot}2H_2O$), from Greek *erion*, wool, and *chalkos*, copper, referring to this mineral's composition and woolly appearance.

Erythrite [$Co_3(AsO_4)_2{\cdot}8H_2O$], from Greek *erythros*, red, referring to this mineral's color.

Eschynite [$(Ln,Ca,Fe,Th)(Ti,Nb)_2(O,OH)_6$], from Greek *aischyne*, shame, referring to the inability of chemists, at the time of its discovery, to separate the constituents of this mineral. A.k.a. **aeschynite, blomstrandine**, after the Swedish chemist Christian Wilhelm Blomstrand (1826–1897), and **priorite**, after the British mineralogist George Thurland Prior (1862–1936).

Euclase [$BeAl(SO_4)(OH)$], from Greek *euklasis*, good cleavage.

Eulytite [$Bi_4(PO_4)_3$], from Greek *eulytos*, easily liquefiable, referring to this mineral's relatively low melting point.

Fluorapatite [$Ca_5F(PO_4)_3$], from fluorine and apatite.

Fluorite (CaF_2), from Latin *fluere*, to flow, referring to its readily fusible nature and use as a flux in smelting.

Fuller's earth (bentonite or palygorskite), from fuller, from Old French *fuler*, to cleanse and thicken cloth.

Glaucodot [$(Co,Fe)AsS$], from Greek *glaukos*, blue green, and *doter*, giver, from *didonai*, to give, referring to this mineral's use in the manufacture of dark blue glass.

Grossular [$Ca_3Al_2(SiO_4)_3$], named after the gooseberry (*Ribes grossularia*), referring to its pale green color. The specific epithet is from French *groseille*, gooseberry.

Halotrichite [$Fe^{2+}Al_2(SO_4)_4{\cdot}22H_2O$], from Greek *hals, hal(o)-*, salt, and *thrix, tricho-*, hair, referring to its silky hairlike habit.

Harmotome ($BaAl_2Si_6O_{16}\cdot 6H_2O$), from Greek *harmos*, joint, and *tome*, section, from *temnein*, to cut, referring to details of this mineral's crystal structure, i.e. an octahedron dividing parallel to the plane that passes through the terminal edges.

Helvite [$Be_3Mn_4(SiO_4)_3S$], from Latin *helvus*, light-bay colored, referring to this mineral's color.

Hematite (Fe_2O_3), from Greek *haima*, blood, referring to this mineral's sometimes reddish brown or red color.

Hessonite [$Ca_3Al_2(SiO_4)_3$], from Greek *hesson*, inferior, less, comparative of *heka*, slightly, referring to the fact that this grossular has a lower hardness and lower density than most other garnet species varieties. A.k.a. **cinnamon stone**.

Hexacelsian [$Ba(AlSiO_4)_2$], with contraction, from hexagonal celsian.

Hornblende [$\square Ca_2(Mg,Fe^{2+})_4(Al,Fe^{3+})(Si_7Al)_{22}(OH)_2$], a German name, from German *Horn*, horn, and *Blende*, deceiver, referring to these minerals being found together with metalliferous ores, but containing no valuable metal.

Hyacinth ($ZrSiO_4$), from Greek *hyakinthos*, precious stone, flowering plant, referring to this mineral's yellow, orange, or red color.

Hyalophane [$(K,Ba)Al(Si,Al)_3O_8$], from Greek *hyalos*, glass, and *phainein*, to show, referring to its transparent crystals.

Hydroxylapatite [$Ca_5(PO_4)_3(OH)$], named as the hydroxyl-bearing end-member of the apatite group. A.k.a. **durapatite**, from Latin *durus*, hard, and apatite.

Hypersthene [$(Fe,Mg)SiO_3$], from hyper- and -sthene, referring to the fact that this mineral is harder than hornblende with which it is often confused.

Ianthinite ($UO_2\cdot 5UO_3\cdot 10H_2O$), from Greek *ianthinos*, violet, referring to its color.

Kainite [$KMg(SO_4)Cl\cdot 3H_2O$], from Greek *kainos*, hitherto unknown, referring to the fact that this mineral was the first found to contain both sulfate and chloride.

Kamacite [α-(Fe,Ni)], from Greek *kamax, kamak-*, shaft, lath, referring to the crystal shape.

Kermesite (SbS_2O), from kermes, ultimately from Arabic *qirmiz*, vivid red, referring to this mineral's red color. A.k.a. **red antimony**.

Kosmochlor ($NaCr^{3+}Si_2O_6$), from Greek *kosmos*, cosmos, universe, and *chloros*, green, referring to this mineral's color and occurrence in a meteorite.

Kyanite (Al_2SiO_5), from Greek *kyanos*, blue, referring to this mineral's most common color.

Lepidolite [$K(Li,Al)_3(Al,Si,Rb)_4O_{10}(F,OH)_2$], from Greek *lepis, lepid-* scale, referring to this mineral's scaly appearance.

Lizardite [$Mg_3Si_2O_5(OH)_4$], from lizard, referring to this mineral's often green or green-blue color. Lizard is ultimately from Latin *lacerta*, lizard, from *lacertus*, the mobile.

Maghemite [$Fe^{2+}(Fe^{3+})_2O_4$], a contraction of magnetic hematite.

Magnetite [$Fe^{2+}(Fe^{3+})_2O_4$], named as natural magnet.

Meionite [$Ca_4Al_6Si_6O_{24}(CO_3)$], from meio-, referring to this mineral's less acute pyramidal shape compared to vesuvianite.

Melilite [$(Ca,Na)_2(Al,Mg,Fe^{2+})(Si,Al)_2O_7$], from Greek *meli, melit-*, honey, referring to its honey-yellow color.

Mellite {$Al_2[C_6(COO)_6]\cdot 16H_2O$}, as melilite.

Mimetite [$Pb_5(AsO_4)_3Cl$], from Greek *mimetes*, imitator, referring to its resemblance to pyromorphite.

Monazite [$(Ce,La,Th,Y)PO_4$], from Greek *monazein*, to live alone, referring to the isolated occurrence and rarity of this mineral.

Nacrite [$Al_2Si_2O_5(OH)_4$], from French *nacre*, mother-of-pearl, ultimately from Arabic *naqqarah*, bowl, referring to its iridescent luster.

Nekoite ($Ca_3Si_6O_{15}\cdot 7H_2O$), anagram of okenite, q.v., referring to these two minerals' similarity.

Nepheline [$(Na,K)AlSiO_4$], from Greek *nephele*, cloud, referring to the cloudy appearance which develops when this mineral is immersed in strong acid.

Nickeline (NiAs), from German *Kupfernickel*, copper bogey.

Okenite ($Ca_{10}Si_{18}O_{46}\cdot 18H_2O$), earlier ockenite, named after the German naturalist Lorenz Ocken (1779–1851).

Ozokerite (a waxy mineral mixture of hydrocarbons), from Greek *oze*, stench, and *keros*, wax, referring to this mineral's appearance and smell. A.k.a. **ozocerite**.

Paracelsian ($BaAl_2Si_2O_8$), from Greek *para*, near, and celsian, referring to the dimorphic relationship between the two minerals. Thus not named in honor of Paracelsus.

Peristerite ($NaAlSi_3O_8$), from Greek *peristera*, dove, pigeon, referring to this mineral's resemblance to the iridescent feathers of a pigeon's neck.

Picromerite [$K_2Mg(SO_4)_2\cdot 6H_2O$], from picr(o)- and -mer, referring to this mineral's bitter taste.

Pharmacolite ($CaHAsO_4 \cdot 2H_2O$), from Greek *pharmakon*, poison, referring to its arsenic content.

Phenakite (Be_2SiO_4), from Greek *phenax, phenak-*, deceiver, because this mineral was easily mistaken for quartz.

Phlogopite [$K(Mg,Fe)_3(Si_3Al)_{10}(F,OH)_2$], from Greek *phlogopos*, fiery looking, from *phlox, phlog-*, flame, and *ops*, eye, face, referring to this mineral's fiery appearance.

Plagioclase [$(Na,Ca)Al(Al,Si)Si_2O_8$], from Greek *plagios*, oblique, and *klasis*, fracture, from *klan*, to break, referring to the small obliquity of the angle between the [001] and [010] cleavages.

Plessite (Fe,Ni), from plesi(o)-, referring to the intergrowth of kamacite and taenite in this mineral.

Polybasite ($Ag_{12}Cu_4Sb_2S_{11}$), this name refers to the base metals in this mineral.

Polycrase [$(Y,Ca,Ce,U,Th)(Ti,Nb,Ta)_2O_6$], from poly- and -crase, referring to the large number of elements occurring in this mineral.

Polydymite (Ni_3S_4), from poly- and Greek *didymos*, twin, referring to this mineral's numerous twinned cystals.

Portlandite [$Ca(OH)_2$], this name refers to the fact that this mineral is a common product of the hydration of portland cement.

Purpurite ($MnPO_4$), from purpur(o)-, referring to this mineral's purple red color.

Pyrargyrite (Ag_3SbS_3), from Greek *pyr*, fire, and *argyros*, silver, referring to its red color and silver content.

Pyrochlore [$(Na,Ca)_2Nb_2O_6(OH,F)$], from pyr(o)- and [1]chlor(o)-, referring to this mineral's turning green in the blowpipe analysis.

Pyrolusite (MnO_2), from Greek *pyr*, fire, and *lousis*, a washing, from *louein*, to wash, referring to the use of this mineral in glass-making to remove brown and green tints. A.k.a. **polianite**, from Greek *poliainesthai*, to become white with foam, from *polios*, gray, pale.

Pyrope [$Mg_3Al_2(SiO_4)_3$], from Greek *pyropos*, fiery-eyed, referring to its characteristic red color.

Pyrophyllite [$Al_2Si_4O_{10}(OH)_2$], from Greek *pyr*, fire, and *phyllon*, leaf, referring to its exfoliation upon strong heating.

Pyroxferroite [$(Fe^{2+},Mn^{2+},Ca)SiO_3$], from pyroxene and Latin *ferrum*, iron. First found on the Moon, but later also on Earth.

Rhodonite [$(Mn^{2+},Fe^{2+},Mg,Ca)(SiO_3)_3$], from Greek *rhodon*, rose, probably from Old Iranian **urda-*, rose, referring to this mineral's color.

Safflorite [$(Co,Fe)As_2$], from German *Safflor*, ultimately from Latin *sapphiri color*, sapphire color, referring to its use as the source of *Zaffer*, zaffer, a mixture of cobalt oxides used as a blue pigment for coloring glass and enamel.

Sanidine ($KAlSi_3O_8$), from Greek *sanis*, board, referring to this mineral's flat crystals.

Saponite [$(Ca,Na)_{0.3}(Mg,Fe^{2+})_3(Si,Al)_4O_{10}\cdot 4H_2O$], from sapo(n)-, referring to this mineral's soapy consistency.

Sapphirine ($Mg_7Al_{18}Si_3O_{40}$), from sapphire, referring to this mineral's sapphire-blue color.

Scolecite ($CaAl_2Si_3O_{10}\cdot 3H_2O$), from Greek *skolex, skolek-*, worm, referring to this mineral's reaction to the blowpipe flame.

Scorodite ($FeAsO_4\cdot 2H_2O$), from Greek *skorodon*, garlic, referring to the garlic odor (of arsane) generated upon strong heating.

Sepiolite [$Mg_4Si_6O_{15}(OH)_2\cdot 6H_2O$], from Greek *sepion*, cuttlebone, referring to this mineral's resemblance to cuttlebone. A.k.a. **meerschaum**, from German *Meerschaum*, from *Meer*, sea, and *Schaum*, foam, referring to the fact that this mineral floats in water.

Smaltite [$(Co,Fe,Ni)As_2$], named after its use in smalt, ground cobalt glass used in glass production.

Sphalerite [$(Zn,Fe^{2+})S$], from Greek *sphaleros*, mistaken, because this mineral was often mistaken for galena, but yielded no lead.

Sphene [$CaTiO(SiO_4)$], from Greek *sphen*, wedge, referring to this mineral's wedge-shaped crystal habit. A.k.a. **titanite**, from titanium.

Spinel ($MgAl_2O_4$), from Latin *spina*, thorn, referring to its sharp octahedral crystals.

Spodumene ($LiAlSi_2O_6$), from Greek *spodumenos*, burnt to ashes, referring to the opaque, ash-grey appearance of this mineral when refined for use in industry.

Staurolite [$Fe^{2+},Mg,Zn)_{3-4}Al_2\square_2Al_{16}Si_8O_{48}H_{2-4}$], from Greek *stauros*, cross, referring to the penetration twinning of the crystals.

Stilpnomelane [$K(Fe^{2+},Mg,Fe^{3+})_8(Si,Al)_{12}(O,OH)_{27}$], from Greek *stilpnos*, shining, and *melas*, black, referring to this mineral's appearance.

Synchysite [$Ca(Ce,La)(CO_3)_3F$], from Greek *synchys*, confounding, referring to early misidentification of this mineral as parisite.

Syngenite [$K_2Ca(SO_4)_2\cdot H_2O$], from syn- and -gen, referring to this mineral's similarity with polyhalite.

Tacharanite ($Ca_{12}Al_2Si_{18}O_{51}\cdot 18H_2O$), from Gaelic *tacharan*, changeling, because this mineral breaks down into other compounds if left standing in air.

Tachyhydrite ($CaMg_2Cl_6\cdot 12H_2O$), from Greek *tachys*, quick, and *hydor*, water, referring to this mineral's ready deliquescence.

Taenite [γ-(Ni,Fe)], from Greek *tainia*, ribbon, referring to the thin platy habit of this mineral.

Tephroite (Mn_2SiO_4), from Greek *tephros*, ash-colored, referring to this mineral's color.

Tetradymite (Bi_2Te_2S), with contraction, from Greek *tetra*, four, and *didymos*, twin, referring to this mineral's usual appearance as fourling twins.

Tetrahedrite [$(Cu,Fe)_{12}Sb_4S_{13}$], referring to the tetrahedral shape of the crystals.

Thaumasite [$Ca_3Si(OH)_6(CO_3)(SO_4)\cdot 12H_2O$], from Greek *thaumazein*, to be surprised, referring to the unusual presence of hydroxysilicate, carbonate, and sulfate in one mineral.

Tincalconite [$Na_2B_4O_5(OH)_4\cdot 3H_2O$], from tincal and Greek *konia*, ashes, dust, powder, referring to this mineral's composition and powdery appearance.

Tridymite (SiO_2), from Greek *tridymos*, triplet, referring to its common twinning as trillings.

Vermiculite [$(Al,Fe,Mg)_3(Al,Si)_4O_{10}(OH)_2\cdot 4H_2O$], from Latin *vermiculus*, diminutive of *vermis*, worm, referring to the fact that this mineral's scales, when heated, open out into wormlike forms.

Xenotime [$(Y,Yb)PO_4$], irregularly from Greek *kenos*, in vain, seemingly, and *timé*, value, price, referring to the fact that the yttrium contained in this mineral was initially mistaken to be a new element.

Zorite [$Na_2Ti(Si,Al)_3O_9\cdot nH_2O$], from Russian *zoria*, the rosy hue of the sky at dawn, referring to its color.

13.3.4 Names based on chemistry

Allophane [$Al_2O_3(SiO_2)_{1.3-2}\cdot(H_2O)_{2.5-3}$], from Greek *allophanes*, appearing otherwise, referring to the change this mineral undergoes in a blowpipe flame.

Anhydrite ($CaSO_4$), named as anhydrous calcium sulfate.

Allargentum ($Ag_{1-x}Sb_x$), from Greek *allos*, other, and Latin *argentum*, silver.

Aluminite [$Al_2SO_4(OH)_4\cdot 7H_2O$], from aluminum.

Arsenolite (As_4O_6), from German *Arsen*, arsenic.

Atheneite [$(Pd,Hg)_3As$], after the Greek goddess of war, arts, fertility, and wisdom Pallas Athena, referring to this mineral's content of palladium which was named after this goddess.

Aventurine (SiO_2), from French *aventure*, chance, referring to this mineral's chance discovery.

Boracite ($Mg_3B_7O_{13}Cl$), from borax, referring to its boron content.

Bromargyrite (AgBr), from Greek *argyros*, silver.

Cafetite [$(Ca,Mg)(Fe,Al)_2Ti_4O_{12}\cdot 4H_2O$], from Ca, Fe, Ti.

Calcite (β-$CaCO_3$), from Latin *calx, calc-*, limestone.

Cerussite ($PbCO_3$), from Latin *cerussa*, white lead, cerussite, possibly of Greek origin.

Chalcomenite ($CuSeO_3\cdot 2H_2O$), from chalc(o)- and Greek *mene*, moon, referring to this mineral's content of copper and selenium, which has been named after the Greek moon goddess Selene.

Chlorargyrite (AgCl), from chlorine and Greek *argyros*, silver.

Chlorocalcite ($CaKCl_3$), from chlorine and calcium, coined in the erroneous belief that this mineral only consists of these two elements.

Digenite (Cu_9S_5), from di- and -gen, referring to the fact that this mineral contains both copper(I) and copper(II) ions.

Geocronite [$Pb_{14}(Sb,As)_6S_{23}$], from Greek *ge, geo-*, Earth, and *kronos*, Saturn, the alchemistic symbol for lead.

Germanite [$Cu_3(Ge,Fe)S_4$], after this mineral's content of germanium.

Glucine [$Ca_2Be_8(PO_4)_4(OH)_8\cdot H_2O$], after its content of beryllium, from the obsolete name glucinium for beryllium.

Hafnon ($HfSiO_4$), from Hf, by analogy with zircon.

Iodargyrite (AgI), from iodine and Greek *argyros*, silver. A.k.a. **iodyrite**.

Liberite (Li_2BeSiO_4), from Li, Be.

Lithiophilite ($LiMnPO_4$), from lithium and phil(o)-.

Lithosite ($K_6Al_4Si_8O_{25}\cdot 2H_2O$), from Greek *lithos*, stone, referring to the fact that this mineral consists of the most abundant elements of the Earth's crust. Literally stonestone.

Makatite [$Na_2Si_4O_8(OH)_2 \cdot 4H_2O$], from Maasai *emaka*, soda, referring to this mineral's high sodium content.

Miargyrite ($AgSbS_2$), from Greek *meion*, less, and argyr(o)-, referring to this mineral's relatively low silver content.

Molybdomenite ($PbSeO_3$), derived from Greek *molybdos*, lead, and *mene*, moon, referring to this mineral's content of lead and selenium, the latter having been named after Selene, the Greek moon goddess.

Palarstanide [$Pd_5(Sn,As)_2$], from Pd, As, Sn.

Phosgenite [$Pb_2(CO_3)Cl_2$], named after phosgene ($COCl_2$), whose components, carbon, chlorine, and oxygen, are present in this mineral.

Polyhalite [$K_2Ca_2(SO_4)_4 \cdot 2H_2O$], from poly- and Greek *hals, hal(o)-*, salt, referring to the many salt components in this mineral.

Polylithionite [$KLi_2AlSi_4O_{10}(F,OH)_2$], from poly- and lithium, referring to this mineral's high lithium content.

Polymignite [$(Ca,Fe,Y,Th)(Nb,Ti,Ta,Zr)O_4$], from Greek *polys*, much, many, and *migninai*, to mix, referring to this mineral's complex composition.

Siderazot (Fe_5N_2), from Greek *sideros*, iron, and French *azote*, nitrogen.

Siderite ($FeCO_3$), from Greek *sideros*, iron, named for its iron content.

Sinoite (Si_2N_2O), from Si, N, and O.

Stibnite (Sb_2S_3), from Latin *stibium*, antimony.

Stronalsite ($SrNa_2Al_4Si_4O_{16}$), from Sr, Na, and Al.

Sulfohalite [$Na_6(SO_4)_2FCl$], from sulfate and halide.

Sulvanite (Cu_3VS_4), from S and V.

Tellurite (TeO_2), from Te.

Thorianite (ThO_2), from Th.

Thorite [$(Th,U)SiO_4$], from Th.

Triphylite ($LiFePO_4$), from Greek *tria*, three, and *phylon*, family. A misnomer since this name was chosen in the erroneous belief that this mineral contained three cations, i.e. Fe, Li, and Mg.

Trona [$Na_2CO_3 \cdot NaHCO_3 \cdot 2H_2O$], from Swedish or Spanish *trona*, trona, from Arabic *natrun*, ultimately from Ancient Egyptian *ntry*, native soda.

Uricite [uric acid, 7,9-dihydro-1*H*-purine-2,6,8(3*H*)-trione], from the chemical name. One of the few organic minerals.

Wolframite [$(Fe,Mn)WO_4$], from German *Wolfram*, wolframite, probably from *Wolf*, wolf, and *Rahm*, froth, dirt, also cream, akin to Medieval Latin *spuma lupi*, wolframite, literally wolf's froth. These names appear to stem from medieval miners' observation that wolframite during its extraction consumed large amounts of molten tin the same way as a wolf eats a sheep.

13.3.5 Class names

Amphibole (a group of inosilicates), from Greek *amphibolos*, ambiguous, referring to these minerals' varying shape.

Apatite (phosphate minerals), from Greek *apatein*, to deceive, to be misleading, referring to the fact that these minerals are often mistaken for others.

Axinite (borosilicates), from Greek *axine*, axe, referring to these minerals' axe-shaped crystals.

Cyclosilicate, from Greek *kyklos*, ring, so named after these minerals' ring structure.

Eudialyte (silicates), from Greek *eudialytos*, easily decomposable, from *eus, eu-*, good, and *dialyein*, to dissolve, an allusion to these minerals' ready dissolution in acids.

Feldspar (tectosilicates), from German *Feldspat*, from *Feld*, field, and *Spat*, a rock or mineral readily breaking into flakes. The corresponding adjective is feldspathic.

Inosilicate, from Greek *is, ino-*, sinew, fiber, referring to these minerals' chain structure.

Mica (phyllosilicates), from Latin *mica*, crumb, maybe also influenced by Latin *micare*, to glitter. The corresponding adjective is **micaceous**.

Nesosilicate, from Greek *nesos*, island, referring to these minerals' structure with isolated silicate tetrahedra.

Olivine (nesosilicates), from olive, referring to these minerals' color.

Phyllosilicate, from Greek *phyllon*, leaf, referring to these minerals' sheet structure.

Pyrite (sulfide minerals such as FeS_2), from Greek *pyrites*, of fire, in fire, from *pyr*, fire, referring to the fact that these minerals give off sparks when struck against steel.

Pyroxene (a group of inosilicates), from Greek *pyr*, fire, and *xenos*, stranger, referring to the fact that these crystals were often found embedded in volcanic glass.

Smectite, clay minerals, The name is from Greek *smektris*, a kind of fuller's earth, from *smechein*, to clean, probably an allusion to these minerals' soapy feel.

Sorosilicate, from Greek *soros*, heap, mound, referring to these minerals' structure with isolated double silicate tetrahedra.

Tectosilicate, from Greek *tektonikos*, pertaining to building, from *tekton*, builder, carpenter, woodworker, referring to these minerals' three-dimensional framwork of silicate tetrahedra.

Zeolite (aluminosilicates), from Greek *zein*, to boil, and Greek *lithos*, stone, referring to these minerals' expulsion of absorbed water upon heating.

13.3.6 Formulas

While the IUPAC recommendations for empirical formulas of inorganic compounds [8] are adequate when one deals with stoichiometric compounds, many minerals are characterized by variable and non-stoichiometric proportions of the elements contained. Therefore mineralogists use notations which can cope with non-stoichiometry.

13.3.6.1 Stoichiometric minerals

In mineralogy molecular formulas for stoichiometric minerals are written according to the IUPAC rules [8] except in cases where the same element appears in more than one oxidation state where the oxidation states are explicitly shown. Thus, while the formula Fe_3O_4 is appropriate for triiron tetraoxide, the corresponding formula for the mineral magnetite is $Fe^{2+}(Fe^{3+})_2O_4$.

13.3.6.2 Non-stochiometric minerals

In the molecular formulas of nonstoichiometric minerals the element and group symbols are written in descending order of occurrence, not in alphabetical order as in stoichiometric compounds.

Thus, the formula $(K,H_3O)(Al,Mg,Fe)_2(Si,Al)_4O_{10}[(OH)_2,(H_2O)]$ of illite shows that the relative amounts of potassium and oxonium cations decrease in the order shown as do the relative amounts of aluminum, magnesium, and iron cations as well as those of silicon- and aluminum-containing anions, respectively.

13.3.6.3 Minerals with point defects

While IUPAC recommends the elaborate Kröger-Vink notation for empirical formulas involving crystal lattice point defects [8] mineralogists employ a simplified system as in the molecular formula of the inosilicate anthophyllite, $\square(Mg,Fe^{2+})_7Si_8O_{22}(OH)_2$, where the symbol □ stands for a point defect.

References

[1] https://en.wikipedia.org/List_of_minerals_(complete) (retrieved August 2, 2018).
[2] www.mindat.org (retrieved August 2, 2018).
[3] Webmineral.com (retrieved August 2, 2018).
[4] https://www.mineralatlas.eu (retrieved August 2, 2018).
[5] http://www.ima-mineralogy.org (retrieved August 2, 2018).
[6] https://web.archive.org/web/20061017105058/http://geo.web.ru/IgMetAtlas/ (retrieved August 2, 2018).
[7] *https://en.wikipedia.org/wiki/List_of_diamonds* (retrieved August 2, 2018).
[8] 'The Red Book', N. G. Connelly, T. Damhus, R. M. Hartshorn, and A. T. Hutton, *Nomenclature of Inorganic Chemistry – IUPAC Recommendations 2005*, Royal Society of Chemistry, Cambridge, UK, 2005, ISBN 0-85404-438-8; free online version, https://www.iupac.org/cms/wp-content/uploads/2016/07/Red_Book_2005.pdf (retrieved August 2, 2018). Lists with corrections of misprints and errors are available as http://www.qmul.ac.uk/HYPERLINK "http://www.qmul.ac.uk/sbcs/iupac/bibliog/RBcorrect.html"sbcs/HYPERLINK "http://www.qmul.ac.uk/sbcs/iupac/bibliog/RBcorrect.html"iupac/bibliog/RBcorrect.html (retrieved March 14, 2019) and http://www.qmul.ac.uk/HYPERLINK "http://www.sbcs.qmul.ac.uk/iupac/bibliog/RBcorrect2.html"sbcs/HYPERLINK "http://www.sbcs.qmul.ac.uk/iupac/bibliog/RBcorrect2.html"iupac/bibliog/RBcorrect2.html (retrieved March 14, 2019) .

14 International non-proprietary names for drugs and excipients

For obvious reasons neither the pharmaceutical nor the medical community could operate smoothly on the basis of the often unwieldy systematic names for drugs. Since the structures of the active substances are not at issue in their day-to-day business an independent, more convenient, naming concept is in use in the shape of International Non-Proprietary Names for Drugs (INN), issued by the World Health Organization (WHO) [1] since 1953. INNs are not only issued in English, but also in Latin, French, Russian, Spanish, Arabic, and Chinese, special pains being taken to make these names mutually intelligible. The names are first issued as proposed INN (pINN) and then as recommended INN (rINN) [2]. A considerable number of INNs have been issued to drug candidates which then failed to pass the clinical tests required for drug registration. National naming systems such as Australian Approved Name (AAN), British Approved Name (BAN), Japanese Accepted Name (JAN), Japanese Pharmacopeia (JP), United States Accepted Name (USAN), and United States Pharmacopeia (USP) have been used as well.

INNs must be unique and consist of a stem, placed either first or, more frequently, last in the name, combined with an arbitrarily chosen name part [3]. These INNs identify a given pharmaceutical product which is typically sold under a freely chosen additional brand name. Thus, the brands Celexa, Celapram, and Citrol contain the same active substance, i.e. citalopram. Some branded preparations may be composed of two active substances such as Lemsip which contains paracetamol and phenylephrine.

It should be noted that the letter y is not allowed in INNs. Where it would otherwise appear it is replaced by i. INN stems typically refer to the therapeutic class which the drug belongs to or its chemical class. Examples are shown below.

14.1 INN stems implying chemical class

-actide, as in alsactide, for synthetic polypeptides with corticotropin-like activity.

Arte-, as in artemether, for artemisinin-derived antimalarials; the sesquiterpene lactone artemisine has been named after the genus name *Artemisia* (mugwort), ultimately after Artemis, the Greek goddess of forests and hills, whose name is of uncertain etymology.

-ase, as in alteplase, for enzymes; the suffix ‘ase’, used for the systematic naming of enzymes, is derived from the name of the enzyme diastase, from Greek *diastasis*, separation.

https://doi.org/10.1515/9783110612714-014

-azepam, as in diazepam, for benzodiazepines.

Cef-, as in cefalexin, for cephalosporins; the natural fungal cephalosporins were named after the genus *Cephalosporium*, from Greek *kephale* (head) and *spora* (seed), from *speirein* (to sow, to strew).

-cillin, as in ampicillin, for derivatives of 6-aminopenicillanic acid.

Cort-, as in fluocortin, for corticosteroids with the exception of prednisolone analogs.

-coxib, as in celecoxib, for COX-2 inhibitors, from the enzyme name prostaglandin-endoperoxide synthase 2 (EC 1.11.99.1).

-cycline, as in doxycycline, for antibiotics derived from tetracycline.

Erg-, as in methysergide, for derivatives of ergot alkaloids.

Estr-, as in mestranol, for estrogens.

Io-, as in iobenguane, for iodine-containing radiopharmaceuticals.

-mab, as in infliximab, for monoclonal antibodies.

-micin, as in calicheamicin, for antibiotics produced by *Micromonospora* bacteria.

-mycin, as in carbomycin, for antibiotics *produced* by *Streptomyces* actinobacteria.

-one, as in oxycodone, for ketones.

-onide, as in budesonide, for steroids to be used externally.

Prost-, as in naxaprostene, for prostaglandins.

-relin, as in somatorelin, for hormone-release stimulating peptides.

Vin-, as in vincristine, for *Vinca* alkaloids or related drugs; the plant genus *Vinca* (periwinkle) was later renamed *Cataranthus*.

14.2 INN stems implying therapeutic class

-ac, as in ibufenac, for anti-inflammatory drugs.

-adol, as in tramadol, for analgetics.

-anib, as in pazopanib, for angiogenesis inhibitors.

-anserin, as in mianserin, for serotonin receptor antagonists.

-arit, as in lobenzarit, for antiarthritic agents.

-astine, as in azelastine, for antihistamines.

-bactam, as in sulbactam, for β-lactamase inhibitors.

Bol-, as in boldenone, for anabolic steroids.

-bol, as in clostebol, for anabolic steroids.

-caine, as in procaine, for local anesthetics, from cocaine; the name of the alkaloid cocaine is after the species *Erythroxylon coca* (coca). The specific epithet is ultimately from Quechua *kuka* (coca).

Calci-, as in calcipotriol, for vitamin D analogs.

-dipine, as in nifedipine, for calcium antagonists.

Gli-, as in glimepiride, for antidiabetic agents.

-icam, as in piroxicam, for antiinflammatory drugs.

-navir, as in darunavir, for antiretroviral protease inhibitors.

-olol, as in atenolol, for beta blockers, derived from the names of the first agents in this class, propanolol and pronethalol.

-pramine, as in imipramine, for antidepressants.

-pril, as in enalapril, for angiotensin-converting-enzyme inhibitors, from the name of the first agent in this class, captopril, the 'capto' part of this name, short for 'mercapto', referring to the presence of a sulfanyl group.

-profen, as in ibuprofen, for nonsteroidal anti-inflammatory drugs (NSAID).

-sartan, as in losartan, for angiotensin receptor II antagonists.

-terol, as in fenoterol, for bronchodilating drugs.

-tidine, as in cimetidine, for H_2 receptor antagonists.

-tinib, as in imatinib, for tyrosine kinase inhibitors.

-trexate, as in methotrexate, for folic acid antagonists.

-triptyline, as in amitriptyline, for tricyclic antidepressants.

-vastatin, as in simvastatin, for 3-hydroxy-3-methylglutaryl-coenzyme A reductase (EC 1.1.1.88 and EC 1.1.1.34) inhibitors.

-vec, as in alipogene tiparvovec, for gene therapy vectors.

-vir, as in aciclovir, for antivirals.

14.3 INNs for acids and bases used in drug modification

Parent drugs can be modified by conversion to salts or esters, cf. Section 14.4.

14.3.1 Acids (as anions)

To name a salt the name of the acid anion is added to the INN of the parent base, for instance:

Besilate, benzenesulfonate ($C_6H_5SO_3^-$), a contraction of the systematic name.

Closilate, 4-chlorobenzenesulfonate ($4\text{-}ClC_6H_4SO_3^-$), cf. besilate.

Edisilate, ethane1,2-disulfonate ($^-O_3SCH_2CH_2SO_3^-$), cf. besilate.

Embonate, 4,4′-methylenebis(3-hydroxynaphthalene-2-carboxylate), also known as pamoate, from embonic acid, q.v.

Hyclate, ethanol—water—hydrogen chloride (1/1/2), a contraction of hydrogen chloride and 'ate'.

Mesilate, methanesulfonate ($CH_3SO_3^-$), cf. besilate.

Tosilate, 4-methylbenzene-1-sulfonate ($4\text{-}CH_3C_6H_4SO_3^-$), cf. besilate.

14.3.2 Bases

To name a salt the name of the base is added to the INN of the parent acid, for instance:

Erbumine, 2-methylpropan-2-amine [$(CH_3)_3CNH_2$]; the INN is a contraction of *tert*-butylamine.

Olamine, 2-aminoethanol ($H_2NCH_2CH_2OH$), from 'ol' and amine.

14.3.3 Ester components

To name an ester the name of the ester component is added to the parent INN.

14.3.3.1 Alkyl components for the esterification of acids

To name the ester the name of the ester component is added to the INN of the parent acid.

Axetil, *rac*-1-(acetyloxy)ethyl, a radical contraction of the systematic name.

Cilextil, *rac*-1-[[(cyclohexyloxy)carbonyl]oxy]ethyl, cf. axetil.

Mofetil, 2-(morpholin-4-yl)ethyl, cf. axetil.

Pivoxetil, *rac*-1-[(2-methoxy-2-methylpropyl)oxy]ethyl, cf. axetil.

Proxetil, *rac*-1-[[[(propan-2-yl)oxy]carbonyl]oxy]ethyl, cf. axetil.

Soproxil, [[(propan-2-yl)carbonyl]oxy]methyl, cf. axetil.

14.3.3.2 Acyl components for the esterification of alcohols

To name such an ester the name of the acyl component is added to the INN of the parent alcohol, for instance:

Ecamate, *N*-ethylcarbamate, a contraction of the systematic name.

Etabonate, ethyl carbonate, cf. examate.

Xinafoate, 1-hydroxynaphthalene-2-carboxylate, an arbitrary name.

14.4 International nonproprietary names modified

Special attention has been given to the naming of drugs obtained by modification of a parent drug such as by formation of a salt or an ester. They receive International Nonproprietary Drug Names Modified (INNM) [4].

Typical INNMs are: abanoquil mesilate, cefpodoxim proxetil, doxycycline hyclate, magnesium valproate, mofetil mycophenolate, mupirocim calcium, perindopril erbumine, and sulfadiazine silver.

14.5 International nonproprietary names for racemic and chiral versions of a drug

After a chiral drug has been assigned an INN for its racemate it might later become necessary to issue another INN for an enantiomer. The following prefixes can be added to the INN of the racemate: dextro- or dex- for the (+)-isomer, levo- or lev- for the (–)-isomer, ar- for the (*R*)-isomer, and es- for the (*S*)-isomer, as shown in the following examples:

Amlodipine (racemate), **levamlodipine** [(–)-isomer].

Amphetamine (racemate), **dextroamphetidine** [(+)-isomer], an irregular name construction. Amphetamine is a contraction of α-methyl(phenylethylamine) (1-phenyl propan-2-amine).

Bupivacaine (racemate), **levobupivacaine** [(−)-isomer].

Formoterol (racemate), **arformoterol** [(*R*)-isomer].

Ketoprofen (racemate), **dexketoprofen** [(+)-isomer].

Omeprazole (racemate), **esomeprazole** [(*S*)-isomer].

Tetramisole (racemate), **levamisole** [(−)-isomer], an irregular name construction.

14.6 Drug family naming

With the parallel therapeutic use of native enzymes, such as the insulins, and a considerable number of modifications thereof, a corresponding portfolio of INNs is required, for instance:

Insulin bovine ($C_{254}H_{377}N_{65}O_{75}S_6$).

Insulin human ($C_{257}H_{383}N_{65}O_{77}S_6$).

Insulin porcine ($C_{256}H_{381}N_{65}O_{76}S_6$).

Insulin aspart ($C_{256}H_{237}N_{65}O_{79}S_6$), this name refers to the substitution of human insulin's amino acid B28, L-proline, with L-aspartic acid.

Insulin degludec ($C_{274}H_{411}N_{65}O_{81}S_6$), this name refers to the deletion of human insulin's amino acid B30, L-tryptophan, and conjugation to hexadecanedioic acid, via a γ-L-glutamyl spacer, to amino acid B29, L-lysine.

Insulin detemir ($C_{267}H_{402}N_{64}O_{76}S_6$), this name refers to the substitution of L-alanine in position B30 of porcine insulin with L-lysine and the acylation of L-lysine in position B29, with a myristyl (tetradecanoyl) group.

Insulin glargine ($C_{267}H_{404}N_{72}O_{78}S_6$), this name, with contraction of glycine and arginine, refers to the substitution of glycine for L-asparagine in position A21 and the addition of two L-arginines at the carboxy terminal of the B chain of native human insulin.

Insulin glulisine ($C_{258}H_{384}N_{64}O_{78}S_6$), this name, with contraction of glutamic acid and lysine, refers to the substitution of L-lysine for L-asparagine in position B3 and of L-glutamic acid for L-lysine in position B29 in human insulin.

Insulin lispro ($C_{257}H_{389}N_{65}O_{77}S_6$), this name, with contraction of lysine and proline, refers to the relocation of L-lysine from position B29 to position B28 and of L-proline from position B28 to position B29 of human insulin.

An important insulin preparation is **NPH insulin**, containing insulin, protamine, and zinc. NPH stands for Neutral Protamine Hagedorn, after the Danish physician and industrialist Hans Christian Hagedorn (1888–1971); NPH insulin is also called **isophane insulin**, isophane meaning a ratio of protamine to insulin equal to that in a solution made by mixing equal parts of a solution of the two in which all the protamine precipitates and a solution of the two in which all the insulin precipitates.

References

[1] http://www.who.int/en/ (retrieved August 2, 2018).
[2] World Health Organization, *Lists of Recommended and Proposed INNs*, http://www.who.int/medicines/publications/druginformation/inn/lists/en/
[3] World Health Organization, *The use of stems in the selection of International Nonproprietary Names (INN) for pharmaceutical substances*, 2018, https://www.who.int/medicines/services/inn/StemBook_2018.pdf?ua=1 (retrieved April 5, 2019).
[4] http://www.who.int/medicines/services/inn/INNMreview%20paperWkDoc167_Feb06_3_.pdf (retrieved August 2, 2018).

15 ISO common names for pesticides and other agrochemicals

The challenges for the naming of agrochemicals and the methods used to name them parallel those known from the naming of drugs, cf. Chapter 14. The International Organization for Standardization (ISO) [1, 2] approves ISO names in the shape of standards [3]. As of January, 2019 ISO had approved 1125 names completely and 110 names provisonally [4].

15.1 Pesticide classes

Pesticides are categorized in the following classes which can be further subdivided into chemical classes. Some of these classes, such as bactericides and fungicides, coincide or overlap with similar drug classes. The pesticide classes are acaricides, algicides, antifeedants, avicides, bactericides, bird repellents, chemisterilants, fungicides, herbicides, herbicide safeners, insect attractants, insect repellents, mating disrupters, molluscicides, nematicides, nitrification inhibitors, plant activators, plant growth regulators, rodenticides, synergists, and virucides.

15.2 ISO common names

No ISO common names are assigned to well-known generic substances with simple names such as the following two molluscicides:

Copper sulfate ($CuSO_4$).

Pentachlorophenol (C_6Cl_5OH).

An ISO common name is most often a contraction of the compound's chemical name or completely arbitrary. This can be seen if one chooses a few herbicides as examples:

Cafenstrole {*N*,*N*-diethyl-3-[(2,4,6-trimethylphenyl)sulfonyl]-1*H*-1,2,4-triazole-1-carbo xamide}, an arbitrary name.

Cyclopyranil, a contraction of the systematic name 1-(3-chloro-4,5,6,7-tetrahydro pyrazolo[1,5-*a*]pyridin-2-yl)-5-[(cyclopropylmethyl)amino]-1*H*-pyrazole-4-carbonitrile.

https://doi.org/10.1515/9783110612714-015

Isocarbamid, a contraction of *N*-isobutyl-2-oxoimidazolidine-1-carboxamide [*N*-(2-methylpropyl)-2-oxoimidazolidine-1-carboxamide]. In common chemical usage isocarbamide would be a synonym for isourea (carbamimidic acid).

MCPB, an initialism for methylchlorophenoxybutyric acid [4-(4-chloro-2-methylphenoxy)butanoic acid].

Prodiamine, a drastic contraction of 2,6-dinitro-N^1,N^1-dipropyl-4-(trifluoromethyl) benzene-1,3-diamine.

15.3 Recommended stems for ISO common names

For the construction of new names certain stems are recommended. However, a stem might already be in use for unrelated existing pesticides. Examples are:

-alin, as in trifluralin, for 2,6-dinitroanilines.

-azine, as in atrazine, for chloro-1,3,5-triazines.

Carb-, -carb, -carb-, as in carbofuran, for carbamates.

-conazole, as in penconazole, for chlorophenyl substituted imidazoles or 1,2,4-triazoles.

Coum-, -coum, as in coumatetralyl, for coumarins.

Din-, as in dinoterb, for dinitrophenols and their esters.

-fop, as in fluazifop, for phenoxypropionic acids.

Fos-, -fos-, -fos, as in quintiofos, for organic phosphorus compounds.

-quat, as in paraquat, for quaternary ammonium compounds.

-thrin as in permethrin, for pyrethroids.

-tryn, as in simetryn, for methylsulfanyl substituted 1,3,5-triazines.

-uron, as in linuron, for urea derivatives.

15.4 ISO common names for similar compounds

Highly analogous compounds are named in a way that expresses their close relationship as shown in the following examples:

Ethirimol, 5-butyl-2-(ethylamino)-6-methylpyrimidin-4-ol.

Dimethirimol, 5-butyl-2-(dimethylamino)-6-methylpyrimidin-4-ol.

Chlorotoluron, *N'*-(3-chloro-4-methylphenyl)-*N*,*N*-dimethylurea.

Chloretoluron, *N'*-(3-chloro-4-ethoxyphenyl)-*N*,*N*-dimethylurea.

15.5 ISO common names for esters and salts of active substances

Metsulfuron, 2-[[(4-methoxy-6-methyl-1,3,5-triazin-2-yl)carbamoyl]sulfamoyl]benzoic acid.

Metsulfuron-methyl, methyl 2-[[(4-methoxy-6-methyl-1,3,5-triazin-2-yl)carbamoyl] sulfamoyl]benzoate.

Metam, methylcarbamodithioic acid.

Metam-sodium, sodium methylcarbamodithioate.

15.6 ISO common names for enantiomers of racemic active substances

Once a racemic substance has been assigned an ISO name the name for a given enantiomer will be the same, with an added -P for (+)- or -M for (−)-. Thus, these expanded names are based on the sign of the optical rotation, not on absolute configuration, as shown in the following examples:

Mecoprop, (*RS*)-2-(4-chloro-2-methylphenoxy)propanoic acid.

Mecoprop-P, (2*R*)-2-(4-chloro-2-methylphenoxy)propanoic acid.

Flamprop, *N*-benzoyl-*N*-(3-chloro-4-fluorophenyl)-DL-alanine.

Flamprop-M, *N*-benzoyl-*N*-(3-chloro-4-fluorophenyl)-D-alanine.

References

[1] www.iso.org/home.html (retrieved August 2, 2018).
[2] https://www.iso.org/standards.html (retrieved August 2, 2018).
[3] https://www.iso.org/standard/4158.html (retrieved August 2, 2018).
[4] http://www.alanwood.net/pesticides/index.html (retrieved February 23, 2019).

16 Initialisms and acronyms

The following examples of initialisms and acronyms are intended to illustrate the way in which such terms are usually constructed. Many of them, such as DDT and dopa, are based upon obsolete, non-systematic compound names.

Some of the more elaborate acronyms, such as INADEQUATE, TEMPO, and WALTZ, seem to have been proposed tongue-in-cheek.

AAS, atom absorption spectroscopy.

ABA, abscisic acid {(2*Z*,4*E*)-5-[(1*S*)-1-hydroxy-2,6,6-trimethyl-4-oxocyclohex-2-en-1-yl]-3-methylpenta-2,4-dienoic acid}.

ABC, atomic, biological, chemical (warfare).

ACC, 1-aminocyclopropane-1-carboxylic acid.

ACE, angiotensin-converting enzyme (EC 3.4.15.1).

ACES, aminoacetylaminoethanesulfonic acid (2-[(2-amino-2-oxoethyl)amino]ethanesulfonic acid).

AchE, acetylcholinesterase (EC 3.1.1.7).

ACM, alkyl acrylate copolymer.

ACP, acyl carrier protein.

ACTH, adrenocorticotropic hormone.

ACV, δ-(L-α-aminoadipyl)-L-cysteinyl-D-valine.

[1]**ADA**, adenosine deaminase (EC 3.5.4.4).

[2]**ADA**, acetamidoiminodiacetic acid, *N*-(carbamoylmethyl)iminodiacetic acid.

[3]**ADA**, azodicarbonamide (diazene-1,2-dicarboxamide).

[1]**ADH**, alcohol dehydrogenase (EC 1.1.1.1).

[2]**ADH**, antidiuretic hormone.

ADI, acceptable daily intake.

ADP, adenosine 5′-diphosphate.

AES, atomic emission spectrometry.

AFM, atomic force microscopy.

AFS, atomic fluorescence spectroscopy.

https://doi.org/10.1515/9783110612714-016

AIBN, azoisobutyronitrile {2-[(2-cyanopropan-2-yl)diazen-1-yl]-2-methylpropanenitrile}.

ALA, δ-aminolevulinic acid (5-amino-4-oxopentanoic acid).

AMP, adenosine 5′-monophosphate.

amu, atomic mass unit.

Anot, 3-amino-5-nitro-*o*-toluamide (3-amino-2-methyl-5-nitrobenzamide).

ANTU, α-naphthylthiourea [*N*-(1-naphthyl)thiourea].

ATPase, adenosine 5′-triphosphatase (EC 3.6.1.3).

AO, atomic orbital.

AOX, adsorbable organic halides. The letter X stands for halogen.

APS, appearance potential spectroscopy.

ASS, acetylsalicylic acid (2-acetoxybenzoic acid).

ATP, adenosine 5′-triphosphate.

A 4, *S*-[2-[(di(propan-2-yl)amino]ethyl] *O*-ethyl methylphosphonothioate, a US military code. A.k.a. **VX**.

BAL, British antilewisite, dimercaprol (2,3-disulfanylpropan-1-ol).

BART, biological activity reaction test.

BET, Brunauer-Emmett-Teller (method), cf. Chapter 12.

BHA, *tert*-butylhydroxyanisole [4-methoxy-2-(2-methylpropan-2-yl)phenol].

BHC, benzene hexachloride (1,2,3,4,5,6-hexachlorocyclohexane), i.e. a misnomer. A.k.a. **HCH** and **lindane**.

BHT, butylated hydroxytoluene [4-methyl-2,6-bis(2-methylpropan-2-yl)phenol]. A.k.a. **DBP**.

BiAS, bismuth active substance.

BINAP, [1,1′-binaphthalene]-2,2′-diylbis(diphenylphosphane).

BIRC5, baculoviral inhibitor of apoptosis repeat-containg 5.

BOD, biological oxygen demand, cf. COD.

^{1}**BTX**, benzene, toluene, xylene.

^{2}**BTX**, bibrotoxin.

^{3}**BTX**, botulinus toxin.

^{4}**BTX**, brevetoxin.

CAF, calcium-activated factor.

CAIR, carboxyaminoimidazole ribotide [5-amino-1-(5-*O*-phosphono-β-D-ribofuranosyl)-1*H*-imidazole-4-carboxylic acid].

cAMP, cyclic adenosine 3′,5′-monophosphate.

CARS, coherent anti-Stokes Raman spectroscopy.

CCC, chlorocholine chloride (2-chloro-*N*,*N*,*N*-trimethylmethanaminium chloride).

CCK, cholecystokinin.

cCMP, cyclic cytidine 3′,5′-monophosphate.

CCN, cement chemist notation.

CD, circular dichroism.

cDNA, complementary deoxyribonucleic acid.

CDP, cytidine 5′-diphosphate.

CE, capillary electrophoresis.

^{1}**CI**, chemical ionization.

^{2}**CI**, configuration interaction.

C.I., colour index.

CIDEP, chemically induced dynamic electronic polarization.

CIDNP (pronounced kidnap), chemically induced dynamic nuclear polarization.

cIMP, cyclic inosine 3′,5′-monophosphate.

CIP, Cahn-Ingold-Prelog, cf. Chapter 12.

CLA, conjugated linoleic acid(s).

CLP, Classification, Labelling and Packaging of Substances and Mixtures.

^{1}**CMC**, carboxymethylcellulose.

^{2}**CMC**, critical micelle concentration.

^{1}**CN**, nitrocellulose, a.k.a. **NC**.

^{2}**CN** (a US military code), ω-chloroacetophenone (2-chloro-1-phenylethan-1-one).

CNDO, complete neglect of differential overlap.

COD, chemical oxygen demand, cf. BOD.

COMT, catecholamine *O*-methyltransferase (EC 2.1.1.6).

COSY, correlated spectroscopy.

COT, cycloocta-1,3,5,7-tetraene.

COX, cyclooxygenase (EC 1.14.99.1).

CPK, Corey-Pauling-Koltun, cf. Chapter 12.

CR, a US military code, dibenzo[*b,f*][1,4]oxazepin.

CRF, corticotropin-releasing factor.

CRH, corticotropin-releasing hormone.

cRNA, complementary ribonucleic acid.

CS (a US military code), (2-chlorobenzylidene)malononitrile {[(2-chlorophenyl) methylidene]propanedinitrile}.

CSF, colony-stimulating factor.

CT, charge transfer.

CTP, cytidine 5′-triphosphate.

^{1}**CTX,** charybdotoxin.

^{2}**CTX**, ciguatoxin.

CVD, chemical vapor deposition.

DABCO, 1,4-diazabicyclo[2.2.2]octane.

DAHP, 3-deoxy-D-*arabino*-hept-2-ulosonic acid 7-(dihydrogen phosphate).

DANTE, delayed alternation with nutations for tailored excitation.

DAP, diallyl phthalate [di(prop-2-en-1-yl) benzene-1,2-dicarboxylate].

DAST, (diethylamino)sulfur trifluoride [*N*-ethyl-*N*-(trifluoro-λ^4-sulfanyl)ethan-1-amine].

DBN, 1,5-diazabicyclo[4.3.0]non-5-ene.

^{1}**DBP**, di-*tert*-butylmethylphenol [2-(2,2,4,4-tetramethylpentan-3-yl)phenol].

^{2}**DBP**, dibutyl phthalate (dibutyl benzenedicarboxylate).

DBU, 1,8-diazabicyclo[5.4.0]undec-2-ene.

DCC, *N,N′*-dicyclohexylcarbodiimide (dicyclohexylmethanediimine).

DDQ, 2,3-dichloro-5,6-dicyano-*p*-benzoquinone (4,5-dichloro-3,6-dioxocyclohexa-1,4-diene-1,2-dicarbonitrile).

DDT, dichlorodiphenyltrichloroethane [1,1′-(2,2,2-trichloroethane-1,1-diyl)bis(4-chlorobenzene].

DDVP, *O*,*O*-dimethyl *O*-(2,2-dichlorovinyl) phosphate [(2,2-dichloroethen-1-yl) dimethyl phosphate].

DEA, diethanolamine, (2,2′-iminobisethan-1-ol).

deet, *N*,*N*-diethyl-*m*-toluamide (*N*,*N*-diethyl-3-methylbenzamide).

DEHP, di(2-ethylhexyl) phthalate [bis(2-ethylhexyl) benzene-1,2-dicarboxylate].

DEP, diethyl phthalate (diethyl benzen-1,2-dicarboxylate).

DHA, (4*Z*,7*Z*,10*Z*,13*Z*,16*Z*,19*Z*)-docosa-4,7,10,13,16,19-hexaenoic acid.

DHEA, dehydroepiandrosterone (3β-hydroxyandrost-5-en-17-one).

DIBOA, 2,4-dihydroxy-2*H*-1,4-benzoxazin-3(4*H*)-one.

DIDP, diisodecyl phthalate [bis(8-methylnonyl) benzene-1,2-dicarboxylate].

DINA, diisononyl adipate [bis(7-methyloctyl) hexanedioate].

^{1}**DINP**, diisononyl phthalate [bis(7-methyloctyl) benzene-1,2-dicarboxylate].

^{2}**DINP**, diisononyl phthalate [bis(3,5,5-trimethylhexyl) benzene-1,2-dicarboxylate].

DIOP, diisooctyl phthalate [bis(6-methylheptyl) benzene-1,2-dicarboxylate].

DIPT, diisopropyl D-tartrate [di(propan-2-yl) (2*S*,3*S*)-2,3-dihydroxybutanedioate].

DLVO, Deryagin-Landau-Verwey-Overbeek, cf. Chapter 12.

DM (a US military code), 10-chloro-5,10-dihydrophenarsazine.

DMA, *N*,*N*-dimethylacetamide.

DMF, *N*,*N*-dimethylformamide.

DMP, dimethyl phthalate (dimethyl benzene-1,2-dicarboxylate).

DMSO, dimethyl sulfoxide [(methanesulfinyl)methane].

DMT, dimethyl terephthalate (dimethyl benzene-1,4-dicarboxylate).

DNA, deoxyribonucleic acid.

DNOC, 4,6-dinitro-*o*-cresol (2-methyl-4,6-dinitrophenol).

DNOP, di-*n*-octyl phthalate (dioctyl benzene-1,2-dicarboxylate), a.k.a. **DOP**.

DOA, dioctyl adipate (dioctyl hexanedioate).

DOC, dissolved organic carbon.

DOM, 2,5-dimethoxy-4-methylamphetamine [1-(2,5-dimethoxy4-methylphenyl)propan-2-amine), a.k.a. **STP**.

DOP, di-*n*-octyl phthalate (dioctyl benzene-1,2-dicarboxylate), a.k.a. **DNOP**.

Dopa, an acronym of 3,4-dioxyphenylalanine (3,4-dihydroxyphenylalanine).

Dopamine, 2-(3,4-dihydroxyphenyl)ethan-1-amine, a contraction of dopa and amine.

DPM, dipivaloylmethane (2,2,6,6-tetramethylheptane-3,5-dione).

DSC, differential scanning calorimetry.

DSIP, delta sleep-inducing peptide.

DTA, differential thermal analysis.

DTPA, diethylenetriamine-*N*,*N*,*N*′,*N*′,*N*″,*N*″-pentaacetic acid {*N*,*N*-bis(2-[bis(carboxymethyl)amino]ethyl]glycine}. A.k.a. **pentetic acid**.

DU, Dobson unit, cf. Chapter 12.

E, ecstasy [3,4-methylenedioxymethamphetamine, 1-(1,3-benzodioxol-5-yl)-*N*-methylpropan-2-amine]. A.k.a. **Adam** and **XTC**.

^{1}**EC**, electron capture.

^{2}**EC**, also E.C., Enzyme Commission.

ECD, electron capture detector.

ECOIN, European Core Inventory.

ED, Entner-Doudoroff pathway, cf. Chapter 12.

EDAX, energy dispersive analysis of X-rays.

EDTA, ethylenediaminetetraacetic acid, *N*,*N*′-ethane-1,2-diylbis[*N*-(carboxymethyl) glycine], 2,2′,2″,2‴-(ethane-1,2-diyldinitrilo)tetraacetic acid. Here IUPAC offers no less than three valid systematic names without preference for any of them.

e.e., enantiomeric excess.

EEDQ, *N*-ethoxycarbonyl-2-ethoxy-1,2-dihydroquinoline [ethyl 2-ethoxyquinoline-1 (2*H*)-carboxylate].

EELS, electron energy loss spectroscopy.

EGCG, epigallocatechol gallate [(2*R*,3*R*)-5,7-dihydroxy-2-(3,4,5-trihydroxyphenyl)-3,4-dihydro-2*H*-1-benzopyran-3-yl 3,4,5-trihydroxybenzoate].

EGF, epidermal growth factor. A.k.a. **urogastrone**.

EGTA, ethylene glycol bis(β-aminoethyl ether)-*N*,*N*,*N*′,*N*′-tetraacetic acid [3,12-bis (carboxymethyl)-6,9-dioxa-3,12-diazatetradecane-1,14-dioic acid].

EHT, extended Hückel theory, cf. Chapter 12.

EI, electron impact.

EIA, enzyme immunoassay.

EINECS, European Inventory of Existing Commercial Chemical Substances.

ELINCS, European List of Notified Chemical Substances.

ELISA, enzyme-linked immunosorbent assay.

ELS, electron loss spectroscopy.

EMA, electron microprobe analysis.

EMIT, enzyme-multiplied immunoassay technique.

EMP, Embden-Meyerhof-Parnas (path), cf. Chapter 12.

EPA, (5*Z*,8*Z*,11*Z*,14*Z*,17*Z*)-eicosa-5,8,11,14,17-pentaenoic acid [(5*Z*,8*Z*,11*Z*,14*Z*,17*Z*)-icosa-5,8,11,14,17-pentaenoic acid], a.k.a. **IPA**.

EPMA, electron probe microanalysis.

EPO, erythropoietin. The name refers to its stimulation of erythropoiesis, i.e. red blood cell production.

EPR, electron paramagnetic resonance.

ESCA, electron spectroscopy in chemical analysis, a.k.a. **IEE**.

ESD, electron-stimulated desorption.

ESR, electron spin resonance.

ET, electron transfer.

EXAFS, extended X-ray absorption fine structure.

FAAS, flame atom absorption spectroscopy.

FAB, fast atom bombardment.

FAD, flavine adenine dinucleotide.

$FADH_2$, reduced flavine adenine dinucleotide.

FAS, fasciculin (toxic proteins).

FCC, fluid catalytic cracking.

FD, field desorption.

FDP, fructose 1,6-diphosphate (1,6-di-*O*-phosphono-D-fructose).

FES, flame emission spectroscopy.

FFF, field flow fractionation.

FFS, flame fluorescence spectroscopy.

FH_2, dihydrofolic acid.

FH_4, tetrahydrofolic acid.

FI, field ionization.

[1]**FIA**, flow injection analysis.

[2]**FIA**, fluorescence immunoassay.

FID, flame ionization detector.

FIM, field ion microscopy.

FIMS, field ion mass spectrometry.

FIR, far infrared.

FMN, flavin mononucleotide.

$FMNH_2$, dihydroflavin mononucleotide.

FPC, fermentation-produced chymosin.

FSGO, floating spherical Gauss orbital, cf. Chapter 12.

FSH, follicle-stimulating hormone.

FT, Fourier transform.

GA (a US military code), tabun [(*RS*)-ethyl *N*,*N*-dimethylphosphoramidocyanidate}.

GABA, γ-aminobutyric acid (4-aminobutanoic acid).

GB (a US military code), sarin (propan-2-yl methylphosphonofluoridate).

GC, gas chromatography.

GD (a US military code), soman (3,3-dimethylbutan-2-yl methylphosphonofluoridate).

GDP, guanosine 5′-diphosphate.

GF (a US military code), cyclosarin (cyclohexyl methylphosphonofluoridate).

GFC, gel filtration chromatography.

GGG, gadolinium-gallium garnet.

GGT, γ-glutamyl transferase (EC 2.3.2.2).

GH, growth hormone. A.k.a. **somatotropin**.

GH-RIF, growth hormone release inhibitory factor.

GHS, Globally Harmonized Systems for Classification and Labelling of Chemicals.

GIP, gastric inhibitory polypeptide.

GLC, gas-liquid chromatography.

GLP, good laboratory practice.

glyme, glycol dimethyl ether (1,2-dimethoxyethane).

[1]**GMP**, good manufacturing practice.

[2]**GMP**, guanosine 5′-monophosphate.

GO, Gauss orbital, cf. Chapter 12.

GOT, glutamate oxaloacetate transaminase (EC 2.6.1.1).

GPC, gel permeation chromatography.

GPD, glucose 6-phosphate dehydrogenase (EC 1.1.1.49).

G-protein, a contraction of GTP-binding protein.

GPT, glutamate pyruvate transaminase (EC 2.6.1.2).

GRAS, generally recognized as safe.

GRF, gonadotropin-releasing hormone.

GRP, gastrin-releasing peptide.

GSC, gas-solid chromatography.

GSH, glutathione [*N*-(L-γ-glutamyl-L-cysteinyl)glycine].

GSSG, oxidized glutathione.

GTO, Gaussian-type orbital.

GTP, guanosine 5′-triphosphate.

HA, hyaluronic acid.

HCB, hexachlorobenzene.

HCH, hexachlorocyclohexane (1,2,3,4,5,6-hexachlorocyclohexane). A.k.a. **BHC** and **lindane**.

HDI, hexamethylene diisocyanate (1,6-diisocyanatohexane), a.k.a. **HMDI**.

HDPE, high-density polyethylene.

HEC, hydroxyethyl cellulose.

HEDTA, *N*-(2-hydroxyethyl)ethylenediaminetriacetic acid {*N*-[2-[bis(carboxymethyl) amino]ethyl]-*N*-(2-hydroxyethyl)glycine}.

HEED, high-energy electron diffraction.

HEIS, high-energy ion scattering.

HERON, heteroatom rearrangements on nitrogen.

5-HETE, (5*S*,6*E*,8*Z*,11*Z*,14*Z*)-5-hydroxyicosa-6,8,11,14-tetraenoic acid.

HETP, height equivalent to a theoretical plate.

HETS, height equivalent to a theoretical stage.

HFS, hyperfine structure.

HGH, human growth hormone.

HLAD, horse liver alcohol dehydrogenase (EC 1.1.1.1).

HMDI, hexamethylene diisocyanate (1,6-diisocyanatohexane), a.k.a. **HDI**.

HMDS, hexamethyldisilazane.

HMO, Hückel molecular orbital, cf. Chapter 12.

HMPA, hexamethylphosphoric triamide, a.k.a. **HMPT**.

HMPT, hexamethylphosphoric triamide, a.k.a. **HMPA**.

HMX (a UK military code), 1,3,5,7-tetranitro-1,3,5,7-tetrazocane. A contraction of high-molecular weight research department explosive.

HNE, (2*E*)-4-hydroxynon-2-enal.

HOESY, heteronuclear Overhauser effect spectroscopy, cf. Chapter 12.

HOMO, highest occupied molecular orbital.

HON, 5-hydroxy-4-oxo-L-norvaline.

HOPO, 2-hydroxypyridine-*N*-oxide (pyridin-2-ol 1-oxide).

HPC, hydroxypropyl cellulose.

5-HPETE, (6*E*,8*Z*,11*Z*,14*Z*)-5-hydroperoxyicosa-6,8,11,14-tetraenoic acid.

HPLC, high-pressure liquid chromatography, high-performance liquid chromatography.

HPMC, 2-hydroxypropyl methylcellulose.

HPTLC, high-pressure thin-layer chromatography, high-performance thin-layer chromatography.

HQNO, 2-heptylquinolin-4-ol 1-oxide.

HREELS, high-resolution electron energy loss spectroscopy. A.k.a. **HRELS**.

HRELS, high-resolution electron energy loss spectroscopy. A.k.a. **HREELS**.

HSA, human serum albumin.

HSAB, hard and soft acids and bases.

5-HT, 5-hydroxytryptamine [3-(2-aminoethyl)-1*H*-indol-5-ol]. A.k.a. **serotonin**.

5-HTP, 5-hydroxy-L-tryptophan.

[1]**IC**, internal conversion.

[2]**IC**, ion chromatography.

ICP, inductively coupled plasma.

ICR, ion cyclotron resonance.

IDP, inosine 5′-diphosphate.

[1]**IEC**, ion exchange chromatography.

[2]**IEC**, ion exclusion chromatography.

IEE, induced electron emission, a.k.a. **ESCA**.

IEF, isoelectric focusing.

IEP, isoelectric point, a.k.a. **IP**.

IESS, inelastic ion scattering spectroscopy.

IETS, inelastic ion tunneling spectroscopy.

IFA, immunofluorescence analysis.

IFN, interferon.

Ig, immunoglobulin.

IL, interleukin.

ILS, ionization loss spectroscopy.

IMMA, ion microprobe mass analysis.

IMP, inosine 5′-monophosphate.

INAA, instrumental neutron activation analysis.

INADEQUATE, incredible natural-abundance double quantum transfer experiment.

INDO, intermediate neglect of differential overlap.

INDOR, internuclear double resonance.

INEPT, insensitive nuclei enhanced by polarization transfer.

INH, isonicotinic hydrazide (pyridine-4-carbohydrazide).

INN, International Nonproprietary (drug) Name.

INNM, International Nonproprietary (drug) Name Modified.

^{1}**INS**, inelastic neutron scattering.

^{2}**INS**, ion neutralization spectroscopy.

IP, isoelectric point, cf. IEP.

IPA, (5*Z*,8*Z*,11*Z*,14*Z*,17*Z*)-icosa-5,8,11,14,17-pentaenoic acid, a.k.a. **EPA**.

IPC, ion pair chromatography.

IPMA, ion pair microanalysis.

IPP, isopentenyl pyrophosphate (3-methylbut-3-en-1-yl diphosphate).

IR, infrared (spectroscopy).

IRMA, immunoradiometric assay.

ISC, intersystem crossing.

ISS, ion scattering spectroscopy.

ITP, inosine 5′-triphosphate.

IUB, International Union of Biochemistry.

IUCr, International Union of Crystallography.

IUPAC, International Union of Pure and Applied Chemistry.

IUPAP, International Union of Pure and Applied Physics.

JSTX, joro spider toxin.

LAFS, laser atomic fluorescence spectroscopy.

LAL, limulus amebocyte lysate.

LAMES, laser microemission spectroscopy.

LAMMA, laser microprobe mass analysis.

LAMMS, laser microprobe mass spectroscopy.

LAMS, laser ablation mass spectrometry.

Laser, light amplification by stimulated emission of radiation.

LC, liquid chromatography.

LCAO, linear combination of atomic orbitals.

LDH, lactate dehydrogenase (EC 1.1.1.27).

LDL, low-density lipoprotein.

LDPE, low-density polyethylene.

LEAFS, laser-excited atomic fluorescence spectroscopy.

LEC, liquid exclusion chromatography.

LEED, low-energy electron diffraction.

LEI, laser-enhanced ionization.

LEIS, low-energy ion scattering.

LF, ligand field.

LFER, linear free-energy relationship.

LFSE, ligand field stabilization energy.

LH, luteinizing hormone.

LH-RH, luteinizing hormone-releasing hormone.

LIBS. laser-induced breakdown spectroscopy.

LIF, laser induced fluorescence.

LIMA, laser induced mass analysis.

LLC, liquid-liquid chromatography.

LLDPE, linear low-density polyethylene.

LLPC, liquid-liquid partition chromatography.

Ln, element symbol for unspecified lanthanoids.

LNG, liquefied natural gas.

LOI, limiting oxygen index.

lox, liquid oxygen.

LPES, laser photodetachment electron spectroscopy.

LPG, liquefied petroleum gas.

LR, Lawesson's reagent [2,4-bis(4-methoxyphenyl)-1,3,2,4-dithiadiphosphetane-2,4-dithione].

LSC, liquid-solid chromatography.

LSD, lysergic acid *N,N*-diethylamide (*N,N*-diethyl-6-methyl-9,10-didehydroergoline-8β-carboxamide).

LT, leukotriene.

LTA, lead tetraacetate [lead(IV) acetate].

LUMO, lowest unoccupied molecular orbital.

MAO, monoamine oxidase (EC 1.4.3.4).

MARCKS, myristoylated alanine-rich C kinase substrate.

MAS, magic-angle spinning.

Maser, microwave amplification by stimulated emission of radiation.

MBAS, methylene blue active substance.

MBK, methyl butyl ketone (hexan-2-one).

MC, methylcellulose.

MCR, multi-component reaction.

MCT, medium-chain triglycerides.

MDA, 3,4-methylenedioxyamphetamine [1-(1,3-benzodioxol-5-yl)propan-2-amine].

MDE, 3,4-methylenedioxyethamphetamine [1-(1,3-benzodioxol-5-yl)-*N*-ethylpropan-2-amine], a.ka. **MDEA**.

MDEA, 3,4-methylenedioxyethamphetamine [1-(1,3-benzodioxol-5-yl)-*N*-ethylpropan-2-amine], a.k.a. **MDE**.

MDI, methylene diphenylene diisocyanate [1,1′-methylenebis(4-isocyanatobenzene)].

MDMA, 3,4-methylenedioxymethamphetamine [1-(1,3-benzodioxol-5-yl)-*N*-methylpropan-2-amine]. A.k.a. **Adam, E, ecstasy**, and **XTC**.

MEK, methyl ethyl ketone (butanone).

MES, morpholine-4-ethanesulfonic acid [2-(morpholin-4-yl)ethane-1-sulfonic acid].

mesna, German *2-Mercaptoethansulfonsäurenatriumsalz* (sodium 2-sulfanylethane-1-sulfonate).

MFO, mixed-function oxidase (EC 1).

MIBK, methyl isobutyl ketone (4-methylbutan-2-one).

MINDO, modified intermediate neglect of differential overlap.

MMA, methyl methacrylate (methyl 2-methylprop-2-enoate).

MMB, 3-mercapto-3-methylbutan-1-ol (3-methyl-3-sulfanylbutan-1-ol).

^{1}**MMT**, (methylcyclopentadienyl)manganese tricarbonyl [tricarbonyl(methyl-μ^5-cyclopentadienyl)manganese].

^{2}**MMT**, monomethoxytrityl chloride {1-[chloro(diphenyl)methyl)-4-methoxybenzene]}.

MNDO, modified neglect of diatomic overlap.

MO, molecular orbital.

MOPS, 3-(morpholin-1-yl)propane-1-sulfonic acid.

MPIC, mobile-phase ion chromatography.

MPS, mucopolysaccharide polysulfate.

MPTP, 1-methyl-4-phenyl-1,2,3,6-tetrahydropyridine.

MS, mass spectrometry.

MSG, monosodium L-glutamate.

MSH, melanocyte-stimulating hormone.

MTB, methyl *tert*-butyl ether (2-methoxy-2-methylpropane), a.k.a. **MTBE**.

MTBE, methyl *tert*-butyl ether (2-methoxy-2-methylpropane), a.k.a. **MTB**.

MTG, methanol to gasoline (process).

MTX, maitotoxin.

NAA, neutron activation analysis.

NAD, nicotinamide adenine dinucleotide. A.k.a. **cozymase**, a contraction of coenzyme and zymase.

NAD$^+$, oxidized nicotinamide adenine dinucleotide. A.k.a. **cozymase**, a contraction of coenzyme and zymase.

NADH, reduced nicotinamide adenine dinucleotide (1,4-dihydronicotinamide-adenine dinucleotide), a.k.a. **NADPH** and **NADPH**$_2$.

NADP$^+$, nicotinamide adenine dinucleotide phosphate.

NADPH, reduced nicotinamide adenine dinucleotide, 1,4-dihydronicotinamide-adenine dinucleotide phosphate, a.k.a. **NADH** and **NADPH**$_2$.

NADPH$_2$, reduced nicotinamide adenine dinucleotide, 1,4-dihydronicotinamide-adenine dinucleotide phosphate, cf. **NADH** and **NADPH**.

NANA, *N*-acetylneuraminic acid.

Napalm, a contraction of naphthenic acid and palmitic acid.

NBS, *N*-bromosuccinimide (1-bromopyrrolidine-2,5-dione).

NC, nitrocellulose, a.k.a. ^{1}**CN**.

NDDO, neglect of diatomic differential overlap.

NIR, near infrared (spectroscopy).

NMDA, *N*-methyl-D-aspartic acid.

NMN, nicotinamide mononucleotide.

^{1}**NMP**, *N*-methyl-2-pyrrolidone (1-methylpyrrolidin-2-one).

^{2}**NMP**, nucleoside 5′-monophosphate.

NMR, nuclear magnetic resonance.

NOE, nuclear Overhauser effect.

NOEL, no observed effect level.

NOESY, nuclear Overhauser enhancement spectroscopy, cf. Chapter 12.

NQR, nuclear quadrupole resonance.

nRNA, nuclear ribonucleic acid.

NTA, nitrilotriacetic acid (2,2′,2″-nitrilotriacetic acid).

OBA, optical brightening agent.

OE, Overhauser effect, cf. Chapter 12.

OES, optical emission spectroscopy.

OMP, orotidine 5′-monophosphate.

OMPA, *N,N,N′,N′,N″,N″,N‴,N‴*-octamethyldiphosphoric tetramide.

ORD, optical rotatory dispersion.

ORTEP, Oak Ridge thermal ellipsoid plot, named after the Oak Ridge National Laboratory, Oak Ridge, TN, USA.

OSPE, octahedral site preference energy.

PAF, platelet-activating factor.

PAGE, polyacrylamide gel electrophoresis.

PAH, polycyclic aromatic hydrcarbon(s).

2-PAM, 2-pyridinealdoxime methiodide {2-[(*E*)-(hydroxyimino)methyl]-1-methylpyridinium iodide}.

^{1}**PAN**, peroxyacetyl nitrate [1-(nitroperoxy)ethan-1-one].

^{2}**PAN**, 1-(2-pyridylazo)-2-naphthol {1-[2-[(*E*)-pyridin-2-yl]diazen-1-yl]naphthalen-2-ol}.

PAS, *p*-aminosalicylic acid (4-amino-2-hydroxybenzoic acid).

PBB, polybrominated biphenyl(s).

PC, paper chromatography.

PCA, protocatechuic acid (3,4-dihydroxybenzoic acid).

PCB, polychlorinated biphenyl(s).

PCDD, polychlorodibenzodioxin(s).

PCDF, polychlorodibenzofuran(s).

PCILO, perturbative configuration interaction using localized orbitals.

PCP. pentachlorophenol.

PCR, polymerase chain reaction.

PED, photoelectric diffraction.

PEG, poly[ethylene glycol] {poly[ethene-1,2-diol]}.

PEP, phosphoenolpyruvate [2-(phosphonooxy)prop-2-enoic acid].

PET, positron emission tomography.

PETN, pentaerythritol tetranitrate {2,2-bis[(nitrooxy)methyl]propane-1,3-diyl dinitrate}.

PG, prostaglandin.

pH, New Latin *potentia hydrogenii*, strength of hydrogen.

PHB, *p*-hydroxybenzoic acid (4-hydroxybenzoic acid).

PhIP, 1-methyl-6-phenylimidazo[4,5-*b*]pyridin-2-amine.

PID, photoionization detector.

PIES, Penning ionization electron spectroscopy, cf. Chapter 12.

pINN, proposed International Nonproprietary (drug) Name.

PIS, Penning ionization spectroscopy, cf. Chapter 12.

pK, from German *Konstante*, a constant, patterned after pH.

PLNM, principle of least nuclear motion.

PLP, pyridoxal 5′-phosphate [(4-formyl-5-hydroxy-6-methylpyridin-3-yl)methyl phosphate, vitamin B_6].

PLZT, the alloy (Pb,La,Zr,Ti).

PM, pyridoxamine [4-(aminomethyl)-5-(hydroxymethyl)-2-methylpyridin-3-ol, vitamin B_6]

PMP, pyridoxamine 5′-phosphate {4-[(aminomethyl)-5-hydroxy-6-methylpyridin-3-yl]methyl dihydrogen phosphate, vitamin B_6}.

PMR, proton magnetic resonance.

PN, pyridoxine [4,5-bis(hydroxymethyl-2-methylpyridin-2-ol, vitamin B_6].

PNDO, partial neglect of differential overlap.

POC, particulate organic carbon.

POM, polyoxometalate.

POP, persistent organic pesticide(s).

PPA, polyphosphoric acid.

PQ, plastoquinone.

PQQ, pyrroloquinoline quinone {4,5-dioxo-4,5-dihydro-1*H*-pyrrolo[2,3-*f*]quinoline-2,7,9-tricarboxylic acid}. A.k.a. **methoxatin**, after its source, a methanol-oxidizing bacterium.

PRL, prolactin.

PS, photoelectron spectroscopy.

PSE, periodic system of the elements.

PT, phosphinothricin.

PTC, phase transfer catalysis.

[1]**PTX**, palytoxin.

[2]**PTX**, pectenotoxin.

[3]**PTX**, pumiliotoxin.

PTZ, phenothiazine.

PU, polyurethane, a.k.a. **PUR**.

PUFA, polyunsaturated fatty acid(s).

PUR, polyurethane, a.k.a. **PU**.

Purex process, plutonium-uranium recovery by extraction.

PVC, poly[vinyl chloride] {poly[1-chloroethene]}.

PVD, physical vapor deposition.

Py-Phe, *N*-[4-(pyren-1-yl)butanoyl]-L-phenylalanine.

[1]**PZT**, the alloy (Pb,Zr,Ti).

[2]**PZT**, the alloy (Pb,La,Zr,Ti).

P5P, pyridoxine 5′-phosphate {5-hydroxy-4-(hydroxymethyl)-6-methylpyridin-3-yl] methyl dihydrogen phosphate, vitamin B_6}.

QCSID, quantum configuration interaction singles and doubles.

QSAR, quantitative structure-activity relationship(s).

RBS, Rutherford backscattering spectroscopy, cf. Chapter 12.

rDNA, ribosomal deoxyribonucleic acid.

RDX, research department explosive (1,3,5-trinitro-1,3,5-triazinane). A.k.a. **cyclonite** and **hexogen**, referring, with contraction, to the presence of six nitrogen atoms in this compound.

RE, rare earth.

REACH, Registration, Evaluation, Authorisation and Restriction of Chemical Substances.

Redox, reduction-oxidation.

REE, rare earth element(s).

[1]**REM**, rare earth metal(s).

[2]**REM**, reflection electron spectroscopy.

REMPI, resonance enhanced multiphoton ionization.

RHEED, reflection high-energy electron diffraction.

RIA, radioimmunoassay.

RIBS, Rutherford ion backscattering, cf. Chapter 12.

RIMS, resonance ionization mass spectrometry.

rINN, recommended International Nonproprietary (drug) Name.

RIS, resonance ionization spectroscopy.

RIST, radioimmunosorbent test.

RLCC, rotating locular countercurrent chromatography.

RN, registry number.

RNA, ribonucleic acid.

RNase, ribonuclease (EC 2.7 and EC 3.1).

ROESY, rotational frame nuclear Overhauser effect spectroscopy, cf. Chapter 12.

ROMP, ring-opening metathesis polymerization.

RPC, reversed-phase chromatography.

RRKM, Rice-Ramsperger-Kassel-Marcus (theory), cf. Chapter 12.

rRNA, ribosomal ribonucleic acid.

RTECS, Registry of Toxic Effects of Chemical Substances.

RTX, resiniferatoxin.

Rubisco, ribulose 1,5-bisphosphate carboxylase (1,5-di-*O*-phosphonopent-2-ulose, EC 4.1.1.39).

SAICAR, *N*-succino-5-aminoimidazole-4-carboxamide ribonucleotide {*N*-[[5-amino -1-(5-*O*-phosphono-β-D-ribofuranosyl)-1*H*-imidazol-4-yl]carbonyl]-L-aspartic acid}.

Salatrim, short- and long-acyl triglyceride molecule.

Salen, *N,N′*-ethylenebis(salicylideneimine) {2,2′-(ethane-1,2-diyl)bis[nitrilo(*E*)methylidene)diphenol}.

[1]**SAM**, α-adenosyl-L-methionine.

[2]**SAM**, scanning Auger spectroscopy, cf. Chapter 12.

SAS, secondary alkanesulfonate.

SBR, styrene-butadiene rubber.

SCF, self-consistent field (method).

SCP, single-cell protein.

SDS, sodium dodecyl sulfate, a.k.a. **SLS**.

SEC, size exclusion chromatography.

SEM, scanning electron microscopy.

SEMINA, SEMUT editing of INADEQUATE ^{13}C NMR spectra. Here we have a second-order acronym based on two first-order acronyms.

SEMUT, subspectral editing using a multiple quantum trap.

Serpin (a protein), serine proteinase inhibitor.

SERRS, surface-enhanced resonance Raman scattering spectroscopy, cf. Chapter 12. A.k.a. **SERS**.

SERS, surface-enhanced resonance Raman scattering spectroscopy, cf. Chapter 12. A.k.a. **SERRS**.

SET, single-electron transfer.

SEXAFS, surface-extended X-ray absorption fine structure.

SFC, supercritical fluid chromatography.

SFE, supercritical fluid extraction.

SI, French *Système International d'Unités*, international system of units.

SINDO, symmetrically othogonalized intermediate neglect of differential overlap.

SLS, sodium lauryl sulfate (sodium dodecyl sulfate), cf. SDS.

SMB, simulated moving bed (chromatography).

SMEAH, sodium bis(2-methoxyethoxy)aluminum hydride {sodium dihydrido[bis(-2-methoxyethan-1-olato-κ*O*)]aluminate(1−)}. A.k.a. **Red-Al**.

SNG, substitute natural gas, synthetic natural gas.

SNOBS, sodium 4-(nonanoyloxy)benzenesulfonate.

snRNA, small nuclear ribonucleic acid.

SOD, superoxide dismutase (EC 1.15.1.1). A.k.a. **cytocuprein** and **erythrocuprein**.

SOMO, singly-occupied molecular orbital.

SP, substance P.

SPE, solid-phase extraction.

SPME, solid-phase microextraction.

sRNA, soluble ribonucleic acid.

SRPES, synchrotron radiation photoelectron spectroscopy.

SRS-A, slow-reacting substances of anaphylaxis.

SRTX, sarafotoxin.

SRXRF, synchrotron radiation X-ray fluorescence spectroscopy.

STEM, scanning transmission electron microscopy.

STM, scanning tunneling microscopy.

STO, Slater-type orbital.

^{1}**STP**, standard temperature and pressure.

^{2}**STP**, initialism of serenity, tranquillity, peace; 2,5-dimethoxy-4-methylamphetamine [1-(2,5-dimethoxy-4-methylphenyl)propan-2-amine). A.k.a. **DOM**.

STX, saxitoxin.

SUMO, small ubiquitin-related modifier.

T_3, 3,3′,5-triiodo-L-thyronine

T_4, L-thyroxine.

TAED, *N,N,N′,N′*-tetraacetylethane-1,2-diamine.

TAPS, *N*-[tris(hydroxymethyl)methyl]-3-aminopropane-1-sulfonic acid.

TBP, tributyl phosphate.

TCD, thermal conductivity detector.

TCDD, tetrachlorodibenzodioxin {2,3,6,7-tetrachlorodibenzo[*b*,*e*][1,4]dioxin}.

TCDF, tetrachlorodibenzofuran {2,3,7,8-tetrachlorodibenzo[*b*,*d*]furan}.

TCNE, tetracyanoethene.

TCNQ, tetracyanoquinodimethane [2,2′-(cyclohexa-2,5-diene-1,4-diylidene) propanedinitrile].

TDA, thermodilatometry.

[1]**TDI**, tolerated daily intake.

[2]**TDI**, toluene 2,4-diisocyanate (2,4-diisocyanato-1-methylbenzene).

TDP, thymidine 5′-diphosphate.

TEA, triethanolamine (2,2′,2″-nitrilotriethanol).

TEELS, transmission electron energy loss spectroscopy.

TEL, tetraethyllead (tetraethylplumbane).

TEM, transmission electron microscopy.

TEMPO, 2,2,6,6-tetramethylpiperidin-1-yloxidanyl, 2,2,6,6-tetramethylpiperidin-1-yloxyl. a radical.

TEPP, tetraethyl pyrophosphate (tetraethyl diphosphate).

TES, *N*-[tris(hydroxymethyl)methyl]-2-aminoethane-1-sulfonic acid.

TETD, *N*,*N*,*N*′,*N*′-tetraethylthiuram disulfide {1,1′,1″,1‴-[disulfanediylbis(carbonothioylnitrilo)]tetraethane}.

TFA, trifluoroacetic acid.

TG, thermogravimetry.

TGA, thermogravimetric analysis.

TGS, triglycine sulfate [sulfuric acid—glycine (1/3)].

THAM, tris(hydroxymethyl)aminomethane [2-amino-2-(hydroxymethyl)propane-1,3-diol]. A.k.a. **tris** and **trometamol**.

THC, tetrahydrocannabinol {(6a*R*,10a*R*)-6,6,9-trimethyl-3-pentyl-6a,7,8,10a-tetrahydro-6*H*-dibenzo[*b*,*d*]pyran-1-ol}.

THEED, transmission high-energy electron diffraction.

THF, tetrahydrofuran.

THIP, 4,5,6,7-tetrahydroisoxazolo[5,4-*c*]pyridin-3(2*H*)-one {4,5,6,7-tetrahydro[1,2] oxazolo[5,4-*c*]pyridin-3(2*H*)-one}.

TIC, total inorganic carbon.

TLC, thin-layer chromatography.

TMD, *trans*-1,1,10-trimethyl-2-decalol [(2*S*,4a*S*,8a*R*)-1,1,4a-trimethyldecahydronaphthalen-2-ol].

TMP, thymidine 5′-monophosphate.

TMPD, tetramethyl-*p*-phenylenediamine (N^1,N^1,N^4,N^4-tetramethylbenzene-1,4-diamine).

TMS, tetramethylsilane.

TMTD, *N*,*N*,*N*′,*N*′-tetramethylthiuram disulfide {1,1′,1″,1‴-[disulfanediylbis(carbonothioylnitrilo)]tetramethane}.

TMU, *N*,*N*,*N*′,*N*′-tetramethylurea.

TNAZ, 1,3,3-trinitroazetidine.

TNF, tumor necrosis factor.

TNT, 2,4,6-trinitrotoluene (1-methyl-2,4,6-trinitrobenzene). A.k.a. **trotyl.**

TNX, 2,4,6-trinitro-*m*-xylene (1,3-dimethyl-2,4,6-trinitrobenzene).

TOAC, 2,2,6,6-tetramethylpiperidine-*N*-hydroxy-4-amino-4-carboxylic acid (4-amino -1-hydroxy-2,2,6,6-tetramethylpiperidine-4-carboxylic acid).

TOC, total organic carbon.

TOD, total oxygen demand.

TOF, time-of-flight (mass spectrometry).

tPA, tissue plasminogen activator.

TPE, transplutonium element(s).

TPMPA, (1,2,3,6-tetrahydropyridin-4-yl)methylphosphinic acid [methyl(1,2,3,6-tetrahydropyridin-4-yl)phosphinic acid].

TPN, triphosphopyridine nucleotide.

TPP, thiamine pyrophosphate.

TRH, thyrotropin-releasing hormone.

tRNA, transfer ribonucleic acid.

TRNOE, transferred nuclear Overhauser effect, cf. Chapter 12.

TRX, thioredoxin.

TRXRF, total reflection X-ray fluorescence spectroscopy.

TSH, thyroid-stimulating hormone.

TTF, 1,4,5,8-tetrathiafulvalene.

TTP, thymidine 5′-triphosphate.

TTX, tetrodotoxin.

TX, thromboxane.

UDP, uridine 5′-diphosphate.

UMP, uridine 5′-monophosphate.

UPS, ultraviolet photoelectron spectroscopy.

UTP, uridine 5′-triphosphate.

UV, ultraviolet (spectroscopy).

UV/VIS, ultraviolet/visible (spectroscopy).

VB, valence bond.

VEELS, vibrational electron energy loss spectroscopy.

VIP, vasoactive intestinal polypeptide.

VIS, visible (spectroscopy).

VLDL, very low-density lipoprotein.

VOC, volatile organic compound(s).

VSC, volatile sulfur compound(s).

VSEPR, valence shell electron pair repulsion.

VX (a US military code), *S*-[2-[(di(propan-2-yl)amino]ethyl] *O*-ethyl methylphosphonothioate. A.k.a. **A 4**.

WALTZ, wide-band alternating-phase low-power technique for zero residual splitting.

WLN, Wiswesser line notation, cf. Chapter 12.

XAES, X-ray induced Auger electron spectroscopy, cf. Chapter 12. Cf. XEAES.

XANES, X-ray absorption near edge.

XDP, xanthosine 5′-diphosphate.

XEAES, X-ray induced Auger electron spectroscopy, cf. Chapter 12. Cf. XAES.

XMP, xanthosine 5′-monophosphate.

XOD, xanthine oxidase (EC 1.17.3.2).

XPES, X-ray photoelectron spectroscopy, a.k.a. **XPS**.

XPS, X-ray photoelectron spectroscopy, a.k.a. **XPES**.

XRD, X-ray diffraction.

XRF, X-ray fluorescence (spectroscopy).

XRS, X-ray spectroscopy.

XTC, ecstasy [1-(1,3-benzodioxol-5-yl)-*N*-methylpropan-2-amine]. A.k.a. **Adam** and **E**.

XTP, xanthosine 5′-triphosphate.

YAG, yttrium-aluminum garnet.

YIG, yttrium-iron garnet.

ZDO, zero differential overlap.

ZEKE, zero kinetic energy.

ZINDO, Zerner intermediate neglect of differential overlap, cf. Chapter 12.

ZSM, zeolite Socony Mobil. Socony is an acronym for Standard Oil Company of New York.

17 The vocabulary of chemistry

Authoritative and comprehensive compilations of general and special chemical terms and their definitions have been provided by IUPAC [1].

17.1 Examples of chemical terms

IUPAC's Gold Book [2] is an excellent source of terminology (with definitions) in general chemistry.

17.1.1 Valid chemical terms

Absorption, from Latin *ab*, away from, and *sorbere, sorpt-*, to swallow up.

Acid, from Latin *acidum*, acid, from *acer, acr-*, sharp.

Acidulation, from Latin *acidulus*, diminutive of *acidus*, sour, from *acer*, sharp.

Adiabatic, from Greek *adiabatos*, which cannot be passed, from a(n)-, di(a)-, and *bainein*, to walk.

Adsorption, from Latin *ad*, to, and *sorbere, sorpt-*, to swallow up.

Agostic, from Greek *agostos*, hooked up.

Aliquot, from Latin *aliquo*t, a few, from *alius*, other, and *quot*, as many as.

Allihn condenser, after the German chemist Felix Richard Allihn (1854–1915).

Allomerism, from all(o)- and -mer, referring to variability of chemical constitution without variation in crystalline form.

Allotropy, from all(o)- and -tropy.

Alloy, ultimately from Latin *alligare*, to bind together, from *ad*, to, and *ligare*, to bind.

Amalgam, ultimately from Greek *malagma*, soft mass, from *malassein*, to soften.

Ambident, from Latin *ambo*, both, and *dens, dent-*, tooth.

Amphoteric, from Greek *amphoteros*, each of two.

Anaphoresis, from ana- and electrophoresis.

Anchimeric, from Greek *anchi*, near, and -mer. A.k.a. **synartetic**, from Greek *synartan*, to join.

https://doi.org/10.1515/9783110612714-017

Anion, from anode and ion.

Annulation, from Latin *anulus*, diminutive of *anus*, ring. A.k.a. **annelation**.

Anode, from Greek *anhodos*, way up, from *ana*, up, and *hodos*, way.

Aprotic, from a(n)- and proton.

Aptamer, from hapto- and -mer.

Atom, from Greek *atomos*, indivisible, ultimately from a(n)- and *temnein*, to cut.

Aufbau, from German *Aufbau*, build-up, from *bauen*, to build.

Autoclave, from aut(o)- and Latin *clavis*, key, from *claudere*, to lock, i.e. named as self-locking apparatus.

Auxochrome, from aux(o)- and -chrome.

Azeotrope, from a(n)-, Greek *zein*, to boil, and *tropos*, turn.

Babo funnel, after the German chemist Lambert Heinrich Joseph Anton Konrad Freiherr von Babo (1818–1899).

Base, ultimately from Greek *basis*, that on which on steps or stands, from *bainein*, to go, to walk, to step. In the chemical sense 'compound substance which unites with an acid to form a salt' introduced in 1754 by the French chemist Guillaume-François Rouelle (1703–1770). Since most of the acids known at the time were liquids a base was seen as a substance which imparted solidity to the salt formed, hence the name base.

Beckman pH meter, after the Amerian chemist Arnold Orville Beckman (1900–2004).

Beckmann thermometer, after the German chemist Ernst Otto Beckmann (1853–1923).

Berthollide (non-stoichiometric compound), after the French chemist Claude Louis Comte de Bethollet (1748–1822).

Bidentate, tridentate, etc., from bi-, tri-, etc. and Latin *dens, dent-*, tooth.

Bifilar, trifilar, etc. from, bi-, tri-, etc. and Latin *filum*, thread.

Binary, from Latin *binaris*, consisting of two.

Bromatology, from Greek *broma, bromat(o)-*, food.

Büchner funnel, after the German chemist Ernst Büchner (1850–1924).

Bunsen burner, after the German chemist Robert Wilhelm Eberhard Bunsen (1811–1899).

Burette, from French *burette*, diminutive of *buire*, pitcher.

Calcination, from Late Latin *calcina*, lime, from Latin *calx, calc-*, lime.

Capto-dative, from Latin *capere, capt-*, to catch, and *dare, dat-*, to give. A.k.a. **pull-push**.

Catalysis, from Greek *kata*, down, against, and lysis.

Cataphoresis, from cat(a)- and electrophoresis.

Cathode, from Greek *kata*, down, against, and *hodos*, way.

Cation, from cathode and ion.

Caustic, from Greek *kaustikos*, capable of burning, corrosive, from *kaiein*, to burn.

Caviplex, from cavitand and complex.

Cavitand, from Latin *cavum*, cavity, hole.

Centri-, from Latin *centrum*, center.

Ceramic, from Greek *keramikos*, pottery, earthenware, from *keramos*, clay, possibly akin to *kerannynai*, to mix.

Chiron, a contraction of chiral synthon.

Chromatography, from [1]chrom(o)- and Greek *graphein*, to write. This name refers to the very first chromatographic separations which involved colored substances whose movements could be followed visually.

Chromophore, from [1]chrom(o)- and -phore.

Cine-substitution, from Greek *kinein*, to move, to stimulate, to set in motion.

Claisen adapter, after the German chemist Ludwig Claisen (1851–1930). A.k.a. **Claisen flask**.

Claisen flask, as Claisen adapter.

Clusius-Dickel column, after the German chemists Klaus Paul Alfred Clusius (1903–1963) and Gerhard Dickel (1913–2017).

Colloid, from Greek *kolloeides*, glue-like, from *kolla*, glue.

Comproportionation, from Latin *cum, con-, com-* with, together with, and proportion.

Condis crystal, a contraction of conformational disorder crystal.

Crucible, from Medieval Latin *crucibulum*, earthen pot for melting metals, of obscure etymology, possibly derived from Latin *crux, cruc-*, cross.

Cryptand; cryptate, from Greek *kryptos*, hidden, from *kryptein*, to hide.

Daltonide (stoichiometric compound), named after the British chemist, physicist, and meteorologist John Dalton (1766–1844). A.k.a. **proustide**, after the French chemist Joseph Louis Proust (1754–1826).

Davies condenser, after James Davies, director of the Gallenkamp Company.

Dean-Stark apparatus, after the American chemists Ernest Woodward Dean (1888–1959) and David Dewey Stark (1893–1979).

Denticity, cf. bidentate.

Dephlegmator, from phlegm, ultimately from Greek *phlegma*, flame, inflammation, mucus, humor caused by heat, from *phlegein*, to burn.

Desiccator, from Latin *desiccare*, to dry out, from *siccus*, dry.

Desorption, from Latin *de*, away from, and *sorbere, sorpt-*, to swallow up.

Dewar flask, after the British chemist James Dewar (1842–1923).

Diafiltration, a contraction of dialysis and ultrafiltration.

Dialysis, from di(a)- and lysis.

Diaphragm, from di(a)- and Greek *phragma*, fence, screen, from *phrassein*, to enclose, to fence in.

Dichroism, from Greek *dichroos*, of two colors.

Digestion, from Latin *digerere, digest-*, to separate, to divide, to arrange.

Dimroth condenser, after the German chemist Otto Dimroth (1872–1940).

Dispersion, from dis- and Latin *spergere, spers-*, to scatter.

Disproportionation, from disproportion.

Distillation, from Latin *distillatio, distillation-*, distillation, from *distillare*, trickle down in minute drops, from *stilla*, little drop, diminutive of *stiria*, drop.

Döbereiner lamp, after the German chemist Johann Wolfgang Döbereiner (1780–1849).

Docimasy, also **docimacy**, assay of metallic ores, from Greek *dokimasia*, examination, scrutiny, test, from *dokimazein*, to assay, to test, to approve, from *dokimos*, approved, assayed, from *dokein*, to seem good, to seem, to think.

d-Orbital (atomic orbital), named after the 'diffuse' subseries in the line spectra of the alkali metals.

Dreiding model, after the Swiss chemist André Samuel Dreiding (1919–2013).

Eigenvalue, from German *eigen*, own, proper.

Electrophile, from electron and Greek *philos*, dear, friendly.

Electrophoresis, from electric and Greek *phoresis*, act of carrying, from *phorein*, to carry, to bear, frequentative of *pherein*, to carry, to bear.

Element, from Latin *elementum*, of unknown etymology. Originally used for the 'four elements', earth, air, fire, and water.

Emulsion, from Latin *emulsus*, milked out, from *mulgere*, to milk.

Enthalpy, from Greek *en*, in, and *thalpos*, heat. The symbol H for enthalpy was chosen to represent heat.

Entropy, from Greek *entrepein*, to tun around, from *trepein*, to turn. It has been suggested that the symbol S for entropy was chosen by the German physicist and mathematician Rudolf Julius Emanuel Clausius (1822–1888) to honor the French military engineer and physicist Sadi Carnot (1796–1832).

Erlenmeyer flask, after the German chemist Erwin Erlenmeyer (1825–1909).

Eudiometer, from Greek *eudia*, fair weather, from *eudios*, quiet, and -meter.

Euosmophore, from eu-, ^{2}osm(o)-, and -phore, patterned after chromophore.

Eutectic, from eu- and -tectic.

Fenske rings, after the American engineer Merrell Robert Fenske (1904–1971).

Filter, from Medieval Latin *feltrum*, felt, from Gothic *felta*, felt.

Fischer projection, after the German chemist Emil Hermann Fischer (1852–1919).

Fission, from Latin *fissio, fission-*, cleavage, from *findere, fiss-*, to split, to cleave.

Fluorescence, a contraction of fluorspar and opalescence.

f-Orbital (atomic orbital), named after the 'fundamental' subseries in the line spectra of the alkali metals.

Fricke dosimeter, after the Danish-American chemist Hugo Fricke (1892–1972).

Friedrichs condenser, after the German chemist Fritz Walter Paul Friedrichs (1882–1958).

Fug-, -fuge, from Latin *fugere*, to flee.

Fugacity, from Latin *fugax, fugac-*, apt to flee, from *fugere*, to flee.

Fusion, from Latin *fusio, fusion-*, a pouring, a casting, from *fundere, fus-*, to pour, to cast.

Galenical, after the Greek-Roman physician Galen (Galenos) (129–199).

Galvanization, after the Italian physiologist Luigi Galvani (1737–1798).

Gel, from gelatin.

Getter, from to get, a substance introduced into a vacuum tube to remove traces of gas.

Golay column, after the Swiss-American mathematician, physicist, and information theorist Marcel Jules Edouard Golay (1902–1989).

Gooch crucible, after the American chemist Frank Austin Gooch (1852–1929).

g-Orbital (atomic orbital), next-higher orbital after an f-orbital. The name was coined by proceeding alphabetically from f with the omission of i and j.

Gravimetry, from Latin *gravis*, heavy.

Haworth projection, after the British chemist Walter Norman Haworth (1883–1950).

Hempel pipet, after the German chemist Walther Matthias Hempel (1851–1916).

Hershberg stirrer, after the American chemist Emanuel Benjamin Hershberg (born 1908).

Hydron (H^+), from hydrogen, patterned after proton. A natural-abundance mixture of $^1H^+$ and $^2H^+$, thus different from **proton** ($^1H^+$).

Hydrophilic, from Greek *hydor, hydr-*, water, and *philos*, dear, friendly.

Hydrophobic, from Greek *hydor, hydr-*, water, and *phobos*, fear.

Hygroscopic, back formation from hygroscope, from hygr(o)-, an instrument measuring humidity and in which hygroscopic substances are used.

Hypergolic, as in hypergolic fuel, from hyper-, Greek *ergon*, work, and 'ol' in the sense of oil, fuel.

Hypsochrome, from Greek *hypsos*, height, and -chrome.

Hysteresis, from Greek *hysteresis*, shortcoming, deficiency, need, from *hysterein*, to come late, to lag, from *hysteros*, later.

Idiochromatic, from Greek *idios*, own, particular, personal, and *chroma, chromat-*, color.

In statu nascendi, Latin, in the state of being generated, literally in the state of being born.

Ion, from Greek *ion*, going, from *ienai*, to go, referring to the mobility of ions in solution.

Ipso-substitution, from Latin *ipse*, self.

Isochore, from is(o)- and Greek *choros*, empty space, room.

Isomer, from Greek *isos*, equal, and -mer.

Isotachophoresis, from is(o)-, Greek *tachos*, speed, and -phore.

Katharometer, from Greek *katharos*, clean, pure, and -meter, referring to the use of this apparatus for the analysis of gas mixtures. A.k.a. **thermal conductivity detector** and **TCD**.

Kofler bench, after the Austrian pharmacognosist Ludwig Kofler (1891–1951).

Kutscher-Steudel extraction apparatus, after the German chemists Friedrich Kutscher (1866–1942) and Hermann Steudel (1871–1967).

Landolt vessel, after the Swiss chemist Hans Heinrich Landolt (1831–1910).

Liebig condenser, after the German chemist Justus von Liebig (1803–1873), but earlier described by the German chemist Johann Friedrich August Göttling (1753–1809), the French physician and chemist Pierre Poissionner (1720–1798), the German scientist Christian Ehrenfried Weigel (1748–1831), and the Finnish chemist Johan Gadolin (1760–1852) [3].

Lipophilic, from lipid and Greek *philos*, dear, friendly.

Lipophobic, from lipid and Greek *phobos*, fear.

Liquidus curve, from Latin *liquidus*, liquid, from *liquere*, to be liquid.

Lithergolic, as in lithergolic propellant, from lith(o)-, Greek *ergon*, work, and 'ol' in the sense of oil, fuel.

Lysis, from Greek *lysis*, a loosening, dissolution, remission of fever, from *lyein*, to loosen.

Maceration, from Latin *macerare*, to soften, to steep.

Magic angle (54.74°), referring to this angle's importance in solid-state NMR spectroscopy.

Magic numbers (2, 8, 20, 28, 50, 82, 126), referring to the exceptional stability of nuclides where the number of protons and/or neutrons corresponds to such a magic number or a sum of such magic numbers.

Manometer, from Greek *manos*, sparse, loose, rare, and -meter.

Meker burner, after the French chemist George Méker (1875–1975).

Membrane, from Latin *membrana*, fine skin, literally that which covers the members of the body.

Mercerization, after the British chemist John Mercer (1791–1866).

Mesoionic, from meso- and ion.

Mesomerism, from meso- and -mer.

Meso-substitution, from meso-.

Metalepsis (substitution), from Greek *metalepsis*, exchange, transposition.

[1]**Metalloid**, from metal and -oid, an element intermediate in properties between the typical metals and nonmetals.

[2]**Metalloid**, from metal and -oid, a nonmetal that can combine with a metal to form an alloy.

Metathesis, ultimately from Greek *metatithenai*, to transpose.

Mineral, from Medieval Latin *mineralis*, of the ore, of the mine, from *minera*, ore. mine.

Mole, a contraction of gram-molecule.

Molecule, from Latin *molecula*, diminutive of *moles*, mass.

[1]**Mortar** (material), from Latin [1]*mortarium*, [1]mortar, of obscure etymology. Another Latin word for [1]mortar is *arenatum*.

[2]**Mortar** (tool), from Latin [2]*mortarium*, [2]mortar, of obscure etymology.

[3]**Mortar** (short cannon), from [2]mortar.

Narcissistic reaction, after the Greek mythical figure Narkissos (Narcissus), a beautiful youth who fell in love with his own image, referring to the fact that the product of such a reaction is the mirror image of the starting material.

Newman projection, after the American chemist Melvin Spencer Newman (1908–1993).

Nucleophile, from nucleus and Greek *philos*, dear, friendly.

Opalescence, from opal.

Orsat apparatus, after the French engineer Louis Hengist Orsat (1837–1882).

Osmosis, from Greek *osmos*, act of thrusting, from *othein*, to push.

Parr bomb, after the American chemist Samuel Wilson Parr (1857–1931).

Pasteurization, after the French biologist, microbiologist, and chemist Louis Pasteur (1822–1895).

Pasteur pipet, as pasteurization.

Peristaltic, as in peristaltic pump, from peri- and Greek *stalsis, stalt-*, compression, constriction, from *stellein*, to place, to send.

Peri-substitution, from Greek *peri*, around, about, from *peran*, to pass through.

Permeable, from Latin *permeabilis*, from *permeare*, to pass through, from *meare*, to go, to pass.

Pestle, from Latin *pistillum*, diminutive of *pilum*, pestle, javelin, from *pinsere*, to pound, to crush.

Petri dish, after the German microbiologist Julius Richard Petri (1852–1921).

Pfeffer cell, after the German botanist Wilhelm Friedrich Philipp Pfeffer (1845–1920).

[1]**Plasma** (the fluid part of blood, lymph, etc.), from Greek *plasma*, something molded, from *plassein*, to mold.

[2]**Plasma** (a collection of charged particles), from Greek *plasma*, something molded, from *plassein*, to mold.

Platforming, from platinum (the catalyst) and reforming.

Podbielniak column, after the American industrialist Walter Joseph Podbielniak (1899–1978).

p-Orbital (atomic orbital), named after the 'principal' subseries in the line spectra of the alkali metals.

π-Orbital (molecular orbital), derived from p-orbital, referring to the common symmetry of π- and p-orbitals.

Precipitation, from Latin *praecipitatus*, past participle of *praecipitare*, to throw or dive headlong, from *praeceps*, steep, headlong, headfirst, from *prae*, before, forth, and *caput, capit-*, head.

Prosthetic group, from Greek *prosthetikos*, adding, furthering, from *prostithenai*, to add.

Quantum satis, as much as necessary, from Latin *quantum*, how much, and *satis*, sufficient, satisfied.

Ramsay grease, after the British chemist William Ramsay (1852–1916).

Raschig rings, after the German chemist Friedrich August Raschig (1863–1928).

Rectification, from Medieval Latin *rectificare*, to rectify, from Latin *rectus*, right, and, ultimately *facere, fact-*, to make, referring to repeated or fractional distillation.

Resin, ultimately from Greek *rhetine*, pine resin.

Rose crucible, after the German mineralogist and analytical chemist Heinrich Rose (1795–1864).

Rotamer, from rotation and -mer.

Saponification, from sapo(n)-.

Schellbach burette, after the German mathematician and physicist Karl Heinrich Schellbach (1804–1892).

Schlenk tube, after the German chemist Wilhelm Schlenk (1879–1943).

Scintillation, from Latin *scintilla*, spark, of obscure etymology.

Seger cone, after the German chemist August Seger (1839–1893).

Septum, from Latin *saeptum*, wall, enclosure, from *saepire*, to hedge in.

Siemens ozonizer, after the German-British engineer Charles William Siemens (1823–1883).

Söderberg electrode, after the Swedish-Norwegian engineer Carl Wilhelm Söderberg (1876–1955).

Sol, from Latin *solutio, solution-*, solution.

Solvent, from Latin *solvens, solvent-*, dissolving, from *solvere*, to loosen, to release, to accomplish, to fulfill.

Soxhlet extraction, after the German chemist Franz Ritter von Soxhlet (1848–1926).

s-Orbital (atomic orbital), named after the 'sharp' subseries in the line spectra of the alkaline metals.

σ–Orbital (molecular orbital), derived from s-orbital, referring to the common symmetry of σ- and s-orbitals.

Stauffer grease, after the American chemist and industrialist John Stauffer, Sr. (1862–1940).

Stearopten (the solid part of a partially solidified natural oil), from Greek *stear*, fat, and *ptenos*, volatile, evaporated.

Stoichiometry, from Greek *stoicheion*, element, ultimately from *stoichos*, row, and *metron*, measure.

Stuart-Briegleb molecular models, after the Swiss-German physicist Herbert Arthur Stuart (1899–1974) and the German chemist Günther Briegleb (1905–1991).

Sublimation, from Medieval Latin *sublimatio, sublimation-* refinement, ultimately from *sublimis*, lofty, high, exalted, eminent, distinguished, literally sloping up to the lintel, from *limen, limin-*, lintel, threshold, sill.

Synthon, from sythesis and ^{3}on-.

Tautomer, from Greek *to auton*, the same, and -mer. A.k.a. **tropomer.**

Teclu burner, after the Romanian-Austrian chemist Nicolae Teclu (1839–1916).

Tele-substitution, from Greek *tele*, far, far away.

Thixotropy, from Greek *thixis*, act of touching, from *thinganein*, to touch, and -tropy.

Titer, from French *titre*, title, qualification, fineness of gold or silver, ultimately from Latin *titulus*, inscription, label, ticket, placard, heading, honorary appellation, title of honor.

Titration, from titer.

Topochemistry (chemistry taking place in the solid state), from topo-.

Topomerism, from topo- and isomerism.

Traube cell, after the German chemist Moritz Traube (1826–1894).

Trituration, from Latin *triturare*, to thresh, to grind, from *terere*, to rub.

Umpolung, from German *Umpolung*, pole reversal, from *umpolen*, to reverse poles, from *Pol*, pole.

Venturi tube, after the Italian physicist Giambattista Venturi (1746–1822).

Vigreux column, after the French glassblower Henri Vigreux (1869–1951).

Vinylogy, a contraction of vinyl homology.

Vulcanization, named for Vulcanus, the Roman god of fire and metalworking, probably of Etruscan origin.

Weinhold vessel, after the German physicist Adolf Ferdinand Weinhold (1841–1917).

Weisz ring oven, after the Hungarian-Austrian chemist Herbert Weisz (1922–2018).

Weston element, after the British chemist Edward Weston (1850–1936).

Widmer column, after the Swiss chemistry student Gustav Widmer.

Winkler burette, after the German chemist Clemens Winkler (1838–1904).

Winkler generator, after the German chemist Fritz Winkler (1888–1950).

Woulfe flask, after the British chemist Peter Woulfe (1727–1803).

Zwitterion, from German *Zwitter*, hermaphrodite, hybrid, from *zwei*, two, and ion. A.k.a. **betaine**.

17.1.2 Obsolete and/or discredited chemical terms

An overview of alchemy and its language is available online [4].

Absolute, as in absolute alcohol. An obsolete term meaning 100% or anhydrous.

Air, an alchemistic name for gas.

Alchemy, derived from Arabic *al-kimiya*, alchemy, ultimately from Greek *chemeia*, chemistry, from *chymos*, juice, from *chein*, to pour.

Alcohol sulphuris (carbon disulfide), an alchemistic New Latin name, referring to this compound's volatility and sulfur content.

Alembic, a distillation vessel used by alchemists. From Arabic *al-anbiq*, distilling flask, via Persian from Greek *ambix*, cup, of unknown, possibly Semitic, origin.

Alkahest, an alchemistic name of obscure etymology. It was first used by the Swiss-German alchemist and physician Philippus Aureolus Paracelsus (Theophrastus Bombast von Hohenheim) (1493–1541) for an ill-defined iatrochemical remedy. The Flemish alchemist Jan Batista van Helmont (1577–1644) used the same name or New Latin *ignis aqua*, fire's water, for the ill-defined and disputed hypothetical universal solvent and universal remedy.

Alkaline air (ammonia), an alchemistic name for ammonia.

Aqua fortis (nitric acid), New Latin, literally strong water. The name refers to the fact that this liquid can dissolve the noble metal silver.

Azoth, ultimately from Arabic *al-za'uq*, the mercury. In alchemy mercury, regarded as the principle of all metals, but also as a universal remedy.

Black lead (graphite), an obsolete name for graphite which was often confused with lead.

Blue vitriol [copper(II) sulfate pentahydrate), cf. vitriol.

Bologna phosphorus (barium sulfide), named for the city of Bologna, Italy, the home of this phosphorescent preparation's discoverer, the Italian alchemist Vincenzo Casciorola (1571–1624). A.k.a. **lapis solis**, New Latin stone of the sun.

Brimstone (sulfur), from Middle English *brinnen*, to burn, and *ston*, stone.

Bromide (potassium bromide), in the sense of sedative refers to this obsolete use of potassium bromide.

Butter of antimony [antimony(III) chloride], referring to its physical appearance.

Cadmia (zinc oxide), as opposed to the modern meaning cadmium oxide.

Cadmia botrytis (zinc oxide), zinc oxide as obtained by sublimation in furnaces, named by its shape, from Greek *botrys*, bunch of grapes.

Cadmia calamitis (zinc oxide), zinc oxide as obtained by sublimation in furnaces, named by its shape, from Greek *kalamitis*, reed-like, tubular, from *kalamos*, reed, quill.

Cadmia capnitis (zinc oxide), zinc oxide as obtained by sublimation in furnaces, ultimately from Greek *kapnitis*, smoky, from *kapnos*, smoke.

Cadmia fornacum (zinc oxide), zinc oxide as obtained by sublimation in furnaces, from Latin *fornax, fornac-*, furnace.

Cadmia ostracitis (zinc oxide), zinc oxide as obtained by sublimation in furnaces, named by its shape, ultimately from Greek *ostrakon*, shell, tile, potsherd.

Cadmia placitis (zinc oxide), zinc oxide as obtained by sublimation in furnaces, named by its shape, ultimately from Greek *plakitis*, scaly, from *plax, plak-*, plate.

Caput mortuum, New Latin literally dead head, alchemistic term for the residue after distillation or sublimation or for dross.

Caput mortuum vitrioli, New Latin, dead head of metal sulfate, an alchemistic term for an iron oxide pigment obtained by calcination of iron sulfate.

Carbinamine (methanamine), an obsolete name for methanamine, from carbine.

Carbine (methyl), an obsolete name for the methyl radical.

Carbinol (methanol), an obsolete name for methanol, from carbine.

Carburet, an obsolete verb meaning 'to cause to combine with carbon'.

Carburetted hydrogen (methane), obsolete name for methane. A.k.a. **light carburetted hydrogen**.

Carbylamine (isocyanide), an obsolete name for isocyanide, from carbon and amine. A.k.a. **carbamine**.

Carmot, the matter of which the mythical philosopher's stone was thought to be composed, of uncertain origin and etymology.

Chrome (chromium), an obsolete name for chromium, preserved in trivial names like chrome alum (potassium chromium sulfate), chrome iron ore (dichromium iron tetraoxide), chrome green (dichromium dioxide), and chrome yellow [lead(II) chromate].

Chrysopoiea, from Greek *chrysos*, gold, and *poiein*, to make, i.e. gold making. A.k. a. **transmutation**. Cf. philosopher's stone.

Cream of tartar (potassium hydrogen tartrate). From tartar, referring to the fact that this crystalline salt is the most valuable part of the crude tartar deposited in wine casks.

Cremor tartari (potassium hydrogen tartrate), New Latin, cream of tartar.

Dephlogisticate, from phlogiston.

Dephlogisticated marine acid air (dichlorine), obsolete name for chlorine gas.

Dephlogisticated air (dioxygen), obsolete name for oxygen gas.

Desoxy- (deoxy-), from German *Desoxy-*, from Latin *de-*, *des-*, off, away from, and oxygen.

Dutch liquid (1,2-dichloroethane), referring to the Dutch botanist Nicolaas Bondt (1765–1796), the German-Dutch physician Jan Rudolph Deiman (1743–1808), the Dutch chemist Anthoni Lauwerenburg (1758–1820), and the Dutch merchant Adriaan Paets van Troostwyck (1752–1837), who formed the *Gezelschap der Hollandsche Scheikundigen* (Society of Dutch Chemists) and first prepared it.

Eau de Javelle (potassium hypochlorite solution), named after the town of Javelle, France.

Eau de Labarraque (sodium hypochlorite solution), named after the French pharmacist Antoine Germain Labarraque (1777–1850).

Elixir, a substance held capable of changing base metals into gold or to prolong life indefinitely, from Arabic *al-iksir*, elixir, ultimately probably from Greek *xerion*, desiccative powder, from *xeros*, dry.

Emanation, from Latin *emanatio, emanation-*, outflow, from *manare*, to flow.

Empyreuma, the peculiar odor of organic substances burnt in closed vessels, from Greek *empyreumata*, live coal covered with ashes, from *empyreuein*, to light a fire.

Essential earth (metal oxide), before the discovery of oxygen regarded as an element.

Fixed air (carbon dioxide).

Flos ferri (calcium carbonate), from Latin *flos, flor-*, flower, and *ferrum*, iron, referring to the occurrence of this mineral in beds of iron ore.

Flowers of antimony [antimony(III) oxide].

Fluoro acid air (silicon tetrafluoride, tetrafluorosilane).

Fool's gold [iron(II) disulfide]), referring to its gold-like appearance.

Green vitriol [iron(II) sulfate heptahydrate).

Hepar (any liver-colored metal sulfide), from Latin *hepar, hepat-*, liver.

Hepar sulfuris (potassium oligosulfides), from hepar and sulfur.

Hepatic air (hydrogen sulfide), from hepar.

Hydrol (secondary alcohol), as in benzhydrol (diphenylmethanol) and benzhydryl- (diphenylmethyl-), from [2]hydr(o)- and 'ol'. So named to show the presence of a hydrogen atom at the hydroxyl group-bearing carbon atom.

Iatrochemistry, from Greek *iatros*, physician, from *iasthai*, to heal.

Inflammable air (dihydrogen).

Glucina (berylla, beryllium oxide), after glucinium (beryllium).

Kipp aparatus, after the Dutch apothecary, chemist, and instrument maker Petrus Jacobus Kipp (1808–1864).

Lana philosophica (zinc oxide), the threads of zinc oxide formed over melted zinc; New Latin, literally mysterious wool. A.k.a. **philosopher's wool** and **nix alba**.

Lapis infernalis (silver nitrate), New Latin, hellish stone, referring to its corrosive action on skin. A.k.a. **lapis** and **lunar caustic**.

Lapis solaris (barium sulfide), New Latin solar stone, referring to this compound's phosphorescence. Cf. Bologna phosphorus.

Laudanum (alcoholic solution of opium), probably from Greek *ladanon*, gum resin, of Semitic origin.

Lithos magnetis (magnetite), from [1]Magnesian stone.

Lead fume [lead(II) oxide].

Luna cornea (silver chloride), horn silver, from Latin *luna*, moon, the alchemistic name for silver, and *corneus*, of horn, from *cornu, cornut-*, horn.

Lunar caustic (silver nitrate), from Latin *luna*, moon, the alchemistic name for silver. A.k.a. **lapis** and **lapis infernalis**.

Lupi spuma [wolframite, $(Fe,Mn)WO_4$], Latin wolf's foam. A.k.a. **spuma lupi**.

Magisterium, alchemistic term for precipitate, likely from New Latin *magisterium* in the sense of masterpiece, referring to the precipitate as the ultimate result of a series of alchemical operations.

Magisterium bismuthi [bismuth nitrate oxide—water (1/1)]. A.k.a. **bismuthum subnitricum**.

[1]**Magnes** (triiron tetraoxide), a corruption of [1]Magnesian stone.

[2]**Magnes** [manganese(IV) oxide]), a corruption of [2]Magnesian stone.

[1]**Magnesia** (triiron tetraoxide), derived from [1]Magnesian stone.

[2]**Magnesia** [manganese(IV) oxide], a corruption of [2]Magnesian stone.

[3]**Magnesia** (magnesium oxide), derived from magnesia alba.

Magnesia alba (magnesium carbonate), from [3]Magnesian stone and from Latin *albus*, white.

Magnesia nigra [manganese(IV) oxide], from [2]Magnesian stone and from Latin *niger, nigr-*, black.

[1]**Magnesian stone** (triiron tetraoxide), an alchemistic name for magnetite, ultimately from Greek *Magnetis lithos*, Magnesian stone. The toponym Magnesian refers either to the ancient Greek province Thracian Magnesia, the ancient town of Magnesia ad Maeandrum, Asia Minor, the ancient town of Magnesia ad Sipylum, Asia Minor, or to the ancient shepherd Magnes who according to Nicander (as cited by Pliny the Elder) discovered this magnetic mineral on Mount Ida, Crete, Greece.

[2]**Magnesian stone** [manganese(IV) oxide], an alchemistic name for manganese dioxide, coined by confusion of [2]Magnesian stone with [1]Magnesian stone because of their similar appearance.

[3]**Magnesian stone** (magnesium carbonate), from Greek *Magnetis lithos*, Magnesian stone. The toponym Magnesian refers either to the ancient Greek province Thracian Magnesia, the ancient town of Magnesia ad Maeandrum, Asia Minor, the ancient town of Magnesia ad Sipylum, Asia Minor.

Magnesia usta (magnesium oxide), from [3]Magnesian stone and from Latin *ustus*, burnt, from *urere, ust-*, to burn.

Manganes (manganese dioxide), ultimately from Greek *manganizein*, to purify, referring to manganese dioxide's ability to decolorize iron-containing green glass.

Marine acid (hydrochloric acid).

Marine acid air (hydrogen chloride).

Marine alkali (sodium carbonate).

Marsh gas (methane).

Mephitic acid (carbon dioxide), from Latin *mephitis*, noxious exhalation from the earth.

[1]**Mephitic air** (carbon dioxide), from Latin *mephitis*, noxious exhalation from the earth, with air in the alchemistic sense of gas.

[2]**Mephitic air** (dinitrogen), as [1]mephitic air.

Mercurius praecipitatus {mercury(II) oxide], red mercury(II) oxide.

Microcosmic salt (ammonium sodium hydrogenphosphate), referring to its isolation from human urine.

Mosaic gold [tin(II) sulfide].

Muriate (chloride), from muriatic acid.

Muriatic acid (hydrochloric acid), from Latin *muria*, brine, referring to its source.

[1]**Nitre** (sodium carbonate), used in antiquity.

[2]**Nitre** (potassium nitrate). This use of the name developed during the Middle Ages.

Nitrous air (nitric oxide, nitrogen monoxide).

Nitrum flammans (ammonium nitrate), New Latin, flaming nitre.

Nix alba (zinc oxide), New Latin, literally white snow. A.k.a. **lana philosphica** and **philosopher's wool**.

Oil of vitriol (sulfuric acid). A.k.a. **spirit of vitriol**.

Oily carburetted hydrogen (ethane). A.k.a. **heavy carburetted hydrogen**.

Oxymuriatic acid (dichlorine), from oxygen and muriatic acid, q.v.

Parachor, from para- and Greek *choros*, space.

Pearl white [chloridohydroxidobismuth(III)].

Philosopher's stone, a legendary alchemical substance capable of turning base metals into gold or silver and also to provide longevity or even immortality [5].

Philosopher's wool (zinc oxide). A.k.a. **Lana philosophica** and **nix alba**, New Latin, literally white snow.

Phlegma, an alchemistic term for distillation residue, from Greek *phlegma, phlegmat-*, flame, inflammation, from *phlegein*, to burn.

Phlogisticate, from phlogiston.

Phlogisticated air (dinitrogen), from phlogiston.

Phlogisticated nitrous air (dinitrogen oxide), from phlogiston.

Phlogiston, from Greek *phlogistos*, inflammable, from *phlogizein*, to set on fire, from *phlox, phlog-*, flame [6].

Powder of Algaroth [antimony(III) chloride oxide], after the Italian physician Vittorio Algarotti (1533–1604).

Purple of Cassius (colloidal gold), after the German physician and alchemist Andreas Cassius (1600–1676).

Pseudomerism (tautomerism), from pseudo- and -mer.

Pyroligneous acid (acetic acid, methanol, water), a technical product obtained by destructive distillation of wood, on occasion used as substitute for vinegar. From pyr(o)- and lign(o)-. A.k.a. **wood vinegar** and **wood acid.**

Prussic acid (hydrocyanic acid), referring to its formation upon treatment of Prussian blue with sulfuric acid.

Quinta essentia Saturni, alchemistic New Latin, quintessence of lead, a name given to acetone by the German physician and chemist Andreas Libavius (Andreas Libau) (1555–1616), who obtained it by heating lead(II) acetate.

Quintessence, from Latin *quinta essentia*, the fifth (and finest) essence or element (after fire, earth, air, and water), from Greek *pempte oysia*, fifth substance. Believed to be the matter celestial bodies were made of.

Red vitriol [cobalt(II)sulfate heptahydrate].

Regulus, literally diminutive of Latin *rex, reg-*, king; the more or less impure mass of metal formed beneath the slag in smelting and reducing ores. Regulus was also a name for metallic antimony because it readily formed an alloy with gold, the king of metals.

Resin of copper [(copper(I) chloride].

Retort, from re- and Latin *tortus*, twisted, from *torquere, tort-*, to twist.

Sal ammoniac (ammonium chloride). New Latin, from Latin *sal ammoniacus*, salt of Ammon.

Sal anglicum (magnesium sulfate heptahydrate), New Latin, English salt. A.k.a. **Epsom salt.**

Sal digestivus Sylvii (potassium chloride), New Latin, Sylvius's digestive salt, after the Dutch physician, chemist, physiologist, and anatomist Franciscus Sylvius (Franz de le Boë) (1614–1672).

Sal mirabilis (sodium sulfate), New Latin, miraculous salt. A.k.a. **Glauber's salt.**

Sal petrae (potassium nitrate), saltpeter, New Latin, salt of Petra, Syria.

Sal sedativum (boric acid), New Latin, sedative salt. However, not a salt.

Sal volatile (ammonium carbonate), New Latin, volatile salt.

Spirit of hartshorn (ammonia).

Spirit of salt (hydrochloric acid).

Spiritus fumans Libavii (tin tetrachloride), New Latin, fuming liquid of Libavius, after its rediscoverer, the German physician and chemist Andreas Libavius (Andreas Libau) (1555–1616). The first mention of tin tetrachloride was by the German Franciscan frater Ulmannus in his 'Buch der Heiligen Dreifaltigkeit' (book of the holy trinity), published 1410–1419.

Spiritus vini (ethanol), New Latin, wine's spirit.

Sublimate [mercury(II) chloride]. A.k.a. **corrosive sublimate.**

Sugar of lead [lead(II) acetate], referring to its sweet taste.

Sulfuretted hydrogen (hydrogen sulfide).

Sweet oil of vitriol (diethyl ether, ethoxyethane).

Tartar (potassium hydrogen tartrate), ultimately from Persian *durd*, lees. A.k.a. **argol**, ultimately from Latin *argilla*, argil.

Transmutation, from Latin *transmutare*, to change over. A.k.a. **chrysopoiea**. Cf. Philosopher's Stone.

Tutty (zinc oxide), impure zinc oxide, ultimately from Sanskrit *tutha, tutaka*, tutty.

Vegetable alkali (potassium carbonate).

Vinyl alcohol (ethanol), from spiritus vini. Not to be confused with ethenol.

Vis vitalis, life's power, from Latin *vis*, power, and *vitalis*, pertaining to life, from *vita*, life, referring to the obsolete belief that carbon compounds could only be produced by living organisms. Hence the term organic, from Greek *organon*, bodily organ.

Vitriol, alchemistic name for metal sulfates of glassy appearance, from Medieval Latin *vitriolum*, something glassy, from Latin *vitrum*, glass, from *vitrum*, woad, referring to the blue-green color of antiquity's glass.

Vitriolic acid (sulfuric acid), from vitriol.

Vitriolic acid air (sulfur dioxide), from vitriolic acid.

Vitriolum, New Latin, vitriol.

Volatile alkali (ammonia).

White arsenic [arsenic(III) oxide].

White lead [lead(II) carbonate].

White vitriol (zinc sulfate).

17.2 Samples from the vocabulary of stereochemistry

IUPAC provides extensive information about this terminology [7].

Achiral, from Greek *a(n)-*, not, and chiral.

Alloisomerism, from allo- and isomerism, an obsolete name for *cis-trans*-isomerism.

Anomer, from ano- and -mer.

Antarafacial, from Sanskrit *antara*, the other, and facial.

Anti- (a), from ant(i)-, referring to a torsion angle between ±90° and 180°.

Anticlinal (ac), from ant(i)- and Greek *klinein*, to lean, referring to a torsion angle between 90° and 150° or between −90° and −150°.

Antiperiplanar (ap), from ant(i)-, peri-, and planar, referring to a torsion angle between ±150° and 180°.

Asymmetry (lack of all symmetry elements other than a one-fold axis of symmetry), from a(n)- and symmetry.

Atropisomer, from Greek *atropos*, without turn, inevitable, and isomer.

Chiral, from Greek *cheir*, hand.

Clinal (c), from Greek *klinein*, to lean, referring to a torsion angle between 30° and 150° or between −30° and −150°.

Diastereomer, from Greek *dia-*, *di-*, through, apart, and stereomer.

Dissymmetry (chirality), from Latin *dis-*, apart, asunder, away, utterly, and symmetry.

e.e., enantiomeric excess.

Enantiomer, from Greek *enantios*, opposite, and -mer.

Epimer, from epi- and -mer.

Meso compound, from Greek *mesos*, middle.

Mutarotation, from Latin *mutare*, to change, to exchange.

Periplanar (p), from peri- and planar, referring to a torsion angle between 0° and ±30° or between ±150° and 180°.

Prochiral, from Greek and Latin *pro*, for, forward, instead of, and chiral.

Racemate, from racemic acid.

Rotamer, from rotation and -mer.

Stereoisomer, from Greek *stereos*, solid, and isomer.

Suprafacial, from Latin *supra*, above, over, and facial.

Syn- (s), from syn-, referring to a torsion angle between 0° and ±90°.

Synclinal (sc), from syn- and Greek *klinein*, to lean, referring to a torsion angle between 30° and 90° or between −30° and −90°. A.k.a. **gauche**, from French *gauche*, left, akward, and **skew.**

Synperiplanar (sp), from syn-, peri-, and planar, referring to a torsion angle between 0° and ±30°.

17.3 Samples from the vocabulary of polymer chemistry

IUPAC provides extensive information about this terminology [8].

17.3.1 General terminology

Atactic, from Greek *a(n)-*, not, and -tactic.

Copolymer, from Latin *cum, co-*, with, together with, and -mer.

Elastomer, from elastic and -mer.

Eutactic, from Greek *eus, eu-*, good, and -tactic.

Heterotactic, from Greek *heteros*, other, different, and -tactic.

Inifer, a contraction of initiator and transfer reagent.

Iniferter, a contraction of initiator, transfer reagent, and terminator.

Initer, a contraction of initiator and terminator.

Isotactic, from Greek *isos*, equal, and -tactic.

Living polymerization, referring to the absence of chain transfer and chain termination reactions.

Monomer, from Greek *monos*, alone, and -mer.

Natta projection, after the Italian chemist Giulio Natta (1903–1979).

Oligomer, from Greek *oligos*, few, scanty, and -mer.

Plastic, from Greek *plastos*, molded, formed, from *plassein*, to mold.

Plasticizer, from plastic.

Polymer, from Greek *polys*, much, many, and Greek *meris*, part, share.

Polynosic, with contraction, from French *polymère non-synthétique*, non-synthetic polymer.

Syndiotactic, from Greek *syndio-*, two together, and -tactic.

-tactic, from Greek *taktikos*, of or relating to arrangement or order.

Tacticity, from -tactic.

Taxogen, from Greek *taxis*, arrangement, order, and -gen.

Telogen, from tel(o)- and -gen.

Telomer, from Greek *telos*, ultimate end, from *tellein*, to accomplish, and -mer.

Terpolymer, from Latin *ter*, thrice, and polymer.

Topotactic, from topo- and -tactic.

Ziegler-Natta catalyst (complexes of titanium, zirconium, or hafnium) after the German chemist Karl Waldemar Ziegler (1898–1973) and the Italian chemist Giulio Natta (1903–1979). A.k.a. **Ziegler catalyst**.

17.3.2 Trivial names for synthetic polymers

Bakelite (thermosetting resins), after the Belgian-American chemist Leo Hendrik Baekeland (1863–1944).

Cellophane (regenerated cellulose), from cellulose and Greek *diaphanes*, translucent.

Celluloid (a thermoplastic composed of cellulose nitrate and camphor), from cellulose and -oid.

Chardonnet silk (nitrocellulose), after the French chemist Louis Marie Hilaire Bernigaud de Grange, Count de Chardonnet (1839–1924).

Collodion (nitrocellulose, dissolved in diethyl ether and ethanol), from Greek *kolloides*, glue-like, from *kolla*, glue.

Dacron {poly[ethylene terephthalate]}, an invented American name for terylene.

Nitrocellulose, not a nitro compound, but a nitric acid ester. A.k.a. **gun cotton** and **pyroxylin**, from pyr(o)- and xyl(o)-.

Nylon (thermoplastic polyamides), an invented name, patterned after rayon. According to an urban legend the acronym NYLON stands for 'Now, you lousy old Nipponese!'.

Orlon (acrylic textile fiber), an invented name, patterned after rayon.

Perlon (polyamides), an invented name, patterned after rayon.

Polyurethane. A.k.a. **PU** and **PUR**.

Rayon (regenerated cellulose), from ray, referring to its shiny appearance. The ending -on patterned after cotton.

Silicone, from silicon and ketone, referring to the formal, though not chemical, resemblance between poly[diphenylsiloxane] and benzophenone.

Teflon {poly[tetrafluoroethene]}, a contraction of the systematic name, patterned after rayon.

Terylene {poly[ethylene terephthalate]}, an anagramamtic contraction of the systematic name.

Thiokol (polysulfide polymers), from 'thio' and Greek *kolla*, glue.

Viscose (regenerated cellulose), from viscous and cellulose.

17.4 Samples from the vocabulary of biochemistry

This section only lists relatively few terms from cell biology to illustrate their etymology.

Aerobic, from Greek *aer*, air, and *bios*, life.

Agglutination, from Latin *agglutinare*, to adhere, from *gluten*, glue.

Agglutinin, a class name, from agglutination.

Agonist, a class name, from Greek *agonistes*, combattant, from *agonizesthai*, to contend, from *agon*, contest.

Alexin, a class name, from Greek *alexein*, to defend.

Ames test, after the American biochemist Bruce Nathan Ames (born 1928).

Anabolism, with contraction, from Greek *ana-*, up, against, back, and metabolism.

Anaerobic, from a(n)- and aerobic.

Antagonist, a class name, from ant(i)- and agonist.

Antibody, a class name.

Antigen, a class name, from antibody and -gen.

Antitoxin, a class name, from anti- and toxin.

Apoptosis, from Greek *apoptosis*, a falling off.

-bacter, from bacterium, from Greek *bakterion*, diminutive of *bakteria*, staff.

Bacteri(o)-, from bacterium.

Bacteriolysin, a class name, from bacterium and lysis.

Basophile, from base and -phile.

Cadherin, a contraction of calcium-dependent cell adhesion molecule.

Calvin cycle, after the American chemist Melvin Calvin (1911–1997).

Capsid, from Latin *capsa*, box, capsule.

Capsomer, from Latin *capsa*, box, capsule, and -mer.

Catabolism, from Greek *katabole*, throwing down.

Centriole, from New Latin *centriolum*, diminutive of Latin *centrum*, center.

Chloroplast, from chlorophyll and Greek *plastos*, formed, molded.

Chromosome, from Greek *chroma, chromat-*, color, and *soma*, body, referring to the ease with which it can be stained.

Chromatin, from Greek *chroma, chromat-*, color, referring to the ease of staining this substance.

Coenzyme, from co- and enzyme.

Complement. This name refers to a group of thermolabile proteins in normal blood serum and plasma that in combination with antibodies causes the destruction of particulate antigens.

Cytoplasm, from Greek *kytos*, container, receptacle, body, and *plasma*, something molded or formed.

Endoplasmic reticulum, from Greek *endon*, within, plasma, something molded or formed, and Latin *reticulum*, diminutive of *rete*, net.

Enzyme, cf. Section 10.2.

Erythrocyte, from Greek *erythros*, red, and *kytos*, hollow vessel.

Extremozyme, a contraction of extremoenzyme.

Gene, after German *Gen*, derived from -gen, from Greek *-genes*, born, produced.

Genome, with contraction, from gene and chromosome.

Golgi body, after the Italian biologist and pathologist Camillo Golgi (1843–1926). A. k.a. **Golgi apparatus** and **Golgi complex**.

Hatch-Slack cycle, after the Australian biochemist and physiologist Davidson Hatch (born 1932) and the British-Australian biochemist and plant physiologist Charles Roger Slack (born 1937).

Hemolysin, a class name, from Greek *haima*, blood, and lysis.

Homochirality, from homo- and chirality.

Krebs cycle, after the German-British biochemist Hans Adolf Krebs (1900–1981).

Krebs-Kornberg cycle, as Krebs cycle and after the German-British biochemist Hans Leo Kornberg (born 1928).

Leukocyte, from Greek *leukos*, white, and *kytos*, hollow vessel.

Lynen cycle, after the German biochemist Feodor Felix Konrad Lynen (1911–1979).

Lysis, from Greek *lysis*, act of loosening, dissolution, from *lyein*, to loosen, to dissolve.

Lysosome, from lysis, from Greek *lysis*, a loosening, and *soma*, body.

Meiosis, from Greek *meiosis*, a lessening.

Metabolism, from Greek *metabole*, change.

Metabolome, from metabolism and -ome, from Greek *-oma*, a suffix forming neuter nouns.

Michaelis-Menten equation, after the German-American biochemist Leonor Michaelis (1875–1949) and the Canadian biochemist Maud Leonora Menten (1879–1960).

Mitochondrion, from Greek *mitos*, thread, and *chondrion*, diminutive of *chondros*, grain, corn.

Mitosis, from Greek *mitos*, thread.

Moore-Stein analysis, after the American biochemists Stanford Moore (1913–1982) and William Howard Stein (1911–1980).

Neutrophile, from neutral and -phile.

Nucleolus, from Latin *nucleolus*, diminutive of *nucleus*, kernel, diminutive of *nux, nuc-*, nut.

Nucleus, from Latin *nucleus*, kernel, diminutive of *nux, nuc-*, nut.

Opsonin, a class name, ultimately from Greek *opsonein*, to purchase victuals, from *opson*, food, and *oneisthai*, to buy. The name refers to proteins that bind to foreign particles and cells, making them more susceptible to the action of phagocytes. A.k.a. **fibronectin**, from Latin *fibra*, fiber, and *nectere*, to tie, to bind.

Organelle, from New Latin *organella*, diminutive of Latin *organum*, organ.

Osborne fractions, after the American biochemist Thomas Burr Osborne (1859–1929).

Ouchterlony's test, after the Swedish microbiologist Örjan Ouchterlony (1914–2004).

Pasteur effect, after the French biologist, microbiologist, and chemist Louis Pasteur (1822–1895).

Phagocyte, from Greek *phagein*, to eat, and *kytos*, hollow vessel.

Pheromone, with contraction, from Greek *pherein*, to bear, to carry, and hormone.

Phytoalexin (a class name), from phyt(o)- and Greek *alexein*, to defend.

Precipitin (a class name), referring to an antibody which precipitates when it unites with its antigen.

Prion, a contraction of proteinaceous infectious particle.

Proenzyme, from pro- and enzyme. A.k.a. **zymogen**.

Ribosome, from ribose and Greek *soma*, body.

Semiochemical, from Greek *semeion*, signal, and chemical.

Sideramine (a class name), from Greek *sideros*, iron, referring to these compounds iron-binding properties.

Toxin, ultimately from Greek *toxikon pharmakon*, arrow poison, from *toxon*, bow, and *pharmakon*, drug.

Warburg-Dickens-Horecker pathway, after the German biochemist Otto Heinrich Warburg (1883–1979), the British biochemist Frank Dickens (1899–1986), and the American biochemist Bernard Leonard Horecker (1914–2010).

Zymogen, from zym(o)- and -gen. A.k.a. **proenzyme**, from pro- and enzyme.

References

[1] https://www.qmul.ac.uk/sbcs/iupac/ (retrieved April 9, 2019).
[2] https://www.qmul.ac.uk/sbcs/iupac/bibliog/gold.html (retrieved April 9, 2019).
[3] https://en.wikipedia.org/wiki/List_of_examples_of_Stigler%27s_law (retrieved August 18, 2018).
[4] https://en.wikipedia.org/wiki/Alchemy (retrieved April 9, 2019).
[5] https://en.wikipedia.org/wiki/Philosopher%27s_stone (retrieved August 26, 2018).
[6] https://en.wikipedia.org/wiki/Phlogiston_theory (retrieved August 26, 2018).
[7] https://www.qmul.ac.uk/sbcs/iupac/stereo/ (retrieved April 9, 2019).
[8] R. G. Jones, J. Kahovec, R. Strepto, E. C. Wilks, M. Hess, T. Kitayama, W. Val Metanomski, *Compendium of Polymer Terminology and Nomenclature, IUPAC Recommendations* 2008, RSC Publishing, Cambridge, UK, 2009, ISBN 978-0-85404-491-7; cf. G. P. Moss (Ed.), *Brief Guide to Polymer Nomenclature, IUPAC Technical Report*, https://www.qmul.ac.uk/sbcs/iupac/BriefGuide/polymer.html (retrieved March 22, 2019).

18 A sampler of miscellaneous prefixes and suffixes occurring in scientific terms

Aci-, as in acichromism, from acid.

Acanth(o)-, as in acanthite, from Greek *akantha*, thorn, spine.

Acar(o)-, as in acaricide, from Greek *akari*, mite, from *keirein*, to cut off.

[1]Actin(o)-, as in actinium, from Greek *aktis, aktin-*, ray.

[2]Actino-, as in actinobolin, after the bacterial genus *Actinomyces*. The genus name is from [1]actino- and Greek *mykes*, fungus.

Ad- (ac-, af-, al-, am-, ap-, ar-, as-, at-, i.e. depending on the following consonant), as in adrenaline, from Latin *ad*, to.

Aden(o)-, as in adenine, from Greek *aden*, gland.

Adren(o)-, as in adrenochrome (3-hydroxy-1-methylindoline-5,6-dione), from Latin *ad*, to, at, and *ren*, kidney, referring to the adrenal gland.

Aer(o)-, as in aerosol, from Greek *aer*, air.

Agr(o)-, as in agroclavine, from Latin *ager, agr-*, field.

-ain(e), as in lidocaine, an arbitrary suffix patterned after cocaine.

Alb(o)-, as in albite, from Latin *albus*, white.

All(o)-, as in allocinnamic acid, from Greek *allos*, other.

Amar(o)-, as in amarolide, from Latin *amarus*, bitter.

Ambi-, as in ambident, from Latin *ambo*, both.

Ampel(o)-, as in ampelopsin, from Greek *ampelos*, grapevine.

Amphi-, as in amphoteric, from Greek *ampho*, both,

Amyl(o)-, as in amylopectin, from Greek *amylon*, starch, literally not ground; starch was obtained from unground grain.

Andr(o)-, as in androstane, from Greek *aner, andr-*, male person, stamen.

Angi(o)-, as in angiotensin, from Greek *angeion*, diminutive of *angios*, vessel.

A(n)-, as in anhydride, from Greek *a(n)-*, non-.

-an, as in glycan, from anhydride.

-an(e), as in alkane, a suffix derived from Latin *-anus*, denoting a relationship.

https://doi.org/10.1515/9783110612714-018

Ana-, as in anabolism, from Greek *ana*, up.

-and, as in cryptand, from Latin -*andus*, to be acted upon.

Ano-, as in anomer, from Greek *ano*, upward, above.

Anorth(o)-, as in anorthite, from Greek *anorthos*, not upright.

Antara-, as in antarafacial, from Sanskrit *antara*, the other.

Anth(o)-, as in anthocyanin, from Greek *anthos*, flower.

[1]**Anthra-**, as in anthranthrene, from anthracene.

[2]**Anthra-**, as in anthracite, from Greek *anthrax*, coal.

[3]**Anthra-**, as in anthranilic acid, referring to carbon, derived from [2]anthra-.

Anti-, as in antiaromatic, from Greek *anti*, against, opposite.

Api-, as in apiose, from Latin *apium*, parsley, celery, utimately from *apis*, bee.

Apo-, as in apomorphine, from Greek *apo*, off, away from.

Argill-, as in argillite, from Latin *argilla*, clay.

Argyr(o)-, as in argyrodite, from Greek *argyros*, silver.

Asco-, as in ascomycin, from New Latin *ascus*, bladder.

-ase, as in endonuclease, from diastase, from Greek *diastasis*, separation.

Aspid(o)-, as in aspidospermine, from Greek *aspis*, shield.

Atto- (10^{-18}), as in attosecond, from Danish and Norwegian *atten*, eighteen.

Aure(o)-, auri-, auro-, as in auramine, from Latin *aureus*, golden yellow, from *aurum*, gold.

Aut(o)-, as in autoclave, from Greek *autos*, self.

Aux(o)-, as in auxin, from Greek *auxanein*, to make grow, from *auxein*, to increase.

Avi-, as in avicide, from Latin *avis*, bird.

Azi-, as in azibenzil (2-diazo-1,2-diphenylethan-1-one), from az(o)-.

-bacter, bacteri(o)-, as in bactericide, from bacterium, from Greek *bakterion*, diminutive of *bakteria*, staff, akin to *baktron*, stick.

Bary-, as in barium, from Greek *barys, baryt-*, heavy.

Bathmo-, as in bathmometry, from Greek *bathmos*, step.

Bath(o)-, bathy-, as in bathochromic, from Greek *bathos*, depth.

Batrach(o)-, as in batrachotoxin, from Greek *batrachos*, frog.

Bet(a)-, as in betaine, after the genus *Beta* (beets). The genus name is from Latin *beta*, beet, of Celtic origin.

Bi-, as in bidentate, from Latin *bis*, twice.

Bil(i)-, as in 21*H*-biline, from Latin *bilis*, bile.

Bio-, as in biotin, from Greek *bios, biot-*, life.

Blast(o)-, as in blastomycin, from Greek *blastos*, bud, germ.

Bry(o)-, as in bryostatin, from Greek *bryon*, moss.

Bufo-, as in bufotenine, from Latin *bufo, bufon-*, toad.

Cacodyl- (dimethylarsanyl-), as in cacodylic acid, from Greek *kakodes*, evil smelling, from *kakos*, bad.

Caffe-, as in caffeine, ultimately from Arabic *qahwa*, coffee.

[1]**Calci-**, as in calcium, from Latin *calx, calc-*, lime.

[2]**Calci-**, as in calcite, from calcium.

Cal(o)-, as in calomel, from Greek *kalos*, beautiful.

Capr(o)-, as in capric acid, from Latin *caper, capr-*, goat.

Carb(o)-, as in carbohydrate, from carbon.

Carcin(o)-, as in carcinogen, from Greek *karkinos*, crab, cancer.

Card(io)-, as in cardiac glycoside, from Greek *kardia*, heart.

Carce(r)-, as in carcerand, from Latin *carcer*, prison.

Carmin-, as in carminic acid, ultimately from Arabic *qurmiz*, kermes, ultimately from Sanskrit *krmi-ja*, red dye produced by a worm, and Latin *minium*, cinnabar.

Carnos-, as in carnosine, from Latin *carnosus*, fleshy, from *caro, carn-*, flesh.

Carph(o)-, as in carpholite, from Greek *karphos*, dry stalk, straw.

Carp(o)-, **-carp**, as in podocarpic acid, from Greek *karpos*, fruit.

Cat(a)-, as in catalysis, from Greek *kata*, down, against.

Catena-, as in catenane, from Latin *catena*, chain.

Caul(o)-, as in caulophylline, from Greek *kaulos*, stem, stalk.

Cedr(o)-, as in cedrane, from Greek *kedros*, cedar.

Celest-, as in celestine, from Latin *caelestis*, of the sky, from *caelum*, sky.

Cell(o)-, as in cellulase, from cellulose.

Centi- (10^{-2}), as in centiliter, from Latin *centesimus*, one hundredth, from *centum*, hundred.

Cephal(o)-, as in cephalosporanic acid, from Greek *kephale*, head.

Cerebr(o)-, as in cerebrose, from Latin *cerebrum*, brain.

[1]**Cer(o)-**, as in cerite, from cerium.

[2]**Cer(o)-**, as in cerotin, from Greek *keros*, wax.

Cerul(o)-, caerul(o)-, as in cerulenin, from Latin *caeruleus*, dark blue.

Cet(o)-, as in cetyl alcohol, from Latin *cetus*, whale.

Chalc(o)-, as in chalcopyrite, from Greek *chalkos*, copper.

Chel(e)-, as in chelate, from Greek *chele*, claw.

Chen(o)-, as in chenodeoxycholic acid, from Greek *chen*, goose.

Chiast(o)-, as in chiastolite, from Greek *chiastos*, cross.

Chim(o)-, as in chimaphilin, from Greek *cheima*, winter.

Chi(ono)-, as in chiolite, from Greek *chion*, snow.

Chit(o)-, as in chitin, from Greek *chiton*, tunic, of Semitic origin.

[1]**Chlor(o)-**, as in chlorine, from Greek *chloros*, yellow green.

[2]**Chlor(o)-**, as in chlorapatite, from chlorine.

Chol(e)-, as in choline, from Greek *chole*, bile.

Chondr(o)-, as in chondroitin, from Greek *chondros*, cartilage, grain.

Chori(o)-, as in choriomammatropin, from chorion, from Greek *chorion*, skin.

[1]**Chrom(o)-, chromato-**, -**chrome**, as in chromium, from Greek *chroma, chromat-*, color.

[2]**Chrom(o)-**, as in chromite, from chromium.

Chron(o)-, as in chronometer, from Greek *chronos*, time.

Chrys(o)-, as in chrysotile, from Greek *chrysos*, gold.

Chym(o)-, as in chymotrypsin, from Greek *chymos*, juice.

-cide, as in pesticide, from Latin -*cida*, killer, from *caedere*, to cut down, to kill.

-cillin, as in penicillin, from Latin *penicillium*, diminutive of *penis*, tail, male member.

Ciner(o)-, as in cinerin, from Latin *cinis, ciner-*, ashes.

Cino-, as in cinobufotalin, from Latin *cinus*, Chinese.

-**clase**, as in euclase, from Greek *klasis*, a breaking, from *klan*, to break.

Clavi(c)-, as in agroclavine, ultimately from Latin *clava*, club.

Clin(o)-, as in clinoenstatite, from Greek *kline*, couch, from *klinein*, to bend, to cause to slope.

Clito-, as in clitocine, from Greek *klitos*, slope.

Cole(o)-, from Greek *koleos, koleon-*, sheath.

Coll(a)-, as in colloid, from Greek *kolla*, glue.

Col(o)-, as in colocynthin, from Greek *kolon*, large intestine.

Con- (**coll**-, **com**-, **corr**-, i.e. depending on the following consonant), as in complex, from Latin *cum, con-*, with.

Copr(o)-, as in coprostane, from Greek *kopros*, dung.

Coron-, as in coronene, from Latin *corona*, crown.

Corr-, as in corrinoid, from core (of vitamin B_{12}).

Cort(ico)-, as in cortisone, from adrenal cortex. Cortex is from Latin *cortex*, bark.

Coryn(o)-, as in corynan, from Greek *koryne*, club.

-crase, as in polycrase, from Greek *krasis*, a mixing, mixture, from *kerannynai*, to mix.

Cre(o)-, creat(o)-, as in creatine, from Greek *kreas, kreat-*. flesh.

Croc(o)-, as in croconic acid, from Greek *krokos*, saffron, crocus, of Semitic origin.

Cry(o)-, as in cryolite, from Greek *kryos*, icy cold, frost.

Crypt(o)-, as in cryptopyrrole, from Greek *kryptos*, hidden, from *kryptein*, to hide.

Cyan(o)-, as in cyanide, from Greek *kyanos*, dark blue.

-cyclin, as in doxycyclin, from tetracyclin.

Cyst(o)-, as in cysteine, from Greek *kystis*, bladder.

Cyt(o)-, as in cytidine, from Greek *kytos*, cell, hollow vessel.

Daidz-, as in daidzin, from Japanese *daidzu*, soy bean.

Deci- (10^{-1}), as in deciliter, from Latin *decima pars*, one tenth, from *decimus*, the tenth, from *decem*, ten.

Dendr(o)-, as in dendrimer, from Greek *dendron*, tree.

Deca- (10), as in decane, from Greek *deka*, ten.

Deci- (10^{-1}), as in deciliter, from Latin *decima pars*, one tenth, from *decimus*, tenth, from *decem*, ten.

-dent, as in ambident, from Latin *dens, dent-*, tooth.

-dentate, as in tridentate, from Latin *dentatus*, toothed, from *dens, dent-*, tooth.

Depsi-, as in depsipeptide, from Greek *depsein*, to tan.

Derm(o)-, as in dermostatin, from Greek *derma*, skin.

De(s)-, as in desiccator, from Latin *de*, away from.

Desmo(o)-, as in desmolase, from Greek *desmos*, bond, from *dein*, to bind.

Deuter(o)-, as in deuterium, from Greek *deuteros*, the second.

Dextr(o)-, as in dextrose, from Latin *dexter, dextr-*, right.

Di(a)-, as in dialysis, from Greek *dia, di-*, through, apart.

Dicty(o)-, as in dictyopterene, from Greek *diktyon*, net, from *dikein*, to throw.

Dis-, as in disrotatory, from Latin *dis-*, apart, asunder.

Dry(o)-, as in dryophantin, from Greek *drys*, tree, oak.

Dvi-, as in dvi-actinium, from Sanskrit *dvi*, two.

Dyn(amo)-, as in dynorphin, from Greek *dynamis*, power.

Dy(o)-, as in dyotropic, from Greek *dyo*, two.

Dys-, as in dysprosium, from Greek *dys-*, hard, bad, difficult.

E- (**ec**-, **ef-**, **ex-**, i.e. depending on the following letter), as in excipient, from Latin *e (x)*, out of.

Ec-, **ex-**, as in ecdysone, from Greek *ek, ex*, out of.

Ecgon(o)-, as in ecgonine, from Greek *ekgonos*, born of, sprung from.

Ect(o)-, as in ectoderm, from Greek *ektos*, out of, out.

Eka-, as in eka-actinium, from Sanskrit *eka*, one.

Ele(o)-, **elai(o)-**, as in eleostearic acid, from Greek *elaion*, olive oil.

Electr(o)-, as in electrophoresis, from Greek *elektron*, amber, referring to the discovery of static electicity by wiping an amber staff with a piece of cloth.

Elu(o)-, as in eluotropic, from Latin *eluere*, to wash out.

En- (**el-**, **em-**, i.e. depending on the following letter), as in energy, from Greek *en*, in, on.

Enanti(o)-, as in enantiomer, from Greek *enantios*, opposite.

End(o)-, as in endothermic, from Greek *endon*, within.

En(o)-, as in enanthic acid, from Greek *oinos*, wine.

Enter(o)-, as in dysentery, from Greek *enteron*, intestine.

Ep(i)-, as in epimer, from Greek *epi*, beside, on, upon, after.

Equ(i)-, as in *equiline*, from Latin *equus*, horse.

Erem(o)-, as in eremophilane, from Greek *eremos*, lonely, solitary, or *eremos*, desert.

Erg(o)-, as in ergone, a contraction of erg(o)- and hormone, from Greek *ergon*, work.

Eri(o)-, as in eriochrome, from Greek *erion*, wool.

Erythr(o)-, as in erythrin, from Greek *erythros*, red.

Estr(o)-, as in estrogen, from Greek *oistros*, gadfly, frenzy.

Etio-, aetio-, as in etioporphyrin, from Greek *aitia*, cause; a prefix used in the naming of degradation products of undetermined structure, derived from known natural products.

Eu-, as in euclase, from Greek *eu*, well, from neuter of *eys*, good.

Ex(o)-, as in exothermic, from Greek *exo*, outside.

Femto- (10^{-15}), as in femtosecond, from Danish and Norwegian *femten*, fifteen.

-fer, as in conifer, from Latinn *ferre*, to bear, to carry.

Ferr(o)-, as in ferritin, from Latin *ferrum*, iron.

Fibr(o)-, as in fibrin, from Latin *fibra*, fibre.

Flav(o)-, as in flavin, from Latin *flavus*, light yellow.

Fol-, as in folic acid, from Latin *folium*, leaf.

-fuge, as in nucleofuge, from Latin *fugere*, to flee.

Fulmin-, as in fulminic acid, from Latin *fulmen, fulmin-*, lightning.

Fulv(o)-, as in fulvene, from Latin *fulvus*, tawny

Fusc(o)-, as in fuscin, from Latin *fuscus*, dark, tawny, brownish orange.

Fus(i)-, fus(a)-, as in fusaric acid, from Latin *fusus*, spindle.

Gal(a)-, galact(o)-, as in galactose, from Greek *gala, galakt-*, milk.

Gall(o)-, as in gallacetophenone, from gallic acid (3,4,5-trihydroxybenzoic acid).

Gam(o)-, as in gamete, from Greek *gamos*, marriage.

Gastr(o)-, as in gastrin, from Greek *gaster*, stomach.

Geisso-, as in geissoschizoline, from Greek *geis(s)on*, cornice, tile, eaves of a roof.

-gen, as in oxygen, from Greek *-genes*, of a certain kind, maker, from *gennaein*, to generate.

Ge(o)-, as in geosmin, from Greek *ge*, earth, land.

Giga- (10^{12}), as in gigabyte, from Greek *gigas, gigant-*, giant.

Glauc(o)-, as in glaucine, from Greek *glaukos*, blue green.

Gli(o)-, as in gliadin, from Middle Greek *glia*, glue.

-globulin, as in immunoglobulin, from Latin *globulus*, diminutive of *globus*, globe.

[1]**Gluc(o)-**, as in glucose, from Greek *gleukos*, must, sweet wine, akin to *glykys*, sweet.

[2]**Gluc(o)-**, as in glucuronic acid, from glucose.

Glut-, as in glutamine, from Latin *gluten*, glue.

[1]**Glyc(o)-**, as in glycerol, from Greek *glykys*, sweet.

[2]**Glyc(o)-**, as in glycoside, from glycose (monosaccharide).

-gon, as in hexagon, from Greek *gonia*, angle.

Gonad(o)-, as in gonadotropin, from gonad.

Gymn(o)-, as in gymnemic acid, from Greek *gymnos*, naked.

Gyn(o)-, as in gynogamone, from Greek *gyne*, woman.

Gyr(o)-, as in gyrase, from Greek *gyros*, ring.

[1]**Hal(o)-**, as in halogen, from Greek *hals, halo-*, salt.

[2]**Hal(o)-**, as in haloform, from halogen.

Hapl(o)-, as in haplophytine, from Greek *haploos*, single, simple.

Hapt(o)-, as in hapticity, from Greek *haptein*, to touch, to adhere.

Hecto- (10^2), as in hectoliter, from Greek *hekaton*, one hundred.

Heli(co)-, as in helicene, from Greek *helix, helik-*, spiral.

Heli(o)-, as in helium, from Greek *helios*, sun.

Helminth(o)-, as in anthelmintic, from Greek *helmis, helminth-*, intestinal worm, parasitic worm.

Helo-, as in helodermin, from Greek *helos*, stake, palisade.

Hemi-, as in hemiacetal, from Greek *hemisys*, half.

Hem(o)-, haem(o)-, as in hemoglobin, from Greek *haima*, blood.

Hen-, as in henicosane, from Greek *heis, hen-*, one.

Herb(i)-, as in herbicide, from Latin *herba*, grass, herb.

Hercyn-, as in hercynite, from *Hercynia Silva* (Hercynian Forest), the Romans' name for the wooded regions east of the Rhine.

Heter(o)-, as in heterocycle, from Greek *heteros*, other, different.

Hipp(o)-, as in hippuric acid, from Greek *hippos*, horse.

Hist(o)-, as in histone, from Greek *histos*, web.

Hol(o)-, as in holoenzyme, from Greek *holos*, complete.

[1]**Homo-**, as in homogenous, from Greek *homos*, same.

[2]**Homo-**, as in homocysteine, a contraction of homologous.

Homodetic, from [1]homo- and Greek *detos*, bound, linked.

Homoleptic, from [1]homo- and -leptic.

Hydn(o)-, as in hydnocarpic acid, from Greek *hydnon*, tuber, truffle, tubercle.

Hyal(o)-, as in hyalic acid, from Greek *hyalos*, glass.

[1]**Hydr(o)-**, as in hydrogen, from Greek *hydor, hydro-*, water.

[2]**Hydr(o)-**, as in decahydronaphthalene, from hydrogen.

Hygr(o)-, as in hygroscopic, from Greek *hygros*, wet, moist.

Hy(o)-, as in hyodeoxycholic acid, from Greek *hys, hyo-*, swine.

Hyper-, as in hyperconjugation, from Greek *hyper*, over, above, beyond.

Hyph(o)-, as in '*hypho*', from Greek *hyphe*, web.

Hypn(o)-, as in hypnone, from Greek *hypnos*, sleep.

Hyp(o)-, as in hypochlorous acid, from Greek *hypo*, under, below, from below.

Hyps(o)-, as in hypsochrome, from Greek *hypsos*, height.

Iatr(o)-, as in iatrochemistry, from Greek *iatros*, physician, from *iasthai*, to heal.

[1]**-ide**, as in actinide, from Greek *-oeides*, -like, -shaped.

[2]**-ide**, as in chloride, an arbitrary suffix derived from oxide.

[3]**-ide**, as in glycoside, an arbitrary suffix derived from anhydride.

Idi(o)-, as in idiochromatic, from Greek *idios*, own, particular, personal.

-il(e), as in benzil, ultimately derived from Latin *-ilis*, a suffix denoting function or relationship.

Imbibition, from in- and Latin *bibere*, to drink.

-in(e), as in quinine, from Latin *-inus*, a suffix meaning belonging to or being similar to.

Infra-, as in infrared, from Latin *infra*, below, within.

In(o)-, as in inosin, from Greek *is, in-*, fiber, muscle, sinew, tendon.

Inter-, as in interleukin, from Latin *inter*, among, between.

Intercalation, from Latin *intercalare*, to insert, from inter- and ultimately *calendae, kalendae*, the first day of the month, literally to insert an extra day (or month) into the calendar.

Intra-, intro-, as in intramolecular, from Latin *intra*, within, inside.

[1]**Iod(o)-**, as in iodopsin, from Greek *ioeides*, purple, violet colored.

[2]**Iod(o)-**, as in iodic acid, from iodine.

Ion(o)-, as in ionophore, from ion.

Ipso-, as in ipso-substitution, from Latin *ipse*, self.

[1]**Irid(o)-**, as in iridium, from Greek *iris*, rainbow.

[2]**Irid(o)-**, as in iridocene, from iridium.

Juxta-, as in juxtaposition, from Latin *iuxta*, next to, from *iungere, iunct-*, to join.

Kilo- (10^3), as in kilogram, from Greek *chilioi*, one thousand.

Kin(e)-, as in kinetics, from Greek *kinein*, to move, to stimulate, to set in motion.

Lacc(a)-, as in laccaic acid, from New Latin *lacca*, lac, ultimately from Sanskrit *laksa*, lac.

[1]**Lact(o)-**, as in lactose, from Latin *lac, lact-*, milk.

[2]**Lact(o)-**, as in lactone, from lactic acid.

[3]**Lact(o)-**, as in lactase, from lactose.

Lampr(o)-, as in lamprophyllite, from Greek *lampros*, bright.

Lan(o)-, as in lanolin, from Latin *lana*, wool.

Leio-, lio-, as in leiocarpic acid, from Greek *leios*, smooth.

Lepid(o)-, as in lepidocite, from Greek *lepis, lepid-*, flake, scale.

-leptic, as in homoleptic, from Greek *lepsis*, a seizing, from *lambanein*, to seize, to take.

Lept(o)-, as in leptin, from Greek *leptos*, peeled, thin, slender, from *lepein*, to peel.

Leuk(o)-, leuc(o)-, as in leucine, from Greek *leukos*, white.

Lev(o)-, laev(o)-, as in levulose, from Latin *laevus*, left.

Lign(o)-, as in lignoceric acid, from Latin *lignum*, wood.

Lip(o)-, as in lipophilic, from Greek *lipos*, bacon, fat.

-lite, -lith, lith(o)-, as in zeolite, from Greek *lithos*, stone.

Loph(o)-, as in lophotoxin, from Greek *lophos*, crest.

Lumi(n)-, as in luminol, from Latin *lumen, lumin-*, light.

Lute(o)-, as in lutein, from Latin *luteus*, yellow, from *lutum*, dyer's rocket.

Lyc(o)-, as in lycopene, from Greek *lykos*, wolf.

Ly(o)-, lys(o)-, as in lysine, from lysis.

-lyte, as in electrolyte, from lysis.

Macro-, as in macrolide, from Greek *makros*, long, large.

Mega- (10^9), as in megaton, from Greek *megas*, huge, powerful.

Meio-, as in meionite, from Greek *meion*, less.

Mela(no)-, as in melanin, from Greek *melas*, black.

Mel(li)-, as in mellite, from Greek *meli*, honey.

Meri-, mero-, -mer, as in polymer, from Greek *meris, meros*, part, share.

Mes(o)-, as in mesoxalic acid, from Greek *mesos*, middle.

Meta-, as in *m*-chlorobenzoic acid, from Greek *meta*, between, with, after.

-meter, as in manometer, from Greek *metron*, measure.

-micin, as in gentamicin, from myc(o)-, for antibiotics produced by *Micromonospora* bacteria.

Micro- (10^{-6}), as in microsecond, from Greek *mikros*, small.

Milli- (10^{-3}), as in millisecond, from Latin *millesimus*, one thousandth, from *mille*, one thousand.

[1]**Morph(o)-**, as in morphology, from Greek *morphe*, form, shape.

[2]**Morph(o)-**, as in morphine, from Morpheus, the Greek god of dreams, from Greek *morphe*, form, shape.

Muc(o)-, as in mucic acid, from Latin *mucus*, nasal mucus.

Murex-, as in murexide, from Latin *murex*, purple shell, from Greek *myax, myak-*, sea mussel.

-mycin, as in streptomycin, from myc(o)-, for antibiotics produced by *Streptomyces* actinobacteria.

Myc(o)-, as in mycomycin, from Greek *mykes*, fungus.

Myel(o)-, as in myelin, from Greek *myelos*, marrow.

My(o)-, as in myoglobin, from Greek *mys, myo-*, muscle.

Myx(o)-, as in myxalamide, from Greek *myxa*, mucus.

Nano- (10^{-9}), as in nanoplankton, from Greek *nanos*, dwarf.

Nemat(o)-, as in nematic, from Greek *nema, nemat-*, thread.

Neo-, as in neon, from Greek *neos*, new.

Nephel(o)-, as in nephelauxetic effect, from Greek *nephele*, cloud.

Neur(o)-, as in neuraminic acid, from Greek *neuron*, nerve, sinew.

Nigr(o)-, as in nigrosin, from Latin *niger, nigr-*, black.

-oid, as in alkaloid, from Greek *-oeides*, -like, from *eidos*, form, shape.

Omm(ato)-, as in ommochrome, from Greek *omma, ommat(o)-*, eye.

[1]**on-**, as in argon, derived from the Greek adjective neuter suffix *-on*.

[2]**on-**, in the names boron and silicon expressing analogy with carbon.

[3]**on-**, in the names of small particles by analogy with proton and electron.

O(o)-, as in ooporphyrin, from Greek *oon*, egg.

Ophi(o)-, as in ophiobolane, from Greek *ophis*, snake.

-opsis, as in *Ampelopsis*, from Greek *opsis*, sight.

Ortho-, as in *o*-chlorobenzoic acid, from Greek *orthos*, straight, right, true.

[1]**Osm(o)-**, as in osmosis, from Greek *osmos*, a thrusting, from *othein*, to push.

[2]**Osm(o)-**, as in osmium, from Greek *osme*, smell, from ozein, to smell.

[3]**Osm(o)-**, as in osmocene, from osmium.

Oste(o)-, as in osteocalcin, from Greek *osteon*, bone.

Ov(o)-, as in ovalbumin, from Latin *ovum*, egg.

Ox(a)-, as in 1,3-dioxole, from oxygen.

Ozo-, as in ozone, from Greek *ozein*, to smell.

Pan(o)-, as in panthetine, from Greek *pas, pant(o)-*, all, every.

Pant(o)-, as in pantolactone, from pantothenic acid, from Greek *pas, pant(o)-*, all, every.

Para-, as in *p*-chlorobenzoic acid, from Greek *para*, beside, similar, near, beyond, irregular, a modification of.

Petro-, as in petroleum, from Greek *petra*, rock, and/or *petros*, stone.

Phae(o)-, phe(o)-, as in phaeophytin, from Greek *phaios*, dusky, grayish brown.

-phan(e), as in tryptophan, from Greek *phaneros*, apparent, visible, from *phainein*, to show.

Phell(o)-, as in phellandrene, from Greek *phellos*, cork, akin to *phloos*, bark.

Phil(o)-, -phile, -philic, as in nucleophile, from Greek *philos*, loving.

Phlog(o)-, as in phlogopite, from Greek *phlox, phlog-*, flame.

Phlor(o)-, as in phloroglucinol, from Greek *phloios, phloos*, bark.

-phobe, -phobic, as in hydrophobic, from Greek *phobos*, fear.

Phorb(o)-, as in phorbol, from Greek *phorbe*, pasture, fodder.

-phore, as in chromophore, from Greek *-phoros*, carrying, from *pherein*, to carry.

Phthi(so)-, as in phthiocol, from Greek *phthisis*, pulmonary tuberculosis, from *phthiein*, to decay, to wane.

Phyc(o)-, as in phycobilin, from Greek *phykos*, seaweed.

Phyll(o)-, -phyll, as in chlorophyll, from Greek *phyllon*, leaf, from *phyein*, to bring forth, to grow, to generate.

Phyl(o)-, as in phylogenetic, from Greek *phylon*, tribe, clan.

Phys(o)-, as in physostigmine, from Greek *physa*, bellows, bladder.

Phyt(o)-, as in phytoene, from Greek *phyton*, plant, from *phyein*, to bring forth, to grow, to generate.

Pico- (10^{-12}), as in picosecond, from Spanish *pico*, small quantity, small number, peak.

Pic(o)-, as in picene, from Latin *pix, pic-*, pitch.

Picr(o)-, as in picric acid, from Greek *pikros*, sharp, bitter.

Piezo-, as in piezoelectric, from Greek *piezein*, to squeeze, to press.

Pil(o)-, as in pilosine, from Greek *pilos*, felt.

Pin(o)-, as in pinonic acid, from Latin *pinus*, pine.

Plast(o)-, as in plastoqinone, from Greek *plastikos*, pliable, from *plassein*, to form, to mold.

Platy-, as in platyphyllin, from Greek *platys*, flat, broad.

Ple(o)-, as in pleonaste, from Greek *pleion*, more.

Plesi(o)-, as in plessite, from Greek *plesios*, near, from *pelas*, near.

Pleur(o)-, as in pleuromutilin, from Greek *pleura*, side, rib.

Pod(o)- as in podocarpane, from Greek *pous, pod-*, foot.

Poiet(o)-, as in *erythropoietin*, from Greek *poietikos*, productive, formative, from *poiein*, to make, to do, to create, to compose.

Prim(o)-, as in primocarcin, from Latin *primus*, the first, from *prae*, ahead.

Pro-, as in prodrug, from Latin *pro*, for.

Prot(o)-, as in proton, from Greek *protos*, the first.

Psil(o)-, as in psilocybin, from Greek *psilos*, bare, akin to *psen*, to rub.

Psychr(o)-, as in psychrometer, from Greek *psychros*, cold, ultimately from *psychein*, to blow.

-pten(e), as in stearopten, from Greek *ptenos*, winged, volatile, from *petesthai*, to fly.

Pter(o)-, as in pteridine, from Greek *pteron*, feather, wing.

Ptil(o)-, as in ptilolite, from Greek *ptilon*, down, feather.

Ptoma-, as in ptomaine, from Greek *ptoma*, a fall, corpse, from *piptein*, to fall.

Ptyal(o)-, as in ptyalin, from Greek *ptyalon*, saliva, from *ptyein*, to spit.

Ptych(o)-, as in ptychography, from Greek *ptychos*, fold, layer, from *ptyssein*, to fold.

Purpur(o)-, as in purpurogallin, from Latin *purpureus*, purple red, from *purpura*, red color, ultimately from Greek *porphyra*, purple stone.

Py(o)-, as in pyocyanine, from Greek *pyon*, pus.

Pyr(o)-, as in pyruvic acid, from Greek *pyr*, fire.

Pyrr(o)-, as in pyrrole, from Greek *pyrrhos*, reddish, literally flame colored.

Quadr(o)-, as in quadricyclane, from Latin *quadratum*, quadrangle, square, from *quattuor*, four.

Quasi-, as in quasi-element, from Latin *quasi*, as if.

Radio-, as in radium, from Latin *radius*, ray.

Ren(o)-, as in renin, from Latin *ren*, kidney.

Rheo-, as in rheology, from Greek *rheos*, stream, from *rhein*, to flow.

Rhiz(o)-, as in rhizonic acid, from Greek *rhiza*, root.

[1]**Rhod(o)-**, as in rhodanin, from Greek *rhodon*, rose, probably from Old Iranian **urda-*, rose.

[2]**Rhod(o)-**, as in rhodocene, from rhodium.

Rhynch(o)-, as in rhynchophylline, from Greek *rhynchos*, snout, bill, beak.

Rubi-, as in rubicene, from Latin *rubeus*, red, reddish.

Rubid(o)-, as in rubidomycin, from Latin *rubidus*, dark red.

Sacchar(o)-, as in saccharic acid, from Latin *saccharum*, sugar, ultimately from Sanskrit *sarkara*, sugar.

Sapo(n)-, as in saponification, from Latin *sapo, sapon-*, soap.

Sapr(o)-, as in sapropterin, from Greek *sapros*, rotten.

Saxi-, as in saxitoxin, from Latin *saxum*, rock.

Schiz(o)-, as in geissoschizoline, from Greek *schizein*, to split.

Scler(o)-, as in sclerosis, from Greek *skleros*, hard.

Scolec(o)-, as in scolecite, from Greek *skolex, skolek-*, worm.

Sider(o)-, as in siderite, from Greek *sideros*, iron.

Sito-, as in sitosterol, from Greek *sitos*, grain, food.

Somat(o)-, as in somatoliberin, from Greek *soma, somat(o)-*, body.

Sphing(o)-, as in sphingolipid, after the Sphinx, in Greek mythology a creature with a human head and the body of a lion, referring to the enigmatic nature of sphingolipids. Sphinx is from Greek *sphinx*, strangler, from *sphingein*, to bind, to squeeze.

Spiro-, as in spirostan, from Greek *speira*, coil, twist.

Spor(o)-, as in sporidesmin, New Latin, from Greek *spora*, a sowing, from *speirein*, to sow.

-statin, as in somatostatin, from Greek *-states*, one that causes to stand, from *histanai*, to cause to stand.

Staur(o)-, as in staurolite, from Greek *stauros*, cross.

Stear(o)-, steat(o)-, as in stearin, from Greek *stear*, fat, tallow.

Stephan(o)-, as in stephanine, from Greek *stephanos*, crown, from *stephein*, to encircle, to crown.

Sterc(o)-, as in stercobilin, after Sterculius, the Roman god of cultivation and manuring, from Latin *stercus*, manure.

Stere(o)-, as in steroid, from Greek *stereos*, solid.

-sthene, as in hypersthene, from Greek *sthenos*, force.

[1]**Strept(o)-**, as in streptomycin, from Greek *streptos*, twisted, easy to bend, pliant, from *strephein*, to twist, to turn.

[2]**Strept(o)-**, from the bacterial genus *Streptococcus*.

[3]**Strept(o)-**, from the bacterial genus *Streptomyces*.

Sub- (**suc**-, **suf**-, **sup**-, i.e. depending on the following letter), as in carbon suboxide (propadienedione), from Latin *sub*, under.

Suber-, as in suberic acid, from Latin *suber*, cork.

Super-, as in superactinoid (elements 122–153), from Latin *super*, above, over.

Syn- (**syl-**, **sym-**, **sys**-, i.e. depending on the following letter), as in syndiotactic, from Greek *syn*-, with, together with, by means of, at the same time.

Tachy-, as in tachysterol, from Greek *tachys*, swift.

Tauto-, as in tautomer, from Greek *to auton*, the same.

Tax(o)-, as in taxol, from Latin *taxus*, yew.

-taxy, as in epitaxy, from Greek *taxis*, arrangement, from *tassein*, to arrange, to put in order.

-tectic, as in eutectic, from Greek *tektos*, melted, from *tekein*, to melt.

Tel(o)-, as in telomer, from Greek *telos*, end, consummation, completeness.

-**tensin**, as in neurotensin, from Latin *tensio, tension*-, tension, from *tendere*, to tighten.

Tephro-, as in tephroite, from Greek *tephros*, ash gray, from *tephra*, ashes.

Tera- (10^{15}), as in terabyte, from Greek *teras, terat*-, monster.

-thecium, as in *Trichothecium*, from Greek *thekion*, diminutive of *theke*, case.

Therm(o)-, as in thermodynamics, from Greek *thermos*, hot.

[1]**Thym(o)-**, as in thymine, from thymus gland, from Greek *thymos*, thymus gland.

[2]**Thym(o)-**, as in thymol, from Greek *thymon*, thyme (*Thymus*).

Thyr(eo)-, as in thyronine, from thyroid gland, ultimately from Greek *thyreoeides*, shaped like a shield, from *thyra*, door.

-tonin, as in melatonin, from tonic, ultimately from Greek *tonos*, tension, tone, from *teinein*, to stretch.

-topic, topo-, as diastereotopic, from Greek *topos*, place.

Trachy-, as in trachylobane, from Greek *trachys*, rough.

Tribo-, as in triboluminescence, from Greek *tribein*, to rub.

Trich(o)-, as in trichothecane, from Greek *thrix, tricho*-, hair.

-troph(o)-, -**trophic**, from Greek *trephein*, to nourish.

Tropo-, -trope, -tropic, -tropin, as in tropomerism, from Greek *tropikos*, pertaining to a turning, from *trope*, a turning, way, manner, from *trepein*, to turn.

Trypt(o)-, as in tryptophan, from tryptic, ultimately from Greek *tripsis*, friction, from *tribein*, to rub.

Ultra-, as in ultraviolet, from Latin *ultra*, beyond, from *ulter, ultr-*, situated beyond.

Virid(o)-, as in viridin, from Latin *viridis*, grass green.

Xanth(o)-, as in xanthine, from Greek *xanthos*, yellow.

Xen(o)-, as in xenon, from Greek *xenos*, strange.

Xer(o)-, as in xerulin, from Greek *xeros*, dry.

Ze(a)-, as in zeaxanthin, from *Zea* (maize, corn, Indian corn).

Zym(o)-, as in zymase, from Greek *zyme*, yeast.

19 Trivial chemical names with disputed etymology

While a considerable number of compounds have received arbitrary trivial names which have been accepted by the chemical community as a matter of routine, cf. Chapter 20, other apparently arbitrary names, also coined without explanation, were subsequently subjected to various attempts of etymological interpretation, some more plausible than others, but never based on solid evidence.

Barbituric acid [pyrimidine-2,4,6(1*H*,3*H*,5*H*)-trione], a.k.a. **6-hydroxyuracil**, was so named by the German chemist Johann Friedrich Wilhelm Adolf von Baeyer (1835–1917) without explanation. While the latter half of the name without doubt has been taken from uric acid [7,9-dihydro-1*H*-purine-2,6,8(3*H*)-trione], and ultimately from urine, the first part has been subject to four more or less credible interpretations.

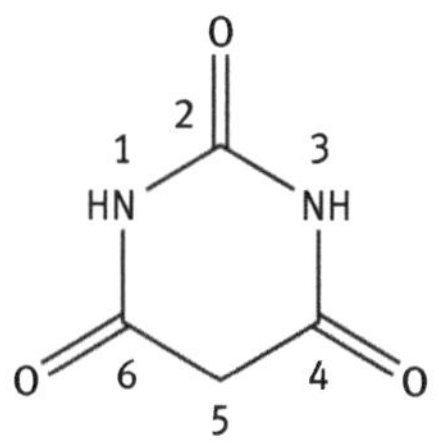

barbituric acid

The first and simplest theory claims that Baeyer chose an acquaintance of his, named Barbara, to become the godmother of this interesting new acid.

Another, far more elaborate story has it, that Baeyer, at a lunch in the company of artillery officers, mentioned that he had prepared a new and still unnamed acid from uric acid on St. Barbara's day (December 4) and that one of the officers then urged him to call this acid barbituric acid. For good measure St. Barbara is also the patron saint of artillerymen. Almost too neat to be true!

A third story, presented by the American chemist Louis Frederick Fieser (1899–1977), but later retracted, claimed that Baeyer regarded barbituric acid as a key to his understanding of the structure of uric acid and its chemistry. Therefore, barbituric acid was supposed to have been named after the German *Schlüsselbart*, the bit of a key, with *Bart*, beard, as required in a scientific context, being first latinized to *barba*, beard.

Also the specific epithet of the lichen species *Usnea barbata* (beard moss, hair lichen, old man's beard, tree moss, witch's broom), from Latin *barbatus*, bearded, from *barba*, beard, has been cited as a possible source of the name barbituric acid, but without proof or rationale.

Parabanic acid (imidazolidine-2,4,5-trione) was first prepared from uric acid [7,9-dihydro-1*H*-purine-2,6,8(3*H*)-trione] by the German chemists Friedrich Wöhler

https://doi.org/10.1515/9783110612714-019

(1800–1882) and Justus von Liebig (1803–1873) and given its trivial name without explanation other than that they found the behavior of this acid quite puzzling.

In correspondence with the Swedish chemist Jöns Jacob Berzelius (1779–1848) Wöhler explained that the name parabanic acid is derived from Greek *parabainein*, to pass over, from *bainein*, to go, but without giving a rationale for this choice.

A possible clue could be the observation by Wöhler and Liebig that parabanic acid is readily hydrolyzed to form another acid, i.e. oxaluric acid [(carbamoylamino)(oxo)acetic acid], and thus constitutes an intermediate stage which is passed on the way from uric to oxaluric acid.

In a different vein it has been claimed that Wöhler and Liebig found parabanic acid in a reaction which they had expected to yield alloxan [pyrimidine-2,4,5,6(1*H*,3*H*)-tetrone]. So para- was used in the sense of 'instead of', 'opposed to', and combined with the ending of alloxan. The letter b was finally inserted for euphonic puroposes or as a residue of 'carb' to express parallelism of its constitution to that of urea (carbamide). This interpretation of the name is based on unlikely assumptions and also ignores Wöhlers own etymology.

Soda (sodium carbonate). This name has been suggested to be derived from Arabic *suwwad*, barilla (the three species *Salsola kali*, *Salsola soda*, and *Halogeton sativus*), the plants whose ashes yield sodium carbonate. Another etymology has as its source Latin *sudanum*, a headache remedy, in turn derived from Arabic *suda*, splitting headache.

20 Some chemical names without known etymology

Ammelide [6-amino-1,3,5-triazine-2,4(1*H*,3*H*)-dione], an arbitrary name for a product of the hydrolysis of melamine.

Ammeline [4,6-diamino-1,3,5-triazine-2(1*H*)-one], an arbitrary name for a product of the hydrolysis of melamine.

Cellone (1,1,2,2-tetrachloroethane), originally coined as a trademark, of unknown etymology.

Cellosolve (2-ethoxyethan-1-ol), originally coined as a trademark, of unknown etymology.

Frenolicin (a group of *Streptomyces* metabolites), so named without any stated reason.

Hadacidin (*N*-formyl-*N*-hydroxyglycine), so named without any stated reason.

Holomycin {*N*-(5-oxo-4,5-dihydro[1,2]dithiolo[4,3-*b*]pyrrol-6-yl]acetamide}, so named without any stated reason.

Isoprene (2-methylbuta-1,3-diene), an arbitrary name created in 1860 by the British chemist Charles Hanson Greville Williams (1829–1910).

Kairine (1-methyl-1,2,3,4-tetrahydroquinolin-8-ol), from Greek *kairos*, right time, opportunity. So named by the German chemist Wilhelm Fischer in 1883, but without explanation.

Lepidine (4-methylquinoline), of obscure etymology. The genus name *Lepidium* (peppercress, peppergrass), has been invoked, but without rationale.

Ligroin (mixture of C_7 and C_8 hydrocarbons), first recorded 1880–1885. Neither the originator nor the intended meaning of this name are known. It has, however, been suggested that the name could be derived from Greek *ligyros*, light, loud, sharp, referring to the fact that this hydrocarbon mixture is colorless.

Lithopone ($BaSO_4{\cdot}ZnS$), from lith(o)- and -pone, the latter of uncertain origin. A possible derivation from Greek *ponos*, labor, toil, grief, has been rejected because it does not make sense.

Melam [N^2-(4,6-diamino-1,3,5-triazin-2-yl)-1,3,5-triazine-2,4,6-triamine], an arbitrary name for a condensation product of ammonium thiocyanate, created in 1834 by the German chemist Justus von Liebig (1803–1873) who subsequently also discovered and named the related condensation products melamine, melem, and

https://doi.org/10.1515/9783110612714-020

melon, explicitly labeling these four names as 'grasped from the air' and thus denying any involvement of Greek *meli*, honey, or *melas*, black.

Melamine (1,3,5-triazine-2,4,6-triamine), an arbitrary name for a substance obtained from melam. The name melamine is unrelated with amine, a compound class which was discovered and named fifteen years later. A.k.a. **cyanuramide**.

Melem (1,3,4,6,7,9,9b-heptaazaphenalene-2,5,8-triamine), an arbitrary name for a condensation product of melamine.

Melon {[4,6-bis(cyanoamino)-1,3,5-triazin-2-yl]cyanamide}, an arbitrary name for a condensation product of melamine.

Naled (1,2-dibromo-2,2-dichloroethyl dimethyl phosphate), so named without any stated reason.

Nosiheptide (a thiopeptide antibiotic), so named without any stated reason. A.k.a. **multhiomycin**.

Oligomycin (a group of antibiotic macrolides), so named without any stated reason.

Soman (3,3-dimethylbutan-2-yl methylphosphonofluoridate), an invented name of undisclosed etymology. A.k.a. **GD, EA 1210, Trilon 144**, and **PFMP**.

Tabun {(*RS*)-[ethyl *N,N*-dimethylphosphoramidocyanidate]}, an invented name of undisclosed etymology. A.k.a. **GA** and **Trilon 38/83**.

Tylosin (a macrocyclic antibiotic), originally coined as a trademark without any stated reason.

Xenyl- {biphenylyl-, [1,1′-biphenyl]-4-yl}, from xen(o)-. Apparently an arbitrary name.

Zarzissine {1*H*-imidazo[4,5-*b*]pyrazin-2-amine}, obtained from the Mediterranean sponge *Anchinoe paupertas*, later renamed *Hymedesmia paupertas*, and named without any stated reason.

Source Language Index

Akkadian, an extinct East Semitic language spoken in Mesopotamia from the mid-3rd millenium BC to the 8th century BC.

American Spanish, Spanish spoken in the Americas.

Ancient Egyptian, an extinct Afro-Asiatic language earlier spoken in Egypt. **Coptic**, the ritual language of the Coptic Orthodox Church of Alexandria, is an offspring of Ancient Egyptian.

Arabic, contemporary Arabic.

Aramaic, a Semitic language spoken in the Near East.

Assyrian, a Semitic language spoken in the Near East.

Bengali, an Indo-Aryan language spoken in South Asia.

Cantonese, the prestige variety of Yue Chinese, spoken in the city of Guangzhou, China and the surrounding region.

Carib, a Cariban language spoken by the Kali'na people of South America.

Celtic, unspecified Celtic language.

Cherokee, a Southern Iroquoian language spoken by the Cherokee people.

Chinese, contemporary Chinese.

Creek Indian, the Muskogee language, spoken in Florida and Oklahoma, USA.

Creto-Minoic, an extinct non-Indo-European language, spoken in Crete, Greece.

Elamite, an extinct language isolate spoken in the Middle East 2800–300 BC.

Ewe, a Niger-Congo language spoken in southeastern Ghana.

Fijian, an Austronesian language spoken in Fiji.

Finnish, contemporary Finnish.

French, contemporary French.

Gaelic, a group of Insular Celtic languages. A.k.a. **Goidelic**.

German, contemporary German.

Gothic, an East Germanic language spoken from the 3rd to the 10th century AD.

https://doi.org/10.1515/9783110612714-021

Greek, Ancient Greek, spoken from the 9th century BC to the 4th century AD.

Hausa, a Chadic language spoken in Central Africa.

Hebrew, contemporary Hebrew.

Hindi, contemporary Hindi.

Iberian, an extinct language earlier spoken in eastern and southeastern Iberia.

Indo-European, an ancient extinct language which became the source of several modern language families.

Irish, contemporary Irish.

Italian, contemporary Italian.

Japanese, contemporary Japanese.

Javanese, an Austronesian language spoken on Java, Indonesia.

Kannada, a Dravidian language spoken in India.

Khoisan, an African language spoken in Namibia, Botswana, and Tanzania.

Kru, a Niger-Congo language spoken in Liberia and Ivory Coast.

Kwa, a Niger-Congo language spoken in Ivory Coast, Ghana, and Togo.

Lappish, the language of the Sami people in Northern Europe.

Late Greek, Greek spoken from the late 2nd to the late 7th century AD.

Late Latin, Latin spoken from the 3rd to the 6th century AD.

Latin, Classical Latin spoken from 75 BC to the 3rd century AD.

Malagasy, an Austronesian language spoken in Madagascar.

Malay, an Austronesian language spoken in Brunei, Indonesia, Malaysia, and Singapore. A.k.a. **Behasa Melayu**.

Malayalam, a Dravidian language spoken in India.

Maasai, an Eastern Nilotic language spoken by the Maasai people in southern Kenya and northern Tanzania. A.k.a. **Maa**.

Malabar, a variant of the Tamil language, spoken in Kerala, India.

Maori, the language of the Maori people of New Zealand.

Mapudungun, formerly called Araucanian, a native language spoken in Chile.

Marquesan, and Austronesian language spoken on the Marquesas Islands.

Medieval French, spoken in northern France from the 8th to the 14th century AD. A.k.a. **Old French**.

Medieval Latin, Latin as used from the 4th to the 14th century AD.

Middle English, spoken from 1150 to 1500.

Middle Low German, spoken from the 12th to the 15th centuries AD in the southern Baltic littoral and the south-eastern North Sea littoral.

Middle Greek, spoken from the 5th to the 6th century AD to 1453.

Modern Greek, contemporary Greek.

Moluccan, an Austronesian language, spoken in parts of Indonesia.

Mongolian, a Mongolic language spoken in Mongolia.

Nahuatl, a Uto-Aztecan language spoken by the Nahua people of Mexico. A.k.a. **Aztec**.

New Latin, a Latin vocabulary created and used for scientific purposes since 1375.

Old English, spoken from the 5th to the 11th century AD.

Old French, spoken in northern France from the 8th to the 14th century AD.

Old Frisian, spoken from the 8th to the 16th century AD.

Old Indo-Aryan, spoken in India from 1500 to 300 BC.

Old Iranian, an Indo-European language spoken until 400 BC.

Old Latin, Latin spoken before 75 BC.

Old Malay, an Austronesian language spoken in the 7th to the 14th century AD.

Old Norse, the language spoken in Scandinavia from the 9th to the 13th century AD.

Old Occitan, an Occitano-Romance language spoken in southwest France from the 8th to the 14th century AD.

Oromo, a Cushitic language spoken in Ethiopia and Kenya.

Osco-Umbrian, an extinct Italic language spoken in Italy from the 1st millenium BC to the 1st millenium AD.

Otomaco, an extinct Amazonian language.

Persian, a Western Iranian language spoken in Iran, Afghanistan, and Tadjikistan. A.k.a. **Farsi**.

Portuguese, contemporary Portuguese.

Proto-Germanic, a reconstructed language, the common ancestor of the Germanic languages, spoken in Northern Europe from the 2nd to the 1st millenium BC.

Proto-Indo-European, the extinct ancestor of the Indo-European languages.

Quechua, an indigenous South American language family. A.k.a. **Runasimi**.

Rapanui, an Eastern Polynesian language spoken on Easter Island, Chile.

Russian, contemporary Russian.

Sanskrit, an old Indo-Aryan language, now mostly used as a ceremonial and ritual language in Hinduism and in some Buddhist practices.

Scythian, an extinct Indo-Iranian language.

Semitic, unspecified ancient Semitic language.

Sinhalese, an Insular-Indic language spoken in Sri Lanka. A.k.a. **Sinhala**.

Somali, a Cushitic language spoken in Somalia.

Spanish, contemporary Spanish.

Swedish, contemporary Swedish.

Tagalog, an Austronesian language spoken in the Philippines.

Tahitian, a Polynesian language spoken on the Society Islands, French Polynesia.

Taino, the extinct Arawakan language of the Taino people of the Caribbean.

Tamil, a Dravidian language spoken in India and Sri Lanka.

Telugu, a Dravidian language spoken in India.

Thai, a Tai-Kadai language spoken in Thailand.

Tongan, an Austronesian language spoken in Tonga.

Tupi, an extinct Tupian language of the Tupi people of Brazil.

Turkish, contemporary Turkish.

Twi, a dialect of the Akan language spoken by the Ashanti people in southern and central Ghana. A.k.a. **Akan Kasi**.

Virginia Algonquian, an extinct language spoken by the Virginia Algonquians. Ak.a. **Powhatan**.

Welsh, contemporary Welsh.

Wolof, the language of the Wolof people in Senegal, the Gambia, and Mauretania.

Yuchi, a near extinct language isolate spoken by the Tsoyaha people in Oklahoma, USA. A.k.a. **Euchee**.

Bei Fragen zur Produktsicherheit wenden Sie sich bitte an:
If you have any questions regarding product safety,
please contact:

Walter de Gruyter GmbH
Genthiner Straße 13
10785 Berlin
productsafety@degruyterbrill.com